ORIGINS:
FROM THE PROTOSUN TO THE FIRST STEPS OF LIFE

IAU SYMPOSIUM 345

COVER ILLUSTRATION:

View of the Earth and Moon from the Galileo spacecraft, 6.2 million kilometers away, on its way to Jupiter in 1992 (credit: NASA). The Earth, filled with water, air, and abundant life, shines brilliantly in stark contrast to its nearest neighbor.

Carl Sagan and collaborators used measurements from this spacecraft as it flew by the Earth to see if they could detect evidence for intelligent life.

(C. Sagan, W. R. Thompson, R. Carlson, D. Gurnett, and C. Hord 1993, "A search for life on Earth from the Galileo spacecraft," Nature, Volume 365, Issue 6448, pp. 715-721)

"There are countless suns and countless earths all rotating round their suns in exactly the same way as the seven planets of our system. We see only the suns because they are the largest bodies and are luminous, but their planets remain invisible to us because they are smaller and non-luminous. The countless worlds in the universe are no worse and no less inhabited than our earth. For it is utterly unreasonable to suppose that those teeming worlds which are as magnificent as our own, perhaps more so, and which enjoy the fructifying rays of a sun just as we do, should be uninhabited and should not bear similar or even more perfect inhabitants than our earth. The unnumbered worlds in the universe are all similar in form and rank and subject to the same forces and the same laws." - Giordano Bruno (1548-1600), as quoted in *The Discovery of Nature* (1965), by Albert W. Bettex, London, Thames and Hudson Publishers.

One of the questions addressed in this symposium is whether the "countless earths" envisioned by Giordano Bruno have any "inhabitants" at all. This question seems much more complex than anyone could have imagined.

INTERNATIONAL ASTRONOMICAL UNION

UNION ASTRONOMIQUE INTERNATIONALE

ORIGINS: FROM THE PROTOSUN TO THE FIRST STEPS OF LIFE

PROCEEDINGS OF THE 345th SYMPOSIUM OF THE INTERNATIONAL ASTRONOMICAL UNION HELD IN VIENNA, AUSTRIA 20–23 AUGUST, 2018

Edited by

BRUCE G. ELMEGREEN
IBM T.J. Watson Research Center, USA

L. VIKTOR TÓTH
Eotvos Lorand University, Budapest, Hungary

and

MANUEL GÜDEL
University of Vienna, Austria

CAMBRIDGE
UNIVERSITY PRESS

CAMBRIDGE UNIVERSITY PRESS
University Printing House, Cambridge CB2 8BS, United Kingdom
1 Liberty Plaza, Floor 20, New York, NY 10006, USA
10 Stamford Road, Oakleigh, Melbourne 3166, Australia

First published 2019

Printed in the UK by Bell & Bain, Glasgow, UK

Typeset in System LaTeX 2ε

A catalogue record for this book is available from the British Library Library of Congress Cataloguing in Publication data

This journal issue has been printed on FSC™-certified paper and cover board. FSC is an independent, non-governmental, not-for-profit organization established to promote the responsible management of the world's forests. Please see www.fsc.org for information.

ISBN 9781108471602 hardback
ISSN 1743-9213

Table of Contents

Preface

Understanding the origin of the Sun, the Earth and how life began on Earth are some of the most challenging and interesting problems facing the science community today. With the goal of addressing many of the complex processes involved, the Scientific Organizing Committee put together this Symposium 345 at the General Assembly of the International Astronomical Union in Vienna, Austria. The meeting, held on 20-24 August 2018, had 10 invited talks, 37 contributed talks, and 151 posters, with 190 registered attendants from 34 countries. The General Assembly also included a Plenary talk on this topic by one of us (MG) and a summary of highlights from the Symposium by a second (LVT) at Division H Days.

The symposium spanned many disciplines in astronomy, planetary science, and astrobiology, following the chain of events that could have been involved in the formation of the Earth and the earliest life forms. Many steps in this process have been inferred from detailed observations and computer models. Stars form in dense interstellar clouds and are usually clustered together. The environment of the early Sun should have been affected by that clustering, as suggested by the isotopic composition of meteorites. Protoplanetary disks that may form Earth-size planets have been studied at high resolution in infrared and radio wavelengths. These observations reveal the chemistry and dust content of the disks, and possible interactions with the growing planets.

While there are no direct observations yet of Earth-size *proto*planets, there is evidence for fully-formed Earth-size planets around other stars, and there are clues to the formation process of the Earth itself from within our Solar System. New models, complemented by observations of exoplanets and solar-system planets, have revealed an important role for the host star's energy output for processing and eroding planetary atmospheres. The conditions for habitability are more difficult to observe and model. How these conditions arose and the nature of the first organisms are topics of intense study.

An accompanying summer school, "Basics of Astrobiology," was held at the University of Vienna on the two days prior to the General Assembly, 17-18 August 2018. There were approximately 100 attendees and 11 speakers. The talks are on www.youtube.com.

– Bruce G. Elmegreen, L. Viktor Tóth, Manuel Güdel, co-chairs SOC, March 2019

Editors

Bruce G. Elmegreen
IBM T.J. Watson Research Center, USA

L. Viktor Tóth
Eötvös Loránd University, Budapest, Hungary

Manuel Güdel
University of Vienna, Austria

Scientific Organising Committee

SOC Co-Chairs

Bruce G. Elmegreen	(IBM T. J. Watson Research Center, USA)
Manuel Güdel	(University of Vienna, Austria)
L. Viktor Tóth	(Eötvös Loránd University, Budapest, Hungary)

SOC Members

Yuri Aikawa	(University of Tsukuba, Tsukuba, Japan)
Muriel Gargaud	(Laboratoire d'Astrophysique de Bordeaux, France)
Lucyna Kedziora-Chudczer	(University of New South Wales, Australia)
Susana Lizano	(UNAM Morelia, Mexico)
Morris Podolak	(University of Tel Aviv, Israel)
Monica Rubio	(University of Chile, Santiago, Chile)

Acknowledgements

We are grateful to the IAU for their partial support of conference participants, to the University of Vienna and the Local Organizing Committee for their exceptional hosting of the General Assembly, to Ms. Madeleine Smith-Spanier in the IAU Paris Office for help with communication, logistics, and the compilation of this book, and to Dr. Nader Haghighipour of the University of Hawaii in Honolulu, USA, for early consultations.

Participants

Abraham, Peter
Abrevaya, Ximena C.
Ackerl, Christine
Aikawa, Yuri
Airapetian, Vladimir
Akimkin, Vitaly
Alfaro, Emilio J.
Alvarado Gomez, Julian David
Alves, Joao
Anglada, Guillem
Apai, Daniel
Artur de la Villarmois, Eizabeth
Arulanantham, Nicole
Audard, Marc
Bagrov, Alexander V.
Bai, Xuening
Berdyugina, Svetlana
Bergin, Edwin
Bertone, Emanuele
Bertrang, Gesa H.-M.
Bisikalo, Dmitry
Bjerkeli, Per
Bochkarev, Nikolai
Bogner, Rebeka
Boro Saikia, Sudeshna
Bosman, Arthur
Breuer, Doris
Broz, Miroslav
Burger, Christoph
Calcutt, Hannah
Chaparro, Germán
Chen, Lei
Chen, Tao
Chouqar, Jamila
Chrenko, Ondrej
Corazza, Lia
Cruz-Saenz de Miera, Fernando
Császár, Anna
Cunningham, Maria
Cuntz, Manfred
de Leon, Jerome
Demidova, Tatiana
Dermott, Stanley
Dionatos, Odysseas
Dorfi, Ernst
Drake, Jeremy
Drozdovskaya, Maria
Dudorov, Alexander
Dwarkadas, Vikram
Dwivedi, Navin Kumar
Dzyurkevich, Natalia
Ehlerová, Sona
Elbakyan, Vardan
Elmegreen, Bruce
Endo, Izumi
Etoh, Yuki
Federrath, Christoph
Fehér, Orsolya
Flock, Mario
Fraser, Helen Jane
Freire da Silva, Danielly
Fujimoto, Yusuke
Garai, Zoltán
Gerják, Tímea
Gregorio-Hetem, Jane
Großschedl, Josefa Elisabeth
Güdel, Manuel
Guenther, Eike
Guinan, Edward
Hacar, Alvaro
Handa, Toshihiro
Hasenberger, Birgit
Herbst-Kiss, Gabor
Hernández-Reséndiz, Patricia
Hetem, Annibal
Higuchi, Aya
Hojaev, Alisher S.
Ipatov, Sergei
Iro, Nicolas
Ismayilov, Nariman
Jatenco-Pereira, Vera
Jensen, Sigurd
JI, Jianghui
Johnstone, Colin
Jorgensen, Jes Kristian
Juvela, Mika
Kadam, Kundan
Kalas, Paul
Kamp, Inga
Kedziora-Chudczer, Lucyna
Khaibrakhmanov, Sergey
Khodachenko, Maxim
Kislyakova, Kristina
Kley, Wilhelm
Koch, Patrick
Kokubo, Eiichiro
Kolomiyets, Svitlana
Kospal, Agnes
Kouwenhoven, M.B.N. Thijs
Kreplin, Alexander
Kristensen, Lars
Krizek, Michal
Kruijssen, Diederik
Kffmeier, Michael
Kumssa, Gemechu
Lai, Dong
Lammer, Helmut
Larreina, Irati
Leitzinger, Martin
Li, Di

Li, Ji
Li, Yaping
Linsky, Jeffrey
Longmore, Steven
Lftinger, Theresa
Makalkin, Andrei
Malhotra, Satish
Mamajek, Eric
Manigand, Sébastien
Marov, Mikhail
Maury, Anaëlle
McIntyre, Sarah
Meingast, Stefan
Melnik, Anna
Mikolaitis, Šarunas
Minamoto, Harumi
Molyarova, Tamara
Monfredini, Thiago
Moor, Attila
Mugrauer, Markus
Murase, Takeru
Nandi, Dibyendu
Nemec, Nina-Elisabeth
Nguyen-Luong, Quang
Nomura, Hideko
Notsu, Shota
Osorio, Mayra
Oya, Yoko
Pacheco, Eduardo
Palous, Jan
Perez, Laura
Perotti, Giulia
Pignatale, Francesco
Pilat-Lohinger, Elke
Postel, Andreas
Pross, Addy
Pye, John
Rab, Christian
Ragulskaya, Mariya
Riaz, Basmah
Rice, Thomas
Rudnitskij, Georgij
Salinas, Vachail
Sargent, Benjamin
Sarid, Gal
Scherf, Manuel
Schlecker, Martin
Shan, Yutong
Shematovich, Valery
Shevchenko, Ivan
Skulskyy, Mykhaylo
Slyusarev, Ivan
Stapelfeldt, Karl
Starchenko, Sergey
Stephens, Ian
Szakats, Robert
Tang, Ya-Wen
Taubner, Ruth-Sophie
Tautvaisiene, Grazina
Tomida, Kengo
Toth, L. Viktor
Tsinganos, Kanaris
UEDA, Takahiro
Valio, Adriana
van den Ancker, Mario
Varga, Jozsef
Varga-Verebelyi, Erika
Vida, Krisztin
Vorobyov, Eduard
Wandel, Amri
Wei, Chen-En
White, Jacob
Wolk, Scott
Yang, Yao-Lun
Yoshida, Kento
Young, Edward
Zahorecz, Sarolta
Zeng, Shaoshan
Zhang, Man Ping
Zinnecker, Hans
Zsidi, Gabriella

Origins: From the Protosun to the First Steps of Life
Proceedings IAU Symposium No. 345, 2018
B. G. Elmegreen, L. Viktor Tóth & M. Güdel, eds.

doi:10.1017/S1743921319003120

Cosmic Pathways to Life: From Interstellar Molecules to the First Traces of Life

Manuel Güdel[1], Bruce G. Elmegreen[2] and L. Viktor Tóth[3]

[1]Department of Astrophysics, University of Vienna, 1180 Vienna, Austria
email: manuel.guedel@univie.ac.at

[2]IBM T. J. Watson Research Center, 1101 Kitchawan Road,
Yorktown Heights, New York 10598 USA email: bge@us.ibm.com

[3]Department of Astronomy, Eötvös Loránd University,
Pázmány Péter sétány 1/A, H-1117 Budapest, Hungary;
Konkoly Observatory of the Hungarian Academy of Sciences, H-1121 Budapest,
Konkoly Thege Miklós út 15-17, Hungary email: l.v.toth@astro.elte.hu

Abstract. The present-day Earth with its innumerable life forms is a product of cosmic evolution starting with the formation of our galaxy and the dense gas clouds within it, and proceeding through the contraction of one of those clouds about 4.6 Gyr ago, first into filaments and then one or more protostellar disks, planets, and central stars, one of which was our Sun. Radioactive debris from a massive nearby star was included. The planets themselves formed through coagulation, accretion, and fragmentation of solid bodies. Habitability depends on a delicate balance between disk accretion by gravity and dispersal by the central star, which determine the size of the planet and its gaseous envelope, combined with a long period of stellar radiation, which has to disperse this envelope but leave a hospitable secondary atmosphere. The final step toward life involves even more complexity as self-replicating bio-molecules form with ever increasing stability.

Keywords. astrobiology, astrochemistry, magnetic fields, Sun: activity, stars: activity, Earth, planets and satellites: formation, solar system: formation, stars: formation

1. Introduction

Observations of local interstellar clouds at infrared to millimeter wavelengths show solar-mass stars and their proto-planetary disks in the process of formation. The nearest disks seen at high resolution contain dust, gas, and ice with structures such as gaps or spirals suggestive of embedded planets and their dynamical processes. The last few years saw major advances in our view of interstellar cloud structures and the differences between low mass and high mass star formation, differences that may in fact be important for the eventual formation of terrestrial planets around solar-type stars.

Optical observations of thousands of stars up to fairly large distances in the Milky Way show periodic dips in brightness up to the percent level, indicating partially occulting planets with measurable sizes. Absorption spectra obtained during planetary transits have given first hints on the compositions of the planetary atmospheres. These advances over the last decade, combined with increasing complexity in computer modeling, bring us closer to visualizing how the Earth might have formed and what the conditions might have been like at its beginning.

The ultimate quest is not only about the origin of the Earth, but also about the origin of life on Earth. There has been progress on this topic also, but most of it has been outside the realm of astronomical observations. Still, there are increasingly sophisticated observations and modeling of possible life-related chemical reactions on other bodies in the Solar System, growing understanding of autonomous activities in RNA, DNA and other bio-molecules, and new physical explanations for what it means to be alive in a world governed by unshakable laws of physics.

This review is based on a plenary lecture by one of the authors (Manuel Güdel) at the General Assembly of the International Astronomical Union in Vienna Austria, delivered on August 22nd, 2018. A series of invited review and contributed talks accompanied this plenary lecture as part of IAU Symposium 345 on "Origins: from the formation of the Sun to the first steps of life." We refer to these other talks here, as published in the present volume. They consider each topic in more detail and cover both observations and numerical modeling at a wide range of resolutions. This plenary talk also benefited from contributions by the Scientific Organizing Committee, as acknowledged below.

The key to life as we know it is the presence of liquid water, presumably on a solid, probably rocky, surface with an atmosphere above it. For the last several billion years, while life on Earth was developing, the conditions in this environment, the water, land and atmospheric pressure, have not changed much, giving the false impression that the Earth might have always been a hospitable planet. However, young solar-type stars are highly variable with strong winds and X-ray emission for their first tens to hundreds of Myr, planetary distances from their stars can change on timescales of 10 to 100 Myr, and outgassing from magma oceans and through volcanism as well as accretion from interplanetary bodies can change the physical conditions on the planet's surface and even the composition of its atmosphere, especially during the first few hundred Myr of a planet's life. The early Earth and other Earth-like planets may not have been so hospitable when proto-life was at its earliest stages. This means that a complete understanding of the origin of the Earth and life as we know it requires detailed astronomical observations of the conditions around other solar-like stars and their Earth-like planets.

2. Formation of the Sun

The Earth and Sun formed at about two-thirds of the present age of the Universe. Figure 1 shows a time line of the cosmic star formation rate (SFR), measured in solar masses per year per cubic Megaparsecs, averaged over all space (from Madau & Dickinson 2014). At early times, the SFR was low because most of the galaxies in which stars form were small. The rate reached a peak at around 10 Gyr ago and has been decreasing slowly ever since because the average density of the universe has been decreasing, and because an increasing fraction of the remaining cosmic gas has become too hot to fall onto the denser parts of galaxies where it can turn into cold clouds and collapse into stars. The peak SFR in our own galaxy, the Milky Way, occurred about 8 Gyr ago, and by the time the Sun and Earth formed, 4.6 Gyr ago, the Milky Way had quieted down by a factor of ~ 3 in its conversion of gas into stars. Life crawled up on land only ~ 0.5 Gyr ago, after living in the ocean for its first ~ 3 Gyr.

Galaxy mergers and cosmic gas infall are important parts of the history of star formation. Because a typical starburst lasts only a few hundred million years, star formation would have stopped early without the continued accretion of new material. It was during the later phase of quiescent accretion when the Solar System and early life formed, after most of the violent collisions with the young Milky Way ended.

Solar-mass stars form in a wide variety of environments, ranging from low-mass regions with only a few other stars to massive young stellar complexes and dense star clusters. The most massive complexes are relatively rare and the nearest examples are fairly distant, making them difficult to study in detail. Nearby examples of solar-mass star formation include the Taurus region at 140 pc (see the Symposium review by Mika Juvela). Figure 2 shows this region with several well-known objects identified, i.e., the

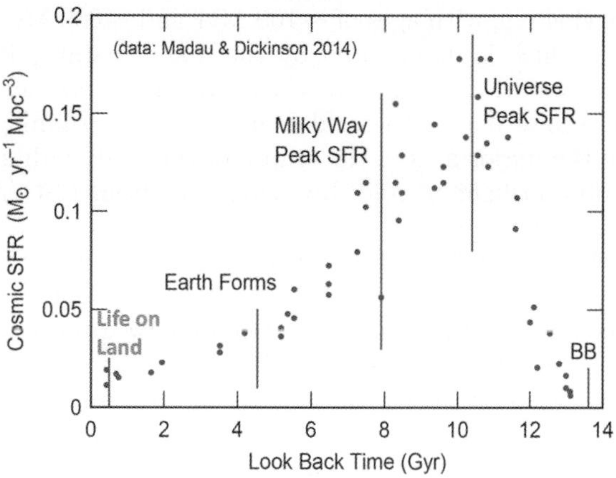

Figure 1. The history of star formation in the universe, averaged over space, from Madau & Dickinson (2014). The universe began with the Big Bang ("BB") around 13.7 Gyr ago, which is on the right-hand side of the plot. The star formation rate increased at first as star-forming galaxies grew in size. Then it decreased as the surrounding gas density decreased with the continued expansion of the Universe. It also decreased as infalling gas heated up in massive dark matter halos and became too hot to accrete onto the enclosed galaxy disks. The Milky Way formed as a small galaxy at first, reaching its peak star formation rate around 8 Gyr ago. The Earth and Sun formed when the Milky Way had quieted down to about one-third of its peak rate.

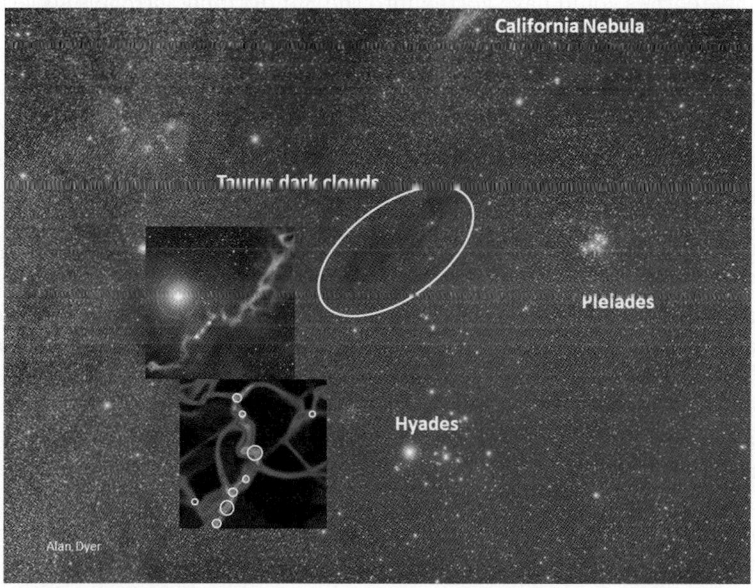

Figure 2. Sky view of the region around the Taurus dark clouds, including the Pleiades and Hyades star clusters. The Taurus clouds are filaments, like most star-forming clouds, with dense knots as the collection points where gravity brings the gas together. The insert on the left is an infra-red view from Hacar *et al.* (2013), and the insert below that is a numerical simulation of a turbulent cloud, from Paolo Padoan (private communication). The background star field is from Alan Dyer (https://amazingsky.net/2011/02/21/the-dark-clouds-of-taurus/).

Pleiades and Hyades clusters, with ages of ~ 100 Myr and ~ 600 Myr, and the California Nebula in the north, which is illuminated by the massive star ξ Persei. Shown inside a white oval are several dark filaments in which low-mass stars are currently forming. An inset on the left shows one of these filaments (the one running to the north-west approximately along the major axis of the ellipse of the oval, with a bright star next to it) glowing in the sub-mm light of cold dust emission (from ESO/APEX; Hacar *et al.* 2013). Several bright spots can be seen from embedded young stars. Also inset is an image from a numerical simulation of a star-forming filament, from Dr. Paolo Padoan (based on work in Padoan & Nordlund 2002). The similar structure results from shock formation in a supersonically turbulent fluid followed by gravitational collapse in the shock-compressed gas (see the Symposium review by Christoph Federrath). Dynamical feedback from young stellar objects maintains an overall low star formation efficiency, although it may be higher where bound stellar clusters form.

While the Taurus clouds are often used as an analogue for solar-type star formation, we now believe the proto-sun formed in a region with more massive stellar companions. The most suggestive evidence for this is the presence of the ^{26}Al decay product, ^{26}Mg, inside Ca and Al-rich inclusions in the most primitive type of meteorite, carbonaceous chondrites (see the Symposium review by Edward Young). ^{26}Al is formed by proton capture in the Mg-Al catalytic cycle of nuclear fusion, which occurs at high temperatures in massive stars. Its half-life is 0.7 Myr, which is so short that the Sun had to form in the midst of these massive stars, either in a massive complex with such stars also forming, or around massive complexes where the gas is contaminated by the semi-continuous production of this radioactive material through on-going star formation (see the Symposium chapter by Yusuke Fujimoto). The ^{26}Al and a few other radio-active elements in meteorites could have gotten into the Sun-forming gas via a supernova explosion that expelled these elements along with a lot of other mass when the massive star exploded, or via the winds from aging massive stars, such as Wolf-Rayet stars, which also carry along proton-capture nucleosynthetic byproducts (see the Symposium chapter by Vikram Dwarkadas). Among these two possibilities, supernovae may be less favored because radioactive iron that should come from a supernova, ^{60}Fe, does not appear to be abundant enough in meteorites.

If the Sun formed along with high mass stars, then regions like the Orion OB association have an increased relevance (see the discussions by Diederik Kruijssen and Michael Küffmeier in this volume.). Large scale surveys of the Galactic cold interstellar medium reveal a complex structure of elongated clouds with connecting striations and cold clumpy filaments, as reviewed by Mika Juvela in this volume. High resolution ALMA-based observations of the filaments resolve smaller filaments, or fibers, inside of them, possibly feeding the star-forming cores. Then the mass of the forming stars could depend on the local "fiber surface density" as suggested by Alvaro Hacar in this volume. In his examples, cluster-forming filaments in Orion present fibers with 2 orders of magnitude higher density than analogous fibers in Taurus. Methods to find and quantify filamentary structures are discussed in the paper by Maria Cunningham in this volume. Observations of dense, star-forming gas throughout the Milky Way are discussed by Quang Nguyen-Luong and gas accretion processes for local protostars are discussed by Ian Stephens in this volume.

Recent astrometric data from the Gaia satellite are now helping us find the former members of dispersed OB associations and understand the kinematics of young stellar clusters (see review by Anna Melnik in this volume). We are also witnessing a significant increase in our chances to link young stellar objects to interstellar clouds where they formed, as shown for Orion clusters by Josefa Elisabeth Grossschedl in this volume.

The remnants of ^{26}Al in carbonaceous chondrites are important not only for tracing the early environment of the Sun, but also for suggesting a significant heat source in

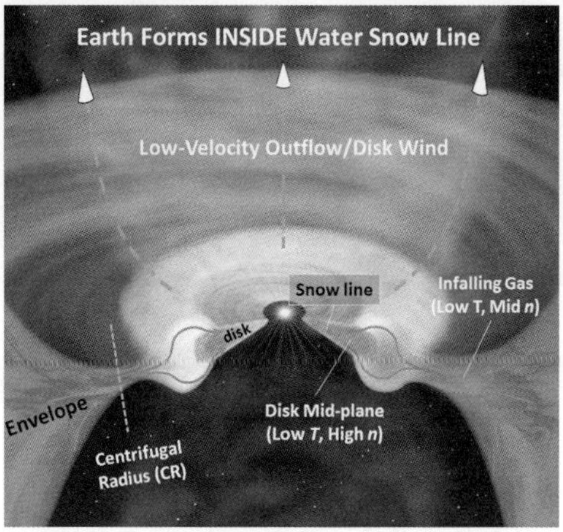

Figure 3. Schematic of a proto-solar disk showing the vertical structure in cross-section (from Yuri Aikawa) (Sakai *et al.* 2017, Credit: RIKEN/Japan). The inner region is warmed by the Sun, but the outer region is cold enough to form ice on grain surfaces. The envelope consists of gas that has fallen in from the surrounding dense cloud. A bi-conical cavity has been cleared out by a wind.

young planetesimals, altering their thermo-mechanical evolution and outgassing volatiles (e.g., Lichtenberg *et al.* 2016). Extrasolar planetesimals without ^{26}Al would have more ice because of this lack of heating, and they would deliver perhaps 10 times more water to their terrestrial planets (Ciesla *et al.* 2015, Lichtenberg *et al.* 2019). The risk of producing water planets (planets with massive, deep water oceans) would be much higher, and such planets are probably much less conducive to forming life. The formation of terrestrial life seems to profit from the availability of surface where water and land meet.

3. Proto-planetary Disks

Figure 3 shows a schematic view of the formation of a low-mass proto-planetary disk like the one that might have formed the Earth. The young star in the center is shown with a small cleared cavity around it (e.g., Andrews *et al.* 2016), and this is surrounded by a disk with an increasing thickness that channels a polar outflow on both sides (for observations of disk outflows, see the discussion by Per Bjerkeli in this volume.). Towards the edge of the disk, the centrifugal force balance for inner circular orbits changes into a more radial force balance between accretion ram pressure and thermal disk gas pressure. The edge of the disk is also heated by this accretion. The figure shows the demarcation between an inner warm part, heated by the star, and a cooler middle part where water ice and snow can form on grain surfaces. Essentially all of these features have been observed, including the snow line (see Symposium reviews by Laura Pérez and Inga Kamp and a discussion of the inner parts of disks around low mass stars by Jozsef Varga). Curiously, the Earth is positioned too close to the Sun to lie within the snow region, making the origin of the Earth's water more indirect. The contribution by Germán Chaparro Molano *et al.* to this Symposium illustrates the many possible types of planetary systems and planet distributions that can come from disk evolution. The complex chemistry of disks and their envelopes is discussed in this volume by Yoko Oya, while disk MHD simulations are discussed by Mario Flock. The effect of stellar flares on disk chemistry is considered in the contribution by Peter Abraham.

Proto-planetary disks are also magnetic. Figure 4 shows the magnetic field morphology of a young stellar object from the observations by Alves *et al.* (2018). The field orientation is shaped by the flow direction, as indicated by the shades of velocity. The magnetic field

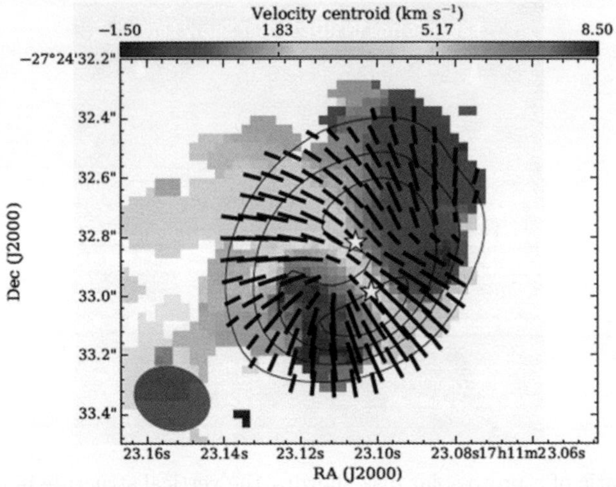

Figure 4. Velocity and magnetic field structure in the proto-solar nebula BHB01-11, from Alves *et al.* (2018). The lines showing the magnetic field orientation are from polarization emission in the far-infrared, rotated by 90°. The velocities are from H_2CO emission and the contours are from Stokes I emission.

morphology can be inferred from the polarized emission of elongated dust grains aligned perpendicular to the magnetic field lines. Dust self-scattering and radiative alignment also produce polarized emission (e.g., Kataoka *et al.* 2017). Azimuthal magnetic fields can stabilize a disk against gravitational instabilities that could form planets (Lizano *et al.* 2010). Observations reported by Anaëlle Maury in this volume suggest an increased importance for magnetic fields from the apparent small sizes of Class 0 protostars.

4. Planetesimal Formation

The heavy elements in a cosmic mixture generally condense into micron-size grains or smaller, and this is the form they most likely had when they entered the solar system along with the gas in a collapsing interstellar cloud. In the high density of the pre-solar nebula, the grains can stick together and accumulate into larger objects, which through a series of collisions and fragmentations, ultimately make rocky or icy planetesimals. Continued collisions combined with ever increasing gravitational forces eventually make the rocky planets like Earth, and for stronger gravity, the larger planets like Jupiter (see the Symposium review by Eiichiro Kokubo, and the discussions in this volume by Francesco Pignatale about the early stages of chondrite formation, and Sergei Ipatov about the formation of the moon). These grain collisions are aided, in part, by their differential drift through the gas, which is caused by a slightly slower orbital velocity of the gas component in the presence of a radial pressure gradient (Weidenschilling 1977). The grains do not feel this full pressure gradient; their attempt to orbit at Kepler speed meets with a headwind, breaking their orbital motion differentially with regard to size so that relative velocities build up and therefore collisions between particles become frequent.

Figure 5 shows how grain-grain interactions depend on their size (from Windmark *et al.* 2012). The tiniest grains can stick together (region "S" in the figure) when they collide, but millimeter and centimeter size grains can stick and bounce ("SB") or bounce ("B") off each other, or transfer mass ("MT") and erode ("E") the largest of the pair. Large equal-size grains can fragment ("F") when they collide. While the fragmentation regime for similarly sized bodies counteracts planetary growth, the "bouncing barrier" (in yellow)

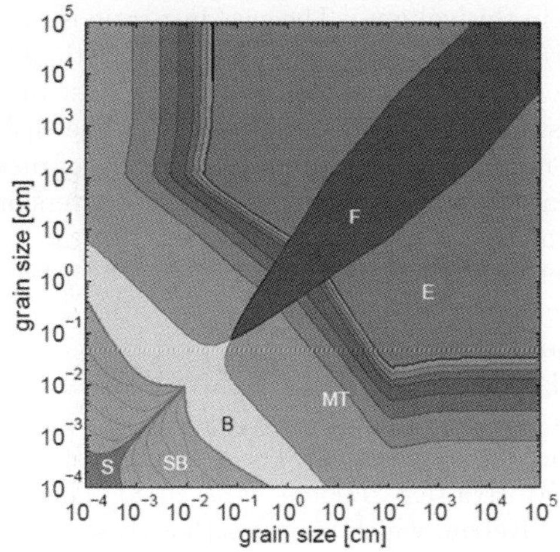

Figure 5. The outcomes of collisions between solid particles are shown by the various shadings in this diagram, as a function of the particle sizes. The letters represent collisions that are sticking (S), stick and bounce (SB), bounce (B), mass transfer (MT), erosion (E), and fragmentation (F), from Windmark *et al.* (2012). The relative velocities between the particles are from a distribution with four sources of grain drift: Brownian motion, turbulence, azimuthal drift and radial drift.

keeps most of the particles small; if only a few bodies cross the barrier by some exceptional conditions, then they can efficiently grow by sweeping up the large number of small bodies without being rapidly fragmented.

The resulting planetesimals that form in icy regions may look like the icy surfaces of comets. Comets that have been photographed by space missions have complex chemical compositions and inhomogeneous surfaces, with important chemicals for life, like water, phosphorus, amino acids and other organics (Altwegg *et al.* 2016).

5. Molecular Emission from Disks

A variety of ring morphologies are found in molecular emission line observations of proto-planetary disks. Ices on grain surfaces tend to occur beyond a certain radius where the temperature is low. There are several ways to probe these snow lines. CO is usually too optically thick, frozen out on dust, or depleted for some other reason to see its full density structure, but its presence and the transition from gaseous CO to frozen CO can be inferred indirectly using N_2H^+ observations. N_2H^+ is destroyed by chemical reactions with CO in the gas phase, so a sudden transition to N_2H^+ emission at some radius could mean the presence of gaseous CO inside that point and frozen CO outside. TW Hya is an example: N_2H^+ is observed only outside a certain radius; that boundary, at a few tens of AU, should be the CO snow line (Qi *et al.* 2013). This radius is large enough to be spatially resolved with ALMA observations. The water snow line is generally much farther in and difficult to resolve in typical disks. When a disk experiences a burst of accretion, however, as in FU Ori type stars, the water snow line extends outwards because of the surge in disk temperature. Around the FU Ori star V883 Ori, Cieza *et al.* (2016) found an abrupt change in dust brightness at 40 AU, which may be caused by dust fragmentation and an increased dust surface area inside the temporary water snow line. Emission of CH_3OH, which has a similar volatility to water, is indeed detected around this transition radius (Lee *et al.* 2019).

There are many ALMA images of dust continuum emission from proto-planetary disks (see review by Laura Pérez in this volume). For planet formation, we also need to know the distribution of the gas, which is the major component in the disk (see reviews by Inga

Kamp and Xuening Bai in this volume). The gas/dust ratio is crucial for some theories of the formation of Jupiter-type planets, which rely on gravitational instabilities or gravity-driven waves (Lichtenberg & Schleicher 2015). The gas/dust ratio is also important for radial migration of dust particles and protoplanets in the disk. Since H_2 is hard to observe, CO and other molecular lines are used as tracers of disk gas. CO depletion by sedimentation of ice-coated dust grains and chemical conversion to other species is a critical issue in the estimation of the gas/dust ratio. H_2O can also be photodissociated to be converted to other molecules. Thus the emission by molecules and the inference of disk properties depends on the interaction between the disk and the stellar UV and ionization by X-rays and decaying ^{26}Al.

While simple molecules such as CO and CN have long been detected in disks, detection of larger molecules became possible only recently thanks to ALMA (see the paper in this Symposium by Jes Jørgensen *et al.*). For example, CH_3OH, CH_3CN and HCOOH have been detected in TW Hya (e.g., Loomis *et al.* 2018). The excitation temperature of CH_3CN is several tens of Kelvin, which indicates that these molecules are non-thermally desorbed from ice. The detection suggests that exoplanet systems could be rich in organics, as was our Solar system (Nakamura-Messinger *et al.* 2006, Altwegg *et al.* 2016). Quantitative comparisons of their abundances with those in comets are of special interest, but not straightforward.

An important question is how the Earth got its water. As mentioned above, the Earth probably formed in a water-dry zone because of its close proximity to the Sun, while water-rich planetesimals formed further out where the temperature was lower. These outer planetesimals could have reached the Earth during a large-scale radial migration of the giant planets, which would have scattered the smaller objects all around (Walsh 2011). In such models, Jupiter-like planets could have had an important role in the origin of life on Earth by indirectly delivering water. The contribution by Mikhail Marov to this volume discusses more about the delivery of water to Earth.

Doris Breuer in this Symposium discusses how water could have come to Earth in two forms: hydrated solids that built up the interior and degassed through volcanic action to supplement the atmosphere, and volatile-rich comets and asteroids that deposit their water directly into the atmosphere, oceans, or surface layers. Plate tectonics on the Earth could have an additional effect on moving water both into and out of the mantle, establishing an equilibrium in surface water content. Planets like Mars without plate tectonics could be continuously losing their water.

6. Gaseous Envelopes around a Young Earth

A young Earth-size planet can accrete a massive H+He envelope from the gas surrounding it in the disk. Because the gaseous part of a protoplanetary nebula is 100 times the mass of the rocky, dusty part, this envelope may contain several Earth-masses of gas, depending on how massive the planetary core grows in the disk. If such an envelope formed and could not be dispersed, then the Earth would be more like a mini-Neptune today. These are not atmospheres in the conventional sense of terrestrial planets. Their surface pressures are huge, and their temperatures high, not supporting habitable conditions. This could be true even for an Earth-like planet in the so-called habitable zone of its star, because that zone is usually defined for planets that contain only light atmospheres like the present-day Earth, Venus or Mars.

There are two solutions to this problem: *First,* the Earth-like planet could form very late, after the nebular gas is gone. One way to limit the lifetime of the proto-planetary disk itself and therefore limit envelope accretion is to remove the gas in a disk is through irradiation by stellar X-rays and/or extreme ultraviolet, which can heat the disk and evaporate it (Alexander *et al.* 2006, Ercolano *et al.* 2008). Once the gas disk is removed, the envelope around a terrestrial planet feels less confining pressure and can expand. According to Stökl *et al.* (2015), low mass terrestrial planets can lose a significant fraction of their envelopes, but high mass planets including those with about an Earth mass, cannot.

Second, if an envelope has already built up when the gas disk disperses, then the irradiation of the envelope by strong stellar X-ray and extreme-ultraviolet (XUV) radiation can heat and progressively evaporate the upper layers of the envelope until it may completely disperse. Under typical circumstances, if the planet avoids accreting up to about one Earth mass already during the disk lifetime of a few million years, its envelope remains thin enough to escape rapidly. If the planet does accrete about two or more Earth masses, it inevitably evolves into a mini-Neptune that will never lose its envelope during the main-sequence life of the host star (Lammer *et al.* 2014.) We see here that a delicate balance between timescales of disk dispersal, planetary solid accretion and stellar activity is crucial for the fate of planetary habitability. Further discussion of atmospheric escape from young planets is in the contribution to this volume by Valery Shematovich and Dmitry Bisikalo. Atmospheric models of habitable planets are in the discussion by Nicolas Iro.

7. Stellar Activity

While the Earth was slowly forming in the proto-solar nebula, the Sun was forming also. Observations of young stars like the Sun show them to be highly active with typically fast rotation, strong variability and a high X-ray flux from flares and other energetic processes (see the discussions in this volume by Orsoly Fehér and John Pye). Based on these observations, we infer that the young Sun had a faster rotation rate than today, and more dynamo action with a stronger magnetic field as a result. A strong magnetic wind then caused the Sun to spin down and lose its high level of XUV and wind activity.

A magnetized wind has so far only been detected from the Sun (the Solar Wind), but the evidence for stellar magnetized winds is the systematic lengthening of the rotation periods of solar and lower mass stars over a half Gyr timescale (Gallet & Bouvier 2015, Johnstone *et al.* 2015). This spin-down is, however, not unique - it in fact depends on the initial rotation period of the star after the proto-planetary disk dispersed. After initial spin-up due to continuing stellar contraction, the star spins down due to angular momentum removal by the wind, but this spin-down is slow if the star started out as a slow rotator, and it is fast if it initially rotated rapidly (up to a hundred times the present solar rotation rate), as both observations of stellar samples and theoretical wind models confirm (Johnstone *et al.* 2015). While this spin-down behavior leads to nearly the same rotation rate after a billion years no matter what the initial rotation rate was, it is interesting that in the time range between around ten and a few hundred Myr stars can take widely differing evolutionary paths with corresponding XUV levels differing by factors of 20 to 40 between fast and slow rotators (Figure 6). By pure accident, this coincides with the time period in which the Earth formed a crust and liquid ocean and presumably outgassed its secondary atmosphere. Such stellar variation is therefore crucial for removing H/He envelopes, but it is also crucial regarding terrestrial-type secondary (outgassed) atmospheres. Too much high-energy irradiation may remove secondary atmospheres, but too low irradiation levels may fail to remove primordial H/He envelopes. Another delicate balance thus becomes evident for the development of habitable conditions, namely between H/He envelope removal and the subsequent preservation of a secondary atmosphere. The main stellar parameter that determines the outcome of this balance is the stellar initial rotation rate, for which nature offers a spread of about two orders of magnitude and therefore a variety of planetary outcomes for any given primordial H/He

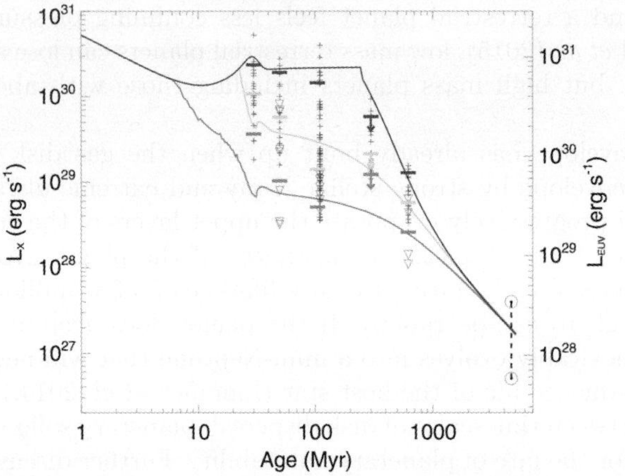

Figure 6. Predicted evolution of L_X (left y axis) and corresponding L_{EUV} (right axis) for three initial rotation rates for a solar-mass star, evolved using a rotation-wind-angular momentum evolution model (Johnstone *et al.* 2015). The blue, green and red curves show the evolutionary tracks for an initially fast, medium, and slow rotator, respectively, for the 10th, 50th, and 90th percentiles of the rotation rate distribution in very young clusters. The individual + symbols are observed values in co-eval cluster samples, inverted triangles mark upper limits. The solid horizontal bars show the 10th, 50th, and 90th percentiles of the observed distributions of L_X at each age. The two solar symbols at 4.5 Gyr show the range of L_X for the Sun over the course of the solar cycle (from Tu *et al.* 2015).

envelope and outgassed atmosphere. Some combinations may lead to habitable conditions, while many may not. More on this topic is in the Symposium contribution by Theresa Lüftinger.

Smaller planets like Mars may lose their atmospheres completely due to stellar activity because their gravity is small. Erkarv *et al.* (2014) showed that Mars should have lost both its H/He envelope and its outgassed H_2O and CO_2 atmospheres in about 10 Myr, with the possibility that its current atmosphere outgassed or was delivered by impacts later.

In summary, these considerations suggest that long-term habitability depends on more than the possibility of liquid water under an Earth-like atmosphere. It requires that the H and He-rich primordial envelope around the planet be removed by solar activity (or that the planet forms very late, after the nearby nebular gas has been removed), and that the planet have enough mass (bigger than Mars) to resist complete solar removal of its atmosphere and be far enough away from the early Sun (further than Venus) to avoid getting too hot.

The long-term evolution of the central star is important, too. The Sun was bolometrically about 10 times brighter when it formed than it is now, and its bolometric luminosity dropped by a factor of ~ 100 in the first 10 Myr. It gradually increased in the next 30 Myr and then, as terrestrial planets accreted to their final mass, stabilized to about a constant output over the remaining 4 Gyr. In another 4 Gyr the Sun's luminosity will increase back up by a factor of about 10 as it turns into a red giant.

Stars with non-solar masses may not have such a favorably long time with a near-constant luminosity. Solar-type stars are ideal as they settle quickly to a steady brightness while the planets are forming and then remain fairly constant for several Gyr, starting with the era when the Earth acquired a habitable, water-rich surface and atmosphere, and sophisticated life formed and developed. More massive stars evolve more quickly at first but burn out sooner. Lower-mass stars settle to their stable states much more slowly.

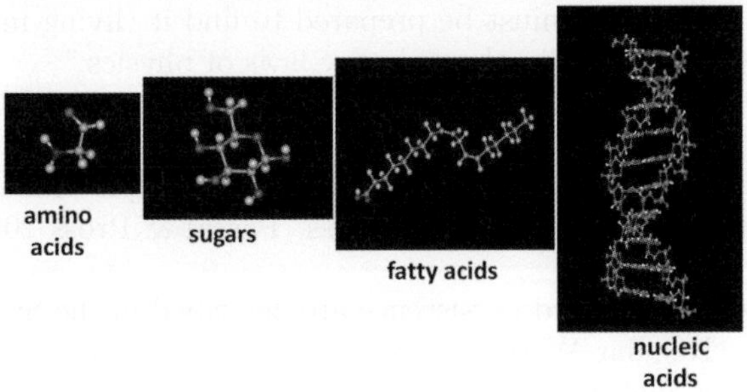

Figure 7. Complex molecules that are the building blocks of life, from the contribution to this volume by Addy Pross.

An M1-type star with about half the mass of the Sun takes ~ 100 Myr before its luminosity stabilizes; before this time, the luminosity drops by a factor of ~ 10 as a power law with time. The situation becomes progressively more serious for later-type M dwarfs, the latest ones declining by a factor of ~ 100 in bolometric luminosity over at least one Gyr. In a case like this, a planet orbiting it at a certain radius may be inside the habitable zone at first, baking in an early hot phase and losing its oceans and atmosphere. The habitable zone then shrinks as the luminosity decreases, passing though the planet's orbit until it lies inside, at which point the planet may be too cold for liquid water (see, e.g., Ramirez & Kaltenegger 2014). Further discussions of the effects of stellar flares on habitability are in the contribution by Adriana Valio in this volume, while Ximena Abrevaya reports experiments on biological reactions to UV radiation.

8. Constraints from Orbital Dynamics

How important are the outer planets for Life on Earth? They seem to be quite important. Pilat-Lohinger *et al.* (2008) and Pilat-Lohinger (2015) considered orbital perturbations affecting the Earth's eccentricity from the outer planets Jupiter and Saturn. They show that if Saturn had ended up at 8.7 au instead of 9.5 au, the eccentricity of the Earth would vary periodically over a few Myr timescale, reaching maximum eccentricities of 0.6–0.7. Then the Earth at perihelion would bake with an equatorial land surface temperature of 80 C, while the Earth at aphelion would freeze with a temperature of near 0 C even at the equator (Williams & Pollard 2002). Stability of the Earth seems to be a key factor in the development of life, and both the central star and the other planets can have a strong influence on that. The implications of planet spin-orbit locking to habitability is discussed in the contribution by Amri Wandel. Observations of Earth-like planets are reviewed by Daniel Apai.

9. Origin of Life

What defines the transition from chemistry to biology? Some life forms such as the cyanobacteria have existed on Earth nearly unchanged in morphology for several billion years. Where did the first biomolecules come from? The amino acids, sugars, fatty acids, and nucleic acids shown in Figure 7 are not life and yet they make up living things. Lipids make vesicles, amino acids make peptides and proteins, nucleotides make RNA and DNA. The basic ingredients of these complex structures are molecules commonly found in interstellar space, H_2, CO, CH_4, NH_3, N_2, HCHO, HCN.

The big unknown is how relatively simple molecular systems with hundreds of atoms assembled into complex biological systems with trillions of atoms, in apparent contradiction to the second law of thermodynamics, which tends to mess things up. To quote

Erwin Schrödinger (1944): "we must be prepared to find it (living matter) working in a manner that cannot be reduced to the ordinary laws of physics."

A general rule governing change in both physical and biological worlds is that unchanging things do not change and changing things do change until they change into things that do not change. Nature may have used this logical principle to make persistent forms of life, with replication the key physical process (Pascal & Pross 2015, Pascal & Pross 2019).

Two evolutionary paths toward persistence are described in the Symposium review by Addy Pross. In the "Regular World", random processes reach thermodynamic equilibrium, and the relevant mathematics is Boltzmann statistical mechanics. In particular, the log of the number of states in a system, which is proportional to the entropy, always increases with time. In the "Replicative World", however, multiplicative processes cause exponential growth. Then the relevant mathematics is Maltusian growth governed by terms involving the sources (replication) and sinks (death) of particular structures. The basic equation for the Replicative World is

$$dX/dt = kXM - gX, \tag{9.1}$$

where population concentration X replicates in proportion to the concentration of resources, i.e., building blocks, M, and the rate, k, and decreases in proportional to a destruction rate, g. When $kM > g$, the growth is exponential until the resources, M, get depleted and $kM = g$, as which point the population reaches equilibrium. Pross calls this "Dynamic Kinetic Stability."

Examples of structures in thermodynamic stability in the Regular World are the chalk cliffs around Dover, UK. They have stability and persistence for 60-100 Myr through a lack of chemical reactivity. Examples of Dynamic Kinetic Stability in the Replicator World are ants, which have persisted for 140-160 Myr, and cyanobacteria, which have persisted for 2000-3000 Myr. Their stability comes from bounded self-replication.

There is no clear evidence for life outside the Earth at the present time. Eduardo Janot Pacheco in this volume considers possible signatures in molecular emission lines for what might be pieces of DNA in interstellar clouds. Possible signs for life on the water-moon Enceladus were discussed by Ruth-Sophie Taubner.

10. Conclusions

There is a sequence of events that have to go right for the formation of life on an Earth-like planet. There has to be a star-formation environment that can make a solar mass star, possibly even with ^{26}Al production if that turns out to be necessary to help melt planetesimals and control the delivery of water. The collapse of the interstellar cloud may need to bring in certain molecules too, such as water, if these molecules cannot be made in the required abundance in the protosolar disk itself (Cleeves *et al.* 2014). Disk chemistry and gas-grain reactions need to put water and other important molecules for life in a usable form, such as in ices on grain surfaces that can be transported to rocky planets. Other organics may need to form in the nebular disk as well. The disk grains then have to grow into planetesimals and eventually Earth-size rocky bodies, which need a light atmosphere to avoid over-heating by the greenhouse effect. This latter step requires either the removal of a massive gaseous envelope of nebular H and He by energetic processes from the young Sun, or a delayed formation of the Earth to a time when there is no local nebular gas. How stellar energetic processes evolve and erode planetary atmospheres depends primarily on the initial stellar rotation period, which in turn is determined by accretion and protostellar disk evolution and dispersal. Water ice and other ices from the cooler regions of the disk, possibly accompanied by organic molecules, have to scatter into the Earth's orbit and eventually to the Earth's surface. Soon after this, the Sun has to be relatively stable for a long time, without large luminosity fluctuations that could throw the Earth into boiling or freezing conditions or destroy its atmosphere.

There could be other environments and pathways to life too, such as through chemical reactions at the rocky bottoms of oceans that are frozen over on the surface. Perhaps some of this diversity can be tested in our own Solar System.

Evidently, life is a product of astrophysical diversity. Astrophysical initial conditions and boundary conditions of protostellar collapse, disk evolution and initial stellar mass and angular momentum as well as planetary properties such as growth rate, final mass and outgassing rate offer many planetary pathways. Apart from initial conditions, the timing of a series of successive events in the star-planet environment and in the planet itself must be right, and timescales matter to fulfil critical conditions for the evolution toward a habitable surface environment. Many pathways will fail to produce habitable conditions even at the "right distance" from the host star because various timescales are "wrong", but some combinations will succeed, as our solar system amply proves.

Acknowledgments

We are grateful to the Scientific Organizing Committee of IAUS 345 for their help in preparing the Plenary talk given by one of us (MG) at the XXXth General Assembly in Vienna on August 22, 2018. This talk was the basis for the present paper. These SOC members were Drs. Yuri Aikawa, Monica Rubio, Susana Lizano, Muriel Gargaud, Morris Podolak, and Lucyna Kedziora-Chudczer.

References

Alexander, R. D., Clarke, C. J., & Pringle, J. E. 2006, *MNRAS*, 369, 216

Altwegg, K., Balsiger, H., Bar-Nun, A., *et al.* 2016, *Science A*, 2, E0285

Alves, F. O., Girart, J. M., Padovani, M., Galli, D., Franco, G. A. P., Caselli, P., Vlemmings, W. H. T., Zhang, Q., & Wiesemeyer, H. 2018, *A&A*, 616, A56

Andrews, S. M., Wilner, D. J., Zhu, Z. *et al.* 2016, *ApJ*, 820, L40

Ciesla, F. J., Mulders, G. D., Pascucci, I., & Apai, D. 2015, *ApJ*, 804, 9

Cieza, L. A., Casassus, S., Tobin, J. *et al.* 2016, *Nature*, 535, 258

Cleeves, L. I., Bergin, E., A., Alexander, C. M. O. D., Du, F., Graninger, D., Öberg, K. I., & Harries, T. J. 2014, *Science*, 345, 1590

Ercolano, B., Drake, J. J., Raymond, J. C., & Clarke, C. C. 2008, *ApJ*, 688, 398

Erkaev, N. V., Lammer, H., Elkins-Tanton, L. T., *et al.* 2014, *P&SS*, 98, 106

Gallet, F., & Bouvier, J. 2015, *A&A*, 577, A98

Hacar, A., Tafalla, M., Kauffmann, J., & Kovács, A. 2013, *A&A*, 554, 55

Johnstone, C. P., Güdel, M., Brott, I., & Lüftinger, T. 2015, *A&A*, 577, A28

Kataoka, A. Tsukagoshi, T., Pohl, A. *et al.* 2017, *ApJL*, 844, L5

Lammer, H., Stökl, A., Erkaev, N. V., Dorfi, E. A., Odert, P., Güdel, M., Kulikov, Yu. N., Kislyakova, K. G., & Leitzinger, M. 2014, *MNRAS*, 439, 3225

Lee, J.-E., Lee, S., Baek, G., *et al.* 2019, *Nature Astron.*, in press, arXiv:1809.00353

Lichtenberg, T. & Schleicher, D. R. G. 2015, *A&A*, 579, A32

Lichtenberg, T., Parker, R. J., & Meyer, M. R. 2016, *MNRAS*, 462, 3979

Lichtenberg, T., Golabek, G. J., Burn, R., *et al.* 2019, Nature Astronomy, Advanced Online Publication

Lizano, S., Galli, D., Cai, M. J., & Adams, F. C. 2010, *ApJ*, 724, 1561

Loomis, R. A., Cleeves, L. I., Öberg, K. I., Aikawa, Y., Bergner, J., Furuya, K., Guzman, V. V., & Walsh, C. 2018, *ApJ*, 859, 131

Madau, P. & Dickinson, M. 2014, *Annu. Rev. Astron. Astrophys.*, 52, 415

Nakamura-Messenger, K., Messenger, S., Keller, L. P., Clemett, S. J., & Zolensky, Michael E. 2006, *Science*, 314, 1439

Padoan, P. & Nordlund, Å. 2002, *ApJ*, 576, 870

Pascal, R. & Pross, A. 2015, *Chem. Commun.*, 5, 16160
Pascal, R. & Pross, A. 2019, *J. Syst. Chem.*, in press
Pilat-Lohinger, E., Süli, Á., Robutel, P., & Freistetter, F. 2008, *ApJ*, 681, 1639
Pilat-Lohinger, E. 2015, *Int. J. Astrobiology*, 14, 145
Qi, C., Öberg, K. I., Wilner, D. J., D'Alessio, P., Bergin, E., Andrews, S. M., Blake, G. A., Hogerheijde, M. R. & van Dishoeck, E. F. 2013, *Science*, 341, 630
Ramirez, R. M., & Kaltenegger, L. 2014, *ApJ*, 797, L25
Sakai, N., Oya, Y., Higuchi, A. E., *et al.* 2017, *MNRAS*, 467, L76
Schrödinger, E. 1944, "What Is Life?" Canto Classics, Cambridge University Press, based on lectures by Erwin Schrodinger at Trinity College Dublin, 1943
Stökl, A., Dorfi, E., & Lammer, H. 2015, *A&A*, 576, A87
Tu, L., Johnstone, C. P., Güdel, M., & Lammer, H. 2015, *A&A*, 577, L3
Weidenschilling, S. J. 1977, *MNRAS*, 180, 57
Williams, D. M., & Pollard, D. 2002, *Int. J. Astrobiology*, 1, 61
Windmark, F., Birnstiel, T., Güttler, C., Blum, J., Dullemond, C. P., & Henning, Th. 2012, *A&A*, 540, A73

Origins: From the Protosun to the First Steps of Life
Proceedings IAU Symposium No. 345, 2018
B. G. Elmegreen, L. Viktor Tóth & M. Güdel, eds.

doi:10.1017/S1743921319001911

Star formation in the solar neighbourhood

Mika Juvela

Department of physics, University of Helsinki, P.O. Box 64 (Gustaf Hällströmin katu 2), FI-00014 University of Helsinki, Finland
email: mika.juvela@helsinki.fi

Abstract. Star formation started as a cosmic process soon after the big bang and still continues in the Milky Way, although at a decreasing rate. The formation of dense interstellar clouds, their fragmentation and eventual collapse lead to the birth of stars. The nearby clouds provide the highest resolution for the study of this process. The progress is closely following the improvement of the infrared and radio-wavelength facilities that enables us to follow even the earliest stages of the star-formation process inside molecular clouds. On the other hand, modern numerical simulations can take into account most of the relevant physics and often provide a more direct access into the general principles of star formation. The comparison of observations and simulations is therefore essential. In this paper, will discuss star formation in the solar neighbourhood, concentrating on the prestellar phases leading up to the formation of protostars.

Keywords. stars: formation, ISM: clouds, ISM: kinematics and dynamics, ISM: structure, ISM: magnetic fields, dust, extinction, infrared: ISM, radio lines: ISM, submillimeter

---

1. Nearby molecular clouds

There are many nearby molecular clouds at distances just beyond 100 pc (see Fig. 1). Because most instruments, for example single-dish radio telescopes, provide an angular resolution better than ~ 1 arcmin, star formation in these clouds can be studied from clouds scales (sizes >1 pc) to clumps ($\sim$1 pc) and down to the structure of gravitationally bound cores (~ 0.1 pc) (Bergin & Tafalla 2007). The nearby clouds have a central role when we try to understand the formation of low and intermediate mass stars.

Many of the best-studied clouds belong to the Gould Belt. This structure is tilted by some 20° with respect to the Galactic plane and was first recognised in the distribution of bright stars in the southern sky. B.A. Gould referred to it already in 1874 (Gould 1874) and wrote in his *Uranometria Argentina*: *"Thus I cannot avoid the conviction that our own system forms part of a small cluster, distinct from the vast organization of that forms the Milky Way..."* (Gould 1879). This is not very far from the current understanding. The belt is not only a stellar structure but also a location of on-going star formation, containing regions such as the Taurus-Perseus-Auriga complex (distances $d \sim$100 - 500 pc), the ρ Ophiuchus cloud ($d \sim$150 pc), and the Orion region ($d \sim$400 pc).

Recent observations have provided a wealth of new information on the nearby clouds. Even now, the data are still revealing entirely new phenomena that are related to the structure of the clouds and the final stages of the prestellar evolution.

2. Observations of nearby clouds

Ground-based observations of molecular lines and dust continuum extend from radio wavelengths (Myers *et al.* 1983, Motte *et al.* 1998) to sub-millimetre (Ward-Thompson *et al.* 2000, Shirley *et al.* 2000). The past two decades have seen a number of space-borne instruments that have complemented these data with infrared and sub-millimetre observations of unprecedented sensitivity. These include satellites such as ISO (Kessler *et al.* 1996; e.g. Tóth *et al.* 2000), Spitzer (Werner *et al.* 2004), AKARI

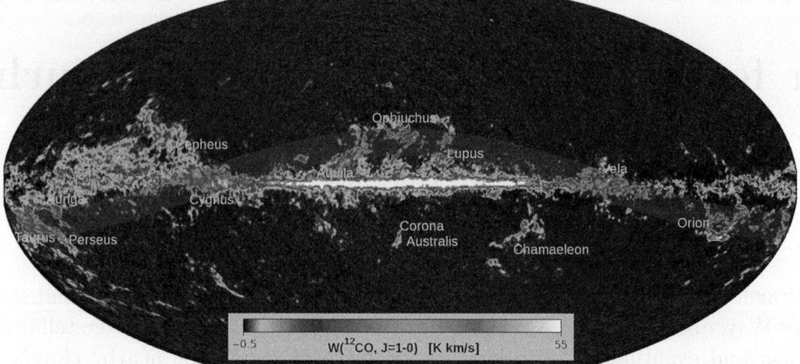

Figure 1. Some nearby molecular clouds marked on the all-sky CO line emission map derived from Planck measurements (Planck Collaboration 2014). These clouds are important targets of star-formation studies, thanks to their typically small distances (100 - 500 pc). Many of the clouds belong to the Gould Belt system (Gould 1879) that is indicated with a cyan band.

(Murakami *et al.* 2007), and WISE (Wright *et al.* 2010). The combination of different instruments makes it possible to follow the full evolution from cold prestellar clouds to protostellar cores, before the young stellar objects (YSOs) again come within the reach of ground-based near-infrared telescopes. Together the ground-based and the space-borne infrared instruments have also made it possible to catalogue YSOs in the nearby clouds and even over the whole sky.

The Herschel Space Observatory (Pilbratt *et al.* 2010) has made a particularly large impact on star-formation studies. It mapped the whole Galactic plane (Molinari *et al.* 2010) and most nearby molecular clouds (André *et al.* 2010) and regions of high-mass star formation (Motte *et al.* 2010). The multi-frequency sub-millimetre continuum data literally shed new light on the star-formation problem by showing the intricate, extended density structures around the star-forming cores. This is in contrast with ground-based continuum surveys where the angular resolution is usually higher but the imaging of extended emission is more challenging (Schuller *et al.* 2009; Ward-Thompson *et al.* 2016). Herschel also carried out heterodyne observations of many key lines that are inaccessible from the ground (e.g. Mottram *et al.* 2014, Goicoechea *et al.* 2015).

One should also not forget extinction measurements from optical to mid-infrared wavelengths (Cambrésy 1999, Bacmann *et al.* 2000, Alves *et al.* 2001). These provide an important point of comparison by being independent of dust temperature. Near-infrared extinction also appears to be little affected by dust evolution compared to the variations seen in the long-wavelength dust opacity (Cardelli *et al.* 1989, Wang & Jiang 2014). With ground-based surveys one can cover everything from diffuse regions (A_V <1 mag) to cores with tens of magnitudes in A_V (Kandori *et al.* 2005, Juvela *et al.* 2008). Nearby clouds have thus been studied combining emission and extinction (Cambrésy *et al.* 2001, Lombardi *et al.* 2006, 2014, Meingast *et al.* 2018) and even all-sky extinction maps exist The ambiguity resulting from the line-of-sight (LOS) mixing of the stellar and interstellar matter distributions is being addressed with 3D extinction mapping methods (Marshall *et al.* 2009, Green *et al.* 2018, Lallement *et al.* 2018).

While space instruments provide improved sensitivity and wavelength coverage, ground-based telescopes still excel in resolution. Radio interferometers, especially ALMA, are revealing new details in the density structure, kinematics, and chemistry of star-forming clouds at the very smallest scales (Kainulainen *et al.* 2017, Hacar *et al.* 2018) and especially at the protostellar phases (e.g. Cox *et al.* 2018).

3. Structure of star-forming clouds

Herschel made cloud filaments a central topic of star-formation studies. Naked-eye observations of dark nebulae go back to William Herschel (Herschel 1785) but the first observations of elongated cloud structures came with astrophotography. E. E. Barnard commented on the *dark lanes*, *streaks*, and *thin threads* visible in his photographs (Barnard 1910). Already in 1894, A.C. Ranyard described the ρ Ophiuchus images as showing *"dark vacant areas or channels... seem to me to be undoubtedly dark structures, or obscuring masses in space, which cut out the light from the nebulous or stellar region behind them"*. Barnard cautiously agreed that these were caused by matter in the interstellar space. The appearance of Pleiades and the Orion nebula was suggestive of a connection between the filamentary nebulosities and stars. Theoretical studies of filament fragmentation and star formation in this context go back at least 70 years (Edgeworth 1946).

Although filaments were later seen in simulations of hydrodynamical and magnetohydrodynamical turbulence (e.g. Padoan *et al.* 2001) they were less evident in observations and only rarely a central topic in star-formation studies (Elmegreen & Elmegreen 1979, Bally *et al.* 1987, Enoch *et al.* 2007). This changed fundamentally with Herschel (the satellite). By imaging large areas with a sensitivity far exceeding that of ground-based instruments (down to below $N(\mathrm{H_2}) = 10^{21}\,\mathrm{cm^{-2}}$), it showed the extensive filamentary networks in both diffuse clouds and actively star-forming regions (André *et al.* 2010, Miville-Deschênes *et al.* 2010, Arzoumanian *et al.* 2011).

The radial profiles of the filaments are observed to be flat compared to the $\rho \sim r^{-4}$ behaviour predicted for isothermal cylinders in hydrostatic equilibrium (Stodólkiewicz 1963, Ostriker 1964). This suggests effects of temperature variations or magnetic fields – or that filaments indeed are not static structures (Kawachi & Hanawa 1998, Nakamura & Umemura 1999, Fiege & Pudritz 2000, Palmeirim *et al.* 2013; Arzoumanian *et al.* 2013). The filaments of nearby clouds also appear to have a narrow distribution of widths, $\sim 0.05 - 0.2\,\mathrm{pc}$ (Arzoumanian *et al.* 2011; Rivera-Ingraham *et al.* 2016). However, in images one can readily see structures at different scales (Fig. 2). Thick filaments may consist of sub-filaments and the width can vary along the filaments, possibly correlated with the column density and core formation (Juvela *et al.* 2012a, Heigl *et al.* 2018). Width measurements are complicated by the varying background, radiative transfer effects, and the finite resolution of the observations (Juvela *et al.* 2012b, Panopoulou 2017, Ossenkopf-Okada & Stepanov 2019). Nevertheless, the observations have lead to attempts to theoretically explain the preferred filament widths (e.g. Fischera & Martin 2012; Hennebelle & André 2013, André *et al.* 2014; Xu & Lazarian 2019).

Molecular line observations revealed in some filaments velocity-coherent substructures that were coined fibers (Hacar & Tafalla 2011). Without full 3D information, detections in the position-position-velocity (PPV) space might suffer from LOS confusion and a PPV structure might not necessarily be identifiable in 3D density (Clarke *et al.* 2018). However, interferometric observations have detected fibers of high column density, narrow width, and almost subsonic velocity dispersion. This suggests that the fibers represent another step closer to the actual star formation (Hacar *et al.* 2018; Monsch *et al.* 2018).

Low-column-density striations are observed perpendicular to the major axis of some more massive structures (Peretto *et al.* 2013; Cox *et al.* 2016). These can be seen also in line emission, for example in Taurus (Heyer *et al.* 2016). The correlation with polarisation suggests a close connection to the magnetic fields. Observations and simulations show that massive filaments tend to be perpendicular to the magnetic field while low-density structures are preferably parallel (Planck Collaboration *et al.* 2016b, Malinen *et al.* 2016, Soler & Hennebelle 2017). Striations could thus represent accretion onto filaments. This is seen in some simulations but is difficult to quantify in observations because of the unknown inclination (Vestuto *et al.* 2003, Palmeirim *et al.* 2013; Chen *et al.* 2017).

Figure 2. Filamentary clouds in the Ophiuchus region. The background shows surface brightness measurements from Planck (Planck Collaboration 2016a) combined with IRAS 100 μm data (credits: ESA/HFI Consortium/IRAS). The zoom-ins show the main ρ Oph cloud (upper box and inset) as seen with Herschel (credit: ESA/Herschel/NASA/JPL-Caltech, R. Hurt 2017), and the Pipe nebula photographed by E.E. Barnard some 110 years earlier (Barnard 1926).

The projected shape of clouds is known accurately from observations but, in the LOS direction, we are only now starting to get constraints from 3D extinction mapping and from YSO and field star distance measurements (Green *et al.* 2018, Galli *et al.* 2018, Großschedl *et al.* 2018). The situation could be better for cloud sub-structures if one could assume that these follow some symmetries. Cores and clumps are often modelled with radial density distributions (Shirley *et al.* 2000, Juvela *et al.* 2018). These may be adequate for quasi-static cores but, because clumps do have sub-structure, spherical models only provide a statistical characterisation of these objects.

Similarly, Herschel filaments are known to have internal structure that is not detected in low-resolution data or without the benefit of velocity information. One may also ask whether they are more cylindrically symmetric objects or sometimes only tangential projections of two-dimensional sheets. The large contrast between the (star-forming) filaments and the background and the quasi-uniform width would seem to argue against the sheet hypothesis. If the structures were extended in the LOS direction, we would also not see such a clear connection between the filaments and the striations. More directly, measurements of high volume densities can show that the mass seen in column density is concentrated in a small volume (Shimajiri *et al.* 2019). Objects such as the Musca cloud contain, or partly even consists of, clumps that have a structure reminiscent of Bonnor-Ebert spheres (Juvela *et al.* 2010). If the filaments were only 2D projections, the alignment between clumps and the filaments would be worse. The same argument applies to the tight spatial correlation with the prestellar and protostellar cores. Filaments may reside in sheets or be the product of sheet collisions but at least the dense ones appear to be closer to the cylinder symmetry. However, the cylindrical shape is only a very crude approximation of the complex structure of the real filaments.

4. Birth of prestellar cores

Cloud fragmentation leads to the formation prestellar and protostellar cores (see the review by C. Federrath in this issue). These are found in regions with background column densities above $N(H_2) \sim 5 \times 10^{21}\,\mathrm{cm}^{-2}$ ($\Sigma \sim 110\,\mathrm{M}_\odot\,\mathrm{pc}^{-2}$) (André *et al.* 2010; Lada *et al.* 2010; Montillaud *et al.* 2015; Marsh *et al.* 2016). Since star formation progresses from atomic clouds to stellar densities, it would seem that almost any "threshold" could be

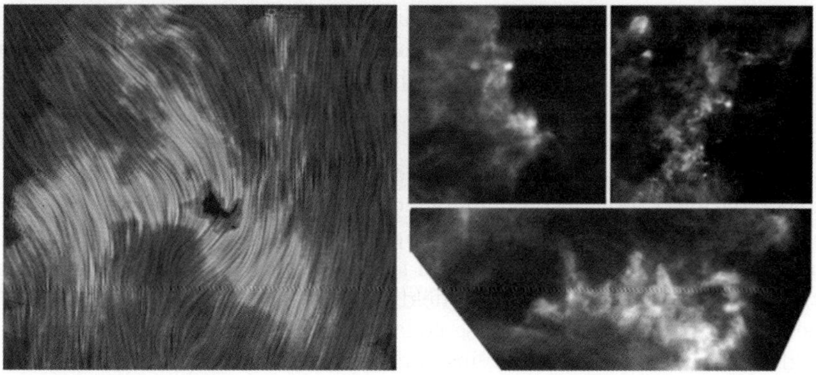

Figure 3. *Left:* Herschel intensity map of the star-forming cloud LDN 1642. The line integral convolution (LIC) pattern shows the plane-of-the-sky magnetic field orientation based on the 353 GHz Planck data (Malinen *et al.* 2016). *Right:* Herschel images of three star-forming clouds where the filamentary structure is strongly modified by external forces (Juvela *et al.* 2012a).

selected by concentrating on a different size scale. However, the coincidence with the threshold of filament stability suggests that this limit is of a more fundamental nature.

The spatial distribution of protostellar sources is strongly associated to super-critical filaments (Könyves *et al.* 2015, Marsh *et al.* 2016, Rivera-Ingraham *et al.* 2017, Bresnahan *et al.* 2018). Isothermal filaments should fragment when the mass per length exceeds the critical value, $M_{\rm crit} = 2\,c_{\rm s}^2/G$, but one should take into account also the effects of internal non-thermal motions, external pressure, magnetic fields, the non-uniform turbulent environment and other external agents (see Fig. 3) – and the dynamical nature of the filaments themselves (Inutsuka & Miyama 1997; Fiege & Pudritz 2000, Myers *et al.* 2009, Heitsch *et al.* 2011, Seifried & Walch 2015, Haugbølle *et al.* 2018).

Cloud and filament properties show systematic trends but each cloud is individual and during its lifetime may be subjected to stellar winds and repeated supernova shocks (Peretto *et al.* 2012; Inutsuka *et al.* 2015). The conditions set by random density and velocity fluctuations affect the structure and the fragmentation. Although, e.g. the observed fragmentation lengths are often attributed to some gravitational or MHD instability, it is not surprising that real clouds do not strictly follow the theoretical predictions (Beuther *et al.* 2015, Roy *et al.* 2015, Contreras *et al.* 2016, Kainulainen *et al.* 2017).

Star formation is characterised by the star-formation efficiency (SFE) and the stellar initial mass function (IMF). To explain the origin of these, one must characterise the preceding step, the prestellar cores, looking at the corresponding core formation efficiency (CFE, the fraction of cloud mass converted into cores) and core mass function (CMF) (Motte *et al.* 1998, Kiss *et al.* 2006, Alves *et al.* 2007). CFE estimates vary in wide range (at least 5 - 35%) between different clouds, probably also depending on the observations used and the chosen extent of the region (Palau *et al.* 2013, Zhang *et al.* 2019, Kong *et al.* 2018). Many studies have commented on the similarity of the IMF and the prestellar CMF after a ∼30% relative shift (efficiency) in masses (e.g. Marsh *et al.* 2016; Sokol *et al.* 2019). Compared to the local clouds, CMF seems to be flatter in high-pressure environments and regions of high-mass star formation (Liu *et al.* 2018). However, it is still unclear how sensitive (or insensitive) the CMF is to the the local cloud conditions.

5. Outlook

Many open question remain concerning star formation in the solar neighbourhood. How are the filaments formed, how do they acquire their final mass, what governs their fragmentation to cores, and what are the lifetimes of all the objects? This should depend on whether evolution is driven mainly by turbulence or gravity, whether external forces (radiation, stellar winds, and supernovae) are important, and what role the still poorly known magnetic fields have (Fig. 3). This complexity makes it difficult to determine the environmental differences that would explain cloud-to-cloud variations of the SFE and the IMF, to the extent they indeed exist. It also limits the generality of the conclusions that can be drawn from any limited cloud sample. This may apply even to the entire sample of nearby clouds, where the influence of massive stars is limited.

Unlike simulations, observations cannot be used to follow the formation of a star from start to finish. We only have a series of snapshots of different environments in different stages of evolution. This has not prevented the determination of many generalities, from the statistics of the large-scale density field (e.g. Kainulainen *et al.* 2009) to the CMF properties. Progress clearly needs close collaboration between observers and modellers. We should also reach a point where, instead of only comparing observations to random model realisations, accurate models are developed for the real individual clouds.

References

Alves, J. F., Lada, C. J., & Lada, E. A. 2001, *Nature* 409, 159

Alves, J., Lombardi, M., & Lada, C. J. 2007, *A&A*, 462, L17

André, Ph., Men'shchikov, A., Bontemps, S., *et al.* 2010, *A&A*, 518, L102

André, P., Di Francesco, J., Ward-Thompson, D., *et al.* 2014, in: H. Beuther, R. S. Klessen, C. P. Dullemond, & T., Henning (eds.), *Protostars and Planets VI*, University of Arizona Press, Tucson, p. 27

Arzoumanian, D., André, Ph., Didelon, P., Könyves, V., Schneider, *et al.* 2011, *A&A* 529, L6

Arzoumanian, D., André, Ph., Peretto, N., Könyves, V. 2013, *A&A* 553, A119

Arzoumanian, D., Shimajiri, Y., Inutsuka, S., Inoue, T., & Tachihara, K. 2018, *PASJ* 70, 96

Bacmann, A., André, P., Puget, J.-L., *et al.* 2000, *A&A* 361, 555

Bally, J., Langer, W. D., Stark, A. A., & Wilson, R. W. 1987, *ApJ* (Letters), 312, L45

Barnard, E. E. 1910, *ApJ* 31, 8

Barnard, E. E., *A Photographic Atlas of Selected Regions of the Milky Way*. Ed. Edwin B. Frost and Mary R. Calvert. Carnegie Institution of Washington, 1927

Beichman, C. A., Neugebauer, G., Habing, H. J., *et al.* 1988, *Infrared Astronomical Satellite (IRAS) Catalogs and Atlases, Vol. 1: Explanatory Supplement* (Washington, DC: GPO)

Bergin, E. A., & Tafalla, M. 2007, *ARAA* 45, 339

Beuther, H., Ragan, S. E., Johnston, K., *et al.* 2015, *A&A*, 584, A67

Bresnahan, D., Ward-Thompson, D., Kirk, *et al.* 2018, *A&A* 615, A125

Cambrésy, L. 1999, *A&A* 345, 965

Cambrésy, L., Boulanger, F., Lagache, G., & Stepnik, B. 2001, *A&A*, 375, 999

Cardelli, J. A., Clayton, G. C., & Mathis, J. S. 1989, *ApJ* 345, 245

Caselli, P., Keto, E., Bergin, E. A., *et al.* 2012, *ApJ* (Letters), 759, L37

Chen, C.-Y., Li, Z.-Y., King, P. K., Fissel, L. M. 2017, *ApJ* 847, 140

Clarke, S. D., Whitworth, A. P., Spowage, R. L., *et al.* 2018, *MNRAS*, 479, 1722

Contreras, Y., Garay, G., Rathborne, J. M., & Sanhueza, P. 2016, *MNRAS*, 456, 2041

Cox, N. L. J., Arzoumanian, D., André, Ph., *et al.* 2016, *A&A* 590, A110

Cox, E. G., Harris, R. J., Looney, *et al.* 2018, *ApJ* 855, 92

Edgeworth, K. E., 1946, *MNRAS* 106, 470

Elmegreen, D. M., & Elmegreen, B. G. 1979, *AJ* 84, 5

Enoch, M. L., Glenn, J., Evans, N. J., II, Sargent, A. I., Young, K. E., & Huard, T. L. 2007, *ApJ*, 666, 982

Fiege, J. D., & Pudritz, R. E. 2000, *MNRAS*, 311, 85

Fischera, J., Martin, P. G. 2012, *A&A* 542, A77

Galli, P. A. B., Loinard, L., Ortiz-Léon, G. N., *et al.* 2018, *ApJ*, 859, 33

Goicoechea, J. R., Teyssier, D., Etxaluze, M., *et al.* 2015, *ApJ* 812, 75

Gould, B.A. 1879, *Resultados del Observatorio Nacional Argentino*, vol. 1, 1

Gould, B.A. 1874, *The American Journal of Science and Arts (Third Series)*, Vol. 8, 325

Green, G. M., Schlafly, E. F., Finkbeiner, D., *et al.* 2018, *MNRAS* 478, 651

Großschedl, J. E., Alves, J., Meingast, S., *et al.* 2018, *A&A*, 619, A106
Hacar, A., & Tafalla, M. 2011, *A&A*, 533, 34
Hacar, A., Tafalla, M., Forbrich, J., *et al.* 2018, *A&A*, 610, A77
Haugbølle, T., Padoan, P., & Nordlund, Å 2018, *ApJ*, 854, 35
Heigl, S., Gritschneder, M., & Burkert, A. 2018, *MNRAS* 2018, 481, L1
Heitsch, F., Naab, T., & Walch, S. 2011, *MNRAS* 415, 271
Hennebelle, P., & André, Ph. 2013, *A&A* 560, A68
Herschel, W., 1785, *Philosophical Transactions*, 75, 213
Heyer, M., Goldsmith, P. F., Yildiz, U. A., *et al.* 2016, *MNRAS* 461, 3918
Inutsuka, S., & Miyama, S. M. 1997, *ApJ* 480, 681
Inutsuka, S., Inoue, T., Iwasaki, K., & Hosokawa, T. 2015, *A&A*, 580, A49
Juvela, M., Pelkonen, V.-M., Padoan, P., & Mattila, K. 2008, *A&A* 480, 445
Juvela, M., Ristorcelli, I., Montier, L. A., *et al.* 2010, *A&A*, 518, L93
Juvela, M., Ristorcelli, I., Pagani, L., *et al.* 2012a, *A&A* 541, A12
Juvela, M., Malinen, J., & Lunttila, T. 2012b, *A&A* 544, A141
Juvela, M., Malinen, J., Montillaud, J., *et al.* 2018, *A&A*, 614, A83
Kainulainen, J., Federrath, C., & Henning, T. 2014, *Science*, 344, 183
Kainulainen, J., Stutz, A. M., Stanke, T., *et al.* 2017, *A&A* 600, A141
Kandori, R., Nakajima, Y., Tamura, M., *et al.* 2005, *AJ* 130, 2166
Kawachi, T. & Hanawa, T. 1998, *PASJ* 50, 577
Kessler, M. F., Steinz, J. A., Anderegg, M. E., *et al.* 1996, *A&A* 315, L27
Kiss, Z. T., Tóth, L. V., Krause, O., *et al.* 2006, *A&A* 453, 923
Kong, S., Tan, J. C., Arce, H. G., *et al.* 2018, *ApJ* (Letters), 855, L25
Könyves, V., André, Ph., Men'shchikov, A., *et al.* 2015, *A&A* 584, A91
Krumholtz, M. 2014, *Phys. Rep.*, 539, 49
Lada, C. J., Lombardi, M., & Alves, J. F. 2010, *ApJ* 724, 687
Lallement, R., Capitanio, L., Ruiz-Dern, L., *et al.* 2018, *A&A* 616, A132
Liu, M., Tan, J. C., Cheng, Y., & Kong, S. 2018, *ApJ* 862, 105
Lombardi, M., Alves, J., & Lada, C. J. 2006, *A&A* 454, 781
Lombardi, M., Bouy, H., Alves, J., & Lada, C. J. 2014, *A&A* 566, A45
Malinen, J., Montier, L., Montillaud, J., *et al.* 2016, *MNRAS* 460, 1934
Marsh, K. A., Kirk, J. M., André, Ph., *et al.* 2016, *MNRAS* 459, 342
Marshall, D. J., Joncas, G., & Jones, A. P. 2009, *ApJ* 706, 727
Meingast, S., Alves, J., & Lombardi, M. 2018, *A&A* 614, A65
Miville-Deschênes, M.-A., Martin, P. G., Abergel, A., *et al.* 2010, *A&A* 518, L104
Molinari, S., Swinyard, B., Bally, J., *et al.* 2010, *A&A* 518, L100
Monsch, K., Pineda, J. E., Liu, H. B. 2018, *ApJ* 861, 77
Montillaud, J., Juvela, M., Rivera-Ingraham, A., *et al.* 2015, *A&A* 584, A92
Motte, F., André, P., & Neri, R. 1998, *A&A* 336, 150
Motte, F., Zavagno, A., Bontemps, S., *et al.* 2010, *A&A* 518, L77
Mottram, J. C., Kristensen, L. E., van Dishoeck, E. F., *et al.* 2014, *A&A* 572, A21
Murakami, H. Baba, H., Barthel, P., *et al.* 2007, *PASJ* 59, 369
Myers, P. C., Linke, R. A., & Benson, P. J. 1983, *ApJ* 264, 517
Myers, P. C. 2009, *ApJ* 700, 1609
Nakamura, F., & Umemura, M. 1999, *ApJ* 515, 239
Ossenkopf-Okada, V., & Stepanov, R. 2019, *A&A* 621, A5
Ostriker, J. 1964, *ApJ* 140, 1056
Padoan, P., Juvela, M., Goodman, A. A., & Nordlund, Åke 2001, *ApJ* 553, 227
Palau, A., Fuente, A., Girart, J. M., *et al.* 2013, *ApJ*, 762, 120
Palmeirim, P., André, Ph., Kirk, J., *et al.* 2013, *A&A* 550, A38
Panopoulou, G. V., Psaradaki, I., Skalidis, R., *et al.* 2017, *MNRAS* 466, 2529
Peretto, N., André, Ph., Könyves, V., *et al.* 2012, *A&A* 541, A63
Peretto, N., Fuller, G. A., Duarte-Cabral, A. *et al.* 2013, *A&A* 555, A112
Pilbratt, G. L., Riedinger, J. R., Passvogel, T., *et al.* 2010, *A&A* 518, L1
Planck Collaboration, 2014, *A&A* 571, A13

Planck Collaboration, 2016a, *A&A* 594, A1
Planck Collaboration 2016b, *A&A* 586, A135
Rivera-Ingraham, A., Ristorcelli, I., Juvela, M., *et al.* 2016, *A&A* 591, A90
Rivera-Ingraham, A., Ristorcelli, I., Juvela, M., *et al.* 2017, *A&A* 601, A94
Roy, A., André, Ph., Arzoumanian, D., *et al.* 2015, *A&A*, 584, A111
Schuller, F., Menten, K. M., Contreras, Y., *et al.* 2009, *A&A*, 504, 415
Seifried, D., & Walch, S. 2015, *MNRAS*, 452, 2410
Shimajiri, Y., André, Ph., Palmeirim, P., *et al.* 2019, *A&A* 623, A16
Shirley, Y. L., Evans, N. J., II, Rawlings, J. M. C., & Gregersen, E. M. 2000, *ApJS* 131, 249
Sokol, A. D., Gutermuth, R. A., Pokhrel, R. *et al.* 2019, *MNRAS* 483, 407
Soler, J. D., & Hennebelle, P. 2017, *A&A* 607, A2
Stodólkiewicz, J. S. 1963, *Acta Astron.* 13, 30
Tóth, L. V., Hotzel, S., Krause, O., *et al.* 2000, *A&A* 364, 769
Vestuto, J. G., Ostriker, E. C., & Stone, J. M. 2003, *ApJ*, 590, 858
Wang, S., & Jiang, B. W 2014, *ApJ* (Letters) 788, L12
Ward-Thompson, D., Kirk, J. M., Crutcher, R. M., *et al.* 2000, *ApJ* (Letters) 537, L135
Ward-Thompson, D., Pattle, K., Kirk, J. M., *et al.* 2016, *MNRAS*, 463, 1008
Werner, M., Roellig, T., Low, F., *et al.* 2004, *ApJS*, 154, 1
Wright, E. L., Eisenhardt, P. R. M., Mainzer, A. K., *et al.* 2010, *AJ* 140, 1868
Xu, S., Ji, S., Lazarian, A. 2019, *ApJ* 878, 157
Zhang, G.-Y., Xu, J.-L., Vasyunin, A. I., *et al.* 2019, *A&A* 620, A163

Discussion

ALFARO: Could you estimate the expected lifetime of filaments?

JUVELA: Filaments are fragmenting to cores and the lifetime of a core is known to be at least ~ 1 Myr. Clouds live at least a few million of years, so the lifetime of a filament must be in the same ball park

KHAIBRAKHMANOV: Filaments are cylinders or sheets? Can we distinguish between these structures?

JUVELA: Regarding cylindrical shape, filaments are likely cylinders. But it is not easy to determine their shape from observations. For some filaments, the volume density has been estimated from molecular lines, which suggest they are cylinders more than sheets.

ZINNECKER: Herschel observations put the filamentary structure of the dense interstellar medium into the limelight. However, in the historical context, who is to be credited with the filamentary nature of dark clouds?

JUVELA: Some credit should certainly be given to Barnard, who drew attention to these elongated structures within the ISM – even if he may not have used the filament word to describe them.

TÓTH: Filaments are usually seen as 2D objects. How much we can trust these are real objects and not just projections?

JUVELA: Simulations have shown that projection effects play a role and especially the filaments seen in continuum are not necessarily single objects in 3D. They may be the superposition of several structures that are not physically connected. However, in the case of such filaments like Musca, it is very likely to be a single object, a real filament.

MAROV: Did you find any correlation between line slopes and fiber/filament structure?

JUVELA: Line profiles can be used to search for signs of collapse - but in the case of fibers, it is just their velocity, they being velocity-coherent in the ppv space.

HACAR: The fibers have narrow lines with clearly subsonic linewidths.

Origins: From the Protosun to the First Steps of Life
Proceedings IAU Symposium No. 345, 2018
B. G. Elmegreen, L. Viktor Tóth & M. Güdel, eds.

doi:10.1017/S1743921319003132

The role of automated methods for filament finding in understanding the complex relationship between filaments, magnetic fields and star formation

Maria R. Cunningham[1], Claire-Elise Green[1], Paul A. Jones[1], Giles Novak[2] and Laura Fissel[3]

[1]School of Physics, University of New South Wales
Sydney, NSW, 2052, Australia
email: maria.cunningham@unsw.edu.au

[2]Department of Physics and Astronomy
North-Western University
2145 Sheridan Road Evanston, IL 60208-3112 USA

[3]NRAO Charlottesville USA

Abstract. The discovery of the ubiquity of filaments in the interstellar medium in the last two decades has begged the question: "What role do filaments play in star formation?" Here we describe how our automated filament finding algorithms can combine with both magnetic field measurements and high-resolution observations of dense cores in these filaments, to provide a statistically large sample to investigate the effect of filaments on star formation. We find that filaments are likely actively accreting mass from the interstellar medium, explaining why some 60% of stars, and all massive stars, form "on-filament".

Keywords. Stars: formation, ISM: molecules.

1. Introduction

The way we do astronomy, particularly radio, millimetre and sub-millimetre astronomy, is rapidly changing. In the not too distant past, a small group of astronomers would apply for time on a telescope, for a particular project with a relatively narrow science focus, travel to the telescope, take the observations, reduce and analyse them, and with some luck produce a refereed journal paper in around three years.

Now, cutting-edge telescopes and observatories, such as the Atacama Large Millimeter Array (ALMA) and the future Square Kilometre Array (SKA) (Sundaram 2015), are funded by many entities and nations. The vast world-wide investment in these telescopes, and the importance of computing power and specialised knowledge in both collecting and reducing the observations, means that many astronomers will not collect the data that they will use for their research. Instead the job of collecting and reducing data is now done by experts in this area.

On the other hand, the quantity and quality of data collected by these new telescopes is unprecedented, leading to a tsunami of high-quality data that is currently ready to be analysed. The amount of data available is far more than any one individual can analyse by hand, and in any case, such an analysis would be a poor use of the large data sets available for exploitation.

So, we now need to automate data analysis processes, so that the large amount of high-quality, reduced data available is exploited to maximise the science.

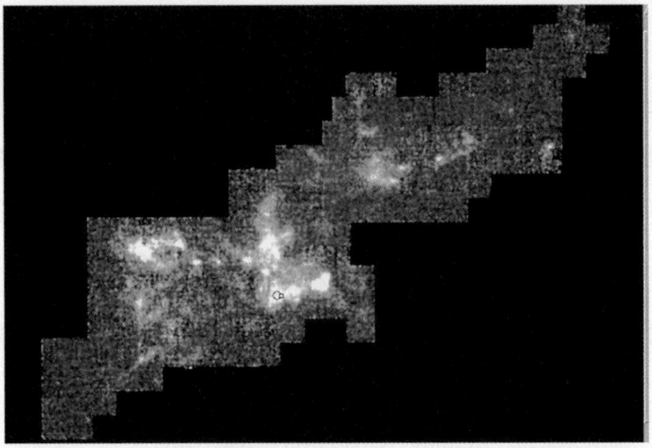

Figure 1. Image of the integrated emission of the HCN(1–0) transition in the Vela C molecular cloud. The filamentary structure is obvious upon visual inspection of this image. However, images of other molecules, and dust such as in the Herschel HOBYS project show similar, but still different, filamentary structure. This highlights the difficulty of defining what is physically a filament. The black outline structure seen in the middle lower part of the image, on the filament described as the "south ridge" shows a particularly dense condensation in this filament (Lowe 2014).

In this paper we describe how automated filament finding in the molecular interstellar medium (ISM) has contributed to our understanding of the relationship between filamentary structure in the ISM, magnetic fields and star formation. We also discuss upcoming surveys which will use automated filament finding and other automated data analysis techniques to extend our analysis, to provide more robust results. It is possible that the relationship between filaments, gravity and turbulence may have different outcomes in differing parts of the Milky Way galaxy (see e.g Luna *et al.* 2006).

2. Filaments in the interstellar medium

The discovery that filaments were significant structures in the ISM cannot be attributed to any one survey, but developed over time (see e.g. Ntormousi *et al.* 2016, and references therein). However, the advent of the Herschel Space telescope cemented the place of filaments as fundamental ISM structures that are important in the star formation process (see e.g. André *et al.* 2010).

So, as Herschel and other large-scale spatial surveys became available, it became important to define exactly what a filament is, and how it was differentiated from the rest of the molecular ISM, in an automated way (Green *et al.* 2017a; 2017b)

Green *et al.* (2017a; 2017b) described a "Goodness-of-Fit" measure that was shown to alleviate visual bias in filament identification, and also automated the determination of the angle between filament elongation and magnetic field direction.

The Vela C molecular line survey of the Mopra telescope†, in Australia, was used to test the filament finding algorithm (Green *et al.* 2017a), along with Herschel HOBYS data of the Vela C molecular cloud (Hill *et al.* 2012). The Mopra Vela C survey is described most completely in Lowe (2014). This survey covered molecular lines from 20 to 116 GHz, detecting some dozens of different molecular transitions. Figure 1 shows the filaments in this region as traced by the Mopra telescope observations of the HCN(1–0) transition. The same data was used to test the automated determination of the angle

† The Mopra telescope, in Coonabarabran, northern NSW, was at the time these observations were taken, run by the CSIRO Australia Telescope National Facility.

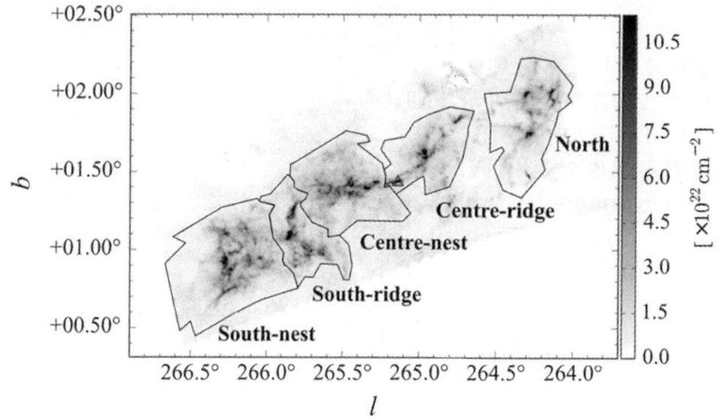

Figure 2. The filamentary regions of Vela C. Presented is the Fissel *et al.* (2016) dust-derived column density map of Vela C overlaid with the different subregions designated by Hill *et al.* (2011). The small triangle marks the position of the RCW 36 IR star cluster (Ellerbroek *et al.* 2013).

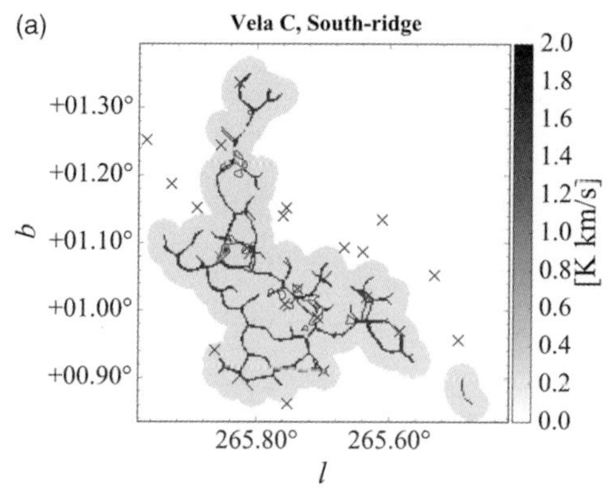

Figure 3. Filamentary structure in Vela C, as determined by the algorithms discussed above. Overlaid on the combined skeletons (in dark grey) are ATCA clumps (small ellipses matching the scale of the colourbar) and the Giannini *et al.* (2012) pre- and protostellar sources, marked by crosses. The light grey represents the average spatial filament extent. See Green 2017 for a high-resolution version of this image.

between filamentary structure and the magnetic field in Vela C, the latter determined from the BLASTPol balloon-borne polarimeter (see e.g Novak 2019).

Figure 1 can be compared with Figure 2, which shows the dust-derived column density map of the Vela C region (Fissel *et al.* 2016). With this comparison, it becomes obvious that different tracers of dense molecular gas give different visual indicators as to the position and extent of filamentary structure.

3. Conclusion and Summary

The automated techniques described in Green *et al.* (2017a, 2017b) can robustly define filaments, using a combination of dust and molecular line data. Figure 3 shows the results of the application of these techniques to the Southern Ridge in Vela C.

Here we summarise the results of the papers mentioned above:

• Filaments are preferred sites for star formation, with ~65% of stars forming on filaments. All massive stars form on filaments.

• Filaments collapse along magnetic field lines

• When stars form on filaments, the direction of collapse is perpendicular to the magnetic fields.

• These results agree with many theoretical predictions (see e.g. Gómez *et al.* 2018).

References

André P., *et al.*, 2010, *A&A*, 518, L102
Ellerbroek L. E., *et al.*, 2013, *A&A*, 558, A102
Fissel L. M., *et al.*, 2016, *ApJ*, 824, 134
Giannini T., *et al.*, 2012, *A&A*, 539, A156
Gómez G. C., Vázquez-Semadeni E., & Zamora-Avilés M., 2018, *MNRAS*, 480, 2939
Green, C.-E., 2017, *PhD Thesis*, University of New South Wales, Sydney, Australia
Green C.-E., Cunningham M. R., Dawson J. R., Jones P. A., Novak G., & Fissel L. M., 2017a, *ApJ*, 840, L17
Green C.-E., Dawson J. R., Cunningham M. R., Jones P. A., Novak G., & Fissel L. M., 2017b, *ApJS*, 232, 6
Hill T., *et al.*, 2011, *A&A*, 533, A94
Hill T., *et al.*, 2012, *A&A*, 548, L6
Lowe V., 2014, *PhDT*, UNSW.
Luna A., Bronfman L., Carrasco L., & May J., 2006, *ApJ*, 641, 938
Novak G., 2019, *AAS*, 233, #127.04
Ntormousi E., Hennebelle P., André P., & Masson J., 2016, *A&A*, 589, A24
Shanmugha Sundaram G., 2015, *IAUGA*, 22, 2225763

Origins: From the Protosun to the First Steps of Life
Proceedings IAU Symposium No. 345, 2018
B. G. Elmegreen, L. Viktor Tóth & M. Güdel, eds.

doi:10.1017/S1743921319001492

3D shape of Orion A with *Gaia* DR2 An informed view on Star Formation Rates and Efficiencies

Josefa E. Großschedl[1], João Alves[1,2], Stefan Meingast[1] and Birgit Hasenberger[1]

[1]Universität Wien, Institut für Astrophysik, Türkenschanzstraße 17, A-1180 Wien
email: josefa.elisabeth.grossschedl@univie.ac.at

[2]University of Vienna, Faculty of Earth Sciences, Geography and Astronomy, Data Science @ Uni Vienna

Abstract. The giant molecular cloud Orion A is the closest massive star-forming region to earth ($d \sim 400$ pc). It contains the rich Orion Nebula Cluster (ONC) in the North, and low-mass star-forming regions (L1641, L1647) to the South. To get a better understanding of the differences in star formation activity, we perform an analysis of the gas mass distribution and star formation rate across the cloud. We find that the gas is roughly uniformly distributed, while, oddly, the ONC region produced about a factor of ten more stars compared to the rest of the cloud. For a better interpretation of this phenomenon, we use *Gaia* DR2 parallaxes, to analyse distances of young stellar objects, using them as proxy for cloud distances. We find that the ONC region indeed lies at about 400 pc while the low-mass star-forming parts are inclined about 70° from the plane of the sky reaching until ∼470 pc. With this we estimate that Orion A is an about 90 pc long filamentary cloud (about twice as long as previously assumed), with its "Head" (the ONC region) being "bent" and oriented towards the galactic mid-plane. This striking new view allows us to perform a more robust analysis of this important star-forming region in the future.

Keywords. methods: statistical, astrometry, stars: distances, stars: pre–main-sequence, ISM: clouds

1. Introduction

The giant molecular cloud Orion A has been extensively studied in the past (e.g., Bally 2008), being the closest massive star-forming region to earth. The distance to the northern part of the cloud, containing the Orion Nebula Cluster (ONC), was, for example, estimated by Menten *et al.* (2007) to be ∼414 pc. Assuming this distance for the whole cloud, leads to a projected length of about 40 pc to 45 pc, when including the high column-density parts of the cloud. The rich ONC is considered a high-mass star-forming region, while only very few of the most massive stars in the cluster's center produce the prominent Orion Nebula (e.g., O'dell 2008). The southern parts of the cloud (L1641 and L1647) produce predominantly low-mass stars, comparable to other nearby star-forming regions. For simplicity we call the ONC region the "Head", and the low-mass star-forming parts the "Tail" of the cloud. The difference in star formation activity in these two parts of the cloud is a well known fact, while we still do not fully understand the origin of the observed phenomenon. To get a better understanding of the cloud properties, we investigate the clouds gas content, star formation rate (SFR), and efficiency (SFE) in Sect. 2, and the distances of young stellar objects (YSOs) using *Gaia* DR2 parallaxes in Sect. 3.

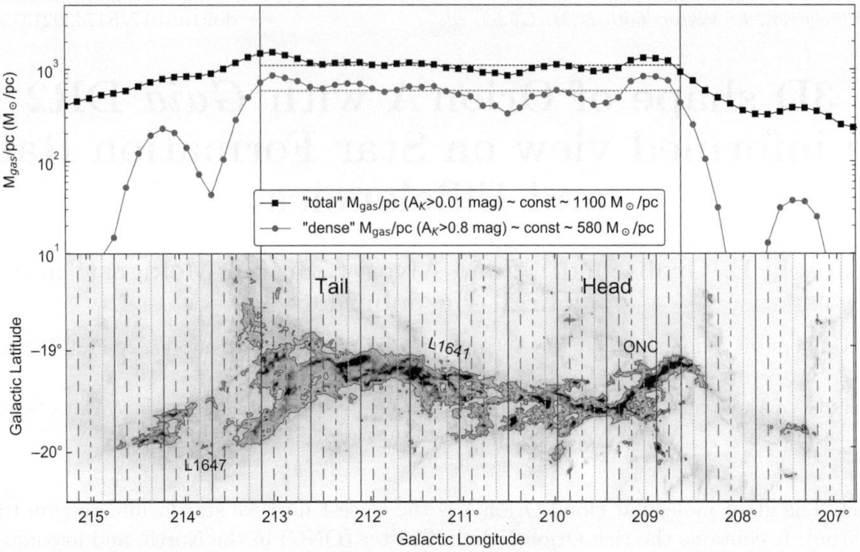

Figure 1. Gas mass distribution across the Orion A cloud. The cloud mass is calculated along l per 0.5° wide bins (~3.6 pc at 414 pc, factor 4 over-sampled), and is given in the top panel in $M_\odot$/pc. The bins are indicated by the vertical lines in the bottom map, while a data point in the top figure is centred above the corresponding bin in the map (e.g., between two dashed lines). Shown separately are the "total" and "dense" gas mass distributions along the cloud. The first is evaluated for pixels with $A_K > 0.01$ mag, which are virtually all pixels in the displayed area, and the second is outlined by the contours at $A_K = 0.8$ mag. The two vertical solid lines in both panels indicate the cloud boundaries, to exclude regions without significant gas mass, while in Galactic latitudes the bins are restricted by the image boundaries ($208.6° < l < 213.2°$, $-20.5° < b < -18.1°$). The two horizontal dashed lines in the top panel show the mean $M_\odot$/pc within the boundaries, as given in the legend.

2. Mass distribution and star formation rate and efficiency

At first guess, one might assume that there is more gas mass available at the ONC region that fuels the relatively high star formation rate. However, when investigating the gas mass distribution across the cloud, we find that it is approximately constant (along Galactic longitudes l). This is highlighted in Fig. 1, where we calculated the gas mass using the dust column-density from a Herschel-Planck-Extinction map (Lombardi *et al.* 2014). For this analysis, we separately look at the gas content within the whole analyzed region ("total") and within higher column-density regions ("dense")†. The latter we define by a star-formation threshold of $A_K > 0.8$mag, as suggested by Lada *et al.* (2010). Extinction A_K is derived from optical depth by $A_K(\text{mag}) = \tau \cdot 3050$ (Meingast *et al.* 2018), and then converted to solar masses for each pixel as follows:

$$M_{\text{gas}} = A_K \times (N_H/A_K \cdot M_p \cdot \mu_{\text{He}} \cdot a_{\text{pix}}) \tag{2.1}$$

with $N_H/A_K = 1.24 \times 10^{22}\ \text{cm}^{-2}\ \text{mag}^{-1}$ (Hasenberger *et al.* 2016), proton mass $M_p = 1.67 \times 10^{-27}$ kg, correction for 10% He abundance $\mu_{\text{He}} = 1.37$, and the pixel area $a_{\text{pix}} = 8.63 \times 10^{33} \text{cm}^2$ (assuming $d = 414$ pc for each 15″ sized Herschel-map pixel). To highlight variations along the cloud we split the cloud into evenly sized bins along l with $\Delta l = 0.5°$ (~3.6 pc at 414 pc) and find that the distribution of the gas mass is roughly constant for both the "total" and "dense" gas mass distributions within the cloud boundaries (Großschedl *et al.* in prep.).

† We do not know the true volume density, but only column-densities, therefore "dense" within parentheses. The "total" gas mass is restricted to the displayed region in Fig. 1.

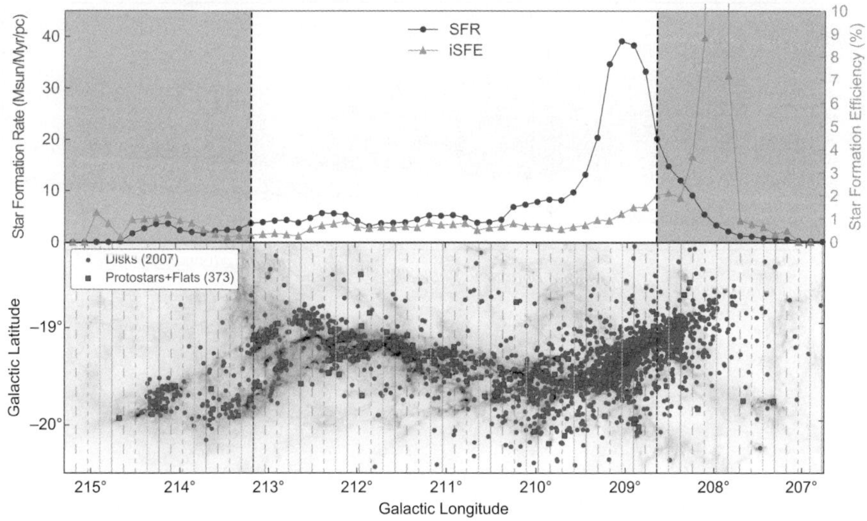

Figure 2. Star formation rate and efficiency across the cloud. The distribution of SFR and instantaneous SFE is calculated per bin across the cloud, similar to Fig. 1. Regions beyond the obvious cloud boundaries should be ignored (gray shaded area, top panel), since they lack significant gas mass (see also Fig. 1). Note the different axes labelling for SFR (left) and SFE (right). The whole YSO sample is used to calculate the SFR, and the younger protostars+flats are used to evaluate the SFE in combination with the gas mass within $A_K > 0.8$ mag.

This seems to be at odds with the observed higher SFR at the Head of the cloud. To get a more quantitative estimate of the SFR, we show its distribution across the cloud (Fig. 2) in a similar manner as the gas mass distribution. To this end we use the YSO catalog of Großschedl *et al.* (2019), which contains a clean sample of 2980 IR-excess YSOs for the whole Orion A region. The 2D distribution of these YSOs is shown in Galactic coordinates in Fig. 2 (bottom), displayed separately for the more evolved Class II YSOs (2607 disks) and the younger Class 0/I and flat-spectrum sources (373 protostars+flats). The SFR per parsec is calculated as follows, using the whole YSO sample:

$$\mathrm{SFR}\,(M_\odot/\mathrm{Myr/pc}) = N_{\mathrm{YSO}} \cdot \bar{M}_{\mathrm{IMF}}/t_{\mathrm{YSO}}/\Delta l \tag{2.2}$$

with the mean IMF mass $\bar{M}_{\mathrm{IMF}} = 0.44\,M_\odot$ (using a Kroupa IMF), the average YSO lifetime $t_{\mathrm{YSO}} \approx 3$ Myrs (e.g., Dunham *et al.* 2015), and the bin size being $\Delta l = 3.6$ pc, assuming a constant distance for the whole cloud. We find that the Head of the cloud produced about a factor of ten more stars in the last 3 to 5 Myrs as compared to the Tail, while, interestingly, the current gas content of these two regions is very similar.

We further calculate the SFE by using only the younger protostars+flats sample in combination with the “dense” gas, since these sources are likely directly associated with the current star-forming gas. This so called instantaneous SFE (iSFE) is calculated as follows:

$$\mathrm{iSFE} = N_{\mathrm{P+F}} \cdot \bar{M}_{\mathrm{IMF}}/(M_{\mathrm{denseGas}} + N_{\mathrm{P+F}} \cdot \bar{M}_{\mathrm{IMF}}) \tag{2.3}$$

The iSFE also shows a roughly constant distribution across the cloud, similar to the gas mass. Variations are at the order of a factor two.

Even though the GMC Orion A is one of the best studied star-forming regions, it is still unclear what caused this burst of star formation at the location of the ONC. This raises the question: How well do we know this cloud?

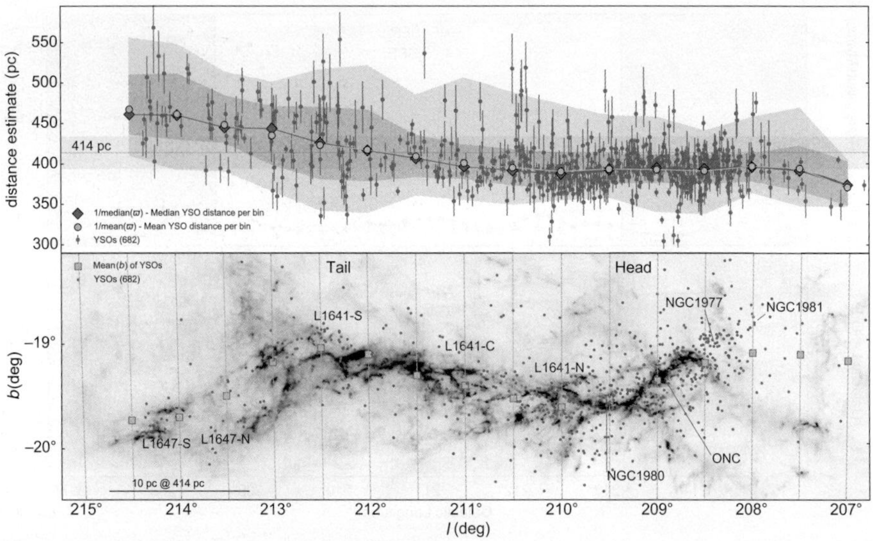

Figure 3. Average distances of IR-YSOs along l. The YSOs parallaxes are averaged within $\Delta l = 1^\circ$ wide bins (factor 2 over-sampled), shown by the circles (mean) and the diamonds (median) in the top panel. The shaded areas in the top pannel show the 1σ and 2σ confidence intervals.

3. Evaluating distance variations in Orion A using *Gaia* DR2

To get potential new insights into this prominent region, we use recent *Gaia* DR2 data (e.g., Gaia Collaboration *et al.* 2016, Gaia Collaboration *et al.* 2018b) to look for distance variations of the IR-excess YSOs connected to Orion A. Due to the youth of these sources they are likely still spatially connected to the cloud regions out of which they formed, or should not have moved too far from these sites. Together with the fact, that they share the same radial velocity as the gas (e.g., Hacar *et al.* 2016), it makes them on average a good proxy for cloud distances. With this, we can estimate the cloud's 3D shape and orientation for the first time. The following methods and results are discussed in detail in Großschedl *et al.* (2018). After cross-matching the YSO catalog from Großschedl *et al.* (2019) with *Gaia* DR2 parallaxes, we end up with roughly 700 IR-YSOs in this region. By averaging YSO distances across the cloud (see Fig. 3), we find that the far end of the Tail is about 70 pc more distant than the Head (ONC), leading to a total length of the cloud of about 90 pc (see Fig. 4). This doubles the previously assumed length of this benchmark cloud. Putting the averaged distances in Galactic cartesian coordinates (Fig. 5) with X pointing toward the Galactic center, we find that the Head seems to be "bent" with respect to the long stretched Tail. Overall, the cloud has a cometary shape pointing toward the Galactic mid-plane.

4. Conclusions and Implications

This striking new view of Orion A will have an impact on some of the observables in this region, like cloud mass, stellar masses, luminosities, or binary separations. The fact that the Tail is differently oriented and more stretched than previously assumed, also calls for a revision of the gas mass distribution and SFR per parsec. This new "image" of the cloud further allows one to speculate on its formation mechanisms.

For example, the 90 pc estimated length of Orion A makes it by far the largest GMC in the solar neighborhood. Together with its aspect ratio (~30:1) and dense gas mass

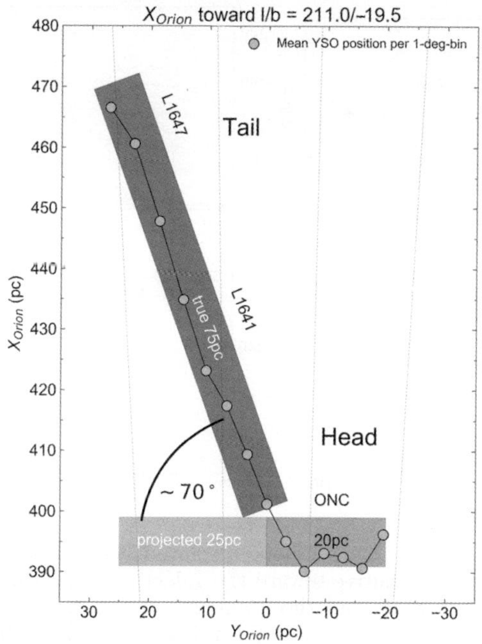

Figure 4. Orientation of the Orion A cloud displayed in a Galactic cartesian coordinate system, with X_{Orion} pointing toward Orion A, at $(l, b) = (211.0, -19.5)$. The sun is at (0,0), Y_{Orion} is similar to l, and Z_{Orion} is similar to b (Z is not shown here). The circles are the averaged YSO distances as derived in Fig. 3. The faint dashed lines are lines of constant l. We find that the Head of the cloud (ONC region) lies almost parallel to the plane of the sky at about 400 pc, while the Tail is inclined about 70° away from the plane of the sky. This leads to an ~75 pc long tail instead of the projected ~25 pc. This makes the Orion A cloud an about 90 pc long filamentary structure, which is twice as long as when assuming a constant distance for the whole region.

fraction (~45%), the Orion A cloud is similar to large-scale Galactic filaments, or so-called "bones" of the Milky Way, as discussed in Zucker *et al.* (2018). However, Orion A lies almost an order of magnitude farther from the Galactic mid-plane as typically found for these structures (about 100 to 130 pc below the plane, assuming $Z_{Sun} \approx +25$ pc). A common suggestion for the origin of the Orion-Monoceros complex is, that it could be the result of an impact of a high-velocity cloud from above the Galactic plane (Franco 1986). However, this seems to be at odds with the orientation of the cloud, pointing toward the Galactic mid-plane from the southern Galactic hemisphere, so more ideas are needed.

External processes might be able to explain the shape of the cloud with the "bent" Head. For example, there is a foreground population near the region of the ONC (e.g., Alves & Bouy 2012; Bouy *et al.* 2014), which could have provided the feedback necessary to bend the head. A Supernova originating from this slightly older population could have taken place in the last about 10 Myrs to 20 Myrs†, triggering the burst of star formation and also shaping this region. Another external event could have taken place, as proposed by Fukui *et al.* (2018). They suggest a cloud-cloud collision at the location of the ONC, being a trigger of star formation. If this event had taken place, it would have had also an impact on the shape of the cloud.

† When considering all ONC members, we find that the cluster's older members (incl. Class III) have ages up to about 10 Myrs when investigating HR-diagrams.

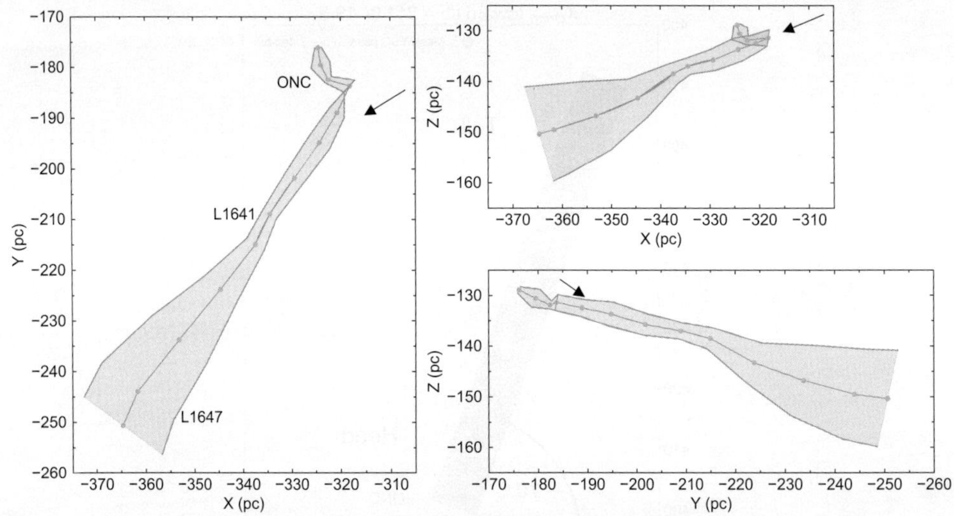

Figure 5. 3D orientation of the Orion A cloud displayed in Galactic cartesian coordinates. The Sun is at (0,0,0), the X-axis is positive toward the galactic center, the Y-axis is positive toward the direction of Galactic rotation, and the Z-axis is positive toward the Galactic North. The dots are the averaged YSO distances as derived in Fig. 3. The grey shaded area shows roughly the cloud extend at $A_K > 0.6$ mag. The arrows indicate the line-of-sight from the Sun pointing toward $(l, b) = (211.0°, -19.5°)$.

Our result illustrates well the new window that is opening with *Gaia* on the structure of the dense ISM. It allows to understand the 3D shape and orientation of star-forming clouds, which will refine other derived properties of these regions.

References

Alves, J., & Bouy, H. 2012, *A&A*, 547, 97

Bally, J. 2008, in: B. Reipurth (eds.), *Handbook of Star Forming Regions, Volume I*, p. 459

Bouy, H., Alves, J., Bertin, E., Sarro, L. M., & Barrado, D. 2014, *A&A*, 564, 29

Dunham, M. M., Allen, L. E., Evans, N. J. II, Broekhoven-Fiene, H., Cieza, L. A., Di Francesco, J., Gutermuth, R. A., Harvey, P. M., Hatchell, J., Heiderman, A., *et al.* 2015, *ApJ*, 220, 11

Franco, J. 1986, *Rev. Mexicana AyA*, 12, 287

Fukui, Y., Torii, K., Hattori, Y., Nishimura, A., Ohama, A., Shimajiri, Y., Shima, K., Habe, A., Sano, H., Kohno, M., *et al.* 2018, *ApJ*, 859, 166

Gaia Collaboration, Prusti, T., de Bruijne, J. H. J., Brown, A. G. A., Vallenari, A., Babusiaux, C., Bailer-Jones, C. A. L., Bastian, U., Biermann, M., Evans, D. W., *et al.* 2016 *A&A*, 595, 1

Gaia Collaboration, Brown, A. G. A., Vallenari, A., Prusti, T., de Bruijne, J. H. J., Babusiaux, C., Bailer-Jones, C. A. L., Biermann, M., Evans, D. W., Eyer, L., *et al.* 2018b *A&A*, 616, 1

Großschedl, J. E., Alves, J., Meingast, S., Ackerl, C., Ascenso, J., Bouy, H., Burkert, A., Forbrich, J., Fürnkranz, V., Goodman, A., *et al.* 2018, *A&A*, 619, 106

Großschedl, J. E., Alves, J., Teixeira, P. S., Bouy, H., Forbrich, J., Lada, C. J., Meingast, S., Hacar, A., Ascenso, J., Ackerl, C., *et al.* 2019, *A&A*, 622

Hacar, A., Alves, J., Forbrich, J., Meingast, S., Kubiak, K., & Großschedl, J. E. 2016, *A&A*, 589, 80

Hasenberger, B., Forbrich, J., Alves, J., Wolk, S. J., Meingast, S., Getman, K. V., & Pillitteri, I. 2016, *A&A*, 593, 7

Lada, C. J., Lombardi, M., & Alves, J. 2010, *ApJ*, 724, 687

Lombardi, M., Bouy, H., Alves, J., & Lada, C. J. 2014, *A&A*, 566, 45
Meingast, S., Alves, J., & Lombardi, M. 2018, *A&A*, 614, 65
Menten, K. M., Reid, M. J., Forbrich, J., & Brunthaler, A. 2007, *A&A*, 474, 515
O'dell, C. R., Muench, A., Smith, N., & Zapata, L. 2008, in: B. Reipurth (eds.), *Handbook of Star Forming Regions, Volume I*, p. 544
Zucker, C., Battersby, C., & Goodman, A. 2018, *ApJ*, 864, 153

Discussion

KRUIJSSEN: In star formation simulations, one side of a cloud often collapses and forms stars first, giving rise to the "kinked" morphology you see in Orion. Especially in a shearing potential, this can persist. Do you have any additional evidence for the idea that this is due to external influences, or could it be secular evolution?

GROSSSCHEDL: Yes, it could also be secular evolution. But this is something we still need to investigate, for example, by comparing the shape and orientation of Orion A within the Galactic disk with simulations of spiral galaxies or collapsing cylindrical clouds.

KHAIBRAKHMANOV: Is there information about kinematics of YSOs in the cloud?

GROSSSCHEDL: Yes, the YSOs radial velocity (RV) is known from APOGEE (e.g., Hacar *et al.* 2016). From this we know that the YSOs share the same RV as the CO gas. With this information, combined with the youth of the YSOs, they are good proxies for cloud distance. So we do not think that they moved significantly far from their birth-sites. But we do not know (yet) the proper motion (PM) of the gas, which can only be measured from stars. We have a running project (VISIONS, Meingast *et al.*) measuring PMs of embedded YSOs in the near infrared. PMs of these sources will also give indirect information of the gas' PM for the first time.

Origins: From the Protosun to the First Steps of Life
Proceedings IAU Symposium No. 345, 2018
B. G. Elmegreen, L. Viktor Tóth & M. Güdel, eds.

doi:10.1017/S1743921319003028

The dense galactic environments of the Milky Way

Quang Nguyen-Luong[1,2,3], Neal Evans[4,2], Kee-Tae Kim[2], Hyunwoo Kang[2] and DEGAMA survey

[1]IBM Canada, 120 Bloor Street East, Toronto, ON, M4Y 1B7, Canada,
[2]Korea Astronomy and Space Science Institute, Yuseoung, Daejeon 34055, Korea,
[3]Visiting researcher at the Graduate School of Natural Sciences, Nagoya City University, Japan
[4]Department of Astronomy, The University of Texas at Austin, 2515 Speedway, Stop C1400, Austin, TX 78712-1205, USA
email: quang.nguyen@ibm.com

Abstract. Star formation takes place in the dense gas phase, and therefore a simple dense gas and star formation rate relation has been proposed. With the advent of multi-beam receivers, new observations show that the deviation from linear relations is possible. In addition, different dense gas tracers might also change significantly the measurement of dense gas mass and subsequently the relation between star formation rate and dense gas mass. We report the preliminary results the DEnse GAs in MAssive star-forming regions in the Milky Way (DEGAMA) survey that observed the dense gas toward a suite of well-characterized massive star-forming regions in the Milky Way. Using the resulting maps of HCO^+ 1–0, HCN 1–0, CS 2–1, we discuss the current understanding of the dense gas phase where star formation takes place.

Keywords. stars: formation, ISM: clouds, ISM: structure, ISM: evolution, Galaxy: evolution

---

1. Introduction

We perform a survey of DEnse GAs in MAssive star-forming regions in the Milky Way (DEGAMA) survey to study the distribution of dense gas in molecular clouds and its role in forming stars. DEGAMA focusses on a larger sample that is sensitive to all gas above a column density threshold of $10^{22}\,\mathrm{cm}^{-2}$ with the goal of improving the understanding of how dense gas is formed and the relation between dense gas and star formation. The threshold column density of $N_{\mathrm{H}_2} \sim 10^{22}\,\mathrm{cm}^{-2}$, which might correspond to a volume density of $1 \times 10^4\,\mathrm{cm}^{-3}$, is chosen because this gas directly builds massive star-forming regions and forms lower mass stars (i.e., Onishi *et al.* 1998, André *et al.* 2010).

At sub-galactic scales, we want to understand how massive star-forming regions can accrete a large amount of mass on a short timescale of 10^5 yr (Peretto *et al.* 2013, Schneider *et al.* 2010) while being inhibited by different types of feedback. Current observations favour the dynamic scenario since it reproduces many aspects of the observational appearance of massive star-forming regions. These regions often show filamentary structure with dense ridges at the centre (Hennemann *et al.* 2010, Hill *et al.* 2011), i.e. the junctions of filamentary networks (Schneider *et al.* 2012), or ridge/hub structures where mass accretion continuously occurs (Liu *et al.* 2015). These locations are very dense with column density $\geqslant 3 \times 10^{22}$ cm^{-2} (Hill *et al.* 2011, Nguyen Luong *et al.* 2011), and often show strong and wide-spread SiO 2–1 emission which is evidence of strong turbulent dissipation (Nguyen-Luong *et al.* 2013).

At the extragalactic scale, we seek to understand the relationship between stellar density and gas density that was put forward by Thackeray (1948) and van den Bergh (1957), and crystallized in a relationship between the observable surface density of SFR, Σ_{SFR}, and the surface density of gas, Σ_{gas}, as: $\Sigma_{SFR} = A\,\Sigma_{gas}^N$ by Schmidt (1959) and Kennicutt (1998). However, this relation is not scale-invariant, the power-law indexes depend on the size scales of the objects, as pointed out using the diffuse cloud tracer CO 1–0 and radio continuum luminosity (Nguyen-Luong *et al.* 2016). These relationships may behave differently if one considers only dense gas tracers, as was suggested for extragalactic environment (Wu *et al.* 2010, Liu *et al.* 2016). The linear relationship manifests in the extragalactic environments (Gao & Solomon 2004) and in the Milky Way environments (Lada *et al.* 2012), but significant gaps remain between extragalactic and galactic star formation laws (Heiderman *et al.* 2010). Nevertheless, the linear relationship between dense gas and SFR shows real scatter in excess of observational uncertainties (Usero *et al.* 2015), so additional factors are at work beyond a simple "more dense gas gives more star formation" model.

With a critical density of $\sim 10^4\,\mathrm{cm}^{-3}$, HCO^+ 1–0 (89.188 MHz) and HCN 1–0 (88.631 GHz), and CS 2–1 (97.980 GHz) is suitable to trace the total dense gas component in star-forming regions. We observed these lines with the Taeduk Radio Astronomy Observatory (TRAO) telescope† in the DEGAMA survey and report its first results in this paper.

2. Observations

TRAO was established in October 1986 with the 13.7 meter Radio Telescope and recently equipped with the 16 pixels 4x4 SEQUOIA receiver. The 2nd IF modules with the narrow band and the 8 channels with 4 FFT spectrometers allow to observe 2 frequencies simultaneously within the 85–100 or 100–115 GHz bands for all 16 pixels of the receiver. We carried out the DEGAMA mapping observations between December 2016 and December 2017. Observations were done in OTF mode and the native velocity resolution is less than 0.1 $\mathrm{km\,s^{-1}}$ (15 kHz) per channel, and their full spectra bandwidth is 60 MHz. The telescope beam size is $\sim 50''$ at 100 GHz, and the main-beam efficiency $\eta_{\rm MB}$ is $\sim 50\%$ at 100 GHz.

Our target is a sample of massive star-forming cloud complexes in the Galaxy that present the entire massive star-forming sequences from quiescent to active and evolved regions. This sample includes the massive star-forming regions that are rather nearby ($d < 5$ kpc), so known star formation rate measurements from direct YSOs counting with Spitzer/WISE or Herschel. The sample includes the following sources that can be categorized in evolutionary stages, from quiescient to more active states.

3. Results

From a global perspective, dense gas tracers such as $L_{\rm HCO^+}$ or $L_{\rm HCN}$ and SFR tracers such as $L_{\rm IR}$ or $L_{\rm H_\alpha}$ are believed to correlate well with each other in log-log space (Gao & Solomon 2004, Bussmann *et al.* 2008, Wu *et al.* 2010). The relations, however, are not identical for different transitions nor for different dense gas tracers (Krumholz & Tan 2007, Narayanan 2008, Juneau *et al.* 2009). Moreover, intrinsic variation in this relationship suggests that simple correlations are inadequate for capturing the full range of star formation behavior (Usero *et al.* 2015).

We calculate the integrated line luminosity using an equation developed in Solomon & Vanden Bout (2005) where

$$\frac{L'_{\rm line}}{\rm K\ km/s\ pc^2} = 23.5 \times 10^{-6} \times \frac{D}{\rm kpc}^{2} \times \frac{A}{''^2} \times \frac{I_{\rm line}}{\rm K\ km/s}\,. \tag{3.1}$$

In DEGAMA survey, integrated line luminosity was derived from a collection of integrated maps of HCO^+ 1–0, HCN 1–0 and CS 2–1. Figure 1 shows an example of

† https://radio.kasi.re.kr/trao/main_trao.php

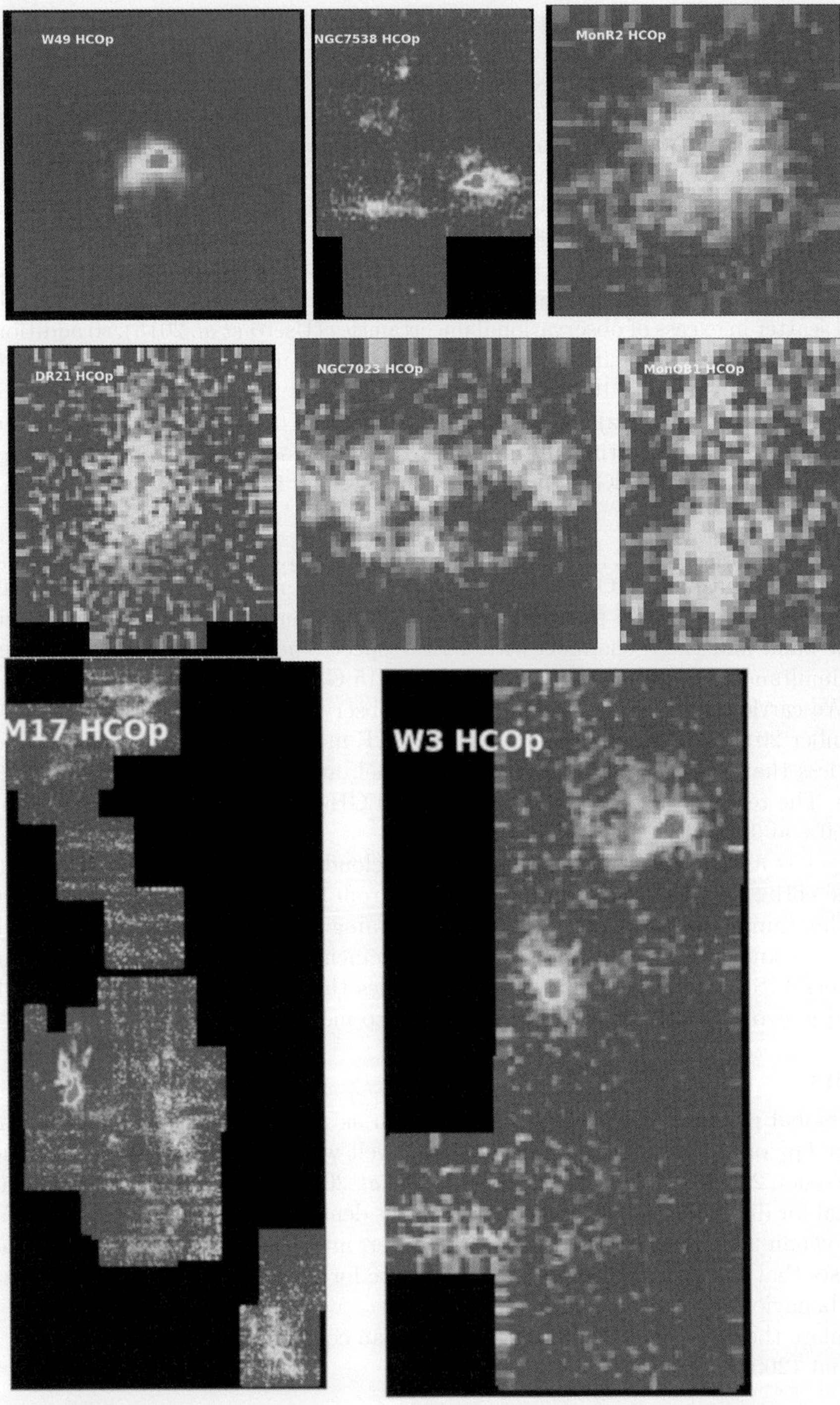

Figure 1. An overview of HCO^+ 1–0 integrated maps of DEGAMA targets.

only HCO^+ 1–0 maps. The ratio of L_{HCO^+}/L_{HCN} from DEGAMA survey shows that HCO^+ varies from cloud-to-cloud and varies around the average extragalactic ratio of 2 (Figure 2a). It might show that although having similar critical density, HCO^+ and HCN might trace quite different types of dense gas. When plotting on the L_{TIR}-L_{HCN}

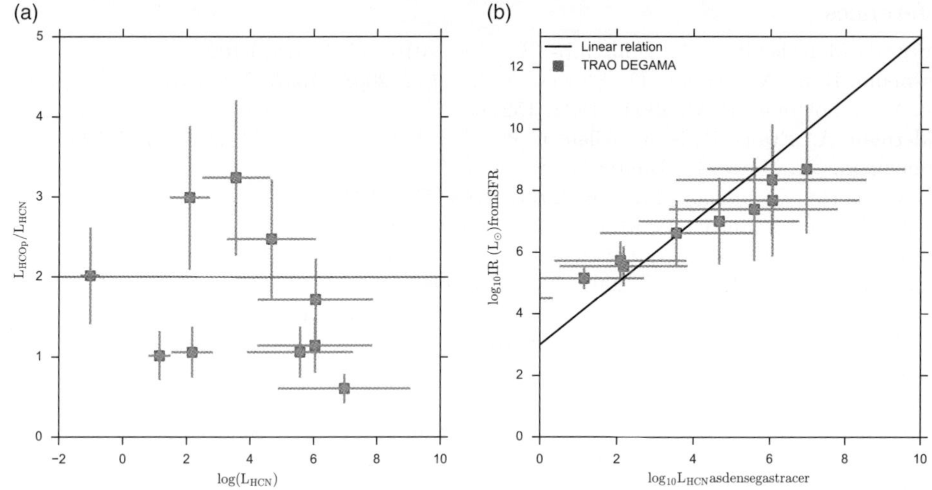

Figure 2. a: Ratio $L_{\mathrm{HCO^+}}/L_{\mathrm{HCN}}$ as function of L_{HCN} luminosity, the horizontal line indicates where the average ratio $L_{\mathrm{HCO^+}}/L_{\mathrm{HCN}} = 2$. **b** L_{TIR} as function of L_{HCN} and the straightline is the linear function $\log(\mathrm{L_{TIR}}) = \log(\mathrm{L_{HCN}}) + 3$ (Gao & Solomon 2004, Wu *et al.* 2010).

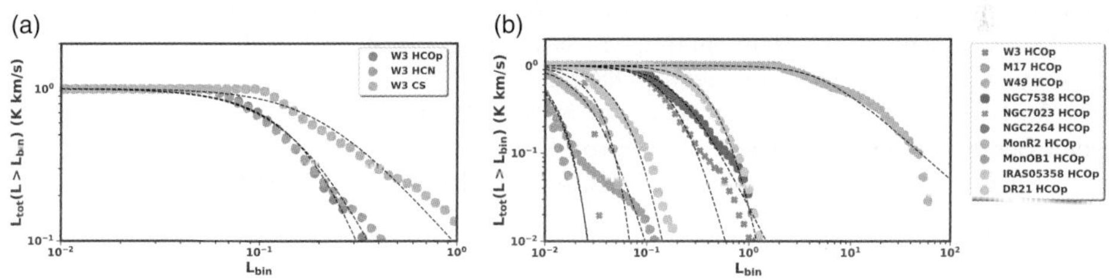

Figure 3. Normalized cumulative integrated line luminosity distributions of different tracers **(a)** and normalized cumulative integrated line luminosity distributions of HCO$^+$ for different DEGAMA targets **(b)**.

plane, there is evidence that Galactic clouds do not follow a simple linear relation as the extragalactic clouds or the scatter is very large (Figure 2b).

Then, we divide the integrated line luminosities into 100 bins and calculate the cumulative distributions (CDs) of the integrated line luminosities for each bin. The results are plotted as functions of luminosity bin in normalized forms (Figure 3). As the CD profiles are different for different sources and different tracers, we model them using a Plummer-like function that describes a flat plateaux and power law decreasing at higher column density as:

$$L(> L_{\mathrm{bin}}) = L_{\mathrm{tot}} \left(\frac{L_{\mathrm{cut}}}{\sqrt{L_{\mathrm{cut}}^2 + L_{\mathrm{bin}}^2}} \right)^p . \tag{3.2}$$

L_{bin} is the luminosity threshold of each bin, L_{tot} is the total luminosity of the cloud, L_{cut} is where the luminosity profile changes from power-law shape to flat plateau, and p is the power-law index of the profile's section at the high luminosity tail. Figures 3 shows that each cloud can be characterized by parameters L_{tot}, L_{cut}, p from the profile in Equation 3.2. This differentiation is potentially applicable to differentiate different types of clouds and also their behaviour in different dense gas tracers. We will explore this topic further in the next paper.

References

André, P., Men'shchikov, A., Bontemps, S., *et al.* 2010, *A&A*, 518, L102
Bussmann, R. S., Narayanan, D., Shirley, Y. L., *et al.* 2008, *ApJL*, 681, L73
Gao, Y., & Solomon, P. M. 2004, *ApJS*, 152, 63
Heiderman, A., Evans, II, N. J., Allen, L. E., Huard, T., & Heyer, M. 2010, *ApJ*, 723, 1019
Hennemann, M., Motte, F., Bontemps, S., *et al.* 2010, *A&A*, 518, L84
Hill, T., Motte, F., Didelon, P., *et al.* 2011, *A&A*, 533, A94
Juneau, S., Narayanan, D. T., Moustakas, J., *et al.* 2009, *ApJ*, 707, 1217
Kennicutt, Jr., R. C. 1998, *ApJ*, 498, 541
Krumholz, M. R., & Tan, J. C. 2007, *ApJ*, 654, 304
Lada, C. J., Forbrich, J., Lombardi, M., & Alves, J. F. 2012, *ApJ*, 745, 190
Liu, H. B., Galván-Madrid, R., Jiménez-Serra, I., *et al.* 2015, *ApJ*, 804, 1
Liu, T., Kim, K.-T., Yoo, H., *et al.* 2016, *ApJ*, 829, 2
Narayanan, D., Cox, T. J., Shirley, Y., *et al.* 2008, *ApJ*, 684, 996
Nguyen Luong, Q., Motte, F., Hennemann, M., *et al.* 2011, *A&A*, 535, A76
Nguyen-Luong, Q., Motte, F., Carlhoff, P., *et al.* 2013, *ApJ*, 775, 88
Nguyen-Luong, Q., Nguyen, Hans V. V., Motte, F., *et al.* 2016, *ApJ*, 833, 23
Onishi, T., Mizuno, A., Kawamura, A., Ogawa, H., & Fukui, Y. 1998, *ApJ*, 502, 296
Peretto, N., Fuller, G. A., Duarte-Cabral, A., *et al.* 2013, *A&A*, 555, A112
Schmidt, M. 1959, *ApJ*, 129, 243
Schneider, N., Motte, F., Bontemps, S., Hennemann, M. *et al.* 2010, *A&A*, 518L, 83S
Solomon, P. M., & Vanden Bout, P. A., *Annu. Rev. Astron. Astrophys.*, 43, 677
Schneider, N., Csengeri, T., Hennemann, M., *et al.* 2012, *A&A*, 540, L11
Thackeray, A. D. 1948, *The Observatory*, 68, 22
Usero, A., Leroy, A. K., Walter, F., *et al.* 2015, *AJ*, 150, 115
van den Bergh, S. 1957, *Zeitschrift fuer Astrophysik*, 43, 236
Wu, J., Evans, II, N. J., Shirley, Y. L., & Knez, C. 2010, *ApJS*, 188, 313

Discussion

HACAR: You are comparing line luminosities but line tracers are subject to excitation, opacity and temperature effects. How do you take into account all these effects in your comparisons?

QUANG: The better way is to compare the column density distributions of different lines if we can have multiple transitions of multiple isotopologue observations of the same molecules. With only lines at single transitions, we can only compare them and examine how excitation, opacity and temperature affect the line luminosity distributions.

MAROV: In modern cosmology there are theories assuming dark matter properties with an existence of multiple galaxies around the Milky Way core. Could you associate dense clouds which you have been talking about with such faint observing features?

QUANG: I couldn't comment on that because there is no observational evidence to distinguish dark matter properties associated with multiple galaxies in the Milky Way vicinity.

Origins: From the Protosun to the First Steps of Life
Proceedings IAU Symposium No. 345, 2018
B. G. Elmegreen, L. Viktor Tóth & M. Güdel, eds.

doi:10.1017/S1743921318008335

Kinematics of OB-associations in the 3-kpc solar neighborhood

Anna M. Melnik[1], **Andrei K. Dambis**[1,2], **Elena V. Glushkova**[2,1] and **Pertti Rautiainen**[3]

[1]Sternberg Astronomical Institute, Lomonosov Moscow State University, Universitetskii pr. 13, Moscow, 119991, Russia
email: anna@sai.msu.ru

[2]Faculty of Physics, Lomonosov Moscow State University, Leninskie Gory 1-2, Moscow, 119991, Russia

[3]Astronomy Research Unit, University of Oulu, P.O. Box 3000, FI-90014 Oulun yliopisto, Finland

Abstract. We use *Gaia* (DR1, DR2) stellar proper motions to study the kinematics of OB-associations. The average one-dimensional velocity dispersion inside 18 OB-associations with more than 10 Gaia DR1 stars is $\sigma_v = 3.9$ km s^{-1}. The median virial and stellar masses of OB-associations are equal to 7×10^5 and 9×10^3 solar masses, respectively. The median star-formation efficiency is $\epsilon = 2.1\%$. We have found the expansion in several OB-associations. Models of the Galaxy with a two-component outer ring R_1R_2 can reproduce the average residual velocities of OB-associations in the Perseus, Sagittarius and Local System complexes.

Keywords. Galaxy: kinematics and dynamics, open clusters and associations: general.

1. Introduction

The distribution of the velocities of young stars in a wide solar neighborhood can tell us a lot about the structure and evolution of the Galaxy. There are several lists of OB-associations in the Galaxy, but the catalog by Blaha & Humphreys (1989) can be considered as the most universal. OB-associations are loose groups of OB stars. The sky-plane sizes of OB-associations from the catalog by Blaha & Humphreys (1989) are very different, but 90% of them are less then 200 pc. The main advantage of OB-associations over open clusters in kinematical studies is that OB-associations contain more stars with known line-of-sight velocities and proper motions, so their average velocities are more reliable.

Star formation proceeds with different intensity in different regions of the Galactic disk. Efremov & Sitnik (1988) identified several regions in the Galaxy containing a lot of molecular and neutral gas as well as young open clusters and OB-associations. These regions were named stellar-gas complexes. Figure 1a shows the distribution of young open clusters (Dias *et al.* 2002) and rich OB-associations (Blaha & Humphreys 1989) in the Galactic plane. We can see that young clusters and OB-associations concentrate to the stellar-gas complexes located in the Sagittarius, Carina, Cygnus, Local System and Perseus regions (e.g., Melnik *et al.* 2016).

2. Results and Conclusions

First let us consider the internal motions inside OB-associations. We used *Gaia* Data Release 1 (DR1) to determine the velocity dispersions inside 18 OB-asssociations with

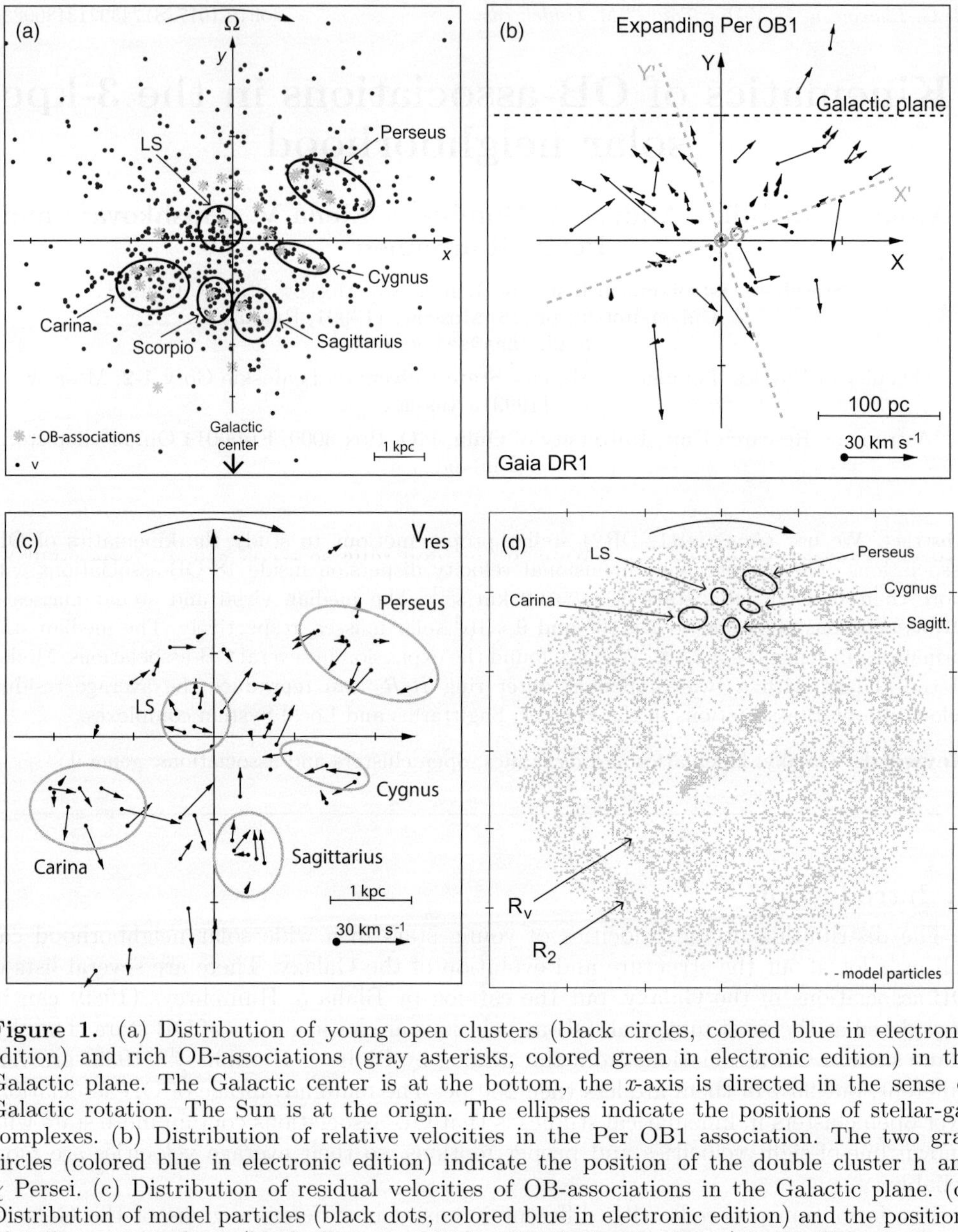

Figure 1. (a) Distribution of young open clusters (black circles, colored blue in electronic edition) and rich OB-associations (gray asterisks, colored green in electronic edition) in the Galactic plane. The Galactic center is at the bottom, the x-axis is directed in the sense of Galactic rotation. The Sun is at the origin. The ellipses indicate the positions of stellar-gas complexes. (b) Distribution of relative velocities in the Per OB1 association. The two gray circles (colored blue in electronic edition) indicate the position of the double cluster h and χ Persei. (c) Distribution of residual velocities of OB-associations in the Galactic plane. (d) Distribution of model particles (black dots, colored blue in electronic edition) and the positions of stellar-gas complexes (black ellipses, colored red in electronic edition) in the Galactic plane.

more than 10 *Gaia* DR1 stars (Gaia Collaboration *et al.* 2016a,b; Michalik, Lindegren & Hobbs 2015). The average one-dimensional velocity dispersion appears to be $\sigma_v = 3.9$ km s^{-1}. Note that *Gaia* DR2 (Brown *et al.* 2018; Lindegren *et al.* 2018) gives a bit larger value of $\sigma_v = 4.8$ km s^{-1} . This small increase in σ_v can be attributed to the contribution of binary systems. However, both estimates of σ_v are considerably smaller than the standard deviation of the velocities of OB-associations from the rotation curve, which is equal to 7.4 km s^{-1}. This fact suggests that OB-associations identified by Blaha & Humphreys (1989) mainly include stars born from the same molecular cloud (see Melnik & Dambis 2017, 2018 for more details).

Giant molecular clouds (GMC) are supposed to be close to their virial equilibrium (Larson 1981; Krumholz, Matzner & McKee 2006) and therefore the velocity dispersions of stars in OB-associations must correspond to the masses of their parent GMC. We can use simple formulae of stellar dynamics to estimate the virial masses of OB-associations:

$$M_{vir} = \frac{5a\sigma_v^2}{G}, \tag{2.1}$$

where a is supposed to be the radius containing 68% of the association member stars. The median value of virial masses of OB-associations with more then 10 *Gaia* DR1 stars appears to be 7×10^5 M$_\odot$, which is consistent with the masses of giant molecular clouds 10^5-10^6 M$_\odot$ obtained by Sanders, Scoville & Solomon (1985).

To estimate the stellar masses of OB-associations we use the multi-component power law of initial mass function (IMF) ($dN \sim M^{\alpha}dM$) by Kroupa (2002) and the assumption that the catalog by Blaha & Humphreys (1989) includes all stars with $M > 20$ M$_\odot$. The median value of stellar masses of OB-associations is 9×10^3 M$_\odot$.

The average efficiency of star formation in a parent GMC is the ratio of the stellar to virial masses of an OB-association. Its median value is 2.1%, which is consistent with other estimates (Myers *et al.* 1986, Evans *et al.* 2009, Garcia *et al.* 2014).

We found the expansion of four OB-associations – Per OB1, Car OB1, Sco OB1 and Ori OB1 – from an analysis of *Gaia* proper motions of their member stars. The expansion of the Per OB1 and Car OB1 associations first was found in the distribution of *Gaia* DR1 proper motions and was then confirmed by *Gaia* DR2 data. The expansion of the Sco OB1 and Ori OB1 associations was obtained with *Gaia* DR2 proper motions only (Melnik & Dambis 2017, 2018).

Figure 1b shows the distribution of relative velocities in the Per OB1 association. The relative velocities are calculated with respect to the center of the OB-association. Also shown is the position of the double cluster h and χ Persei residing practically at the center of the association. We can see that the fastest expansion happens in the direction Y′, which is nearly perpendicular to the main plain of the double cluster. The expansion velocity is 4.7 ± 1.3 km s^{-1}. The mass loss in a gas cloud due to thermal pressure of HII regions can create an expanding group of stars. If mass is ejected from the system within a time comparable to the crossing time, then the system becomes unbound after 50% mass loss (Hills 1980). However, more accurate treatment of relaxation processes shows that the system can form an expanding OB-association with a bound cluster in the center (Kroupa, Aarseth & Hurley 2001). The kinetic and potential energies depend differently on mass, so the mass-loss destroys the virial equilibrium. The gravity of the remaining gas just cannot keep the group inside the original volume.

Now let us consider motions of OB-associations as single entities. Figure 1c shows the distribution of residual velocities of OB-associations in the Galactic plane. The residual velocities are determined as the differences between the observed heliocentric velocities and the velocities due to the galactic differential rotation and the solar motion toward the apex ($V_{res} = V_{obs} - V_{rot} - V_{ap}$). The residual velocities characterize non-circular motions in the Galactic disc. The projections of the residual velocity on the radial and azimuthal directions are designated by V_R and V_T, respectively. We can see that the residual velocities in some complexes have a preferred direction. In the Perseus complex the majority of OB-associations have the radial component V_R of the residual velocity directed toward the Galactic center, while in the Sagittarius region and in the Local System most of objects move away from the Galactic center. This kinematical feature can be reproduced in a model with a two-component outer ring R_1R_2.

Models with analytical Ferrers bars form the outer rings R_1 and R_2 after ~ 1 Gyr from the start of simulations. Figure 1d shows the distribution of model particles and the positions of stellar-gas complexes in the Galactic plane. The outer rings form due to the resonance between the epicyclic motion and the orbital motion with respect to the bar. The outer rings lie near the Outer Lindblad Resonance (OLR) of the bar. Of the two outer rings, R_1 is located a bit closer to the Galactic center than the ring R_2. We can see that the Perseus complex is related to the ring R_2 while other complexes belong to the ring R_1. Models with the outer rings R_1R_2 can reproduce the residual velocities of OB-associations in the Perseus, Sagittarius and LS stellar-gas complexes (Melnik & Rautiainen 2009, 2011; Rautiainen & Melnik 2010, Melnik 2019).

Acknowledgements

This work has made use of data from the European Space Agency (ESA) mission *Gaia* (https://www.cosmos.esa.int/gaia), processed by the *Gaia* Data Processing and Analysis Consortium (DPAC, https://www.cosmos.esa.int/web/gaia/dpac/consortium). Funding for the DPAC has been provided by national institutions, in particular the institutions participating in the *Gaia* Multilateral Agreement.

References

Blaha, C., Humphreys, R. M. 1989, *AJ*, 98, 1598
Brown, A. G. A., Vallenari, A., Prusti, T. *et al.* 2018, *A&A*, 616, A1, *Gaia* DR2
Dias, W. S., Alessi, B. S., Moitinho, A., Lépine, J. R. D. 2002, *A&A*, 389, 871
Efremov, Yu. N., Sitnik, T. G. 1988, *Soviet Astron. Lett.*, 14, 347
Evans, N. J., Dunham, M. M., Jorgensen, J. K. *et al.* 2009, *ApJS*, 181, 321
Gaia Collaboration *et al.* 2016, *A&A*, 595, A1
Gaia Collaboration, *et al.* 2016 *A&A*, 595, A2
Garcia, P., Bronfman, L., Nyman, L.-A., Dame, T. M., Luna, A. 2014, *ApJS*, 212, 2
Hills, J. G. 1980, *ApJ*, 225, 986
Kroupa, P. 2002, *Science*, 295, 82
Kroupa, P., Aarseth, S., Hurley, J. 2001, *MNRAS*, 321, 699
Krumholz, M. R., Matzner, C. D., McKee, C. F. 2006, *ApJ*, 653, 361
Larson, R. B. 1981, *MNRAS*, 194, 809
Lindegren, L., Hernández, J., Bombrun, A. *et al.* 2018, *A&A*, 616, A2, *Gaia* DR2
Melnik, A. M. 2019, 485, 2106
Melnik, A. M., Dambis, A. K. 2017, *MNRAS*, 472, 3887
Melnik, A. M., Dambis, A. K. 2018, submitted to *MNRAS*
Melnik, A. M., Rautiainen, P. 2009, *Astron. Lett.*, 35, 609
Melnik, A. M., Rautiainen, P. 2011, *MNRAS*, 418, 2508
Melnik, A. M., Rautiainen, P., Glushkova, E.V., Dambis, A. K. 2016, *Ap&SS*, 361, 60
Michalik, D., Lindegren, L., Hobbs, D. 2015, *A&A*, 574, 115
Myers, P. C., Dame, T. M., Thaddeus, P. *et al.*, 1986, *ApJ*, 301, 398
Rautiainen, P., Melnik, A. M. 2010, *A&A* 519, 70
Sanders, D. B, Scoville, N. Z., Solomon, P. M. 1985, *ApJ*, 289, 373

Discussion

ALFARO: Did you determine the vertex deviations for these stars? In my opinion this vertex deviation is a more accurate test for comparing with the models.

MELNIK: The vertex deviation approach is relevant for a sample of objects located in a small solar vicinity while residual velocities can tell us a lot about kinematics in a wide solar neighborhood (3 kpc).

Origins: From the Protosun to the First Steps of Life
Proceedings IAU Symposium No. 345, 2018
B. G. Elmegreen, L. Viktor Tóth & M. Güdel, eds.

doi:10.1017/S174392131900173X

Star formation in cloud cores – simulations and observations of dense molecular cores and the formation of solar mass stars

C. Federrath

Research School of Astronomy and Astrophysics, Australian National University, Canberra, ACT 2611, Australia
email: christoph.federrath@anu.edu.au

Abstract. Star formation is inefficient. Recent advances in numerical simulations and theoretical models of molecular clouds show that the combined effects of interstellar turbulence, magnetic fields and stellar feedback can explain the low efficiency of star formation. The star formation rate is highly sensitive to the driving mode of the turbulence. Solenoidal driving may be more important in the Central Molecular Zone, compared to more compressive driving agents in spiral-arm clouds. Both theoretical and observational efforts are underway to determine the dominant driving mode of turbulence in different Galactic environments. New observations with ALMA, combined with other instruments such as CARMA, JCMT and the SMA begin to reveal the magnetic field structure of dense cores and protostellar disks, showing highly complex field geometries with ordered and turbulent field components. Such complex magnetic fields can give rise to a range of stellar masses and jet/outflow efficiencies in dense cores and protostellar accretion disks.

Keywords. ISM: clouds, ISM: jets and outflows, magnetic fields, stars: formation, turbulence

---

1. The big picture

The formation of stars powers the evolution of galaxies and determines the initial conditions for planet formation, and thus, ultimately for life. Star formation happens in the dense cores of filamentary molecular clouds, but many scales and processes are involved, ranging from cosmological initial conditions giving rise to the first galaxies, over the dense clouds within, to finally the protostellar accretion disks where binaries and planets form (see Figure 1). This is an extremely rich physics problem, involving gravity, turbulence, magnetic fields, and feedback via jets, outflows and radiation. We are far from having solved all these problems, but new observations help us map the structure of the turbulent gas and magnetic fields in the dense protostellar cores, and new computer models are reaching the level of detail that enables a quantification of the star formation rate and the initial mass function of stars.

2. The inefficiency of star formation

Star formation is inefficient. Molecular clouds only turn a few percent of their gas mass into stars per freefall time. If gravity were the only thing acting on the clouds, we would expect the star formation rate to be of the order of 100% per freefall time (i.e., the entire cloud would be turned into stars in one freefall time). Thus, physical processes other than gravity must be limiting the collapse and reducing the star formation rate to only a few percent per freefall time.

Figure 2 shows four simulations by that quantify the effects of gravity alone (top-left panel), added turbulence (top-right panel), added magnetic fields (bottom-left panel),

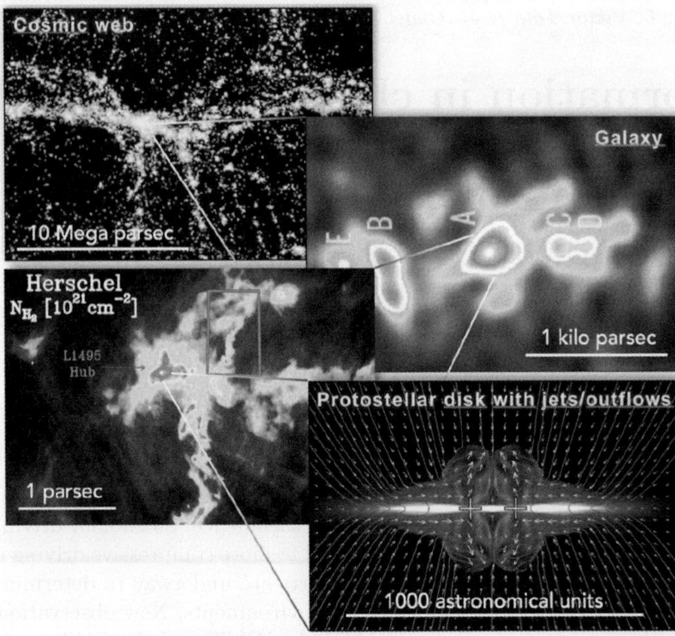

Figure 1. Schematic zoom-in on cosmic star formation. The spatial range covered by star formation is more than a billion: from the cosmic web (size scales of order 10 Mega parsec) to galaxies, interstellar clouds, and finally to the protostellar disks (scales of order 1000 astronomical units) that spawn new stars and planets. Observing, modeling, and understanding the complex interplay of the physical and chemical processes across this huge range of scales from the cosmic web down to stars and planets is one of the biggest challenges in astrophysics. Images adopted from Taylor & Kobayashi (2015), Sharda *et al.* (2018), Arzoumanian *et al.* (2018), and Kuruwita *et al.* (2017).

and added jet/outflow feedback (bottom-right panel). With gravity alone, the star formation rate proceeds close to the maximum freefall rate. Turbulence, magnetic fields, and finally feedback, reduce the star formation rate by a factor of 2–3 in each step, leading to near-realistic (observed) typical star formation rates of a few percent per freefall time. Typical observed values are around 1–2% per freefall time (Krumholz & Tan 2007; Federrath 2013b; Onus *et al.* 2018). The best values achieved in the most complex simulation is a star formation rate per freefall time of $\sim 4\%$, which still overestimates the star formation rate by a factor of ~ 2, most likely because radiation feedback was not included in these simulations, which will also have a profound impact on the initial mass function of stars (Offner *et al.* 2009; Bate 2012; Krumholz *et al.* 2012; Federrath *et al.* 2017a; Cunningham *et al.* 2018; Guszejnov *et al.* 2018).

3. The role of turbulence

Recent theoretical advances allow us to predict the star formation rate based upon four fundamental physical parameters of a molecular cloud (Krumholz & McKee 2005; Padoan & Nordlund 2011; Hennebelle & Chabrier 2011; Federrath & Klessen 2012),

1. the virial parameter $\alpha_{\rm vir} = 2E_{\rm turb}/E_{\rm grav}$ (ratio of turbulent kinetic to gravitational energy),
2. the sonic Mach number $\mathcal{M} = \sigma_v/c_{\rm s}$ (ratio of turbulent velocity dispersion to sound speed),
3. the turbulence driving parameter b, i.e., whether the turbulence is driven solenoidally ($b = 1/3$) or compressively ($b = 1$) (Federrath *et al.* 2008, 2010), and
4. the plasma $\beta = P_{\rm th}/P_{\rm mag}$ (ratio of thermal to magnetic pressure).

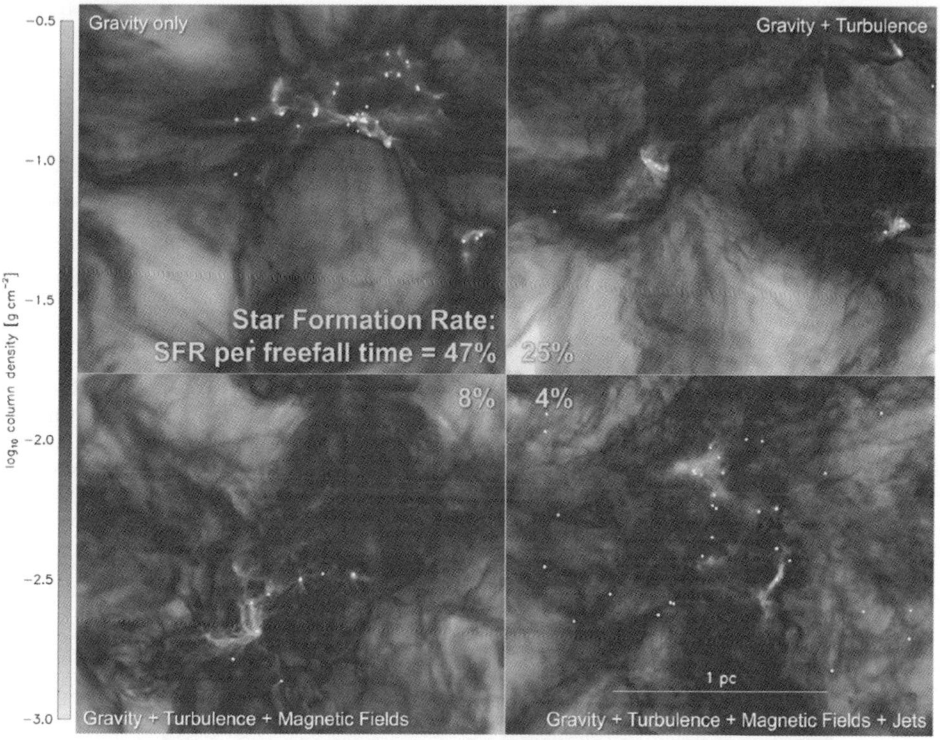

Figure 2. Four numerical simulations of star formation in the same clouds, but with systematically increasing physical complexity (Federrath 2015): Gravity only (top left), Gravity vs. Turbulence (top right), Gravity vs. Turbulence + Magnetic Fields (bottom left), and Gravity vs. Turbulence + Magnetic Fields + Jet/Outflow Feedback (bottom right). Stars are shown as white circles and the resulting star formation rates are indicated in each panel. Only the combination of gravity, turbulence, magnetic fields, and feedback yields realistic star formation rates of a few percent per freefall time, as observed in most real clouds. Simulation movies available: `http://www.mso.anu.edu.au/~chfeder/pubs/ineff_sf/ineff_sf.html`.

These theories rest on the statistics of supersonic magnetized turbulence, its density probability distribution function and power spectrum (Federrath 2013a). Figure 3 shows the theoretical predictions of the star formation rate per freefall time as a function of $\alpha_{\rm vir}$ and $\mathcal{M}$, for solenoidal driving (left-hand panel) and compressive driving (right-hand panel). These can be understood as follows: 1) Large $\alpha_{\rm vir} > 1$ correspond to unbound clouds, such that the star formation rate decreases very quickly with increasing $\alpha_{\rm vir}$. Typical clouds have $\alpha_{\rm vir} \sim 1$ (Kauffmann *et al.* 2013; Hernandez & Tan 2015). 2) Increasing $\mathcal{M}$ produces stronger shocks, and hence a higher dense-gas fraction (Konstandin *et al.* 2012; Kainulainen *et al.* 2013, 2014). 3) More compressive driving of the turbulence also leads to enhanced dense-gas fractions compared to solenoidal driving (Federrath *et al.* 2008, 2010). The latter effect can be very strong, leading to star formation rates that differ by more than a factor of 10 between solenoidal and compressive driving, as confirmed in simulations (Federrath & Klessen 2012).

Given the strong effect of turbulence driving on star formation, a logical next step is to find out which driving mode (solenoidal versus compressive) dominates in different regions of the Galaxy. Figure 4 shows a schematic to illustrate the potentially important effect of environment. For example, spiral-arm compression and cloud-cloud collisions in the galactic disk may be considered compressive drivers, while shear in the Central

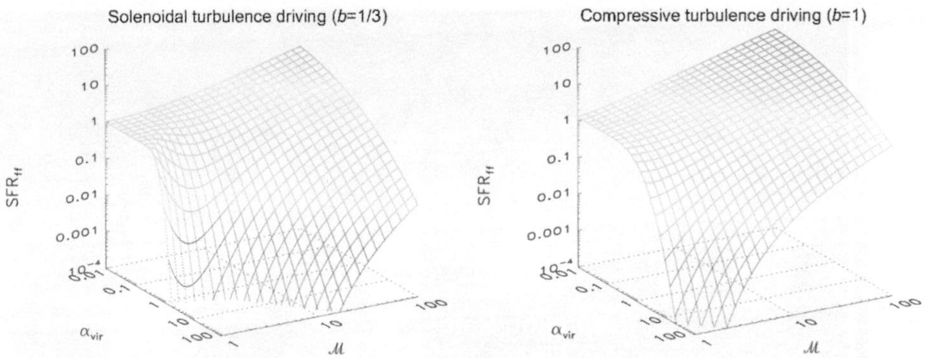

Figure 3. Theoretical predictions of the star formation rate per freefall time as a function of virial parameter $\alpha_{\rm vir}$ and turbulence Mach number $\mathcal{M}$, for solenoidal driving of the turbulence ($b = 1/3$; left-hand panel) and compressive driving ($b = 1$, right-hand panel). The difference in star formation rate is about an order of magnitude in the typical parameter range of observed clouds ($\alpha_{\rm vir} \sim 1$, $\mathcal{M} \sim 10$) between the two driving modes of turbulence, emphasizing the importance of the turbulence driving mechanism in controlling the formation of stars. Figure adopted from Federrath & Klessen (2012)

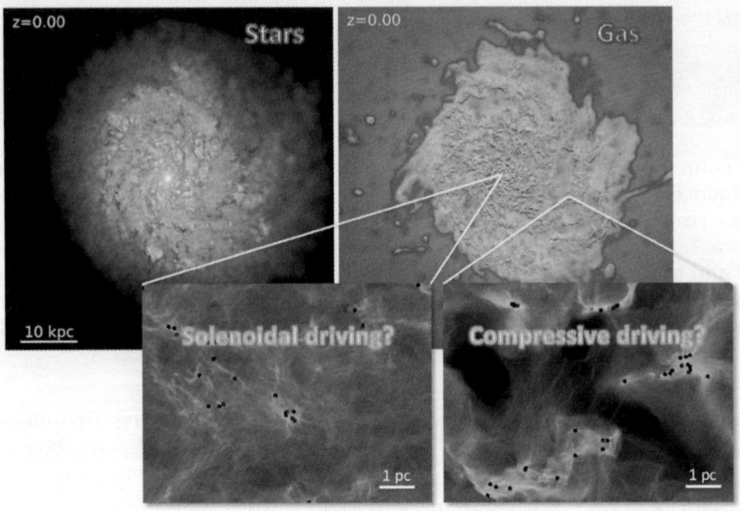

Figure 4. Top panels: stars and gas in a Milky Way type galaxy formed in the FIRE simulations (Hopkins *et al.* 2014). Bottom panels: two high-resolution idealized molecular cloud simulations focussing on small scales (Federrath & Klessen 2012, 2013). Currently the galaxy-scale (top) and cloud-scale (bottom) simulations are completely decoupled from one another. By combining them we can determine the dominant turbulence driving mechanisms in different environments. For example, the central molecular regions of galaxies may be subject to strong shearing motions and hence may be dominated by solenoidal driving of turbulence, while spiral-arm clouds may form via cloud-cloud collisions and spiral-arm compression, which are in the category of compressive drivers.

Molecular Zone is a solenoidal driver (Federrath *et al.* 2016, 2017b). Determining the dominant turbulent driving mode in different clouds and galactic environments is the combined effort of observers and theorists (Brunt & Federrath 2014; Jin *et al.* 2017; Körtgen *et al.* 2017; Orkisz *et al.* 2017).

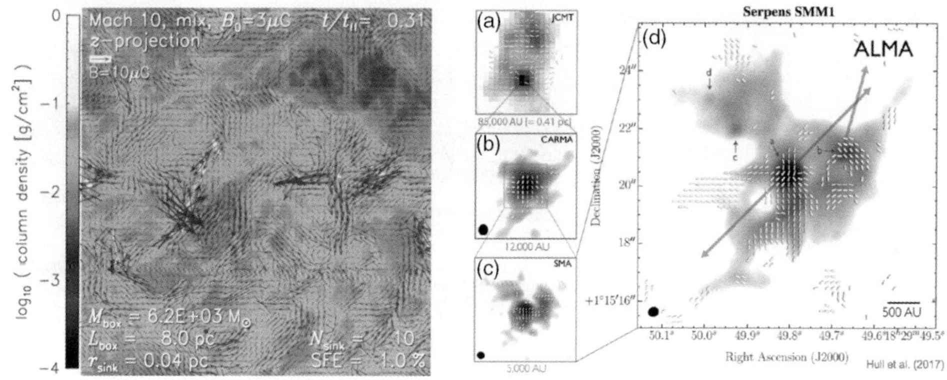

Figure 5. Left: Numerical simulation of a magnetized cloud with magnetic field vectors superimposed on the column density. Right: Zoom-in on a dense core in Serpens, observed with JCMT, CARMA, SMA, and ALMA, probing the magnetic field structure via dust polarization. The cloud breaks up into several individual over-densities (labelled a–d) and jets/outflow may be launched from of these regions (indicated by the arrows). The magnetic field structure in both simulations and observations is highly complex, with a combination of ordered and turbulent field configurations. Images adopted from Federrath & Klessen (2012) and Hull *et al.* (2017).

4. The role of magnetic fields and feedback

The simulations and theoretical models discussed in the previous section have shown that the magnetic field can reduce the star formation rate by a factor of 2–3 compared to purely hydrodynamic cases. Here we want to focus on the effects of the magnetic field configuration. Recent observations of dust polarization with ALMA reveal highly complex magnetic field structures including some ordered, chaotic (turbulent), and rotational components (Hull *et al.* 2017; Lee *et al.* 2018; Zhang *et al.* 2018; Cox *et al.* 2018). Many of these cases differ substantially from the classical hour-glass morphology. The most likely reason for this is the turbulence in the parental molecular cloud, creating flows that twist, tangle and compress the magnetic field lines into equally complex structures as the velocity and density fields of these clouds and cores (Federrath 2016). Figure 5 shows an example of the complexity of the magnetic field in both simulations and observations.

The magnetic field configuration may have profound consequences for star formation, the structure of the disks, the launching of jets and outflows, and ultimately for the formation of planets. Figure 6 shows recent simulation results that aim to isolate the effect of the magnetic field structure on the formation of solar-mass stars. Three identical simulations are compared, only differing in the initial field configuration. The panels from left to right show simulations with initial magnetic field in a uniform (ordered) configuration (aligned with the rotation axis of the core), a partially turbulent case (in which the ordered and turbulent components have the same strength), and a fully turbulent field. The total strength of the initial magnetic field was kept constant at 100 μGauss in all three cases, as were all the other initial parameters (Gerrard *et al.* 2019). The bottom panels of Figure 6 show the gas density when the protostar(s) have reached 0.1 solar masses. Most importantly, we see that the uniform and partially turbulent cases launch jets and outflows along the rotation axis of their disks, while the fully turbulent magnetic field is not capable of launching jets at all. The turbulent field also leads to the formation of 3 stars compared to only a single star in both uniform and partially turbulent field configurations. In summary, the magnetic field structure strongly affects the disk evolution, outflow efficiency and fragmentation, and thus, the mass distribution of stars.

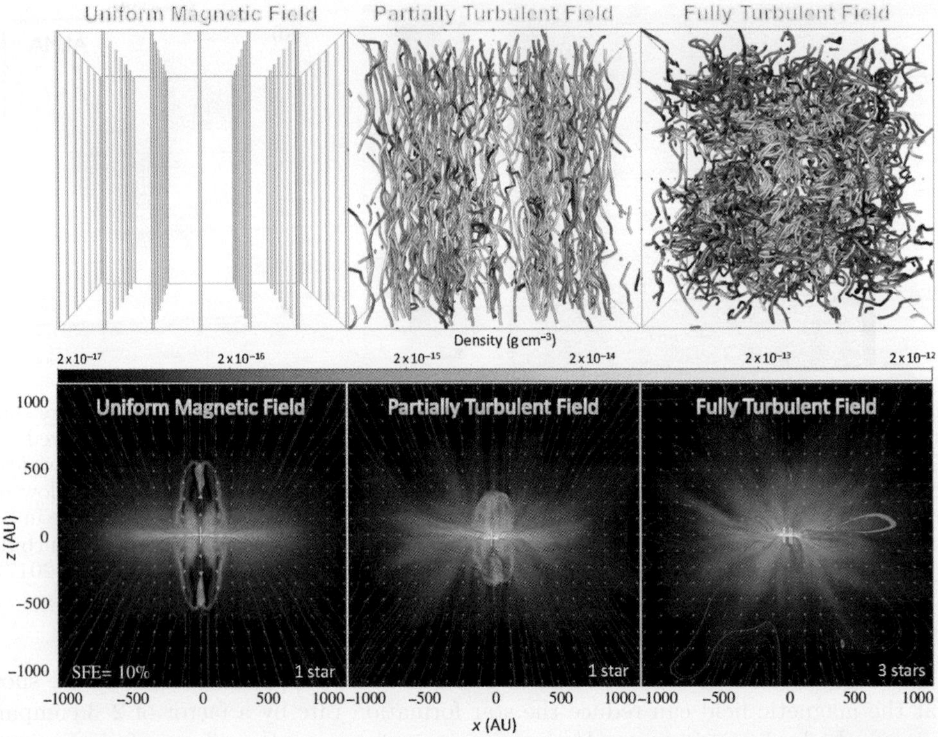

Figure 6. Three simulations of solar-mass star formation. The simulations are identical, except for the magnetic field structure at the start of the simulations. The initial magnetic field configuration is shown in the top panels; from left to right: 1) uniform, 2) partially turbulent, 3) fully turbulent. The bottom panels show the respective accretion disks that form in these three different configurations, when the star(s) has/have accreted 0.1 solar masses. The star(s) (shown as crosses) formed in the disks, seen edge-on in the images. The magnetic field configuration has a strong impact on the outflow/jet strengths (with the uniform and partially turbulent runs launching outflows, while no jets are produced in the fully turbulent case) and on the fragmentation of the core/disk (with the fully turbulent field leading to the formation of 3 stars as opposed to only 1 star each, in the uniform and partially turbulent simulations). Images adopted from Gerrard *et al.* (2019).

5. Summary and conclusions

The main conclusions are

• Star formation is inefficient. The relevant physical processes opposing fast gravitational collapse and making star formation slow and inefficient are: 1) turbulence, 2) magnetic fields, and 3) feedback. Each of these processes reduces the star formation rate by factors of 2–3. In their combination, turbulence, magnetic fields and feedback can yield realistic, observed, low star formation rates per freefall time, of only a few percent.

• Turbulence can reduce the star formation rate by more than a factor of 10 when driven by a solenoidal driver (such as shear), compared to a compressive driver (such as supernova explosions or galactic spiral-arm compression). Solenoidal driving may dominate in the Central Molecular Zone, while compressive drivers may be more relevant in spiral-arm clouds. Determining the mixture of turbulent modes is critical to understand and predict star formation.

• ALMA observations can now probe the magnetic fields in dense protostellar cores down to scales of a few astronomical units. They reveal highly complex field geometries, containing both ordered and turbulent magnetic field components. Simulations with ordered, partially turbulent and completely turbulent fields show that the outflows and jets launched and the stars formed depend significantly on the different field configurations in these dense cores. More observational constraints and simulations that take these complex magnetic field structures into account will be needed to unravel the initial mass function of stars and the origin of binaries and planets.

Acknowledgments

C. F. acknowledges funding by the Australian Research Council (Discovery Projects DP170100603 and Future Fellowship FT180100495), and the Australia-Germany Joint Research Cooperation Scheme (UA-DAAD). We further acknowledge high performance computing resources provided by the Leibniz Rechenzentrum and the Gauss Centre for Supercomputing (grants pr32lo, pr48pi and GCS Large-scale project 10391), the Partnership for Advanced Computing in Europe (PRACE grant pr89mu), the Australian National Computational Infrastructure (grant ek9), and the Pawsey Supercomputing Centre with funding from the Australian Government and the Government of Western Australia, in the framework of the National Computational Merit Allocation Scheme and the ANU Allocation Scheme. The simulation software FLASH was in part developed by the DOE-supported Flash Center for Computational Science at the University of Chicago.

References

Arzoumanian, D., Shimajiri, Y., Inutsuka, S.-i., Inoue, T., & Tachihara, K. 2018, *PASJ*, 70, 96
Bate, M. R. 2012, *MNRAS*, 419, 3115
Brunt, C. M., & Federrath, C. 2014, *MNRAS*, 442, 1451
Cox, E. G., Harris, R. J., Looney, L. W., *et al.* 2018, *ApJ*, 855, 92
Cunningham, A. J., Krumholz, M. R., McKee, C. F., & Klein, R. I. 2018, *MNRAS*, 476, 771
Federrath, C. 2013a, *MNRAS*, 436, 1245
—. 2013b, *MNRAS*, 436, 3167
—. 2015, *MNRAS*, 450, 4035
—. 2016, *JPP*, 82, 535820601
Federrath, C., & Klessen, R. S. 2012, *ApJ*, 761, 156
—. 2013, *ApJ*, 763, 51
Federrath, C., Klessen, R. S., & Schmidt, W. 2008, *ApJ*, 688, L79
Federrath, C., Krumholz, M., & Hopkins, P. F. 2017a, *Journal of Physics Conference Series*, 837, 012007
Federrath, C., Roman-Duval, J., Klessen, R. S., Schmidt, W., & Mac Low, M. 2010, *A&A*, 512, A81
Federrath, C., Rathborne, J. M., Longmore, S. N., *et al.* 2016, *ApJ*, 832, 143
Federrath, C., Rathborne, J. M., Longmore, S. N., *et al.* 2017b, *IAU Symposium*, 322, ed. R. M. Crocker, S. N. Longmore, & G. V. Bicknell, 123–128
Gerrard, I. A., Federrath, C., & Kuruwita, R. L. 2019, *MNRAS*, submitted
Guszejnov, D., Hopkins, P. F., Grudić, M. Y., Krumholz, M. R., & Federrath, C. 2018, *MNRAS*
Hennebelle, P., & Chabrier, G. 2011, *ApJ*, 743, L29
Hernandez, A. K., & Tan, J. C. 2015, *ApJ*, 809, 154
Hopkins, P. F., Kereš, D., Oñorbe, J., *et al.* 2014, *MNRAS*, 445, 581
Hull, C. L. H., Girart, J. M., Tychoniec, Ł., *et al.* 2017, *ApJ*, 847, 92
Jin, K., Salim, D. M., Federrath, C., *et al.* 2017, *MNRAS*, 469, 383
Kainulainen, J., Federrath, C., & Henning, T. 2013, *A&A*, 553, L8
—. 2014, *Science*, 344, 183
Kauffmann, J., Pillai, T., & Goldsmith, P. F. 2013, *ApJ*, 779, 185
Konstandin, L., Girichidis, P., Federrath, C., & Klessen, R. S. 2012, *ApJ*, 761, 149
Körtgen, B., Federrath, C., & Banerjee, R. 2017, *MNRAS*, 472, 2496
Krumholz, M. R., Klein, R. I., & McKee, C. F. 2012, *ApJ*, 754, 71

Krumholz, M. R., & McKee, C. F. 2005, *ApJ*, 630, 250
Krumholz, M. R., & Tan, J. C. 2007, *ApJ*, 654, 304
Kuruwita, R. L., Federrath, C., & Ireland, M. 2017, *MNRAS*, 470, 1626
Lee, C.-F., Li, Z.-Y., Hirano, N., *et al.* 2018, *ApJ*, 863, 94
Offner, S. S. R., Klein, R. I., McKee, C. F., & Krumholz, M. R. 2009, *ApJ*, 703, 131
Onus, A., Krumholz, M. R., & Federrath, C. 2018, *MNRAS*, 479, 1702
Orkisz, J. H., Pety, J., Gerin, M., *et al.* 2017, *A&A*, 599, A99
Padoan, P., & Nordlund, Å. 2011, *ApJ*, 730, 40
Sharda, P., Federrath, C., da Cunha, E., Swinbank, A. M., & Dye, S. 2018, *MNRAS*, 477, 4380
Taylor, P., & Kobayashi, C. 2015, *MNRAS*, 448, 1835
Zhang, Y., Higuchi, A. E., Sakai, N., *et al.* 2018, *ApJ*, 864, 76

Discussion

CUNTZ: Please comment on the effect(s) of metallicity, as the latter is expected to impact the cooling processes.

FEDERRATH: Indeed, the composition (metallicity) determines the thermodynamic response of the gas. This is particularly important when comparing present-day star formation with star formation in the early Universe, i.e., for the formation of the First Stars, where cooling was not as efficient as in solar-metallicity gas.

KHAIBRAKHMANOV: What is the physical mechanism that determines filament formation at the sonic scale?

FEDERRATH: Filaments and cores may form at the sonic scale, because it is the scale, where shocks transition from supersonic to subsonic speeds. Thus, filaments may be the stagnation points of compressive turbulent flows, i.e., filaments forming at the intersection of planar shock waves.

LINSKY: Can you explain the initial mass function of stars? Was Salpeter correct in his prediction of the IMF?

FEDERRATH: The IMF is one of the biggest challenges to explain and understand. It is likely determined by a combination of physical processes, in particular gravity, turbulence, magnetic fields, and feedback (by jets/outflows and radiation). The IMF will be the focus of advanced simulations in the next few years.

Origins: From the Protosun to the First Steps of Life
Proceedings IAU Symposium No. 345, 2018
B. G. Elmegreen, L. Viktor Tóth & M. Güdel, eds.

doi:10.1017/S1743921319002102

ALMA observations of the Orion Integral Filament: evidence for fibers in a massive cloud

Alvaro Hacar[1], Mario Tafalla[2], Jan Forbrich[3] and Josefa Grossschedl[4]

[1]Leiden Observatory, Leiden University, P.O. Box 9513, 2300-RA Leiden, The Netherlands
email: hacar@strw.leidenuniv.nl

[2]Observatorio Astronomico Nacional (IGN), C/ Alfonso XII, 3, E-28014, Madrid, Spain

[3]Centre for Astrophysics Research, University of Hertfordshire, College Lane, Hatfield, AL10 9AB, UK

[4]University of Vienna, Türkenschanzstrasse 17, A-1180 Vienna, Austria

Abstract. The connection between low- and high-mass filaments is a matter of strong debate. In order to bridge these two filamentary regimes, we have investigated the internal structure of the Integral Filament in Orion using ALMA observations of the N_2H^+ (1-0) emission line in Band 3 in combination with previous single-dish data. Our ALMA mosaics, the largest of its kind carried out so-far in local clouds, reveal the presence of multiple sonic-like fibers inside this massive filament. In combination with the identification of fibers in regions such as Taurus, Musca, and Perseus, the first unambiguous detection of fibers in Orion highlights the importance of these gas substructures as the fundamental building blocks of both low- and high-mass filaments.

Keywords. stars: formation, ISM: kinematics and dynamics, ISM: clouds, radio lines: ISM

1. Fibers in low-mass filaments

The unprecedented sensitivity of the Herschel observations has revealed the presence of filaments in low- and high-mass star-forming regions, harbouring most of the cores and stars formed within molecular clouds (see André *et al.* 2014, for a review). The ubiquitous detection of filaments highlights the key role of these gas structures determining the initial conditions for star-formation. However, the detailed origin of stars in both low- and high-mass filaments remains poorly understood.

Recent molecular observations have investigated the internal substructure of several Herschel filaments with exquisite detail. From the analysis of different millimeter line tracers, Hacar *et al.* (2013) demonstrated that the previously described as single B213-L1495 filament in Taurus, is actually a collection of small-scale elongated substructure, referred to as *fibers*. These fibers are characterized by a quiescent velocity structure, showing sonic-like internal motions. They contain all the dense cores within this region, passing on their quiescent properties to these stellar seeds. After their original discovery, fibers have been systematically detected in low-mass star-forming regions such as Musca (Hacar *et al.* 2016) and Perseus (Hacar *et al.* 2017a). Hydrodynamical simulations (e.g., Smith *et al.* 2014) suggest these fibers as the first sonic-like structures decoupled from the turbulent cascade, regulating the transition between the supersonic diffuse gas and the quiescent dense cores prior to the formation of stars.

In a recent work (Hacar *et al.* 2018), our group reported the first direct evidence of fibers in the Integral Shape Filament (ISF; see Johnstone & Bally 1999) in the Orion A cloud. The ISF is the most massive filament in the solar neighbourhood (D$\sim$ 414 pc,

Menten *et al.* 2007) and the only one forming high-mass stars (aka Trapezium). The ISF is usually regarded as a local benchmark of current star-formation theories. The detection of fibers in this unique target can help to unify our current description of the internal structure of both low- and high-mass filaments.

2. ALMA observations in the ISF

We have investigated the Orion ISF using the Atacama Large Millimeter Array (ALMA). We observed the central spine of the ISF filament covering the Orion Nebula Cluster (ONC), including the Trapezium stars and Orion BN/KL source, and both OMC-1 and OMC-2 clouds (see Fig.1, Left panel). Our observations targeted the N_2H^+ (1-0) line (93.173 GHz, Band 3) and combine two 150-pointing interferometric ALMA-12m mosaics with previous IRAM30m (single-dish) observations (Hacar *et al.* 2017a). The full reduction process and combination of single-dish plus interferometric maps is described in Hacar *et al.* (2018). These data are part of the ORION-4D project and are available on-line in a public website (see `https://sites.google.com/site/orion4dproject/`).

We display the total integrated intensity N_2H^+ (1-0) map of the ISF in Fig. 1 (Central panel). Our observations correspond to the largest ALMA mosaic carried out so-far in local star-formation studies. Our final map is convolved to a final beamsize of 4.5", corresponding to a spatial resolution of $\sim$ 2000 AU at the distance of Orion. Our observations cover a total of 2 orders of magnitude in scale, from the total length of $\sim$ 2 pc of the ISF down to 2000 AU. Moreover, the unique combination with our previous IRAM30m observations allow us to recover all relevant scales.

Our maps show the internal substructure of the ISF with unprecedented detail. As density selective tracer, the N_2H^+ (1-0) emission reveals the internal distribution of dense gas ($> 10^5$ cm^{-3}) within this filament. As seen in Fig. 1 (Central panel), the gas in both OMC-1 and OMC-2 regions is strongly organized in multiple elongated gas fibers forming a complex network (see also Sect. 3 for a discussion of their properties). Despite the differences in resolution and sensitivity, the same gas substructure is also observed in the continuum (Fig. 1, Left panel) ruling out contamination, abundance, and/or excitation effects affecting our analysis. The observed ALMA fibers contain most of the dense cores, continuum sources, and protostars identified within the ISF (see also Fig. 1 left panel). For the first time, our new ALMA observations reveal the intrinsic fibrous substructure of the massive Orion ISF.

3. Narrow fibers in Orion

We have characterized the observed ISF substructure using a new hierarchical version of the Friends-In-Velocity algorithm (see Hacar *et al.* 2013), namely, Hi-FIVE (see Hacar *et al.* 2018, for a description). Hi-FIVE simultaneously considers both the spatial and velocity distribution of the cloud in order to extract those velocity-coherent structures found within complex molecular datacubes. From the analysis of our more than 75 000 N_2H^+ (1-0) spectra, Hi-FIVE identifies a total of 55 independent fibers, recovering $\sim$ 90% of the total emission of this molecule detected within our ALMA maps. We show the spatial distribution of fibers in Fig. 1 (Right panel; red segments). These fibers explain the full internal structure of the ISF forming a complex and intertwined network in both OMC-1 and OMC-2 regions.

Our HiFIVE algorithm allows us to statistically investigate the physical properties of the ISF fibers. In Figure 2, we show the cumulative distributions of all the non-thermal motions in units of the local sound speed (Left panel), the mass-per-unit-length (Central panel), and total length (Right panel) of all detected fibers in both OMC-1 (blue) and OMC-2 (red) regions. We compare these two fiber populations with the

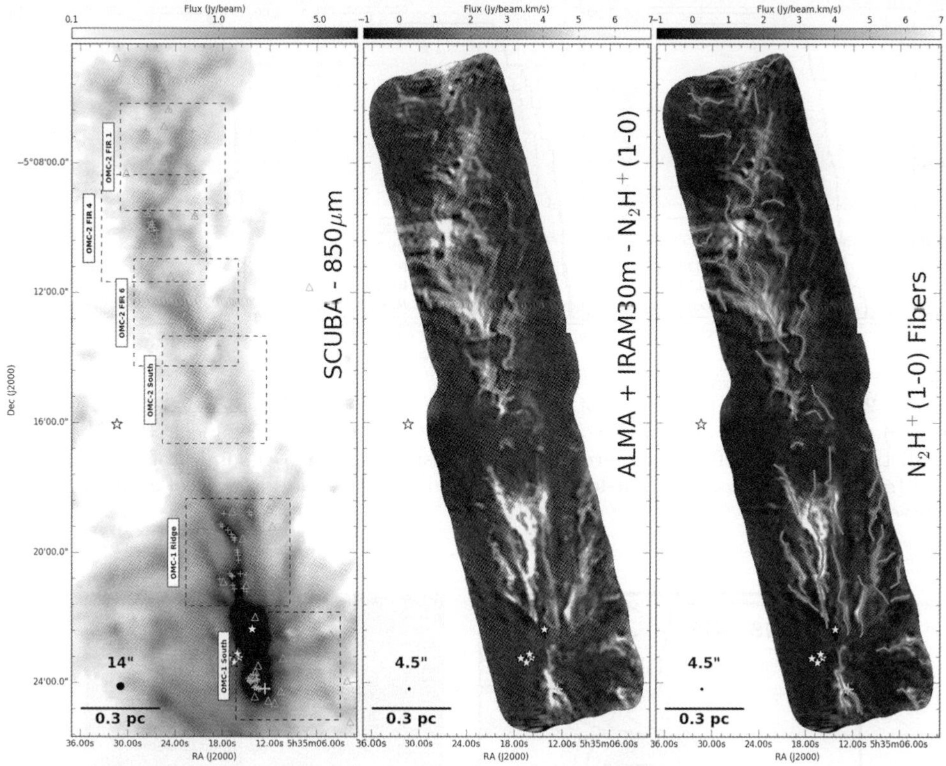

Figure 1. ALMA observations of the Orion ISF (see Hacar *et al.* 2018, for details): (Left) SCUBA-850μm continuum emission (Johnstone & Bally 1999), (Centre) ALMA-12m + IRAM30m N_2H^+ (1-0) integrated emission, and (Right) the 55 velocity-coherent fibers detected in N_2H^+ (main axis indicated by red segments). The location of all the continuum sources (blue crosses) and Spitzer protostars (blue triangles) are indicated in the continuum image. For reference, the position of the Trapezium (four white stars), the Orion BN/KL region (yellow star), and NU Ori star (white isolated star), are indicated in all panels. The synthesized beamsize of our ALMA observations (4.5" or $\sim$ 2000 AU) is indicated in the lower corner of both central & right panels. We note the unprecedented large dynamic range in resolution and sensitivity of our combined ALMA+IRAM30m observations.

observed properties of fibers in clouds such as NGC1333 in Perseus (orange), B213 in Taurus (black), and Musca (green) reported in previous studies. As illustrated by this comparison, fibers show similar properties in terms of internal motions (left panel) and mass distribution close to hydrostatic equilibrium (central panel). Independently of their environment, fibers are recognized as the first sonic structures (i.e., $\sigma_{NT}/c_s \sim 1$) in both low- and high-mass clouds.

While we find similar kinematics, we identify a systematic change in the physical size and width of fibers in regions of increasing mass. In Figure 3, we display two independent fibers observed in the Taurus (left) and Orion (right) clouds, both detected in N_2H^+ (1-0), at the same physical scale. In regions like Taurus, fibers present a typical Full-Width-Half-Maximum (FWHM) of $\sim$ 0.1 pc, consistent with the expected constant filament width observed in Herschel filaments (see Arzoumanian *et al.* 2011). On the contrary, the Orion fibers show much narrower widths on the order of $\lesssim$ 0.03 pc. A similar trend is observed in the reported fiber lengths between Taurus ($\sim$ 0.5 pc) and Orion ($\sim$ 0.15 pc) (see Fig. 2, Right panel). While similar in kinematics, our ALMA results indicate a

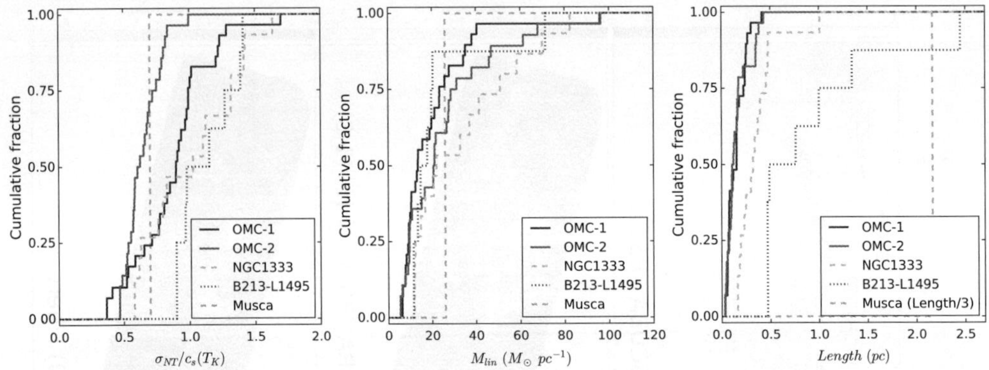

Figure 2. Statistical properties of fibers in the OMC-1 (blue), OMC-2 (red), NGC1333 (orange), B213 (black), and Musca (green) filamentary regions. All panels show the cumulative distribution of (Left) Non-thermal velocity dispersion (σ_{NT}) in units of the local sound speed (c_s), (Centre) mass-per-unit-length (M_{lin}), and (Right) total Length of all fibers detected in these regions.

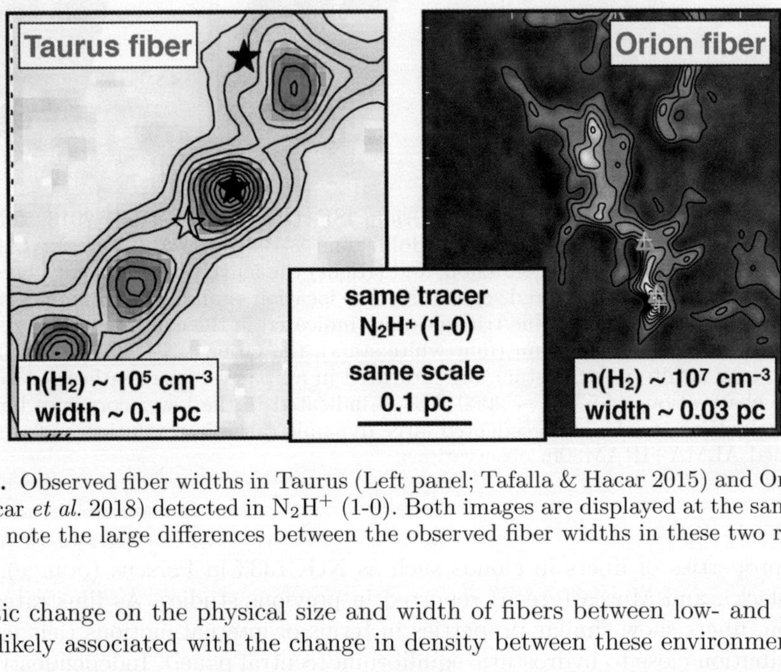

Figure 3. Observed fiber widths in Taurus (Left panel; Tafalla & Hacar 2015) and Orion (Right panel Hacar *et al.* 2018) detected in N_2H^+ (1-0). Both images are displayed at the same physical scale. We note the large differences between the observed fiber widths in these two regions.

systematic change on the physical size and width of fibers between low- and high-mass regions, likely associated with the change in density between these environments.

4. Fibers in low- and high-mass filaments

Originally identified in low-mass filaments like Taurus or Musca, increasing number of studies have suggested the presence of fiber-like structures in regions of increasing complexity such as young proto-clusters (e.g., Fernández-López *et al.* 2014) and Infrared Dark Clouds (e.g., Henshaw *et al.* 2014). Our ALMA results extend the detection of fibers into massive filaments. In particular, the first unambiguous detection of these objects in Orion indicates that fibers are the fundamental building blocks of low- and high-mass filaments. In all environments, fibers regulate the formation of dense cores and stars, playing a pivotal role in the star-formation process. In both low- and high-mass filaments fibers are described by their quiescent internal dynamics and a quasi-stable configuration.

Remarkably, fibers seem to adapt their widths and lengths to the density of their local environment (see Hacar *et al.* 2018, for a full discussion).

References

André, P., Di Francesco, J., Ward-Thompson, D., *et al.* 2014, *Protostars and Planets VI*, 27
Arzoumanian, D., André, P., Didelon, P., *et al.* 2011, *A&A*, 529, L6
Fernández-López, M., Arce, H. G., Looney, L., *et al.* 2014, *ApJL*, 790, L19
Hacar, A., Tafalla, M., Kauffmann, J., & Kovács, A. 2013, *A&A*, 554, A55
Hacar, A., Kainulainen, J., Tafalla, M., Beuther, H., & Alves, J. 2016, *A&A*, 587, A97
Hacar, A., Alves, J., Tafalla, M., & Goicoechea, J. R. 2017a, *A&A*, 602, L2
Hacar, A., Tafalla, M., & Alves, J. 2017b, *A&A*, 606, A123
Hacar, A., Tafalla, M., Forbrich, J., *et al.* 2018, *A&A*, 610, A77
Henshaw, J. D., Caselli, P., Fontani, F., Jiménez-Serra, I., & Tan, J. C. 2014, *MNRAS*, 440, 2860
Johnstone, D., & Bally, J. 1999, *ApJL*, 510, L49
Matthews, B. C., McPhee, C. A., Fissel, L. M., & Curran, R. L. 2009, *ApJS*, 182, 143
Menten, K. M., Reid, M. J., Forbrich, J., & Brunthaler, A. 2007, *A&A*, 474, 515
Smith, R. J., Glover, S. C. O., & Klessen, R. S. 2014, *MNRAS*, 445, 2900
Tafalla, M., & Hacar, A. 2015, *A&A*, 574, A104
Ward-Thompson, D., Pattle, K., Bastien, P., *et al.* 2017, *ApJ*, 842, 66

Discussion

LISEAU: The 1-0 transition of N_2H^+ has an upper level energy less than 5K implying that in Orion considerably higher levels will be predominantly populated, making mass determinations based on (1-0) potentials very uncertain. Furthermore, relative intensities of the main and satellite components indicate the levels to be severely optically thick, making them unsuitable for mass determinations.

HACAR: Under normal conditions, the high column densities observed in Orion would indeed lead into highly optically thick lines of $N_2H^+(J=1-0)$. However, much to our surprise, the high-densities ($\sim 10^7$ cm^{-3}) and relatively higher temperatures ($\sim 20-30$ K) found in the dense gas in Orion help to keep this $J=1-0$ transition optically thin. Using RADEX calculations, in our paper (Hacar *et al.* 2018) we demonstrate that this combination of temperature and density excite many molecules at higher J-states ($J >$ 2...3), depopulating the $J-1-0$ level and making this fundamental level optically thin. Our estimates are also in agreement with our direct measurements of the $N_2H^+(1-0)$ lines and their hyperfine structure showing low line opacities.

KHAIBRAKHMANOV: What is the magnetic field direction in the Orion filament and in subfilament structures?

HACAR:The morphology of the B fields changes significantly along the ISF region. In OMC-1, the B-field shows an hourglass shape (see Ward-Thompson *et al.* 2017) likely created by the gravitational collapse of this region (Hacar *et al.* 2017a). In OMC-2, the B-field morphology is less organized and its origin is less clear (e.g., see Matthews *et al.* 2009).

ZINNECKER: Do the fiber bundles oscillate back and forth hydromagnetically, as proposed in the slingshot theory of the integral-shaped filament in Orion?

HACAR: At large scales, our fibers follow a velocity profile created by the gravitational collapse of the OMC-1 reported in our previous study (Hacar *et al.* 2017a). These results contradict the so-called "slingshot mechanism" as they are collapsing motions rather than oscillations or ejections.

Origins: From the Protosun to the First Steps of Life
Proceedings IAU Symposium No. 345, 2018
B. G. Elmegreen, L. Viktor Tóth & M. Güdel, eds.

doi:10.1017/S1743921319001510

Constraints from zoom-in simulations on the protostellar accretion process

Michael Kuffmeier†

Institute for Theoretical Astrophysics (ITA), Zentrum für Astronomie (ZAH), University of Heidelberg,
Albert-Ueberle-Straße 2, DE-69120, Heidelberg, Germany
email: ru151@uni-heidelberg.de

Abstract. Stars are embedded in different environments of Giant Molecular Clouds during their formation phase. Despite this fact, it is common practice to assume an isolated spherical core as the initial condition for models of individual star formation. To avoid the uncertainties of initial and boundary conditions, we use an alternative approach of zoom-in simulations to account for the environment in which protostars form. Our models show that injections of ^{26}Al from a close-by supernova into the young solar system were highly unlikely. Moreover, we find that the accretion process of protostars is heterogeneous and environment-dependent.

Keywords. stars: formation, (magnetohydrodynamics:) MHD, accretion, accretion disks, solar system: formation, ISM: clouds, (stars:) binaries: general

1. Introduction

Over the last few decades, a picture of isolated star formation has been established: an individual star forms due to the gravitational collapse of a prestellar core (Shu *et al.* 1987). The core is typically assumed to be an isolated entity, approximately spherically in shape often referred to as 'spherical cow' approach. Starting with the assumption of a non-rotating, unmagnetized sphere, over the years models have become increasingly sophisticated by accounting for additional effects such as e.g. rotation, magnetic fields, and turbulence (e.g. Machida *et al.* 2007, Seifried *et al.* 2013, Tomida *et al.* 2015, Vaytet *et al.* 2018).

Regardless of these advances, the fundamental limitations of the spherical cow approach still apply, namely idealized initial and boundary conditions. In fact, already in his pioneering work on protostellar collapse, Richard Larson pointed out: 'For the boundary condition [...] we have in most cases adopted the simplest assumption' and 'For the initial conditions we have again adopted the simplest assumptions' (Larson 1969). 50 years later, it is widely accepted that stars form in different environments of Giant Molecular Clouds (GMCs) and the morphology of prestellar cores deviates from spherical symmetry owing to the connection to the underlying filamentary structure in the interstellar medium (ISM) (Andre *et al.* 2010, Federrath 2016, Rivera-Ingraham *et al.* 2017).

Therefore, models accounting for the protostellar environment provided by the GMC are required, such as has been done in recent 'zoom-in' simulations. In these simulations, the starting point is a turbulent GMC, in which prestellar cores form consistently and where the formation process of stars and disks is studied by applying sufficient adaptive mesh refinement (AMR) around individual protostars. (For a similar approach

† International Postdoctoral Fellow of Independent Research Fund Denmark (IRDF)

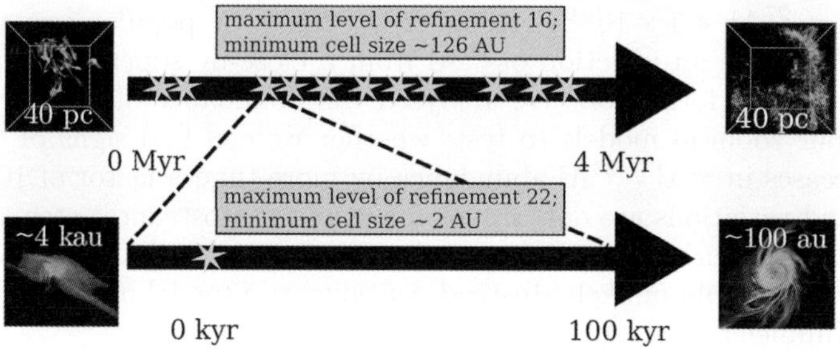

Figure 1. Sketch of the zoom-in technique adopted from Kuffmeier *et al.* (2017). In the first step, the turbulent GMC evolves for about 4 Myr using a minimum cell size of 126 au. Afterwards, sink particles are selected for zoom-ins using a minimum cell size of typically 2 au in the vicinity of the sinks.

using smoothed-particle hydrodynamics (SPH) and without considering magnetic fields see Bate 2018.) In this proceeding, we briefly summarize some major results that were obtained with the zoom-in method. In particular, we used this approach to constrain

- the abundance of short-lived radionuclides such as ^{26}Al,
- the accretion process of stars, and
- the formation process of protostellar disks in different environments.

Moreover, we briefly investigate formation of stellar multiples as well as we discuss the implications of young forming disks for planet formation.

2. Zoom-in method

A detailed presentation of the zoom-in method and the sink prescriptions used for stars is beyond the scope of this contribution, and we refer the reader to Kuffmeier *et al.* (2016) and Kuffmeier *et al.* (2017) for a comprehensive description of the method. Using a modified version (Haugbølle *et al.* 2018) of the adaptive mesh-refinement code RAMSES (Teyssier 2002, Fromang *et al.* 2006), we model a turbulent magnetized Giant Molecular Cloud (GMC) as a (40 pc)^3 cubic box with periodic boundary conditions. As illustrated in Figure 1, we evolve the GMC for about 4 Myr with a minimum cell size of 126 au. The mass of the GMC is $\approx 10^5$ M$_\odot$ and stars more massive than 8 M$_\odot$ explode as supernovae after a mass-dependent lifetime driving turbulence as well as enriching the GMC with short-lived radionuclides ^{26}Al and ^{60}Fe. To gain better insight of the properties during the accretion phase, we then select a number of stars typically in the mass range of 1 to 2 M$_\odot$ and rerun the simulation with a minimum cell size of ≈ 2 au in the vicinity of the stars for time scales of ~ 100 kyr. We applied yet another step of zooming-in to resolve the properties of one disk with a minimum cell size of 0.06 au for ~ 1000 yr at $t \approx 50$ kyr after sink formation.

3. Results

3.1. *Constraints on the ^{26}Al abundance in the early solar system*

Calcium-Aluminum rich Inclusions (CAIs) are the oldest known components of the solar system, probably forming within the first 10^4 yr of its existence (Larsen *et al.* 2011, Haugbølle *et al.* 2017). However, in the young solar system the abundance of the short-lived radionuclide ^{26}Al, produced during the process of supernova explosions, is puzzling to explain. Cosmochemical measurements reveal a discrepancy in relative abundance of ^{26}Al (radioactive) compared to ^{27}Al (stable) in ordinary CAIs with a canonical value of ^{26}Al / ^{27}Al $\sim 5 \times 10^{-5}$ (MacPherson *et al.* 1995) and in ^{26}Al-poor FUN CAIs with values of ^{26}Al / ^{27}Al $< 1 \times 10^{-6}$) (Kööp *et al.* 2018). A popular scenario to explain the difference suggests an injection of ^{26}Al from a close-by supernova during the early phase of solar system formation. We analysed the abundance of ^{26}Al around different protostars in our zoom-in models to test, whether we can find signs of ^{26}Al injections leading to increases in ^{26}Al / ^{27}Al abundance by more than a factor of 10. In short, the answer is no and variations are only minor for young protostellar systems. We draw the conclusion that a scenario of an early supernova injection within the first $t < 10^5$ yr of the solar system is highly unlikely. Instead, we suggest a model of thermal processing of dust grains (Kuffmeier *et al.* 2016).

3.2. *Heterogeneous accretion and disk formation*

To study the effect of the protostellar environment on the accretion process of different sinks, we analyse the accretion profiles of nine sinks for about 10 to 100 kyr. In contrast to the self-similar collapse of a singular isothermal sphere (Shu 1977), the accretion rate is neither spatially homogeneous nor constant in time. Sinks typically have high accretion rates of more than 10^{-5} M$_{\odot}$yr^{-1} in the early collapse phase followed by a decrease in accretion rate. The decrease in accretion rate, however, may differ from object to object. Sinks forming from rather isolated cores show more significant decreases in accretion rates during their evolution than sinks in more perturbed environments, where additional mass infall can raise the accretion profile (for late infall accretion see Padoan *et al.* 2014). For the more embedded sinks, we generally find flatter accretion profiles (Kuffmeier *et al.* 2017). Moreover, we find that the presence of a massive disks can cause increases onto the central sink due to disk instabilities (Kuffmeier *et al.* 2019).

3.3. *Disk formation*

Spherical collapse simulations, demonstrate that rotation of the collapsing prestellar core leads to the formation of a disk around the protostar owed by the conservation of angular momentum. Models that additionally account for magnetic fields by numerically solving the equations of ideal magnetohydrodynamics (MHD) find that magnetic fields efficiently transport angular momentum outwards, and hence, quench disk formation – at odds with observations of disks around protostars even in the earliest stages. The magnetically induced suppression of disks is commonly referred to as ‘magnetic braking catastrophe’. However, catastrophic magnetic braking is circumvented by accounting for non-ideal MHD effects (e.g. Wurster *et al.* 2018) and/or turbulence (e.g. Seifried *et al.* 2013) in more advanced models of spherical collapse. Our models do not account for non-ideal MHD effects, but consistently include the underlying turbulence in the GMC. Out of six runs using a minimum cell size of 2 au, we find three sinks that host rotationally supported disks already after only ~ 10 kyr. In the other three cases, the formation of disks is suppressed or disks only occur intermittently. We find that angular momentum is predominantly transported in radial direction from the disk via Maxwell stresses with turbulence as a counter mechanism preventing disk formation altogether (Kuffmeier *et al.* 2017). Anticipating results with non-ideal MHD effects included, especially ambipolar diffusion (Hennebelle *et al.* 2016, Masson *et al.* 2016), we expect a higher fraction of disks per sink due to further reduction of magnetic braking (Kuffmeier & Nauman 2017). Considering the early presence of disks around young stars with high accretion rates and continuous infall, we emphasize that the mass budget for planet formation may well be larger than in models assuming steady disk masses.

3.4. *Outlook: the origin of stellar multiples*

Initially, we selected sinks that formed from the least perturbed prestellar cores in our simulations to simplify the comparison with models assuming isolated core collapse. However, we have started to look into more perturbed cores that are connected to massive GMC filaments. In particular, we find the formation of a binary companion at a distance of ≈ 1500 au from the primary at $t \approx 36$ kyr after the formation of the primary (Kuffmeier *et al.* 2019). This companion forms inside a filament that is connected to the primary due to compression on time scales of only a few kyr. After its formation, the companion gains more mass and approaches the primary within about 15 kyr to form a binary system of ~ 100 au distance. Based on these findings and against observations of Class 0 protostars of which the separation distribution shows an enhancement at about 100 au (Tobin *et al.* 2018), we suggest the following formation scenario for binary systems: companions form inside filaments at ~ 1000 au distance (Offner *et al.* 2010, Offner *et al.* 2016), move towards the primary due to its gravitational potential on time scales of about $\sim 10^4$ yr, eventually forming a binary system due to conservation of angular momentum with a characteristic separation of ~ 100 au.

4. Summary

We briefly present the major results of studying the accretion process of individual protostars forming in different environments using zoom-in simulations. Regarding the abundance of ^{26}Al in the early solar system, our results demonstrate that an early injection scenario is highly unlikely to be the origin of the discrepancy measured between ordinary CAIs and FUN CAIs. Accounting for the GMC environment additionally shows that the accretion process of protostars is heterogeneous in space, time and differs among protostars. Consistent with these results the properties of disks vary in different environments and young disks can be efficiently replenished with fresh material from the interstellar medium.

References

André, P., Men'shchikov, A., Bontemps, S. *et al.*, 2010, *A&A*, 518, L102
Bate, M. R., 2018, *MNRAS*, 475, 5618
Federrath, C., 2016, *MNRAS*, 457, 375
Fromang, S., Hennebelle, P. & Teyssier, R., 2006, *A&A*, 457, 371
Haugboelle, T., Grassi, T., Frostholm Mogensen, T. *et al.* 2017, *LPI Contributions*, 1975, 2025
Haugbølle, T., Padoan, P. & Nordlund, Å., 2018, *ApJ*, 854, 35
Hennebelle, P., Commerçon, B., Chabrier, G. *et al.*, 2016, *ApJL* 830, L8
Kööp, L., Nakashima, D., Heck, P. R. *et al.*, 2018, *Geochimica et Cosmochimica Acta*, 221, 296
Kuffmeier, M., Frostholm Mogensen, T., Haugbølle, T. *et al.*, 2016, *ApJ*, 826, 22
Kuffmeier, M., Haugbølle, T. & Nordlund, Å., 2018, *ApJ*, 846, 7
Kuffmeier, M. & Nauman, F., 2017, arXiv:1710.11195, *arXiv preprint*
Kuffmeier, M., Calcutt, H. & Kristensen, E. L., 2019, arXiv:1907.02083, arXiv preprint
Larsen, K. K., Trinquier, A., Paton, C. *et al.*, 2011, *ApJL*, 735, L37
Larson, R. B., 1969, *MNRAS*, 145, 271
Machida, M. N., Inutsuka, S.-i. & Matsumoto, T., 2007, *ApJ*, 670, 1198
MacPherson, G. J., Davis, A. M. & Zinner, E. K., 1995, *Meteoritics*, 30, 365
Masson, J., Chabrier, G., Hennebelle, P. *et al.*, 2016, *A&A*, 587, 20
Padoan, P., Haugbølle, T. & Nordlund, Å., 2014, *ApJ*, 797, 32
Offner, S. S. R., Kratter, K. M., Matzner, C. D. *et al.*, 2016, *ApJ*, 725, 1485
Offner, S. S. R., Dunham, M. M., Lee, K. I. *et al.*, 2016, *ApJL*, 827, L11
Rivera-Ingraham, A., Ristorcelli, I., Juvela, M. *et al.*, 2017, *A&A*, 601, A94
Seifried, D., Banerjee, R., Pudritz, R. E. *et al.*, 2013, *MNRAS*, 432, 3320
Shu, F. H., 1977, *APJ*, 214, 488

Shu, F. H., Adams, F. C., & Lizano, S., 1987, *ARAA*, 25, 23
Teyssier, R., 2002, *A&A*, 385, 337
Tobin, J. J., Looney, L. W., Li, Z.-Y. *et al.*, 2018, *ApJ*, 867, 43
Tomida, K., Okuzumi, S., & Machida, M. N., 2015, *ApJ*, 801, 117
Vaytet, N., Commerçon, B., Masson, J. *et al.*, 2018, *A&A*, 615, 5
Wurster, J., Bate, M. R. & Price, D. J., 2018, *MNRAS*, 475, 1859

Discussion

WALLER: You noted that stars do not form in isolation and went on to describe simulations invoking turbulence and accretion. My question is whether you have considered the effects of UV irradiation by stars that have already formed.

KÜFFMEIER: We account for heating induced by UV irradiation by applying the recipe of Franco and Cox (1986). According to this recipe heating is quenched in the densest gas. We plan to implement a full radiative transfer scheme in the future.

MAURY: Are the different disk properties sampled around unequal objects linked to unequal mass-to-flux ratios?

KÜFFMEIER: The prestellar cores in our simulations deviate from the classical scenario of a spherically symmetric core due to the presence of turbulence in the GMC. Therefore, the mass-to-flux ratio is not constant since the amount of enclosed mass and magnetization can change over time. We find that mass-to-flux ratios are in the range of 2-10 with large disks forming in the case of the highest mass-to-flux ratios. However, we also see disks forming at lower values, indicating that disk formation is not only determined by a single mass-to-flux ratio.

Origins: From the Protosun to the First Steps of Life
Proceedings IAU Symposium No. 345, 2018
B. G. Elmegreen, L. Viktor Tóth & M. Güdel, eds.

doi:10.1017/S1743921319001686

The implications of clustered star formation for (proto)planetary systems and habitability

J. M. Diederik Kruijssen[1] and Steven N. Longmore[2]

[1]Astronomisches Rechen-Institut, Zentrum für Astronomie der Universität Heidelberg, Mönchhofstraße 12-14, 69120 Heidelberg, Germany
email: kruijssen@uni-heidelberg.de

[2]Astrophysics Research Institute, Liverpool John Moores University, IC2, Liverpool Science Park, 146 Brownlow Hill, Liverpool L3 5RF, United Kingdom
email: s.n.longmore@ljmu.ac.uk

Abstract. Star formation is spatially clustered across a range of environments, from dense stellar clusters to unbound associations. As a result, radiative or dynamical interactions with neighbouring stars disrupt (proto)planetary systems and limit their radii, leaving a lasting impact on their potential habitability. In the solar neighbourhood, we find that the vast majority of stars form in unbound associations, such that the interaction of (proto)planetary systems with neighbouring stars is limited to the densest sub-regions. However, the fraction of star formation occurring in compact clusters was considerably higher in the past, peaking at $\sim 50\%$ in the young Milky Way at redshift $z \sim 2$. These results demonstrate that the large-scale star formation environment affects the demographics of planetary systems and the occupation of the habitable zone. We show that planet formation is governed by multi-scale physics, in which Mpc-scale events such as galaxy mergers affect the AU-scale properties of (proto)planetary systems.

Keywords. stars: formation, planetary systems: formation, planetary systems: protoplanetary disks, ISM: structure, galaxies: star clusters, galaxies: evolution, galaxies: formation

1. Introduction

The formation of stars and (proto)planetary systems proceeds in a hierarchically structured way (e.g. Elmegreen 2008; Kruijssen 2012; Longmore *et al.* 2014). In the bound substructures, dynamical interactions and external photoionisation are common processes that may disrupt (proto)planetary systems and affect the occupation of the habitable zone within these systems (e.g. Scally & Clarke 2001; de Juan Ovelar *et al.* 2012). These external processes become more efficient towards higher gas and stellar densities.

Observations probing the evolution of star formation in galaxies over cosmic time show that the cosmic star formation rate peaked at redshift $z = 1$–3 (Madau & Dickinson 2014). On the size scale of giant molecular clumps in which the stars are born (0.1–1 kpc, Genzel *et al.* 2011), the interstellar medium in these galaxies is extreme when compared to the conditions in the solar neighbourhood. The gas density, gas pressure, cosmic ray ionisation rate and FUV interstellar radiation field are several orders of magnitude higher than for clouds forming stars in the vicinity of the Sun (Swinbank *et al.* 2011; Tacconi *et al.* 2013). This means that the conditions under which most stars and planetary systems formed were likely very different than the conditions under which we observe the formation of planetary systems today, with plausibly important implications for the demographics of the planet population.

2. Environmental effects on protoplanetary discs

The pre-ALMA suite of protoplanetary disc (PPD) observations available in the literature already demonstrated that PPD properties change as a function of the local stellar environment. Specifically, the disc radii are statistically smaller in high-density environments, dropping below 200 AU for stellar number densities $>5 \times 10^3$ pc^{-2} (de Juan Ovelar *et al.* 2012). Since then, ALMA has revealed this effect in much higher detail (Eisner *et al.* 2018). In de Juan Ovelar *et al.* (2012), we used a simple analytical model to show that the observed truncation may be caused by dynamical perturbations from passing stars, but more recent work shows that external photoevaporation is likely responsible (see below). Importantly, dynamical truncations continue for planetary systems, beyond PPD dispersal. As a result, the time for which the habitable zone may be occupied by planets (the 'HZ lifetime') is a function of stellar mass and ambient stellar density. Even in a low-density cluster with a stellar density $>\{10^2, 10^3\}$ pc^{-2} (or $M > \{10^3, 10^4\}$ M$_\odot$), Earth would have been ejected from the habitable zone by the present day, with HZ lifetimes of $<\{4, 0.3\}$ Gyr, respectively (de Juan Ovelar *et al.* 2012).

Rosotti *et al.* (2014) present hydrodynamical simulations of 50 PPDs in a bound stellar cluster. The PPDs evolve self-consistently due to viscous spreading and dynamical perturbations. They undergo tidally-driven morphological transformations, which in extreme cases can completely destroy disc in a few 10^5 yr. The PPD radii are set by closest dynamical encounter, where disc shrinkage occurs for closest encounter distances less than ~ 6 disc radii. Consistently with the analysis by de Juan Ovelar *et al.* (2012), the simulations show that PPDs in high-density environments are statistically smaller, dropping below 200 AU for projected stellar number densities $>3 \times 10^3$ pc^{-2}.

All observed candidates of environmentally-driven shrinkage in the de Juan Ovelar *et al.* (2012) sample are located in the Orion Nebula Cluster. These discs are well known to be affected by external photoevaporation (Henney & O'Dell 1999; Scally & Clarke 2001). The question thus rises whether disc disruption is dominated by dynamical encounters or photoevaporation. Winter *et al.* (2018) model disc destruction by both processes and show that external photoevaporation dominates in clusters hosting massive stars (affecting PPDs for UV flux densities of $>3 \times 10^3$ G$_0$ or densities >20 pc^{-3}). Dynamical encounters dominate only in low-mass regions without massive stars and require extreme densities of $>10^4$ pc^{-3}. This shows that external UV irradiation generally dominates over PPD dispersal by dynamical encounters, and that (proto)planetary systems in (bound) clusters are systematically subjected to externally-driven disruption.

3. The fraction of planetary systems born in dense stellar clusters

A wide range of recent work shows that the efficiency of cluster formation depends on the local gas density (Elmegreen 2008; Kruijssen 2012). In the interstellar medium, the gas density probability distribution function (PDF) is lognormal, of which the width increases with the gas pressure (Vazquez-Semadeni 1994; Padoan *et al.* 1997; Krumholz & McKee 2005). Towards the high-density tail of the density PDF, the gas has short free-fall times. At a fixed star formation efficiency per free-fall time (e.g. Leroy *et al.* 2017; Utomo *et al.* 2018), this implies high integrated star formation efficiencies, low gas fractions, and, as a result, high bound stellar fractions (Kruijssen 2012). In other words, unbound associations form at the low-density end of the gas density PDF, whereas compact, bound stellar clusters form at the high-density end. Note that the external photoevaporation of PPDs does not strictly require the young stars to reside in a gravitationally *bound* cluster, but the long-term disruption of planetary systems by dynamical perturbations does.

When integrating the gas density PDF, we obtain the fraction of star formation occurring in bound clusters (the cluster formation efficiency or CFE; Bastian 2008; Kruijssen 2012) as a function of the gas pressure. The resulting CFEs range from a few % at low gas pressures (or surface densities, $\sim 10\ M_\odot\ pc^{-2}$), such that few planetary systems are affected by environmental effects, to up to $\sim 50\%$ at high gas pressures (or surface densities, $> 200\ M_\odot\ pc^{-2}$), where many planetary systems are affected by environmental effects. The prediction that the CFE is a function of the gas pressure has been quantitatively confirmed in observational studies of young stellar cluster populations (Adamo *et al.* 2015; Johnson *et al.* 2016; Ward & Kruijssen 2019). For instance, the solar neighbourhood is observed (Lada & Lada 2003) and predicted (Kruijssen 2012) to form 7% of all stars in bound stellar clusters, whereas the CFE is predicted and observed to be about 40% at the high gas pressures near the Galactic Centre (Ginsburg & Kruijssen 2018), which are similar to those at high redshift (Kruijssen & Longmore 2013).

4. Cosmological context

Given the environmental dependence of the CFE, a general statement regarding its impact on the externally-driven dispersal of (proto)planetary systems requires a representative census of the CFE as a function of galactic environment and cosmic time. This is now possible thanks to the E-MOSAICS project (for MOdelling Star cluster population Assembly In Cosmological Simulations within EAGLE; Pfeffer *et al.* 2018; Kruijssen *et al.* 2019), which is a set of hydrodynamical cosmological 'zoom-in' simulations of 25 Milky Way-mass galaxies that include a physically-motivated, sub-grid model for the formation and dynamical evolution of the entire star cluster population. These simulations represent the first time that the formation and evolution of the cluster population can be followed self-consistently across cosmic history.

The E-MOSAICS simulations show that the CFE systematically increases with redshift (Pfeffer *et al.* 2018), such that more (proto)planetary systems are affected by external disruption at earlier cosmic epochs (see Figure 1 for an example). The conditions in the present-day Universe thus represent a lower limit to the fraction of (proto)planetary systems undergoing externally-driven disruption. At the formation time of the Sun, the CFE was typically a factor of 1.5–2 larger than at the present day, whereas the CFE peaked at 30–50% at the peak of the star formation history at $z \sim 2$, corresponding to a lookback time of about 10 Gyr. At lookback times > 8 Gyr, most planetary systems are expected to be affected by environmental effects. In addition, the sharp peaks at redshifts $z = \{0.7, 2, 2.5\}$ correspond to galaxy mergers, which drive an increase of the gas pressure and a corresponding increase of the CFE. As a result, the external disruption of (proto)planetary systems is enhanced at these redshifts, and generally towards earlier cosmic times. This shows that Mpc-scale events can have an important impact on the formation and evolution of AU-scale planetary systems.

5. Conclusion: planetary systems are shaped by galactic environment

In summary, the presented results lead to the following conclusions.

(a) Dynamical interactions and (especially) external photoevaporation affect properties of (proto)planetary discs and planetary systems, mostly limiting the maximum radius where planets reside and shortening disc lifetimes.

(b) These environmental effects can sterilise planetary systems by preventing planet formation or ejecting planets from the habitable zone.

(c) The fraction of planetary systems potentially subject to these influences is environmentally dependent, increasing from a few % at present (with a CFE of 7% in the current solar neighbourhood) to $> 50\%$ when most stars in Milky Way formed.

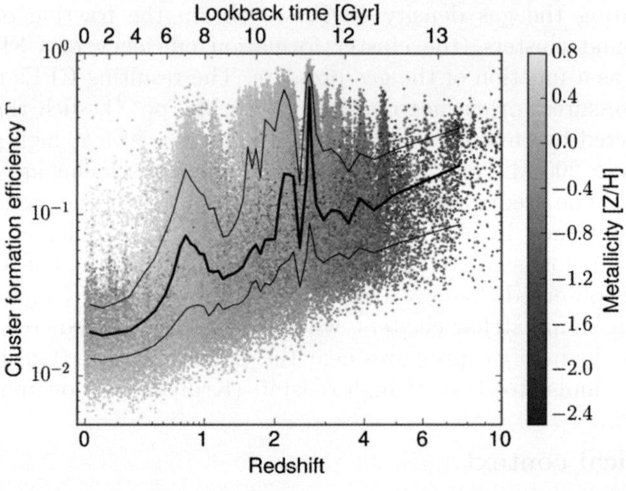

Figure 1. Cluster formation efficiency (fraction of star formation occurring in bound clusters) as a function of redshift for one of the cosmological zoom-in simulations of Milky Way-mass galaxies from the E-MOSAICS project (Pfeffer *et al.* 2018; Kruijssen *et al.* 2019), which include a model for the formation and evolution of the stellar cluster population. The data are colour-coded by metallicity, with the median and standard deviation at each redshift indicated by the thick and thin solid lines, respectively. The peaks at $z = \{0.7, 2, 2.5\}$ correspond to galaxy mergers, which drive an increase of the gas pressure and a corresponding increase of the CFE. As a result, the external disruption of (proto)planetary systems is enhanced at these redshifts, and generally towards earlier cosmic times. This shows that Mpc-scale events can have an important impact on the formation and evolution of AU-scale planetary systems. Figure from Pfeffer *et al.* (2018).

Building on these results, we have derived a model for PPD lifetimes as a function of the local density at which star formation proceeds in the context of the galactic environment (Winter *et al.* 2019). This model will enable a systematic assessment of environmental effects on (proto)planetary systems.

References

Adamo A., Kruijssen J. M. D., Bastian N., Silva-Villa E., & Ryon J., 2015, *MNRAS*, 452, 246
Bastian N., 2008, *MNRAS*, 390, 759
de Juan Ovelar M., Kruijssen J. M. D., Bressert E., Testi L., Bastian N., & Cánovas H., 2012, *A&A*, 546, L1
Eisner J. A. *et al.*, 2018, *ApJ*, 860, 77
Elmegreen B. G., 2008, *ApJ*, 672, 1006
Genzel R. *et al.*, 2011, *ApJ*, 733, 101
Ginsburg A., & Kruijssen J. M. D., 2018, *ApJL*, 864, L17
Henney W. J., & O'Dell C. R., 1999, *AJ*, 118, 2350
Johnson L. C. *et al.*, 2016, *ApJ*, 827, 33
Kruijssen J. M. D., 2012, *MNRAS*, 426, 3008
Kruijssen J. M. D., & Longmore S. N., 2013, *MNRAS*, 435, 2598
Kruijssen J. M. D., Maschberger T., Moeckel N., Clarke C. J., Bastian N., & Bonnell I. A., 2012, *MNRAS*, 419, 841
Kruijssen J. M. D., Pfeffer J. L., Crain R. A., & Bastian N., 2019, *MNRAS*, 486, 3134
Krumholz M. R., & McKee C. F., 2005, *ApJ*, 630, 250
Lada C. J., & Lada E. A., 2003, *Annu. Rev. Astron. Astrophys.*, 41, 57
Leroy A. K. *et al.*, 2017, *ApJ*, 846, 71
Longmore S. N. *et al.*, 2014, *Protostars and Planets VI*, 291
Madau P., & Dickinson M., 2014, *Annu. Rev. Astron. Astrophys.*, 52, 415

Padoan P., Nordlund A., & Jones B. J. T., 1997, *MNRAS*, 288, 145
Pfeffer J., Kruijssen J. M. D., Crain R. A., & Bastian N., 2018, *MNRAS*, 475, 4309
Rosotti G. P., Dale J. E., de Juan Ovelar M., Hubber D. A., Kruijssen J. M. D., Ercolano B., & Walch S., 2014, *MNRAS*, 441, 2094
Scally A., & Clarke C., 2001, *MNRAS*, 325, 449
Swinbank A. M. *et al.*, 2011, *ApJ*, 742, 11
Tacconi L. J. *et al.*, 2013, *ApJ*, 768, 74
Utomo D. *et al.*, 2018, *ApJL*, 861, L18
Vazquez-Semadeni E., 1994, *ApJ*, 423, 681
Ward J. L., & Kruijssen J. M. D., 2019, *MNRAS*, 475, 5659
Winter A. J., Clarke C. J., Rosotti G., Ih J., Facchini S., & Haworth T. J., 2018, *MNRAS*, 478, 2700
Winter A. J., Kruijssen J. M. D., Chevance M., Keller B. W., & Longmore S. N., 2019, *MNRAS* submitted, arXiv:1907.04602

Discussion

PALOUS: What is the role of metallicity in setting the cluster formation efficiency?

KRUIJSSEN: In the model for the cluster formation efficiency that I showed, metallicity could enter through the star formation efficiency or the bound fraction. However, neither of these have strong dependencies on metallicity. So metallicity is largely a tracer of birth environment (galaxy mass, redshift; see the colour coding in Figure 1), but plays no role in setting the cluster formation efficiency directly.

Origins: From the Protosun to the First Steps of Life
Proceedings IAU Symposium No. 345, 2018
B. G. Elmegreen, L. Viktor Tóth & M. Güdel, eds.

doi:10.1017/S1743921319001716

MASSES: An SMA Large Project Surveying Protostars to Reveal How Stars Gain their Mass

Ian W. Stephens[1], Michael M. Dunham[2,1], Philip C. Myers[1], Riwaj Pokhrel[3,1], Tyler L. Bourke[4] and the MASSES team

[1]Harvard-Smithsonian Center for Astrophysics
60 Garden Street, Cambridge, MA, USA
e-mail: `ian.stephens@cfa.harvard.edu`

[2]Department of Physics, State University of New York at Fredonia
280 Central Avenue, Fredonia, NY 14063, USA
email: `michael.dunham@fredonia.edu`

[3]Department of Astronomy, University of Massachusetts, Amherst, MA 01003, USA
[4]SKA Organization, Jodrell Bank
Lower Withington, Macclesfield, Cheshire SK11 9FT, UK

Abstract. Low-mass stars form from the gravitational collapse of dense molecular cloud cores. While a general consensus picture of this collapse process has emerged, many details on how mass is transferred from cores to stars remain poorly understood. MASSES (Mass Assembly of Stellar Systems and their Evolution with the SMA), an SMA large project, has just finished surveying all 74 Class 0 and Class I protostars in the nearby Perseus molecular cloud to reveal the interplay between fragmentation, angular momentum, and outflows in regulating accretion and setting the final masses of stars. Scientific highlights are presented in this proceedings, covering the topics of episodic accretion, hierarchical thermal Jeans fragmentation, angular momentum transfer, envelope grain sizes, and disk evolution.

Keywords. stars: formation, surveys, ISM: jets and outflows, ISM: clouds, ISM: kinematics and dynamics, ISM: molecules, accretion, astrochemistry

1. Introduction

Stars form via the collapse of molecular cloud cores (~0.05 pc in size). The collapse and accretion onto the star itself encompasses a complex interplay of physical processes. To constrain these physical processes at ~200 – 5000 au scales, we embarked on a Submillimeter Array (SMA) large project called the Mass Assembly of Stellar Systems and their Evolution with the SMA (MASSES). The MASSES survey (e.g, Stephens *et al.* 2018) observed all 74 known Class 0 and I protostars within the Perseus molecular cloud to create a complete and unbiased protostellar sample within a single star-forming cloud. The survey included both continuum and spectral line observations at both 230 GHz and 345 GHz. In the SMA's subcompact configuration, we observed at both of these frequencies simultaneously, and in the extended configuration, we observed at 230 GHz only. The lines and continuum frequencies observed are listed in Table 1. Notably, CO(2–1) and CO(3–2) trace the outflows (e.g., Figure 1), while the continuum and $C^{18}O$(2–1) trace the envelopes. The approximate age of each protostar can be assessed via the bolometric temperatures, allowing us to assess the evolution of protostars based on these tracers. More details of the MASSES survey are discussed in Stephens *et al.* (2018).

Table 1. Tracers Probed by MASSES

Tracer	Transition	Frequency (GHz)
1.3 mm cont		231.29
CO	$J = 2-1$	230.53796
^{13}CO	$J = 2-1$	220.39868
C^{18}O	$J = 2-1$	219.56036
N_2D^+	$J = 3-2$	231.32183
850 μm cont		356.410
CO	$J = 3-2$	345.79599
HCO^+	$J = 4-3$	356.73424
$H^{13}CO^+$	$J = 4-3$	346.99835

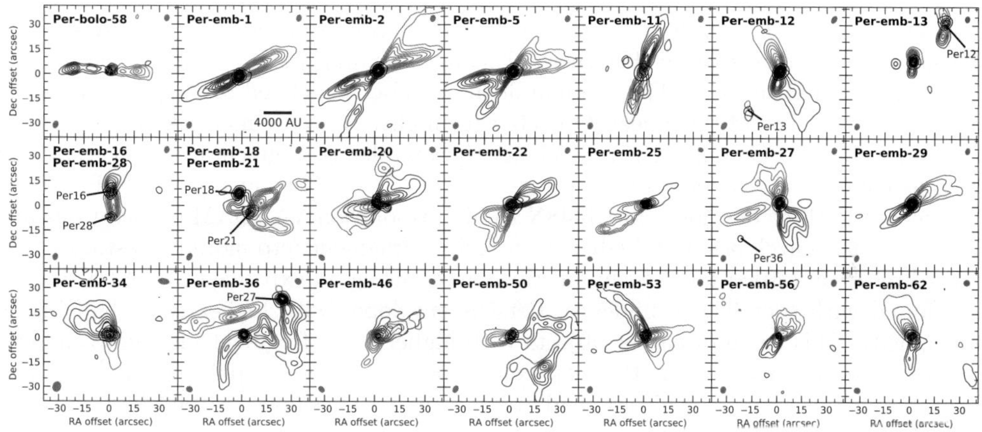

Figure 1. Selected CO(2–1) outflows from Stephens *et al.* (2018), as mapped by the MASSES survey in the SMA's subcompact configuration. Red and blue contours show the red-shifted and blue-shifted outflow lobes, respectively.

The MASSES survey is made more powerful using the ancillary data from the Karl G. Jansky Very Large Array (VLA) from the VLA Nascent Disk and Multiplicity (VANDAM) survey (Tobin *et al.* 2016). This survey imaged the continuum in the Ka-band with resolution up to ∼15 au.

Individuals can go to https://dataverse.harvard.edu/dataverse/MASSES to download the subcompact visibility and image data. The full subcompact+extended data will be released in a forthcoming paper. In this proceedings, we discuss the initial scientific results of the MASSES survey.

2. Scientific Highlights

2.1. *Evidence for Episodic Accretion*

CO and C^{18}O molecules freeze out on dust grains at temperatures between ∼20 and 30 K. Protostellar envelopes have a temperature gradient. Near the protostar, the temperature is expected to be significantly higher than the sublimation temperature, while far from the the protostar, the temperature is expected to be much lower. If one assumes that the bolometric luminosity of the protostar is relatively constant, simple chemical modeling can be used to predict the size of an envelope as measured by C^{18}O(2–1) based on the current bolometric luminosity of the source. Jørgensen *et al.* (2015) observed C^{18}O(2–1) for 16 protostars across 6 different molecular clouds, and found that the C^{18}O(2–1) sizes were typically much larger than that predicted by the current bolometric luminosity of the protostars. Frimann *et al.* (2017) used MASSES data to attempt the same experiment for 21 protostars in Perseus and found the same result.

Episodic accretion may possibly account for the discrepancy between the expected size and the observed size of $C^{18}O$. If there was a sudden burst of accretion in past that greatly increased the luminosity, the $C^{18}O$ size would grow rapidly. When the luminosity goes back to normal, carbon monoxide does not freeze out instantaneously onto dust grains. The freeze-out time scale is about 10,000 yr for densities $\sim 10^6\,\mathrm{cm}^{-3}$ at 10 K (Visser & Bergin 2012). Thus, the larger sizes observed may have been from past accretion bursts. Based on the fraction of sources showing larger than expected $C^{18}O(2\text{–}1)$ sizes, Frimann *et al.* (2017) estimated that accretion bursts occur every 20,000 to 50,000 years.

2.2. *Inefficient Thermal Jeans Fragmentation*

A Jeans Mass, M_J, defines the amount of mass needed to overcome pressure support so that an object can collapse. The thermal Jeans mass, $M_{J,th}$ is the mass needed to overcome the thermal support. The thermal Jeans number is defined as $N_{J,th} = M/M_{J,th}$, where M is the mass of the object. Objects with higher $N_{J,th}$ are more likely to fragment into more sources. Complete (i.e., efficient) thermal Jeans fragmentation suggests that $N_{J,th}$ is equal to the number of fragments. Lee *et al.* (2015) compared MASSES observations of the three envelopes in L1448N to high resolution VANDAM observations and found that the envelopes with higher $N_{J,th}$ indeed fragment into more protostars.

Pokhrel *et al.* (2018) followed up on this survey using an expanded MASSES dataset and found with over 95% confidence, Perseus envelopes with higher $N_{J,th}$ had higher multiplicity. This trend of parent objects with higher $N_{J,th}$ fragment into more children was also true at other scales in Perseus: from clouds (scales of $\sim$10 pc) to clumps ($\sim$1 pc), from clumps to cores ($\sim$0.05 pc), and from cores to envelopes ($\sim$1000 AU). At each scale, complete Jeans fragmentation suggests more fragments than observed. Therefore, fragmentation in Perseus can be described by inefficient thermal Jeans fragmentation. Including non-thermal support based on velocity linewidths predicts no fragmentation on large scales, and has no effect on small scales.

2.3. *Lack of Angular Momentum Transfer from Large to Small Scale*

If a collapsing core transfers its angular momentum directly to smaller scales, one would expect that wide binaries (defined here as protostellar pairs with separations between 1,000 and 10,000 au) would have the same angular momentum axis (i.e., axes are parallel). Lee *et al.* (2016) used MASSES data to analyze wide binaries in the Perseus cloud. The angular momentum axis of each protostar is ascertained from the outflow direction. Lee *et al.* (2016) found that the angular momentum axes actually appear randomly or perpendicularly oriented with each other, even when considering projection effects. Simulations by Offner *et al.* (2016) suggested that this is a signature of turbulent core fragmentation.

A filament may have two kinds of flows: accretion flow onto the filament or flow through the filament. If one of these flows dominates, a vorticity could be induced on the filament. Thus, a protostar could inherit angular momentum from gas flow on or through filaments. Stephens *et al.* (2017) compared the outflow direction (again, as a proxy for the angular momentum axis) to the protostar's filament direction. When considering projection effects, the alignment between outflows and filaments were significantly not parallel or perpendicular. Instead they were consistent with being randomly aligned. However, another possibility is that outflows and filaments are sometimes parallel and sometimes perpendicular, with perpendicular outflows and filaments about 3 times more common than parallel.

2.4. *The Ability to Estimate Protostellar Disk Masses and Evolution from Unresolved Observations*

Jørgensen *et al.* (2009) proposed a model where one can estimate Class 0 and I disk masses using unresolved (~1000 au resolution) interferometric observations along with large-scale single dish observations. The model is a simple radiative transfer model that assumes a density profile and dust opacity index. The accuracy of this method has mostly been untested, as there are very few observations of the disks about embedded Class 0 and I protostars. The VANDAM survey has high enough resolution to resolve disks, and disk masses from this survey were estimated in Segura-Cox *et al.* (2016, 2018) and Tychoniec *et al.* (2018). In Andersen *et al.* (2019), we used the MASSES survey to test whether we can accurately apply the Jørgensen *et al.* (2009) model to estimate disk masses with at ~1000 au resolution. We found that the method works extremely well; the estimated disk MASSES were very similar and strongly correlated to those estimated with high resolution in the VANDAM survey. We find that disks masses are relatively constant with age, but envelope MASSES decrease with age.

References

Andersen, B. C., Stephens, I. W., Dunham, M. M., *et al.* 2019, *ApJ*, 873, 54 (http://adsabs.harvard.edu/abs/2019ApJ...873...54A)
Frimann, S., Jørgensen, J. K., Dunham, M. M., *et al.* 2017, *A&A*, 602, A120
Jørgensen, J. K., van Dishoeck, E. F., Visser, R., *et al.* 2009, *A&A*, 507, 861
Jørgensen, J. K., Visser, R., Williams, J. P., & Bergin, E. A. 2015, *A&A*, 579, A23
Lee, K. I., Dunham, M. M., Myers, P. C., *et al.* 2015 *ApJ*, 814, 114
Lee, K. I., Dunham, M. M., Myers, P. C., *et al.* 2016, *ApJL*, 820, L2
Offner, S. S. R., Dunham, M. M., Lee, K. I., Arce, H. G., & Fielding, D. B. 2016, *ApJL*, 827, L11
Pokhrel, R., Myers, P. C., Dunham, M. M., *et al.* 2018, *ApJ*, 853, 5
Segura-Cox, D. M., Harris, R. J., Tobin, J. J., *et al.* 2016, *ApJL*, 817, L14
Segura-Cox, D. M., Looney, L. W., Tobin, J. J., *et al.* 2018, *ApJ*, 866, 161
Stephens, I. W., Dunham, M. M., Myers, P. C., *et al.* 2017, *ApJ*, 846, 16
Stephens, I. W., Dunham, M. M., Myers, P. C., *et al.* 2018, *ApJS*, 237, 22
Tobin, J. J., Looney, L. W., Li, Z.-Y., *et al.* 2016, *ApJ*, 818, 73
Tychoniec, Ł., Tobin, J. J., Karska, A., *et al.* 2018, *ApJS*, 238, 19
Visser, R., & Bergin, E. A. 2012, *ApJL*, 754, L18

Discussion

ZINNECKER: Why are the outflows randomly directed?

STEPHENS: Simulations by Stella Offner have shown turbulence in a core can change the spin axis of a protostar on short timescales. The outflow direction may also be random from their filament because filament direction could change with time, so the current filament direction may not reflect the natal environment.

KHAIBRAKHMANOV: Can the magnetic field be responsible for inefficient fragmentation?

STEPHENS: Yes, it may be able to provide support, but the field strength is not well constrained at all scales. We also have working models showing that substructure may also reflect why we see inefficient thermal Jeans fragmentation.

Origins: From the Protosun to the First Steps of Life
Proceedings IAU Symposium No. 345, 2018
B. G. Elmegreen, L. Viktor Tóth & M. Güdel, eds.

doi:10.1017/S1743921319001777

The birth environment of the solar system constrained by the relative abundances of the solar radionuclides

Edward D. Young

Department of Earth, Planetary, and Space Sciences,
University of California, Los Angeles
email: eyoung@epss.ucla.edu

Abstract. The relative abundances of the radionuclides in the solar system at the time of its birth are crucial arbiters for competing hypotheses regarding the birth environment of the Sun. The presence of short-lived radionuclides, as evidenced by their decay products in meteorites, has been used to suggest that particular, sometimes exotic, stellar sources were proximal to the Sun's birth environment. The recent confirmation of neutron star - neutron star (NS-NS) mergers and associated kilonovae as potentially dominant sources of r-process nuclides can be tested in the case of the solar birth environment using the relative abundances of the longer-lived nuclides. Critical analysis of the 15 radionuclides and their stable partners for which abundances and production ratios are well known suggests that the Sun formed in a typical massive star-forming region (SFR). The apparent overabundances of short-lived radionuclides (e.g. ^{26}Al, ^{41}Ca, ^{36}Cl) in the early solar system appears to be an artifact of a heretofore under-appreciation for the important influences of enrichment by Wolf-Rayet winds in SFRs. The long-lived nuclides (e.g. ^{238}U, ^{244}Pu, ^{247}Cr, ^{129}I) are consistent with an average time interval between production events of 10^8 years, seemingly too short to be the products of NS-NS mergers alone. The relative abundances of all of these nuclides can be explained by their mean decay lifetimes and an average residence time in the ISM of ~ 200 Myr. This residence time evidenced by the radionuclides is consistent with the average lifetime of dust in the ISM and the timescale for converting molecular cloud mass to stars.

Keywords. ISM, stars: Wolf-Rayet, solar system: formation, Sun: abundances

---

1. Introduction

A worthy goal for studying the origins of the solar system is to establish whether or not the Sun formed in a typical environment and under typical conditions as judged by comparisons with star-forming regions in the Milky Way today. Advantages of making this link between the birth of our solar system and that of other planetary systems in general are twofold. Firstly, a reconstruction of the birth environment of the Sun is relevant to assessing the probability that planetary systems analogous to our solar system exist, have existed, or will exist in the Milky Way Galaxy. Secondly, if we establish the basic characteristics of the solar birth environment, we can test our understanding of star and planet formation in analogous environments using the solar system as the most well understood example.

Among the many physical and chemical characteristics of the solar system, the relative abundances of the radionuclides may be some of the best clues to the solar birth environment. The initial relative abundances of many radionuclides in the solar system have been obtained through painstaking measurements of the isotopes themselves in meteorites, or, in the case of the short-lived radionuclides (SLRs), their decay products in meteorites. Here we present two different paradigms for interpreting the meaning of the relative abundances of the radionuclides and their implications for reconstructing the environment in which our solar system formed. The two views are in effect basis vectors for all models seeking to explain the solar abundances of the radionuclides. The argument is made that initial abundances of SLRs are likely to be typical of massive star-forming regions (SFRs) similar to the Cygnus or Carina regions in our Galaxy today.

2. Overview

Theories for the provenance of the solar-system radionuclides can be divided into two broad categories. In the first category, which we will refer to here as "punctuated delivery", the radiochemistry of the solar system is the result of the mingling of materials from a variety of discrete nucleosynthesis sources that are sometimes attributed to individual stars of specified masses and ages, including AGB stars, supernovae (SNe), and Wolf-Rayet stars (e.g., Wasserburg *et al.* 1996, 2006; Lugaro *et al.* 2014; Dwarkadas *et al.* 2017). The punctuated delivery scenarios emphasize the "granularity" (Wasserburg *et al.* 1996) of the interstellar medium in that a chance encounter with a single stellar source can dominate the inventory of a particular nuclide comprising pre-stellar material. In this type of interpretation, the solar abundances of radionuclides are the result of a sequence of random discrete events (e.g., the explosion of a proximal supernova).

The second category of explanations for the solar abundances of the radionuclides appeals to widespread, quasi-continuous self-enrichment of massive SFRs (Jura *et al.* 2013; Young 2014; Fujimoto *et al.* 2018). In these scenarios, abundances of SLRs represent a steady-state resulting from quasi-continuous production and losses in the SFRs over time. The chemistry of the solar system therefore reflects the global evolution of the SFR in which the Sun formed. Some models are of course a mix of the two categories (e.g., Gounelle & Meynet 2012), representing vectors in the two-dimensional cartesian space spanned by the two theories described here. In what follows the essence of the two distinct types of scenarios are compared using simple mathematical formulations. The results are displayed in plots of the relative abundances of radionuclides normalized for differences in chemistry and differences in stellar production rates against their average lifetimes imposed by radioactive decay. An eventual assessment of which of these two distinct types of scenarios is most likely will place constraints on the events leading up to the formation of the solar system, and thus the solar birth environment.

3. Punctuated Delivery

The punctuated nature, or "granularity" (Wasserburg *et al.* 1996), of stellar nucleosynthesis events that could have seeded parental solar system material with nuclides can be described by an equation based on a geometric series summing individual nucleosynthesis events with an average temporal spacing δt followed by a final actual free decay time Δt (Wasserburg *et al.* 2006; Lugaro *et al.* 2014)

$$\frac{N_{\rm R}}{N_{\rm S}} = \left[\frac{P_{\rm R}}{P_{\rm S}}\frac{\delta t}{T}\left(1 + \frac{{\rm e}^{-\delta t/\tau}}{1-{\rm e}^{-\delta t/\tau}}\right)\right]\exp(-\Delta t/\tau) \tag{3.1}$$

where N_i and P_i are the number and production rates for radionuclides (R) and stable isotope partner (S), respectively, T is the age of the Galaxy (taken to be 7.4 Gyr at the time of solar system formation), Δt is in this case the time interval between the last event (LE) and the birth of the solar system, and τ is the mean life of R against radioactive decay. Dividing $N_{\rm R}$ by $N_{\rm S}$ eliminates the effects of element-specific chemistry when evaluating Equation 3.1 for different radionuclides. $N_{\rm R}/N_{\rm S}$ ratios are well known for the early solar system from precise measurements of the radioactive decay products of the

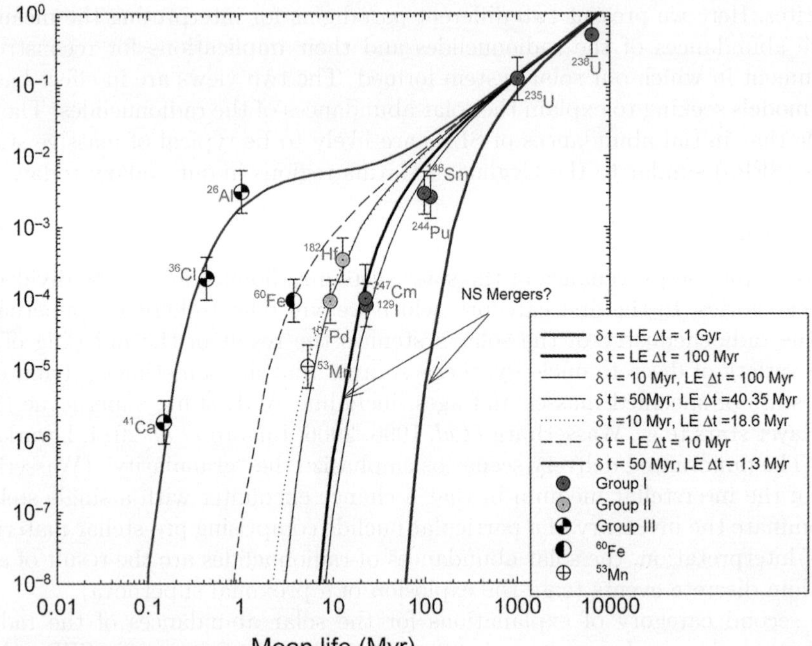

Figure 1. Plot of radionuclide abundances ratioed to their stable nuclide partners ($N_{\mathrm{R}}/N_{\mathrm{S}}$) and production ratios ($P_{\mathrm{R}}/P_{\mathrm{S}}$) (or $\alpha(R)$ as described in the text) vs. mean decay lifetimes for radionuclides present at the birth of the solar system. The values for the abundance ratios, production ratios, and decay lifetimes are from Young (2014) and references therein updated with new production ratios for ^{182}Hf and ^{107}Pd (Lugaro *et al.* 2014) and the ^{247}Cm/^{232}Th ratio and production ratio from Tissot *et al.* (2016) and Goriely & Arnould (2001), respectively. Model curves are from Equation 3.1 fit to the different nuclide groups as indicated in the legend and in the text.

SLRs and the relative abundances of the longer-lived nuclides in meteoritical materials (Huss *et al.* 2009). Comparisons of different radionuclides are further facilitated by dividing the abundance ratio on the left-hand side of Equation 3.1 by the production ratio on the right-hand side of Equation 3.1. Production ratios are obtained from models of stellar nucleosynthesis (e.g., Rauscher *et al.* 2002; Woosley & Heger 2007; Sukhbold *et al.* 2018). The resulting quotient, $\alpha(R) = (N_{\mathrm{R}}/N_{\mathrm{S}})/(P_{\mathrm{R}}/P_{\mathrm{S}})$, for different radionuclides depends on the radioactive mean lifetime τ and the duration of radioactive decay between events, all else equal. Greater time intervals of decay following nucleosynthesis produce steeper trajectories in plots of $\alpha(R)$ vs τ. If the radionuclides inherited by the solar system from its parent molecular cloud were all of similar average age, $\alpha(R)$ should vary only with τ, with the shorter-lived nuclides being much less abundant relative to their production rates than the longer-lived nuclides.

Young (2016) showed that Equation 3.1 can be used to group the solar radionuclides for which production ratios are well known. Figure 1 shows updated curves of $\log[(N_{\mathrm{R}}/N_{\mathrm{S}})/(P_{\mathrm{R}}/P_{\mathrm{S}})]$, or equivalently $\log(\alpha(R))$, versus $\log(\tau)$, defining distinct groups of radionuclides. These groupings succinctly summarize many of the relationships previously described in the literature (e.g., Wasserburg *et al.* 2006; Lugaro *et al.* 2014). For example, the relative concentrations of r-process neutron-capture products ^{129}I, ^{244}Pu, ^{247}Cm, ^{235}U and ^{238}U and the p-process nuclide ^{146}Sm are all explained by Equation (1) using $\delta t = 10$ Myr and $\Delta t = 100$ Myr (Figure 1, where ^{232}Th is used as the nearly stable nuclide to calculate $\alpha(R)$ for the actinides). Here we label these isotopes as Group I.

The value for δt is consistent with the frequency of core-collapse supernova (CCSN) events affecting random positions in the Galactic disk (Meyer & Clayton 2000) and so is appropriate if r-process nuclides form mainly in CCSNe. Note that δt is not the frequency of supernovae anywhere in the Galaxy, which is of order 1 per 50 yrs, but rather the frequency with which a random position in the Galaxy experiences a supernova explosion.

However, mounting evidence suggests that a principal source of r-process nuclides may be kilonovae resulting from mergers of neutron stars (NS-NS mergers) and possibly NS-black hole (NS-BH) mergers (Thielemann *et al.* 2017; Kasen *et al.* 2017; Smartt *et al.* 2017). Assuming similar production ratios for both CCSNe and NS-NS mergers (but see Cote *et al.* 2018), especially among the actinides, Figure 1 provides a constraint on the frequency of NS-NS mergers under the assumption that they were the primary source of solar r-process nuclides. For example, Figure 1 shows two curves where the frequency of nucleosynthesis events is either 100 Myr or 1 Gyr (for simplicity in this case we set $\Delta t = \delta t$). The latter δt of 1 Gyr is in keeping with some estimates that NS-NS mergers are less frequent than CCSNe by a factor of 100 to 1000 (Cote *et al.* 2017; Tsujimoto & Shigeyama 2014). The shorter event frequency δt of 100 Myr fits the data reasonably well while the longer δt of 1 Gyr does not fit the data. The implication is that either kilonovae events were only about ten times less frequent than CCSNe prior to solar system formation, or they were not the primary source of r-process nuclides leading up to the formation the Sun.

Lugaro *et al.* (2014) showed that a single set of values for δt and Δt can explain the relative concentrations of both s-process neutron addition nuclides ^{107}Pd and ^{182}Hf. A value for δt of 50 Myr, appropriate for AGB star encounters, and a similar value for Δt of 40.35 Myr, fits the $\log[(N_R/N_S)/(P_R/P_S)]$ values for ^{107}Pd and ^{182}Hf (Figure 1). We refer to these radionuclides as Group II. ^{53}Mn and ^{60}Fe are treated separately from Group I and II. ^{53}Mn is also fit by the Group II curve but its origin must be distinct as it is a SN product, suggesting a shorter δt interval. ^{60}Fe requires its own Δt (Figure 1). The shortest-lived nuclides, labeled Group III, have $\log[(N_R/N_S)/(P_R/P_S)]$ values that are explained with Equation 3.1 using $\delta t = 50$ Myr and $\Delta t = 1.3$ Myr (Figure 1). The latter model reflects the fact that these short-lived nuclides have no "memory" of events prior to their most recent synthesis. Five distinct models defined by five Δt values (δt values are prescribed *a priori* by astrophysical constraints) represented by five curves are therefore required to explain the radionuclide abundances in Figure 1, one each for Groups I, II, and III and two others for ^{53}Mn and ^{60}Fe.

In general, quantitative assessments of punctuated additions of SLRs to regions of star formation lead to relatively low probabilities for the initial solar abundances of SLRs (Adams 2010; Adams *et al.* 2014). In these scenarios, the solar system is assessed to be atypical in its complement of ^{26}Al, for example.

4. Quasi-continuous Self-enrichment of Star Forming Regions

The quasi-continuous self-enrichment model for the solar birth environment relies on the enhanced production of several short-lived radionuclides emanating from Wolf-Rayet (WR) stars to explain the solar abundances of ^{26}Al, ^{36}Cl, and ^{41}Ca at the time the solar system formed (referred to as the initial abundances). WR stars evolve from massive progenitors ($M_* >\sim 25 M_\odot$). Their large progenitor masses ensure that the WR stage occurs within several Myr of the birth of the star cluster. WR stars are therefore expected to be spatially correlated with star-forming regions since they don't have time to flee their birth environments before they die (Young 2014). Young (2014) showed that the solar radionuclide abundances considered in Figure 1 are explained by a single model based on a two-phase interstellar medium (ISM) composed of a molecular cloud phase (MC)

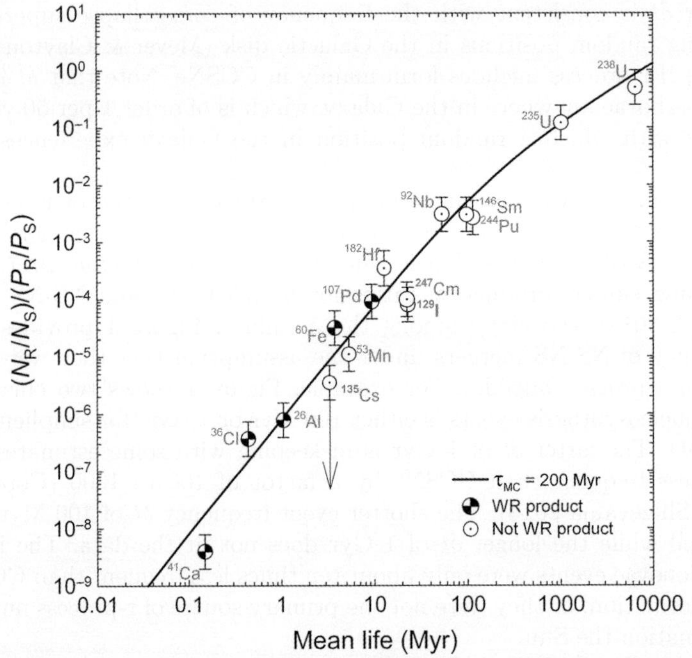

Figure 2. Plot of radionuclide abundances normalized to their stable nuclide partners (N_R/N_S) and production ratios (P_R/P_S) vs. mean decay lifetime for radionuclides in the early solar system. The plot is similar to Figure 1 but with ordinate values calculated using production ratios that include Λ_W/Λ_{SNe} from Equation 4.2, and published WR production terms compiled by Young (2014). The curve is from Equation 4.1 fit to the data using $\Lambda_W/\Lambda_{SNe} = 4000$ and $\tau_{MC} = 200$ Myr with a reduced χ^2 of 0.95. The value for ^{92}Nb and the upper limit for ^{135}Cs were not used to derive the best-fit curve, as discussed in the text.

and a diffuse phase and with a molecular cloud mass fraction x_{MC} of ~ 0.17, equivalent to that today. The model equation (Young 2014; Jacobsen 2005) is

$$\log\left(\frac{N_{R,MC}}{N_{S,MC}}\right) - \log\left(\frac{P_R}{P_S}\right) = 2\log\tau - \log\left[(1 - x_{MC})\tau_{MC} + \tau\right] - \log T \tag{4.1}$$

where the abundances of radionuclides and their stable partners now refer to those in the molecular cloud phase and environs in the SFR ($N_{R,MC}$ and $N_{S,MC}$, respectively) as opposed to the diffuse ISM outside of the SFR. The radionuclide production terms relevant to the molecular cloud setting are cast in terms of production from supernovae (P_R^{SNe}) and production from WR winds (P_R^W):

$$\frac{P_R}{P_S} = \frac{\Lambda_{SNe} P_R^{SNe}}{\Lambda_{SNe} P_S} + \frac{\Lambda_W P_R^W}{\Lambda_{SNe} P_S} \tag{4.2}$$

where Λ_W and Λ_{SNe} are the relative efficiencies for trapping the two sources of nuclides in the star-forming region. WR production values are listed by Young (2014).

Young (2016) showed that the model fits solar-system data using two independent parameters, an enhancement of WR wind production over SN production in SFRs, with $\Lambda_W/\Lambda_{SNe} \sim 4000$, and a sequestration time of nuclides in molecular cloud dust, τ_{MC}, of ~ 200 Myr (Figure 2). The τ_{MC} in Equation 4.1 is the *average* time spent in SFR molecular clouds where they experience radioactive decay with no new additions before incorporation into stars. Because individual clouds exist for shorter time spans than the SFR as a whole, time spent passing from one cloud to another via inter-cloud space in the SFR is included in the residence time. The radionuclide residence time τ_{MC} of ~ 200 Myr is consistent with the lifetime of dust in the interstellar medium (Tielens 2005) and simple estimates for the timescale for converting molecular clouds to stars (Draine 2011). In general, large values for Λ_W/Λ_{SNe} are consistent with the fact that massive stars like WR progenitors apparently do not typically end their lives as energetic SNe but rather collapse by fallback to form black holes directly (Fryer 1999; Smartt 2009). In part for this reason, the products of WR winds are enhanced relative to SNe products in SFRs sufficiently massive as to host a population of large stars (Young 2014). The relatively quiescent ultimate fate of WR stars is supported by calculations showing that the fraction of stars with $M_* > 18M_\odot$ that explode rather than undergoing direct collapse is just 8% (Sukhbold *et al.* 2018).

Numerical simulations (e.g., Young 2014) show that if clusters form every 1 to 1.5 Myr in an SFR, and if mixing is sufficiently efficient, then the abundances of ^{26}Al, ^{36}Cl, and ^{41}Ca will be enhanced in SFRs relative to the average ISM as a whole. Jura *et al.* (2013) appealed to this concept when they pointed out that the initial solar relative abundance of ^{26}Al, usually expressed as $^{26}\text{Al}/^{27}\text{Al} = 5 \times 10^{-5}$, is indistinguishable from that deduced by assuming that the 1.5 to 2.5 $M_\odot$ of ^{26}Al in the Galaxy (Martin *et al.* 2009) resides primarily in regions where stars are forming. These SFRs are represented to first order by the $8.4 \times 10^8 M_\odot$ of H_2 gas in the Galaxy (Draine 2011) and solar-like ratios of heavy elements. A correlation between SFRs and ^{26}Al in the Galaxy is supported by maps of the 1.8 MeV gamma ray emission from ^{26}Al decay in the Galaxy (Diehl *et al.* 2006). The conclusion from this analysis is that the solar initial abundance of ^{26}Al, and by inference those of ^{36}Cl and ^{41}Ca as well, was not extraordinary, but rather typical of star-forming regions as the result of self enrichment by WR stars. A recent numerical simulation at the Galactic scale by Fujimoto *et al.* (2018) leads to a similar conclusion regarding the importance of widespread self enrichment of SLRs in and near giant molecular clouds in the Galaxy.

The earlier work showing that a single curve in Figure 2 explains the solar abundances of radionuclides makes the tacit prediction that as more radionuclide initial abundances are obtained from measurements in meteorites, they too should fall on the same curve defined by the two parameters $\Lambda_W/\Lambda_{SNe} \sim 4000$ and $\tau_{MC} \sim 200$ Myr. This is because these two parameters should be intrinsic properties of the SFR in which the solar system formed. This prediction is put to the test by new data for the initial relative abundances of two additional short-lived radionuclides. These new data include an estimate for the initial abundance of the p-process product ^{92}Nb, yielding $^{92}\text{Nb}/^{93}\text{Nb} = 1.7 \pm 0.6 \times 10^{-5}$ (Iizuka *et al.* 2016), and a new upper limit on the initial abundance of the s-process product ^{135}Cs, yielding $^{135}\text{Cs}/^{133}\text{Cs} \leq 2.8 \times 10^{-6}$ (Brennecka & Kleine 2017). Using the mean production ratio for $^{92}\text{Nb}/^{93}\text{N}$ of 5.65×10^{-3} from Schonbachler *et al.* (2002) and the $^{135}\text{Cs}/^{133}\text{Cs}$ production ratio of 0.8 from Harper (1996), the new data plot within error of the solar-system curve in Figure 2. The new data are thus supportive of the prediction that all radionuclides, excluding those produced by spallation, will plot on the curve defined by $\Lambda_W/\Lambda_{SNe} \sim 4000$ and $\tau_{MC} \sim 200$ Myr. If this proves to be true as more data are acquired, we will have learned about two fundamental parameters that characterized the SFR in which the solar system formed.

5. Model Selection and Concluding Remarks

The two different interpretations of the solar-system radionuclide data portrayed in Figures 1 and 2 each have elements to recommend them. For example, the physical and temporal separation between r-process nucleosynthesis and production of lower-mass nuclides by CCSNe implied by the new kilonova data might be expressed by the different curves for these nuclides in Figure 1. Conversely, all 15 radionuclides for which we have good estimates of $\alpha(R)$ values can be fit with a single curve with 2 fit parameters in Figure 2 with an acceptable reduced χ^2 of near unity. Parsimony would seem to be on the side of the quasi-continuous self-enrichment model. This point has been made using a simple Bayesian analysis (Young 2016). However, parsimony may not be the best arbiter for choosing between the two models described here. Nonetheless, if the self-enrichment model continues to explain more data coming from detailed analyses of meteorite samples, the physical and chemical characteristics of the solar system may prove useful for understanding better the nature of massive star-forming regions in general.

References

Adams, F. C. 2010, *Annu. Rev. Astron. Astrophys.*, 48, 47

Adams, F. C., Fatuzzo, M., & Holden, L. 2014, *ApJ*, 789, 18pp

Brennecka, G. A. & Kleine, T. 2017, *ApJL*, 837, 6pp

Cote, B., Belczynski, K., Fryer, C. L., Ritter, C., Paul, A., Wehmeyer, B., & O'Shea, B. W. 2017, *ApJ*, 836, 20pp

Cote, B., Eichler, M., Arcones, A., Hansen, C. J., Simonetti, P., Frebel, A., Fryer, C. L., Pignatari, M., Reichert, M., Belczynski, K., & Matteucci, F. 2018, arXiv:1809.03525v2

Diehl, R., Halloin, H., Kretschmer, K., Lichti, G., Schonfelder, V., Strong, A., von Kienlin, A., Wang, W., Jean, P., Knodlseder, J., Roques, J., Weidenspointner, G., Schanne, S., Hartmann, D., Winkler, C., & Wunderer, C. 2006, *Nature*, 439, 45

Draine, B. T. 2011, Physics of the Interstellar and Intergalactic Medium (Princeton, N.J.: Princeton University Press)

Dwarkadas, V. V., Dauphas, N., Meyer, B., Boyajian, P., & Bojazi, M. 2017, *ApJ*, 851, 14pp

Fryer, C. L. 1999, *ApJ*, 522, 413

Fujimoto, Y., Krumholz, M. R., & Tachibana, S. 2018, *MNRAS*, 480, 4025

Goriely, S. & Arnould, M. 2001, *A&A*, 379, 1113

Gounelle, M. & Meynet, G. 2012, *A&A*, 545, A4

Harper, Charles L., J. 1996, *AJ*, 466, 1026

Huss, G., Meyer, B. S., Srinivasan, G., Goswami, J. N., & Sahijpal, S. 2009, *Geochemica et Cosmochimica Acta*, 73, 4922

Iizuka, T., Lai, Y.-J., Akram, W., Amelin, Y., & Schonbachler, M. 2016, *Earth and Planetary Science Letters*, 439, 172

Jacobsen, S. 2005, in Chondrites and the Protoplanetary Disk, ed. A. Krot, E. Scott, & B. Reipurth, Vol. 341 *ASP*, 548–557

Jura, M., Xu, S., & Young, E. D. 2013, *ApJ*, 775, L41

Kasen, D., Metzger, B., Barnes, J., Quataert, E., & Ramirez-Ruiz, E. 2017, *Nature*, 551, 80

Lugaro, M., Heger, A., Osrin, D., Goriely, S., Zuber, K., Karakas, A., Gibson, B. K., Dopherty, C. L., Lattanzio, J. C., & Ott, U. 2014, *Science*, 345, 650

Martin, P., Knodlseder, J., Diehl, R., & Meynet, G. 2009, *A&A*, 506, 703

Meyer, B. S. & Clayton, D. D. 2000, *Sp.Sci.Rev.*, 92, 133

Rauscher, T., Heger, A., Hoffman, R., & Woosley, S. 2002, *ApJ*, 576, 323

Schonbachler, M., Rehkamper, M., Halliday, A. N., Lee, D.-C., Bourot-Denise, M., Zanda, B., Hattendorf, B., & Gunther, D. 2002, *Science*, 295, 1705

Smartt, S. J. 2009, *Annu. Rev. Astron. Astrophys.*, 47, 63

Smartt, S. J., Chen, T.-W., Jerkstrand, A., *et al.* 2017, *Nature*, 551, 75

Sukhbold, T., Woosley, S. E., & Heger, A. 2018, *ApJ*, 869, 22pp

Thielemann, F.-K., Eichler, M., Panov, I. V., & Wehmeyer, B. 2017, *Ann. Rev.Nuc.& Part.Sci.*, 67, 253

Tielens, A. 2005, in *The Physics and Chemistry of the Interstellar Medium*, 461–475

Tissot, F. L. H., Dauphas, N., & Grossman, L. 2016, *Science Advances*, 2, 7 pp

Tsujimoto, T. & Shigeyama, T. 2014, *A&A*, 565, 4pp

Wasserburg, G. J., Busso, M., & Gallino, R. 1996, *ApJ*, 466, L109

Wasserburg, G. J., Busso, M., Gallino, R., & Nollett, K. M. 2006, *Nuclear Physics A*, 777, 5

Woosley, S. & Heger, A. 2007, *Phys.Rep.*, 442, 269
Young, E. D. 2014, *Earth and Planetary Science Letters*, 392, 16027
Young, E. D. 2016, *ApJ*, 826, 6pp

Discussion

LISSAUER: Assuming Wolfe-Rayet stars are the primary source of ^{26}Al in the Solar System, what fraction of planetary systems would you expect to have similar excesses?

YOUNG: A consequence of discrete sampling of the stellar initial mass function is that larger star clusters more reliably produce more massive stars. Larger clusters implies formation in higher-mass star-forming regions. The fraction of planetary systems with solar-like relative concentrations of ^{26}Al should therefore correlate with the fraction of systems that form in massive star-forming regions where the high-mass end of the IMF is well sampled and WR stars are virtually guaranteed.

KRUIJSSEN: What is the nature of the molecular cloud "residency timescale" that you use to explain the radioisotope abundance as a function of their lifetime? You seem to use something like the gas depletion time (or Mgas/SFR), but cloud lifetimes are now being found to be 10-30 Myr in our recent work. How would the curve change if you use a shorter timescale?

YOUNG: We consider that characteristic evolutionary timescales correlate with scale, such that of order 10^6 yrs applies to formation of individual clusters, $\sim 3 \times 10^7$ yrs applies for the longevity of giant molecular clouds, and 10^8 yrs applies to the typical lifetime of molecular cloud complexes comprising spiral arms. Of these timescales, the gas depletion time is, arguably, the best measure of the time that a typical nuclide persists from its synthesis to its incorporation into a star and associated planetary system. The relative abundances of the solar-system radionuclides, especially the long-lived nuclides, are telling us that the average interval for isolation and radioactive decay was on the order of 200 Myr leading up to the formation of the solar system. This implies that these nuclides were likely passed from cloud to cloud as a result of inefficient star formation in the clouds. Evolution curves in Figure 2 would be considerably shallower with radioactive free decay times of 10 to 30 Myr, and would not be consistent with the long-lived nuclide data irrespective of the implications for the short-lived nuclides.

Origins: From the Protosun to the First Steps of Life
Proceedings IAU Symposium No. 345, 2018
B. G. Elmegreen, L. Viktor Tóth & M. Güdel, eds.

doi:10.1017/S1743921319002850

Triggered Star Formation inside the Shell of a Wolf-Rayet Bubble as the Origin of the Solar System

Vikram V. Dwarkadas[1], Nicolas Dauphas[1], Bradley Meyer[2], Peter Boyajian[1] and Michael Bojazi[2]

[1]University of Chicago, Chicago, IL, 60637
email: vikram@astro.uchicago.edu

[2]Clemson University

Abstract. A constraint on Solar System formation is the high ^{26}Al/^{27}Al abundance ratio, 17 times higher than the average Galactic ratio, while the ^{60}Fe/^{56}Fe value was lower than the Galactic value. This challenges the assumption that a nearby supernova was responsible for the injection of these short-lived radionuclides into the early Solar System. We suggest that the Solar System was formed by triggered star formation at the edge of a Wolf-Rayet (W-R) bubble. We discuss the details of various processes within the model using numerical simulations, and analytic and semi-analytic calculations, and conclude that it is a viable model that can explain the initial abundances of ^{26}Al and ^{60}Fe. We estimate that 1%-16% of all Sun-like stars could have formed in such a setting.

Keywords. methods: analytical, methods: numerical, Solar System: formation, stars: mass loss, stars: winds, outflows, stars: Wolf-Rayet, ISM: bubbles

---

A critical constraint on Solar System formation is the high ^{26}Al/^{27}Al abundance ratio of 5×10^{-5} at the time of formation, which was $\sim$17 times higher than the average Galactic ratio (Lee *et al.* 1976, Jacobsen *et al.* 2008, MacPherson *et al.* 1995, Tang & Dauphas 2012). The ^{60}Fe/^{56}Fe value measured from meteorites is about 2×10^{-8}, at least an order of magnitude lower than the Galactic background (Tang & Dauphas 2012, Tang & Dauphas 2015). The abundance of ^{26}Al as inferred in meteorites is too high (Tang & Dauphas 2012, Huss *et al.* 2009) to be accounted for by long-term Galactic chemical evolution (Tang & Dauphas 2012, Meyer & Clayton 2000, Wasserburg *et al.* 2006) or early Solar System particle irradiation (Marhas *et al.* 2002, Duprat & Tatischeff 2007).

These results challenge the assumption that a nearby supernova (SN) was responsible for the injection of these short-lived radionuclides into the early Solar System, as had been suggested almost 40 years ago by Cameron & Truran (1977), unless one can explain why a SN injected only ^{26}Al, but did not manage to inject any ^{60}Fe.

Besides SNe, sources of ^{26}Al include AGB stars and Wolf-Rayet (W-R) stars. The probability of an AGB star being found near the Sun is extremely small (Kastner & Myers 1994). Wasserburg *et al.* (2017) showed that an AGB star is unlikely to simultaneously explain the initial Solar System abundance of ^{26}Al, ^{60}Fe, ^{107}Pd and ^{182}Hf.

This leaves W-R stars as the source of ^{26}Al in the early Solar System (Tang & Dauphas 2015, Arnould *et al.* 1997, Arnould *et al.* 2006, Gaidos *et al.* 2009, Gounelle & Meynet

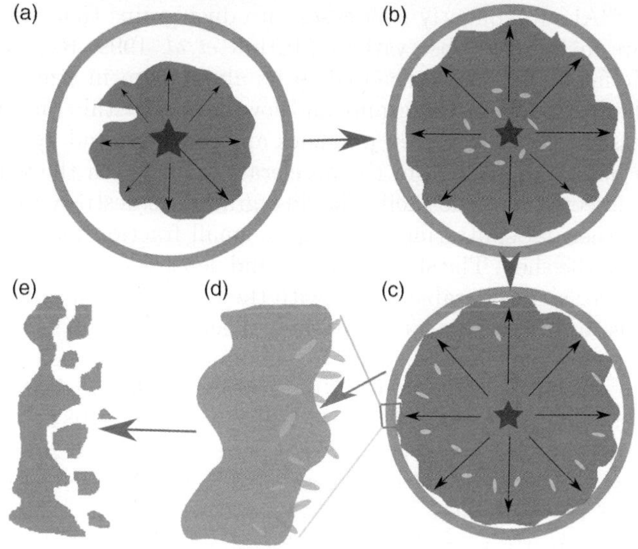

Figure 1. Cartoon version of our model of the formation of the Solar System. (a) Strong winds and ionizing radiation from a massive star build a windblown bubble. Blue region is the windblown region, yellow is the dense shell. (b) The bubble grows with time as the star evolves into the W-R phase. Dust forms in the wind close in to the star, and we assume that the ^{26}Al condenses into dust grains. (c) Dust is carried out by the wind toward the dense shell. (d) Wind is decelerated at the shell, dust detaches from the wind and penetrates the dense shell where it is destroyed, releasing the ^{26}Al. (e) Triggered star formation in the shell will give rise to various solar systems, including ours.

2012, Young 2014, Young 2016). These hot stars form the final phase of post-main-sequence massive stars above $\sim 25\,M_\odot$ (Crowther 2007). W-R stars produce ^{26}Al, which is carried out in their winds, but no ^{60}Fe. Although stellar evolution models differ in the amount of ^{26}Al produced, an analysis of various stellar evolution models (Langer *et al.* 1995, Palacios *et al.* 2005, Limongi & Cieffi 2006, Ekstrom *et al.* 2012, Georgy *et al.* 2012, Georgy *et al.* 2013) suggests that W-R stars with initial mass over about 50 $M_\odot$ can provide the requisite amount of ^{26}Al to seed the early Solar System in most models. The ^{26}Al itself is mainly produced during the main-sequence stage, but is mostly emitted in the strong mass-loss that follows the main-sequence phase, including the W-R phase.

Massive stars have strong supersonic winds, with wind speeds of order 1000-2000 km s^{-1}, and mass loss rates of 10^{-7} to $10^{-4.5}\,M_\odot yr^{-1}$ (Crowther 2007). These winds sweep up the surrounding medium to form large wind-blown bubbles, having a low density hot interior surrounded by a high-density cool shell (Toala & Arthur 2011, Dwarkadas & Rosenberg 2013). The combined action of wind shocks and ionization fronts due to the hot stars can lead to local increases in the density of the shell, leading to collapse of the shell in certain regions, and the formation of a new ensemble of stars (Elmegreen & Lada 1977, Lefloch & Lazareff 1994). This is known as triggered star formation, and is often seen in bubbles around massive stars (Deharveng *et al.* 2003, Deharveng *et al.* 2005, Zavagno *et al.* 2007, Brand *et al.* 2011). At least one case of triggered star formation around a W-R star is known (Liu *et al.* 2012).

In our detailed model of the formation of the Solar System (Dwarkadas *et al.* 2017, Figure 1), the collapse of a wind-blown shell around a W-R star leads to the formation of the dense cores that give rise to the Sun and planets. Aluminium-26 is produced during the H-burning lifetime of the massive star, and is found in the He core and in the H left behind by the receding core at the end of H-burning (Limongi & Cieffi 2006). In W-R stars these layers are ejected mainly through mass-loss, which reaches deep into the interior. The ^{26}Al subsequently condenses into dust grains that have been observed around W-R stars in infrared observations (Tuthill *et al.* 1999, Rajagopal *et al.* 2007, Marchenko & Moffat 2007), and estimated to be about 1μm in size (Marchenko *et al.* 2002). Due to the large size of the grains and low density within the bubble, the dust grains survive passage through the reverse shock and shocked wind, reach the dense shell swept-up by the bubble, detach from the decelerated wind with the wind velocity, and are subsequently injected into the shell. The dust grains are destroyed in the shell due to ablation and non-thermal sputtering, although a small fraction may survive, while the ^{26}Al is released into the shell. The shell thickness and density, as well as the velocity of the dust grains, vary over a considerable range, with the result that the grains will penetrate to different depths in different parts of the shell. Therefore, while some regions may be rich in ^{26}Al, others will not possess significant amounts of ^{26}Al. Such a diversity in ^{26}Al abundance is also seen in meteorites. FUN (with fractionation and unidentified nuclear isotope anomalies) CAIs (Esat *et al.* 1979, Armstrong *et al.* 1984, Macpherson *et al.* 2007) and hibonite rich CAIs are generally ^{26}Al poor (Koop *et al.* 2016), whereas spinel hibonite spherules are consistent with the canonical ^{26}Al/^{27}Al ratio (Liu *et al.* 2009).

The final fate of the W-R star is model dependent. In some models most massive W-R stars collapse directly to a black hole (Sukhbold *et al.* 2016), with no explosive nucleosynthesis accompanied by production of ^{60}Fe. In other models the star explodes as a SN, but the ejecta distribution is asymmetric (Mazzali *et al.* 2005, Maeda *et al.* 2008) making it likely that no ejecta will reach the proto-Solar System. Even if it does, it is unlikely that the hot, fast ejected material can easily mix with the cooler material in the Solar System (Goswami & Vanhala 2000), and therefore that additional ^{60}Fe will be injected into the proto-Solar System. The ^{60}Fe present comes from swept-up Galactic material, that could have been cooling for the several million-year stellar lifetime. Assuming that the IMF of the triggered stars is the same as that of the field stars, we estimate that 1-16% of all Sun-like stars could have been formed via triggered star formation in bubbles.

The model thus explains the abundance of ^{26}Al and ^{60}Fe in the early Solar System. Any such model must also be able to explain other short-lived radionuclides found in meteorites, such as ^{36}Cl, ^{41}Ca, ^{53}Mn etc. These investigations are in progress. Another caveat is that if all the dust grains are not destroyed within the shell, then some dust grains should remain that reflect the W-R composition. So far no dust grains in meteorites have been ascribed to W-R stars, and it is not even clear what a definitive sign of W-R origin would be. Finally, detailed models of ^{26}Al condensation onto dust grains under non-equilibrium conditions are needed to confirm that the ^{26}Al will condense onto dust grains.

References

Armstrong, J. T., Hutcheon, I. D., & Wasserburg, G. J. 1984, *LPI*, 15, 15

Arnould, M., Paulus, G., & Meynet, G. 1997, *A&A*, 321, 452

Arnould, M., Goriely, S., & Meynet, G 2006, *A&A*, 453, 653

Brand, J., Massi, F., Zavagno, A., Deharveng, L., & Lefloch, B. 2011, *A&A*, 527, 62

Cameron, A., & Truran, J. 1977 *Icarus*, 30, 447

Crowther, P. 2007, *ARA&A*, 45, 177

Deharveng, L., Lefloch, B., Zavagno, A., Caplan, J., Whitworth, A. P., Nadeau, D., & Martín, S. 2003, *A&A*, 408, L25

Deharveng, L., Zavagno, A., & Caplan, J. 2005, *A&A*, 433, 565

Duprat, J., & Tatischeff, V. 2007, *ApJ*, 671, 69

Dwarkadas, V. V., Dauphas, N., Meyer, B., Boyajian, P., & Bojazi, M. 2017, *ApJ*, 851, 147

Dwarkadas, V. V & Rosenberg, D 2013, *HEDP*, 9, 226

Elmegreen, B., & Lada, C. 1977, *ApJ*, 214, 725

Ekstrom, S., Georgy, C., Eggenberger, P., Meynet, G., Mowlavi, N., Wyttenbach, A., Granada, A., Decressin, T., Hirschi, R., Frischknecht, U., Charbonnel, C., & Maeder, A. 2012, *A&A*, 537, 146

Esat, T. M., Papanastassiou, D. A., & Wasserburg, G. J. 1979, *LPSC*, 10, 361

Gaidos, Eric, Krot, Alexander N., Williams, Jonathan P., Raymond, Sean N. 2009, *ApJ*, 696, 1854

Georgy, C., Ekstrm, S., Meynet, G., Massey, P., Levesque, E. M., Hirschi, R., Eggenberger, P., Maeder, A. 2012, *A&A*, 542, 29

Georgy, C., Ekström, S., Eggenberger, P., Meynet, G., Haemmerlè, L., Maeder, A., Granada, A., Groh, J. H., Hirschi, R., Mowlavi, N., Yusof, N., Charbonnel, C., Decressin, T., Barblan, F. 2013, *A&A*, 558, 103

Goswami, J. N., & Vanhala, H. A. T. 2000, in: Mannings, V., Boss, A. P., Russell, S. S. (eds.), *Protostars and Planets IV*, (Tucson: University of Arizona Press), p. 963

Gounelle, M. & Meynet, G. 2012, *A&A*, 545, 4

Huss, G. R., Meyer, B. S., Srinivasan, G., Goswami, J. N., Sahijpal, S. 2009, *GeCoA*, 73, 4922

Jacobsen, B., Yin, Q., Moynier, F. , Amelin, Y., Krot, A. N. , Nagashima, K., Hutcheon, I. D., Palme, H. 2008, *E&PSL*, 272, 353

Kastner, J. H., & Myers, P. C. 1994, *ApJ*, 421, 605

Koop, L., Davis, A. M., Nakashima, D., Park, C., Krot, A. N., Nagashima, K., Tenner, T. J., Heck, P. R., Kita, N. T. 2016, *GeoCoA*, 189, 70

Langer, N., Braun, H., & Fliegner, J. 1995, *Ap&SS*, 224, 275

Lee, T., Papanastassiou, D. A., Wasserburg, G. 1976, *GeoRL*, 3, 109

Lefloch, B. & Lazareff, B. 1994, *A&A*, 289, 559

Limongi, M. & Cieffi, A. 2006, *ApJ*, 647, 483

Liu, T., Wu, Y., Zhang, H., & Qin, S-L 2012, *ApJ*, 751, 68

Liu, M.-C., McKeegan, K. D., Goswami, J. N., Marhas, K. K., Sahijpal, S., Ireland, T. R., Davis, A. M. 2009, *GeoCoA*, 73, 5051

MacPherson, G. J., Bullock, E. S., Janney, P. E., Davis, A. M., Wadhwa, M., Krot, A. N. 2007, *LPI*, 38, 1378

MacPherson, G. J., Davis, A. M., Zinner, E. K. 1995, *Metic*, 30, 365

Maeda, K., Kawabata, K., Mazzali, P. A., Tanaka, M., Valenti, S., Nomoto, K., *et al.* 2008, *Science*, 319, 1220

Marchenko, S. & Moffat, A. F. J 2007, *ASPC*, 367, 213

Marchenko, S. V., Moffat, A. F. J., Vacca, W. D., Côté, S., Doyon, R. 2002, *ApJ*, 565, L59

Marhas, K. K., Goswami, J. N., Davis, A. M. 2002, *Sci*, 298, 2182

Mazzali, P. A., Kawabata, K. S., Maeda, K., Nomoto, K., Filippenko, A. V., Ramirez-Ruiz, E. *et al.* 2005, *Sci*, 308, 1284

Meyer, B., & Clayton, D. 2000, in: W. Benz, R. Kallenbach, & G. W. Lugmair (eds.) *From Dust to Terrestrial Planets. Proceedings of an ISSI Workshop*, (Dordrecht, The Netherlands: Kluwer), 133

Palacios, A., Meynet, G., Vuissoz, C., Knödlseder, J., Schaerer, D., Cerviño, M., Mowlavi, N. 2005, *A&A*, 429, 613

Rajagopal, J., Menut, J-L., Wallace, D., Danchi, W. C., Chesneau, O.,Lopez, B.;, Monnier, J. D., Ireland, M., Tuthill, P. G. 2007, *ApJ*, 671, 2017

Sukhbold, T., Ertl, T., Woosley, S. E., Brown, J. M., Janka, H.-T. 2016, *ApJ*, 821, 38

Tang, H., & Dauphas, N. 2012, *E&PSL*, 359, 248

Tang, H., & Dauphas, N. 2015, *ApJ*, 802, 22

Toala, J. A., & Arthur, S. J. 2011, *ApJ*, 737, 100

Tuthill, P. G., Monnier, J. D., & Danchi, W. C. 1999, *Nature*, 398, 487

Wasserburg, G. J., Busso, M., Gallino, R., Nollett, K. M. 2006, *Nuc. Phys. A*, 777, 5

Wasserburg, G. J., Karakas, A. I., & Lugaro, M. 2017, *ApJ*, 836, 126

Young, E. 2014, *E&PSL*, 392, 16

Young, E. 2016, *ApJ*, 826, 129

Zavagno, A., Pomarès, M., Deharveng, L., Hosokawa, T., Russeil, D., Caplan, J. 2007, *A&A*, 472, 835

Discussion

You-Hua Chu: WR stars have three varieties – nitrogen sequence, carbon sequence and oxygen sequence. Dust in stellar winds is found around WC stars, bubbles are found mostly around WN stars. These need to be carefully distinguished, instead of lumped together as WR stars.

Dwarkadas: While dust is mostly seen around WC stars, it is not a problem for our model as long as dust is formed in some stage and present around the star so that ^{26}Al can condense onto it. Wind bubbles are certainly found in all stages, not just in the WN stage. They do not suddenly appear in one stage and disappear in the next. A quick search of the literature reveals several examples of nebulae around WR stars in the WC phase, e.g., Stock & Barlow 2010 MNRAS, 409, 1429; Drissen, Shara and Moffett, 1991, AJ, 101, 1659; Cappa *et al.* 2005, A&A, 436, 155; Marston, A. 1997, ApJ, 475, 188. Marston notes that nebulae around WC stars are larger and could be missed. So there is no problem with having dust in some phase as the bubbles do exist in all phases.

Origins: From the Protosun to the First Steps of Life
Proceedings IAU Symposium No. 345, 2018
B. G. Elmegreen, L. Viktor Tóth & M. Güdel, eds.

doi:10.1017/S1743921319001534

Short-lived radioisotopes in meteorites from Galactic-scale correlated star formation

Yusuke Fujimoto[1], **Mark R. Krumholz**[1] **and Shogo Tachibana**[2]

[1]Research School of Astronomy & Astrophysics, Australian National University, Canberra, Australian Capital Territory 2611, Australia
email: yusuke.fujimoto@anu.edu.au

[2]UTokyo Organization for Planetary and Space Science (UTOPS), The University of Tokyo, 7-3-1 Hongo, Tokyo 113-0033, Japan

Abstract. Meteoritic evidence shows that the Solar system at birth contained significant quantities of short-lived radioisotopes (SLRs) such as ^{60}Fe and ^{26}Al produced in supernova explosions and in the Wolf-Rayet winds. Explaining how they travelled from these origin sites to the primitive Solar system before decaying is an outstanding problem. In this paper, we present a chemo-hydrodynamical simulation of the entire Milky Way to measure for the distribution of ^{60}Fe/^{56}Fe and ^{26}Al/^{27}Al ratios over all stars in the Galaxy. We show that the Solar abundance ratios are well within the normal range. We find that SLRs are abundant in newborn stars because star formation is correlated on Galactic scales, so that ejecta preferentially enrich atomic gas that will subsequently be accreted onto existing GMCs or will form new ones. Thus new generations of stars preferentially form in patches of the Galaxy contaminated by previous generations of stellar feedback.

Keywords. hydrodynamics, methods: numerical, Galaxy: disk, ISM: kinematics and dynamics, meteors, meteoroids

---

1. Introduction

Short-lived radioisotopes (SLRs) are radioactive elements with half-lives ranging from 0.1 Myr to more than 15 Myr that existed in the early Solar system. They were incorporated into meteorites' primitive components when the oldest solids formed in the Solar protoplanetary disc.

Most SLRs form in the late stages of massive stellar evolution, followed by injection into the ISM by stellar winds and SNe (e.g., Huss *et al.* 2009). Explaining how they travelled from these origin sites to the primitive Solar system before decaying is an outstanding problem (e.g., Adams 2010). Proposed mechanisms fall into three broad scenarios. The first scenario is a SN-triggered collapse: a nearby Type II SN injects SLRs and triggers the collapse of the early Solar nebula (e.g., Cameron & Truran 1977; Boss *et al.* 2010; Gritschneder *et al.* 2012). The second scenario is direct pollution: the Solar system's SLRs were injected directly into an already-formed protoplanetary disc by SN ejecta within the same star-forming region (e.g., Chevalier 2000; Ouellette *et al.* 2010). The third scenario is sequential star formation events and self enrichment in a giant molecular cloud (GMC, e.g., Gounelle & Meynet 2012; Young 2014; Kuffmeier *et al.* 2016).

Consensus has not reached yet, and no one has yet investigated galactic-scale SLR distributions. However, one should take account of Galactic-scale chemodynamics for chemical enrichment due to massive stars. Here, we study the Galactic-scale distributions of ^{26}Al and ^{60}Fe produced in stellar winds and supernovae, and propose a new contamination scenario: contamination due to Galactic-scale correlated star formation.

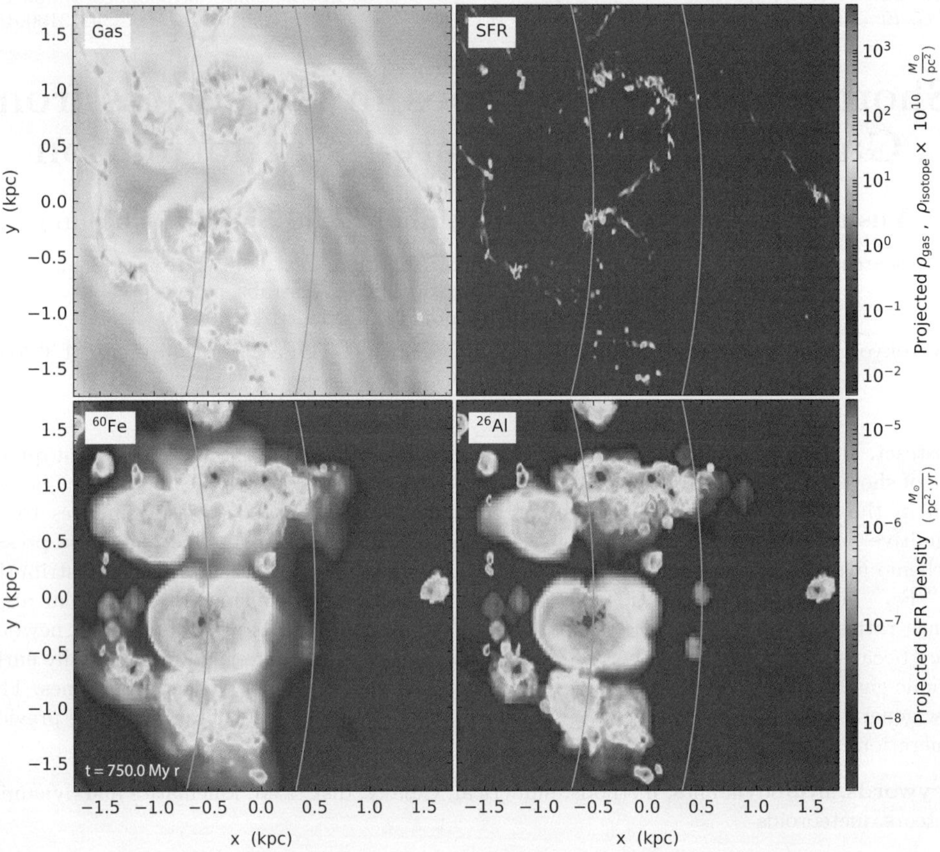

Figure 1. The face-on galactic disc zoomed in on a spot near the Solar Circle (Figure from Fujimoto *et al.* 2018). Panels show the gas (top-left), star formation rate (top-right), ^{60}Fe (bottom-left) and ^{26}Al (bottom-right) surface densities. The two arcs show Galactocentric radii of 7.5 and 8.5 kpc, bounding the Solar annulus.

2. Method

We study the abundances of ^{60}Fe and ^{26}Al in newly-formed stars by performing a high-resolution chemo-hydrodynamical simulation of the interstellar medium (ISM) of a Milky-Way like galaxy. The simulation includes hydrodynamics, self-gravity, radiative cooling, photoelectric heating, stellar feedback in the form of photoionisation, stellar winds and supernovae to represent dynamical evolution of the turbulent multi-phase ISM, and a fixed axisymmetric logarithmic potential to represent the gravity of old stars and dark matter, which causes the galactic-scale shear motion of the ISM in a flat rotation curve. In the simulation, when self-gravity causes the gas to collapse past our ability to resolve, we insert "star particles" that represent stochastically-generated stellar populations drawn star-by-star from the initial mass function (IMF). Each massive star in these populations evolves individually until it produces a mass-dependent yield of ^{60}Fe and ^{26}Al at the end of its life. We subsequently track the transport and decay of these isotopes, and their incorporation into new stars. Further details on our numerical method are given in Fujimoto *et al.* (2018).

3. Results

Fig. 1 shows the distributions of gas and isotopes in the Galactic disc zoomed in on a 3.5 kpc-region centred on the Solar Circle. The distributions of ^{60}Fe and ^{26}Al are

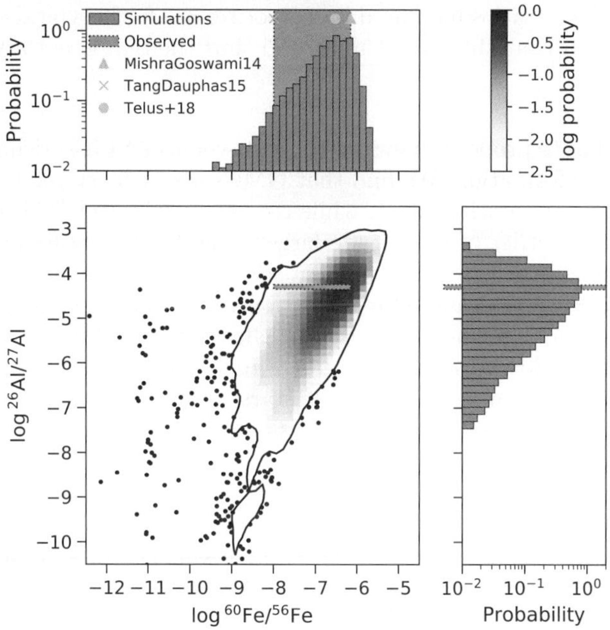

Figure 2. The abundance ratios of short-lived isotopes in newly-formed stars (Figure from Fujimoto *et al.* 2018). The central panel shows the joint PDF of ${}^{60}\mathrm{Fe}/{}^{56}\mathrm{Fe}$ and ${}^{26}\mathrm{Al}/{}^{27}\mathrm{Al}$ from our simulations, with colours showing probability density and black points showing individual stars in sparse regions. The top and right panels show the PDFs of ${}^{60}\mathrm{Fe}/{}^{56}\mathrm{Fe}$ and ${}^{26}\mathrm{Al}/{}^{27}\mathrm{Al}$ individually, with simulations shown in blue. All simulation data are for stars formed from $740-750$ Myr, at Galactocentric radii from 7.5 - 8.5 kpc. Green bands show the uncertainty range of Solar System meteoritic abundances (Lee *et al.* 1976; Mishra & Goswami 2014; Tang & Dauphas 2012; Telus *et al.* 2018); for ${}^{60}\mathrm{Fe}$, due to the wide range of values reported in the literature, we also show three representative individual measurements as indicated in the legend.

strongly correlated with the star-forming regions, which correspond to the highest-density regions (reddish colours) visible in the gas plot. This is as expected, since these isotopes are produced by massive stars, which, due to their short lives, do not have time to wander far from their birth sites.

However, there are important morphological differences between the distributions of ${}^{60}\mathrm{Fe}$, ${}^{26}\mathrm{Al}$, and star formation. The ${}^{60}\mathrm{Fe}$ distribution is the most extended, with the typical region of ${}^{60}\mathrm{Fe}$ enrichment exceeding 1 kpc in size, compared to ~ 100 pc or less for the density peaks that represent star-forming regions. The ${}^{26}\mathrm{Al}$ distribution is intermediate, with enriched regions typically hundreds of pc in scale. The larger extent of ${}^{60}\mathrm{Fe}$ compared to ${}^{26}\mathrm{Al}$ is due to its larger lifetime (2.62 Myr versus 0.72 Myr for ${}^{26}\mathrm{Al}$) and its origin solely in fast-moving SN ejecta (as opposed to pre-SN winds, which contribute significantly to ${}^{26}\mathrm{Al}$).

To investigate abundance ratios of isotopes in newborn stars, whenever a star particle forms in our simulations, we record the abundances of ${}^{60}\mathrm{Fe}$ and ${}^{26}\mathrm{Al}$ in the gas from which it forms, since these should be inherited by the resulting stars. Fig. 2 shows the probability distribution functions (PDFs) for the abundance ratios ${}^{60}\mathrm{Fe}/{}^{56}\mathrm{Fe}$ and ${}^{26}\mathrm{Al}/{}^{27}\mathrm{Al}$; we derive the masses of the stable isotopes ${}^{56}\mathrm{Fe}$ and ${}^{27}\mathrm{Al}$ from the observed abundances of those species in the Sun. The PDF of ${}^{60}\mathrm{Fe}$ peaks near ${}^{60}\mathrm{Fe}/{}^{56}\mathrm{Fe} \sim 3\times10^{-7}$, but is ~ 2 orders of magnitude wide, placing all the meteoritic estimates well within the ranges covered by the simulated PDF. The ${}^{26}\mathrm{Al}$ abundance distribution is similarly broad, but the measured meteoritic value sits very close to its peak, as ${}^{26}\mathrm{Al}/{}^{27}\mathrm{Al} \sim 5\times10^{-5}$.

Clearly, the abundance ratios measured in meteorites are fairly typical of what one would expect for stars born near the Solar Circle, and thus the Sun is not atypical.

4. Conclusion

Our results lead us to propose a new enrichment scenario: SLR enrichment via Galactic-scale correlated star formation. We find that GMCs are at most 100 pc in size and their star forming regions are much smaller, while regions of ^{60}Fe and ^{26}Al contamination due to supernovae are an order of magnitude larger (Fig. 1). The SLRs are not confined to the molecular clouds in which they are born. However, SLRs are nonetheless abundant in newborn stars (Fig. 2) because star formation is correlated on Galactic scales. Thus, although SLRs are not confined, they are in effect pre-enriching a halo of the atomic gas around existing GMCs that is very likely to be subsequently accreted or to form another GMC, so that new generations of stars preferentially form in patches of the Galaxy contaminated by previous generations of stellar winds and supernovae.

References

Adams, F. C. 2010, *ARAA*, 48, 47

Boss, A. P., Keiser, S. A., Ipatov, S. I., Myhill, E. A., & Vanhala, H. A. T. 2010, *ApJ*, 708, 1268

Cameron, A. G. W., & Truran, J. W. 1977, *Icarus*, 30, 447

Chevalier, R. A. 2000, *ApJ* (Letters), 538, L151

Fujimoto, Y., Krumholz, M. R., & Tachibana, S. 2018, *MNRAS*, 480, 4025

Gounelle, M., & Meynet, G. 2012, *A&A*, 545, A4

Gritschneder, M., Lin, D. N. C., Murray, S. D., Yin, Q.-Z., & Gong, M.-N. 2012, *ApJ*, 745, 22

Huss, G. R., Meyer, B. S., Srinivasan, G., Goswami, J. N., & Sahijpal, S. 2009, *Geochimica et Cosmochimica Acta*, 73, 4922

Kuffmeier, M., Frostholm Mogensen, T., Haugbølle, T., Bizzarro, M., & Nordlund, Å. 2016, *ApJ*, 826, 22

Lee, T., Papanastassiou, D. A., & Wasserburg, G. J. 1976, *Geophysical Research* (Letters), 3, 109

Mishra, R. K., & Goswami, J. N. 2014, *Geochimica et Cosmochimica Acta*, 132, 440

Ouellette, N., Desch, S. J., & Hester, J. J. 2010, *ApJ*, 711, 597

Tang, H., & Dauphas, N. 2012, *Earth and Planetary Science* (Letters), 359, 248

Telus, M., Huss, G. R., Nagashima, K., Ogliore, R. C., & Tachibana, S. 2018, *Geochimica et Cosmochimica Acta*, 221, 342

Young, E. D. 2014, *Earth and Planetary Science* (Letters), 392, 16

Discussion

KAMP: Should I read from your diagram that given the lower value of ^{60}Fe, the probability of having an environment with Solar radio-nuclei composition is more like $\sim 1\%$, so in fact not so common in the galaxy as a whole?

FUJIMOTO: The PDF is distributed very wide, almost 4 orders of magnitude. Considering this wide distribution, the Solar abundance is well within a normal range.

JOHNSTONE: Do you find a radial dependence of the ratios in the Galaxy?

FUJIMOTO: We don't find any radial dependence of the abundance ratios.

KOKUBO: How does the radial migration of the sun affect the results?

FUJIMOTO: The radial migration might not affect our results because there is no radial dependence of the abundance.

Origins: From the Protosun to the First Steps of Life
Proceedings IAU Symposium No. 345, 2018
B. G. Elmegreen, L. Viktor Tóth & M. Güdel, eds.

doi:10.1017/S1743921319002060

Episodic accretion in focus: revealing the environment of FU Orionis-type stars

O. Fehér[1,2], Á. Kóspál[1,3], P. Ábrahám[1], M. R. Hogerheijde[4], Ch. Brinch[5] and D. Semenov[3]

[1]Konkoly Observatory, Research Centre for Astronomy and Earth Sciences, Hungarian Academy of Sciences, 1121 Budapest, Konkoly Thege Miklós út, 15–17, Hungary

[2]Institut de Radioastronomie Millimétrique, 38400, Saint-Martin-d'Hères, 300 Rue de la Piscine, Grenoble, France

[3]Max Planck Institute for Astronomy, Königstuhl 17, D-69117 Heidelberg, Germany

[4]Leiden Observatory, Leiden University, Niels Bohrweg 2, 2333 CA Leiden, The Netherlands

[5]Niels Bohr International Academy, The Niels Bohr Institute, University of Copenhagen, Blegdamsvej 17, 2100 Copenhagen Ø, Denmark

Abstract. The earliest phases of star formation are characterised by intense mass accretion from the circumstellar disk to the central star. One group of young stellar objects, the FU Orionis-type stars exhibit accretion rate peaks accompanied by bright eruptions. The occurance of these outbursts might solve the luminosity problem of protostars, play a key role in accumulating the final star mass, and have a significant effect on the parameters of the envelope and the disk. In the framework of the Structured Accretion Disks ERC project, we are conducting a systematic investigation of these sources with millimeter interferometry to examine whether they represent normal young stars in exceptional times or they are unusual objects. Our results show that FU Orionis-type stars can be similar to both Class I and Class II systems and may be in a special evolutionary phase between the two classes with their infall-driven episodic eruptions being the main driving force of the transition.

Keywords. stars: pre–main-sequence, molecular data, circumstellar matter

1. Introduction

Star formation is a complicated interplay of fragmentation and gravitational collapse of interstellar clouds, the mass accretion onto the surface of the newly formed star, turbulence, magnetic fields and environmental effects. These mechanisms play out on multiple scales and compete with each other, thus to understand them, a self-consistent description is needed. One of the important results in the last few years was that protostars prove to be less luminous than theorized (Dunham *et al.* 2014 and references therein). Studies of eruptive young stars e.g. EX Lupi (EXors) and FU Orionis-type sources (FUors) (Herbig 1977; Hartmann & Kenyon 1996) revealed that these Sun-like stars acquire their mass through outburst events. These powerful eruptions can have significant consequences for the luminosity and the final mass of the stars and influence their outflows and the disk structures, see e. g. Kun *et al.* (2011), Mosoni *et al.* (2013), Cieza *et al.* (2016).

FUors are low-mass, pre-main sequence stars, showing 5–6 mag brightness increase in the optical and infrared regime, then slow fading or a brightness plateau over years or decades. The eruptions are caused by enhancement of the accretion rate from the circumstellar disk to the star, from a typical rate of 10^{-7} to $10^{-4}\,M_\odot\,yr^{-1}$. The eruptions were theorised to be triggered by gravitational instability (Armitage *et al.* 2001;

Boley *et al.* 2006), viscous-thermal instability (Bell & Lin 1994), or the effect of a close companion (Bonnell & Bastien 1992). Recent studies suggest that the outbursts are recurring and that infall from an envelope is needed to replenish the material in the disk (Vorobyov & Basu 2006; Vorobyov *et al.* 2013). After a sequence of eruptions the envelope clears out and the star may enter a quiescent phase, first showing smaller magnitude eruptions like EXors, then becoming similar to T Tauri-type stars (Vorobyov & Basu 2015).

Only a few tens of FUors are known so far (Audard *et al.* 2014). Analysis of their infrared emission and 10 μm silicate feature led to an evolutionary sequence proposed by Green *et al.* (2006) and Quanz *et al.* (2007). Category 1 FUors are younger, embedded in dense, dusty envelopes with the silicate feature in absorption, while Category 2 FUors have a remnant envelope, with the silicate feature in emission. FUor outbursts thus seem to happen during the first few million years of star formation and affect the inner disk heavily. Since this is the site and time of terrestrial planet formation, investigation of the process is imperative to describe the stages of star and planet formation.

2. Eruptive young stars through millimeter interferometry

High angular resolution and high sensitivity interferometry offered by the Atacama Large Millimeter/submillimeter Array (ALMA) or the Northern Extended Millimeter Array (NOEMA) allows us to study the morphology and kinematics of FUor disks and envelopes. We measured the emission of ^{13}CO, C^{18}O and the 2.7 mm continuum around 12 FUors with NOEMA, reaching spatial resolutions of around 1000 au: V1057 Cyg, V1515 Cyg, V2492 Cyg and V2493 Cyg in 2012, V733 Cep, V1735 Cyg and RNO 1B/C in 2014, and V899 Mon, V900 Mon, V960 Mon, and V582 Aur in 2017. We also performed observations with the 30 m telescope to detect the CO emission correctly on larger scales and made a spectral survey in the 110 GHz band to search for other molecular species.

The maps revealed extended, fragmented gas clouds around all our targets on a few thousand au spatial scale, with a local peak at or close to the star (Fehér *et al.* 2017). Only this central clump is found at e.g. V1057 Cyg and V1735 Cyg, but several other clumps were detected around e.g. V1515 Cyg and V2492 Cyg. The central clumps are roughly circular with radii of 1500–5000 au and masses of 0.02–0.5 $M_\odot$, often warmer than their surroundings, with average temperatures of 20–50 K. Some of the CO clumps appear in the continuum and other continuum sources can also be seen close to some FUors. A good example is RNO 1B/C (see Fig. 1, to the right). Clump A is located between the two members of the binary FUor and appears in both CO and continuum. Clump B is seen in CO but not seen in the continuum, however, the strong millimeter source to the south does not appear in CO. The other bright continuum source to the north-east appears as a faint local CO maximum. This source was also detected in the far-infrared. The velocity channel maps of V1057 Cyg revealed a velocity gradient in the envelope, indicating rotation. A larger scale gradient was measured in the environment of V2492 Cyg, possibly caused by the shock front that appears as a bright Hα rim close to the star. The ^{13}CO lines around V1735 Cyg have prominent line wings, indicating the presence of an outflow (see Fig. 1, to the left). The velocity channel maps around V1515 Cyg show a feature similar to an expanding bubble around the star. Maps of V582 Aur revealed that the star is located at the edge of an extended cloud structure with two well-identified velocity components and the unresolved continuum source around the FUor is possibly a disk with a total mass of 0.04 $M_\odot$ (Ábrahám *et al.* 2018).

The classification of our targets into categories based on their evolutionary categories faces many issues. In the case of V1057 Cyg the mass of the envelope corresponds to the characteristic mass of Class 0/I objects and it is detected in continuum, implying a significant spatial extent and dust content. However, the silicate feature appears in

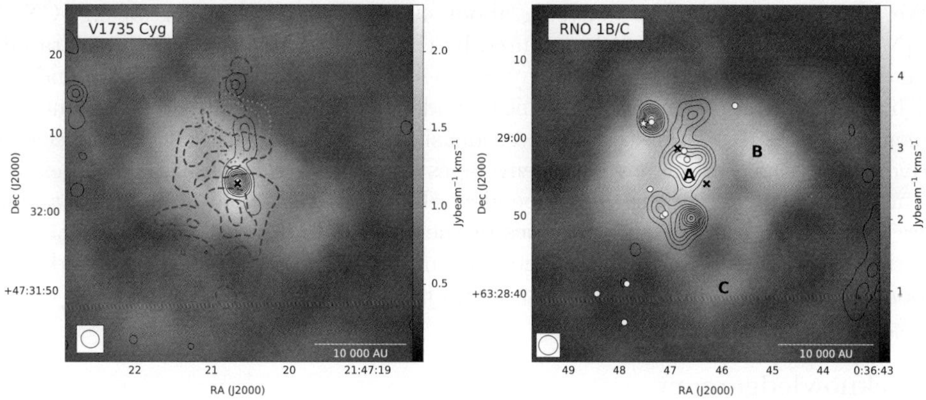

Figure 1. NOEMA CO and continuum maps of northern FUors. Left: ^{13}CO integrated intensity map of V1735 Cyg. The cross marks the optical position of the FUor, the black contours are the 2.7 mm continuum emission in units of Jy beam^{-1} at the 10...90% of the peak. The dashed contour is the blue-shifted (integrated between -0.2–2 km s^{-1}) and the dotted contour is the red-shifted lobe (integrated between 7–9.4 km s^{-1}) of the outflow at 50, 70 and 90% of their peak emission. Right: ^{13}CO integrated intensity map of RNO 1B/C. The black contours are the 2.7 mm continuum emission at the at the 10...90% of the peak. The two crosses are the members of the binary FUor-system (RNO 1B in the center). The letters mark clumps identified on the CO map and circles mark other IR/submm/mm sources in the neighbourhood.

emission. This is only possible if there is no envelope present in the line of sight to absorb most of the 10 μm emission or if the inclination of the system is high enough for the disk to be seen from the direction of an outflow cavity. The same is true for V1515 Cyg, except its envelope is only weakly detected in the continuum. The lowest envelope mass was found at V2493 Cyg and it was not detected in the 1.3 mm or 2.7 mm continuum either, suggesting a very small envelope and more evolved state. No 10 μm spectra was measured for this FUor. For V733 Cep and RNO 1B/C there is no gas clump coinciding with the exact optical position of the star that can be identified as an envelope, but in both cases the silicate feature is detected in absorption. Neighbouring sources and their circumstellar structures might heavily influence the emission received from the targets.

3. Future perspectives

Recent studies by Cieza *et al.* (2018), Li *et al.* (2017) and Liu *et al.* (2017) revealed that most FUors possess smaller disks than T Tauris. To sustain the observed high accretion rates these disks have to be really massive. Submillimeter Array data showed that some targets have very low spectral indices and their millimeter emission can be fitted with an inner, hot, optically thick disk and an outer, colder disk. Thus, a huge amount of mass may be hidden from the observer in the inner few astronomical units of these disks. Answering the question whether these small but massive FUor disks are evolutionary or intrinsic property will be addressed in the analysis of our ALMA Cycle 4 observations.

The detailed analysis of the 2017 NOEMA observations for three targets is in progress and we expect to receive interferometric data of four other FUors (RNO 127, PP 13S, HH 354 and V1184 Tau) in the near future. This time, thanks to the new receiver PolyFiX not only CO but the emission of many other molecular species will be imaged on interferometric spatial resolutions. We will also propose observations for the more extended configuration of NOEMA for more close-by targets., since with an angular resolution of around 1″ we will be able to resolve the envelope-disk interface and its dynamics.

We are in the process of analysing about 40 molecular transitions of e.g. CN, SO, HC_3N, N_2H^+ in the 3 mm band measured by the IRAM 30m telescope. The data allows us to examine circumstellar morphology tracing structures of different density and chemical history. We aim to use the chemical model MUSCLE (Multi Stage CLoud codE, Semenov *et al.* 2010) to fit the observed density of the different molecular species and derive the physical parameters of the envelopes and other circumstellar structures. Using the models we can derive the time-dependent changes of molecular abundances and even infer the time and other characteristics of the FUor outburst. Since none of the FUor eruptions have been observed from their beginning to their end yet, this would give us important predictions for both new observations and the modeling of the FUor eruption process.

4. Acknowledgement

This project has received funding from the European Research Council (ERC) under the European Union's Horizon 2020 research and innovation programme under grant agreement No 716155 (SACCRED, PI: Á. Kóspál).

References

Ábrahám *et al.* 2018 *ApJ*, 853, 28
Armitage *et al.* 2001, *MNRAS*, 324, 705
Audard *et al.* 2014, *Protostars and Planets VI*, 387
Bell & Lin 1994, *ApJ*, 427, 987
Boley *et al.* 2006, *ApJ*, 651, 517
Bonnel & Bastien 1992, *ApJ*, 401, L31
Cieza *et al.* 2016, *Nature*, 535, 258
Cieza *et al.* 2018, *MNRAS*, 474, 4347
Dunham *et al.* 2014, *Protostars and Planets VI*, 195–218
Fehér *et al.* 2017, *A&A*, 607, A39
Green *et al.* 2006, *ApJ*, 648, 1099
Hartmann *et al.* 1996, *ARA&A*, 34, 207
Herbig *et al.* 1977, *ApJ*, 217, 693
Kóspál *et al.* 2017, *ApJ*, 836, 226
Kun *et al.* 2011, *MNRAS*, 413, 2689
Li *et al.* 2017, *ApJ*, 840,72
Liu *et al.* 2017, *A&A*, 612, A54
Mosoni *et al.* 2013, *A&A*, 552, A62
Quanz *et al.* 2007, *ApJ*, 668, 359
Semenov *et al.* 2010, *A&A*, 522, A42
Vorobyov & Basu 2006, *ApJ*, 650, 956
Vorobyov & Basu 2015, *ApJ*, 805, 115
Vorobyov *et al.* 2013, *MNRAS*, 433, 3256

Discussion

KHAIBRAKHMANOV: Are there data on magnetic fields in the investigated objects?

FEHER: As far as we know there is no magnetic field measurements for our targets. For erupting sources we cannot measure the stellar photosphere but we have used Zeeman Doppler Imaging to measure the magnetic field of EX Lupi, which is a quiescent EXor.

Origins: From the Protosun to the First Steps of Life
Proceedings IAU Symposium No. 345, 2018
B. G. Elmegreen, L. Viktor Tóth & M. Güdel, eds.

doi:10.1017/S1743921318008505

Characterizing the youngest protostellar disks with the IRAM-PdBI and ALMA interferometers

Anaëlle Maury

AIM, CEA, CNRS, Université Paris-Saclay, Université Paris Diderot, Sorbonne Paris Cité, F-91191 Gif-sur-Yvette, France
email: anaelle.maury@cea.fr

Abstract. I present our observations and modeling of the 1.3 mm and 3.18 mm dust continuum emission in Class 0 protostars, from the IRAM-PdBI CALYPSO survey. We show that most protostars are better reproduced by models including a disk-like dust continuum component contributing to the flux at small scales, but less than 25% of these candidate protostellar disks are resolved at radii > 60 au, which favors magnetized models of rotating protostellar collapse for disk formation (Maury *et al.* 2019). I also present new ALMA observations of the molecular line emission in the IRAM04191 protostar, suggesting a small counter-rotating disk is detected in this young low-luminosity solar-type protostar. Finally, I show our ALMA observations of the magnetic field topology in the B335 protostar, which when compared to the typical output from protostellar collapse models, suggest the magnetic field might be responsible for constraining the disk size to remain very small in this protostar (Maury *et al.* 2018).

Keywords. stars: formation, ISM: individual (L1448, NGC1333, Serpens South, Serpens Main, L1157, GF9-2, L1527, IRAM04191, B335), ISM: magnetic fields

1. Introduction

Understanding the first steps in the formation of protostars and protoplanetary disks is a great unsolved problem of modern astrophysics. At the simplest level, the formation of circumstellar disks is a natural consequence of the conservation of angular momentum during the collapse of rotating protostellar envelopes in the course of the main accretion phase (Cassen & Moosman 1981, Terebey *et al.* 1984). Hydrodynamic simulations show that in the absence of magnetic fields, rotationally supported disks form and quickly grow to reach large radii > 100 au after a few thousand years (Yorke & Bodenheimer 1999). On the other hand, magnetohydrodynamics (MHD) numerical simulations of magnetized protostellar collapse form much smaller disks, at scales $r \sim 20$ au because of strong magnetic braking (for a review, see Li *et al.* 2014). Observationally, the key to constraining theoretical models for the formation of protostars and disks lies in high-resolution studies of the youngest protostars. Class 0 protostars are representative of the main accretion phase of protostellar evolution: they are observed only $t \lesssim 4 - 9 \times 10^4$ yr after the formation of a central hydrostatic protostellar object (Evans *et al.* 2009; Maury *et al.* 2011) while most of their mass is still in the form of a dense core/envelope. Only the recent advent of powerful interferometric facilities at (sub) millimeter wavelengths has allowed the inner envelopes of Class 0 protostars to be explored with sufficient sensitivities to distinguish envelope emission from resolved disk emission at the relevant scales (20–200 au). The characteristics of Class 0 disks, and the exact ingredients responsible for early disk properties during the main accretion phase, hence remain strongly debated.

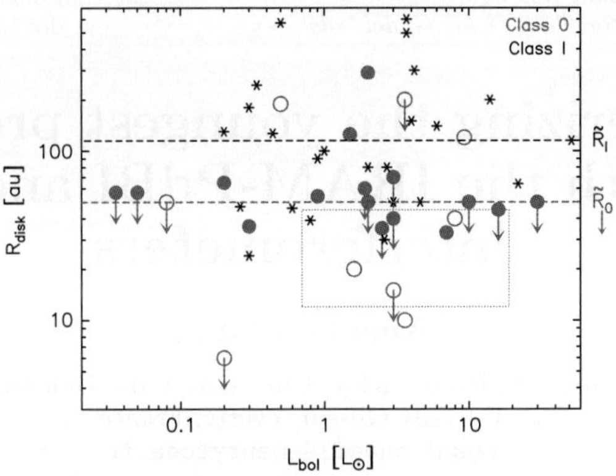

Figure 1. Protostellar disk radii $R_{\rm disk}$ in function of the protostellar bolometric luminosity $L_{\rm bol}$, for all Class 0 (red circles) and Class I (black stars) protostars observed between 0.7 mm and 2.7 mm with a spatial resolution better than 50 au, from both the CALYPSO sample and the literature. See Maury *et al.* (2019) for source names and references. The dashed lines show the median disk radii: in red from the sample of 25 Class 0 disks where $\tilde{R}_0$ <50 au (including upper limit radii), and in black the median radius from the sample of 25 Class I disks is $\tilde{R}_I = 115$ au.

2. Class 0 disks in the CALYPSO sample

Only a survey providing high angular resolution observations for a large sample of Class 0 protostars can shed light on the controversy about the pristine characteristics of protostellar disks and ultimately on the importance of magnetic fields in regulating disk formation during protostellar formation. CALYPSO (see `http://irfu.cea.fr/Projets/Calypso/`) is a survey of 16 Class 0 protostars, carried out with the IRAM Plateau de Bure (PdBI) interferometer at high angular resolution ($\sim 0.3''$). The 16 Class 0 objects of the CALYPSO sample are among the youngest known solar-type protostars (André *et al.* 2000), with $M_{\rm env} \sim 0.5$–10 $M_\odot$, and internal luminosities $L_{\rm int} \sim 0.03$–30 $L_\odot$ (see Table 1 in Maury *et al.* 2019). We analyzed the dust continuum emission datasets, obtained at both 1.37 mm and 3.18 mm: our analysis shows that less than 25% of the CALYPSO Class 0 disk candidates are resolved at radii >60 au. When taking into account all available literature on resolved Class 0 disks (or upper limit sizes obtained with interferometers), we also show that a similar fraction ($< 28\%$) of Class 0 protostars studied so far may harbor large disks with radii > 60 au. Our results suggest that the formation of disks during the Class 0 phase could occur at smaller scales than predicted by hydrodynamical models of rotating protostellar collapse (Maury *et al.* 2010, Maury *et al.* 2019). At the current state of our knowledge, this may either point to (i) the extreme youth of these objects (which may not yet have developed a disk), (ii) an overestimate of the initial angular momentum of the cores in hydrodynamical collapse models, or (iii) a magnetically regulated disk formation scenario. I discussed these scenari and conclude that, because they reduce the centrifugal radius and produce a disk size distribution that peaks at radii < 100 au during the main accretion phase (Li *et al.* 2014, Hennebelle *et al.* 2016), magnetized models of rotating protostellar collapse are favored by our CALYPSO observations.

3. A counter-rotating disk in IRAM 04191 ?

IRAM 04191+1522 is a Class 0 protostar in the Taurus molecular cloud (André *et al.* 1999). Its large envelope mass (1.5 $M_\odot$), low-luminosity (0.1 $L_\odot$) and temperature suggest an age of 1-3 $\times$ 10^4 yr and a stellar mass $\sim$0.05 $M_\odot$, making it one of the youngest protostars in Taurus. Single-dish observations of molecular lines, combined with radiative

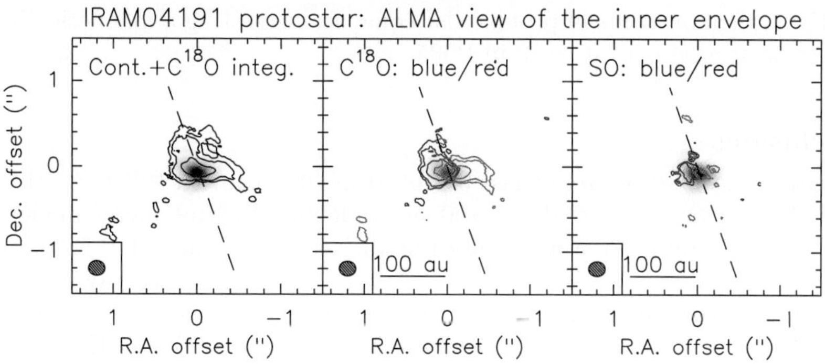

Figure 2. In all panels, the background map shows the ALMA 1.29 mm dust continuum emission. Left: The black superimposed contours show the levels of $C^{18}O$(2–1) integrated emission. Center: The blue and red contours show the levels of respectively blue-shifted and red-shifted $C^{18}O$(2–1) emission (integrated over $\pm$ 1.5 km s^{-1}with respect to the systemic velocity 6.6 ± 0.1 km s^{-1}). Right: Same for the SO(65–54) emission.

transfer modeling of the IRAM 04191 envelope, enabled to constrain the outer envelope kinematics: a steep velocity gradient of ~40 km s^{-1} pc^{-1} was detected along the NW-SE direction, i.e., perpendicular to the axis of the outflow, in the envelope at scales r~4000 au (see Fig. 2 in Belloche *et al.* 2002). This velocity gradient is consistent with differential rotation of the envelope, with the blueshifted emission to the northwest and the redshifted emission to the southeast (Belloche *et al.* 2002, Takakuwa *et al.* 2003, Lee *et al.* 2005). We have observed the dust continuum emission and molecular line emission from IRAM 04191, using ALMA to obtain a synthesized beam $0.18'' \times 0.15''$ ($\sim$ 23au). Our ALMA molecular emission maps reveal an organized velocity gradient at scales $r \lesssim 50$ au. Considering a systemic velocity of the IRAM 04191 core 6.6 ± 0.1 km s^{-1}, the $C^{18}O$(2–1) and SO($6_5 - 5_4$) emission are tracing a velocity gradient perpendicular to the axis of the large-scale outflow (blue to the east and red to the east, see Fig. 2), but in a direction opposite to the one observed at the envelope large-scales. This reversal between the large- and small-scale velocity gradients suggests a counter-rotating disk is forming within the IRAM 04191 protostar (Maury *et al.* in prep). Investigations of the envelope kinematics at intermediate scales are crucially needed to propose a formation scenario for this young disk, and polarimetric observations could shed light on a possible scenario where magnetic braking would be responsible for the kinematics reversal in this pristine protostellar envelope.

4. A magnetically-regulated disk formation scenario in B335

The role of the magnetic field during protostellar collapse is still poorly constrained from an observational point of view, and only few constraints exist that shed light on the magnetic braking efficiency during the main accretion phase. I presented our ALMA polarimetric observations of the thermal dust continuum emission at 1.3 mm, towards the B335 Class 0 protostar (Maury *et al.* 2018). Linearly polarized dust emission is detected at all scales probed by our observations (50 to 1000 au). The magnetic field structure has a very ordered topology in the inner envelope, with a transition from a large-scale poloidal magnetic field, in the outflow direction, to strongly pinched in the equatorial direction. We compared our data to a family of magnetized protostellar collapse models. We show that only models with an initial core mass-to-flux ratio $\mu \sim 5 - 6$ are able to reproduce the observed properties of B335, especially the upper-limits on its disk size, its large-scale envelope rotation β and the pronounced magnetic field lines pinching observed in our ALMA data. In these MHD models, the magnetic field is dynamically relevant to regulate the typical outcome of protostellar collapse, suggesting a magnetically-regulated disk formation scenarios is at work in B335.

5. Conclusions

Observational studies of large Class 0 protostar samples with PdBI, ALMA and VLA found only few large disks at scales r>100 au, while pure hydrodynamic models routinely form rotationally-supported protostellar disks with typical radius ∼150-500 au, as a result of the conservation of the initial core's angular momentum during protostellar collapse. Contrasting with this simple picture, the inclusion of magnetic fields in models of protostellar collapse reduce the disk sizes drastically at the beginning of the main accretion phase, producing a disk size distribution consistent with our observational findings in Class 0 protostars. Our observations of small disks, counter-rotating disks and organized magnetic fields in the youngest star-forming cores question the established paradigm of disk formation as a simple consequence of angular momentum conservation during the main accretion phase: they instead highlight the need to investigate magnetized models in order to unveil the mechanisms responsible for protostellar disk properties. In the future, obtaining accurate disk size and mass distributions, as well as detailed maps of the magnetic field topology in star forming cores, will be of paramount importance to make progress in our understanding of the formation and early evolution of both stars and protoplanetary disks.

Acknowledgements

I thank the CALYPSO and MagneticYSOs collaborations for their help in producing the papers which results are partly shown here. This work has benefited from the support of the European Research Council under the European Union's Seventh Framework Programme (Advanced Grant ORISTARS with grant agreement no. 291294 and Starting Grant MagneticYSOs with grant agreement no. 679937).

References

André, P., Motte, F., & Bacmann, A., *ApJL*, 513, L57 (1999)
André, P., Ward-Thompson, D., & Barsony, M., *Protostars and Planets IV*, 59 (2000)
Belloche, A., André, P., Despois, D. & Blinder, S., *A&A*, 393, 927 (2002)
Cassen, P. & Moosman, A., *Icarus*, 48, 353 (1981)
Evans, N., Dunham, M., Jørgensen, J. *et al.*, *ApJ*, 181, 321 (2009)
Hennebelle, P., Commerçon, B., Chabrier, G., & Marchand, P., *ApJL*,830, L8 (2016)
Lee, C.-F., Ho, P. T. P., & White, S. M., *ApJ*, 619, 948 (2005)
Li, Z.Y., Banerjee, R., Pudritz, R., Jørgensen, J., Shang, S., Krasnopolsky, R., & Maury, A.J., *Protostars and Planets VI*, 173 (2014)
Maury, A.J., André, Ph., Hennebelle, P., Motte, F., Stamatellos, D., Bate, M., Belloche, A., Duchêne, G. & Whitworth, A., *A&A*, 512, A40+ (2010)
Maury, A.J., André, Ph., Men'shchikov, A., Könyves, V., & Bontemps, S., *A&A*, 535, A77 (2011)
Maury, A.J., Girart, J.M., Zhang, Q., Hennebelle, P., Keto, E., Rao, R., Lai, S.P., Ohashi, N., & Galametz, M., *MNRAS*, 477, 2, 2760 (2018)
Maury, A.J., André, Ph., Testi, L. *et al.*, *A&A*, 621, A76 (2019)
Takakuwa, S., Ohashi, N., & Hirano, N., *ApJ*, 590, 932 (2003)
Terebey, S., Shu, F., & Cassen, P., *ApJ*, 286, 529 (1984)
Yorke, H. W. & Bodenheimer, P., *ApJ*, 525, 330 (1999)

Discussion

DWARKADAS: You started by mentioning that the angular momentum in the solar system is mostly locked in Jupiter. Could MHD instabilities in viscous disks be partly responsible for most of the angular momentum being at large radii?

MAURY: Transport of angular momentum in disks is indeed an unsolved problem. The point I wanted to make is that during the formation process of the star, most of the core angular momentum is not transferred to the star and part is found later on in the more evolved protoplanetary disks, in which planets like Jupiter form. Jupiter is the most massive planet so it seems all right that it took most of the disk's angular momentum.

JOHNSTONE: Beautiful observations. What importance do you believe the jet/outflow plays?

MAURY: Probably part of angular momentum is being carried away by jets during the main accretion phase. But from observations of angular momentum in protoplanetary disks, it seems the bulk of the angular momentum must remain in the system to be transferred to the disk later on, so magnetic braking is a prime suspect.

CUNTZ: Can you comment on the role of the central star (especially its mass) regarding the disk structure, in particular the attained distribution of angular momentum.

MAURY: We have no good constraints on the stellar mass yet, because this requires to characterize the Keplerian pattern of these small disks, mostly unresolved with existing molecular line observations.

Origins: From the Protosun to the First Steps of Life
Proceedings IAU Symposium No. 345, 2018
B. G. Elmegreen, L. Viktor Tóth & M. Güdel, eds.

doi:10.1017/S1743921319002278

Formation and Evolution of Protoplanetary Disks: Observations and Modeling of Jets, Disks, and Disk Substructures

Laura M. Pérez

Universidad de Chile, Departamento de Astronomía,
Camino El Observatorio 1515, Las Condes, Santiago, Chile
email: `lperez@das.uchile.cl`

Abstract. Planet formation takes place in the gaseous and dusty disks that surround young stars, known as protoplanetary disks. With the advent of sensitive observations and together with developments in theory, our field is making rapid progress in understanding how the evolution of protoplanetary disks takes place, from its inception to the end result of a fully-formed planetary system. In this review, I discuss how observations that trace both the dust and gas components of these systems inform us about their evolution, mass budget, and chemistry. Particularly, the process of disk evolution and planet formation will leave an imprint on the distribution of solid particles at different locations in a protoplanetary disk, and I focus on recent observational results at high angular resolution in the sub-millimeter regime, which have revealed a variety of substructures present in these objects.

Keywords. stars: formation, planetary systems: protoplanetary disks, instrumentation: high angular resolution, instrumentation: interferometers

---

1. Introduction

Our current view of the process of low-mass star and planet formation is that they occur simultaneously. When protostars are born, they are generally surrounded by circumstellar material whose geometry is envelope-like in the earliest phases of formation, and later becomes disk-like once the envelope is dissipated. Planets are formed out of the same circumstellar material that feeds the protostar, and we generally refer to these planet-forming disks as protoplanetary disks (Williams & Cieza 2011). Eventually, the protostar will reach the main-sequence, perhaps surrounded by a planetary system, while the leftover material from this process may be observed as a debris disk. In the following, I discuss recent ALMA results related to the earliest stages of star formation, when jets and outflows are still present, then I discuss what ALMA observations tells us about the evolution of solids in protoplanetary disks, and finally I present new results on disk substructures observed at high angular resolution with ALMA.

2. Outflows and jets

Jets and outflows become part of the picture in the early stages of the process, playing an important role: they extract excess angular momentum during the protostellar collapse, allowing for accretion of material onto the forming protostar. Outflows/jets are present across a wide range of stellar masses, from low to high-mass stars, and they have an impact over different physical scales: from the disk to the envelope, and all the way to cloud scales (e.g. Aso *et al.* 2015, Yusef-Zadeh *et al.* 2017). ALMA observations in dust continuum and molecular lines can trace the outflow (Louvet *et al.* 2018) or the jet (Lee *et al.* 2017) from young protostars, constraining the launching radii to be quite small. In the case of the edge-on low-mass star HH 330 (Louvet *et al.* 2018) the morphology and kinematics is consistent with a slow disk wind launching at radii < 7au. These observations also reject the wide binary hypothesis that was used to explain the wiggling of the optical jet. For HH 212, Lee *et al.* (2017) are able to measure the rotation of the jet down to 10au from the protostar, inferring a ~ 0.05au launching radius.

Another important aspect of protostellar evolution, which can be traced with detailed studies of outflows/jets, is how acccretion proceeds. Nascent stars are not continuously fed material from the disk but rather have time-variable accretion. It is only recently that theory can be put to the test via observations of jets whose knotty structure may be a signature of episodic protostellar accretion. In a recent study, Vorobyov *et al.* (2018) modeled the accretion history of gravitationally unstable disks and compared the time delay between bursts of luminosity/accretion with the observed episodic ejection, measured as the separation between jet components/knots in the CARMA-7 source studied in Plunkett *et al.* (2015). Such comparisons are finally testing theoretical predictions against observations.

3. The evolution of solids in disks

Inside the circumstellar disks that generally surround young stars is where the ISM dust is converted into planetary systems. At long radio wavelengths (mm to cm) the dust thermal emission mainly arises from macroscopic particles ("pebbles") and thus, observations at this wavelength regime allow for the last direct detection of solids before we can study the end product, planetary systems. Surveys of young star-forming regions have provided an overview of disk evolution, allowing for the comparison with host properties, environment, and age. From IR-excess studies, we know that disk lifetimes are short (Hernández *et al.* 2007) but only recently we have directly observed the evolution of disk mass with time from mm-wave observations (Ansdell *et al.* 2016, Pascucci *et al.* 2016, Barenfeld *et al.* 2016), where the oldest star-forming regions appear to have a larger fraction of lower-mass disks than the younger ones (albeit caveats remain regarding sample completeness, Cieza *et al.* 2019). In any case, grain growth, fragmentation, and the radial drift of macroscopic particles will have an impact on the observed solid mass, and the last two particularly become barriers to the planet formation process (see, e.g. Testi *et al.* 2014). A way to test for the evolution of solids in disks and the presence of these barriers is by inferring their particle-size distribution from multi-wavelength observations. In a few examples we know that there is a radial segregation by particle size, either from constraints on disk size with wavelength (e.g. Tripathi *et al.* 2018) or from the observed variations of the opacity spectral index with radial distance from the star (e.g Tazzari *et al.* 2016). However, nature somehow must overcome these barriers and perhaps a way to promote solid agglomeration and support the planet formation process is by the presence of substructure, since a disk with substructure will concentrate solids needed for planetesimal formation (e.g. Pinilla *et al.* 2015, Birnstiel *et al.* 2013). Substructure can be naturally generated via instabilites in the disk, for example large-scale gravitational instabilities (e.g., Dipierro *et al.* 2015), zonal flows (e.g., Johansen *et al.* 2009), or hydrodynamical instabilities that can concentrate solids in vortices (e.g., Zhu & Stone 2014). On the other hand, the presence of companions, either stellar-mass or planet-mass, will also excite substructure in the disk (e.g. Wolf & D'Angelo 2005) that can then trap the solids and promote its concentration. For many disks, we already knew about substructure: this is the case of transition disks, for which the presence of a cavity/gap was

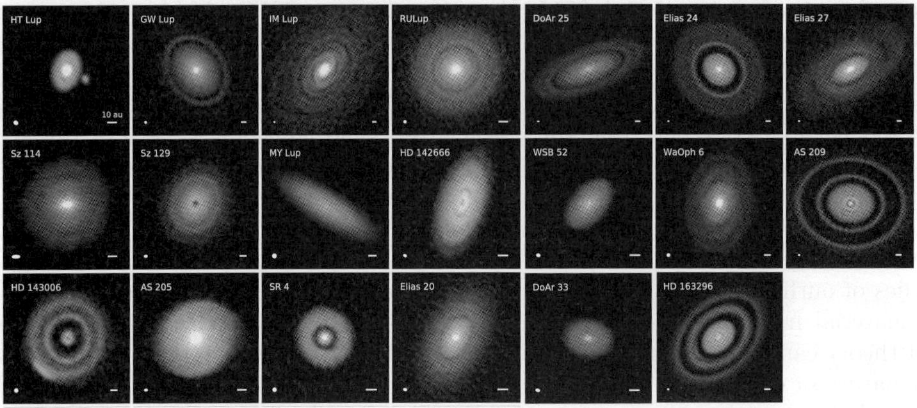

Figure 1. DSHARP observations of the 20 young stars targeted with ALMA high resolution observations at 1.3 mm. Adapted from Andrews *et al.* (2018).

inferred from spectral energy distribution modeling and latter confirmed through imaging (some examples are Brown *et al.* 2009, Geers *et al.* 2007, Andrews *et al.* 2011). With the advent of ALMA, many disks have been targeted with sufficient angular resolution to reveal a whole variety of substructures: gaps and rings, sometimes with asymmetries, and sometimes with spirals (e.g. Fukagawa *et al.* 2013, Pérez *et al.* 2014, ALMA Partnership *et al.* 2015, Tobin *et al.* 2016, Pérez *et al.* 2016, Dong *et al.* 2018). But to understand the prevalence, statistics, and properties of such features, we needed a larger sample from which to study.

4. DSHARP: Disks Substructures at High Angular Resolution Project

The "Disks Substructures at High Angular Resolution Project" (DSHARP) is an ALMA Large Program devoted to study substructures in protoplanetary disks, through observations of high angular resolution at millimeter wavelengths of a large sample of young stars (Andrews *et al.* 2018). The focus of the program is on "classical" Class II disks, since transitional disks already have known substructure and debris disks are more evolved objects. We targeted 20 young stars, two of which were multiple systems, and we managed to image 23 protoplanetary disks in ALMA Band 6 (1.3 mm wavelength) with an angular resolution of few au over the full extent of the disk (Figure 1). Both dust continuum emission and a gas tracer (^{12}CO) were targeted. These observations have revealed a that substructure appears to be ubiquitous (Andrews *et al.* 2018) and that the most common type of substructure is concentric rings/gaps (Huang *et al.* 2018). Rings/gaps can have high and low contrast, can be located close and far from the star, and we find no correlations between the stellar properties and ring properties (Huang *et al.* 2018). Surprisingly, the second most common type of substructure in this sample was spiral arms (Huang *et al.* 2018), which appear to be quite diverse as well in terms of their properties around single-star systems. In multiple systems, we find that the substructure observed is evidence of star-companion-disk interactions (Kurtovic *et al.* 2018). Asymmetries in the disks of this sample were found only in two targets: HD 163296 (Isella *et al.* 2018) and HD 143006 (Pérez *et al.* 2018, see Figure 2), since the majority of the disks observed had very symmetric dust continuum emission, as beautifully demonstrated for the case of AS 209 (Guzmán *et al.* 2018). Several of the observed rings appear to be radially narrow compared to the pressure scale height, indicative of dust trapping taking place in these rings (Dullemond *et al.* 2018). Detailed hydrodynamical simulations

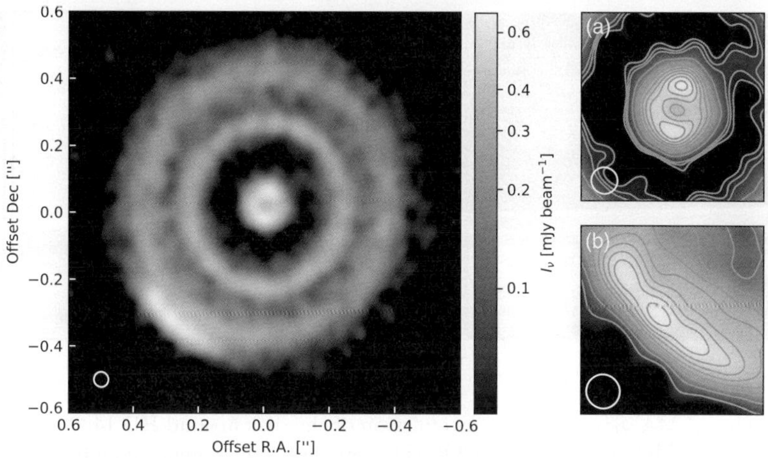

Figure 2. The ALMA view in dust continuum emission of the disk around HD 143006. The left panel shows that the disk resolves intro three concentric rings and an arc of emission outside of the outermost ring. The right panels show that the innermost ring connects to the outer disk with a low-surface brightness bridge-like feature (panel (a)), and that the bright arc of emission resolves into further substructure with three peaks roughly separated by 10au (panel (b)). In panel (a), the contours are at 3, 4, 5σ, then continue being spaced by 5σ, while in panel (b) the contours start and are spaced by 5σ. Note that σ is the rms noise of the image ($\sigma = 14.3$ μJy beam^1). Ellipses in the bottom-left corner of each panel indicate the beam size of 46×45 mas (7.6×7.4 au). Adapted from Pérez *et al.* (2018).

show that several of the observed gaps can be explained with the presence of massive planets (Zhang *et al.* 2018). Finally, opacity calculations needed to interpret in terms of mass the rings observed within the DSHARP collaboration are presented in Birnstiel *et al.* (2018). For a full detailed discussion of all these findings see the Focus Issue from ApJ Letters (Andrews *et al.* 2018, Huang *et al.* 2018, Huang *et al.* 2018, Kurtovic *et al.* 2018, Birnstiel *et al.* 2018, Dullemond *et al.* 2018, Zhang *et al.* 2018, Isella *et al.* 2018, Pérez *et al.* 2018).

5. Observing disk substructures at multiple wavelengths: an example in the T Tauri star HD 143006

Multi-wavelength observations of disks at high spatial resolution can provide critical constraints on the structure and distribution of solids and gas in protoplanetary disks (e.g. Cazzoletti *et al.* 2018). In particular, scattered-light imaging observations are critical as they trace the disk surface layer, they critically depend on the disk structure and stellar parameters, and current instruments can provide high angular resolution images (e.g. SPHERE, GPI, SCExAO, NACO). As an example of what can we learn from studying the same object at multiple wavelengths, we discuss the case of HD 143006, a G7 spectral type young star at a distance of 165 ± 5 pc (Gaia Collaboration *et al.* 2018), with a mass between 1.5-2.0 $M_\odot$, and age between $4 - 12$ Myr old (Andrews *et al.* 2018). Benisty *et al.* (2018) analyzed SPHERE J-band polarized intensity imaging of this star at 40 mas resolution, which shows a strong East-West asymmetry in scattered light, with the East side of the disk being brighter than the West. They show through hydrodynamic simulations and radiative transfer, that a moderately misaligned inner disk (misalignment $< 30°$) can reproduce the set of narrow shadows in the innermost bright ring observed in scattered light, together with the East-West broad shadow observed in the entire disk. Thus, shadows in scattered light can be a different probe of disk substructure, as scattered light observations are very sensitive to the illumination pattern. However, scattered

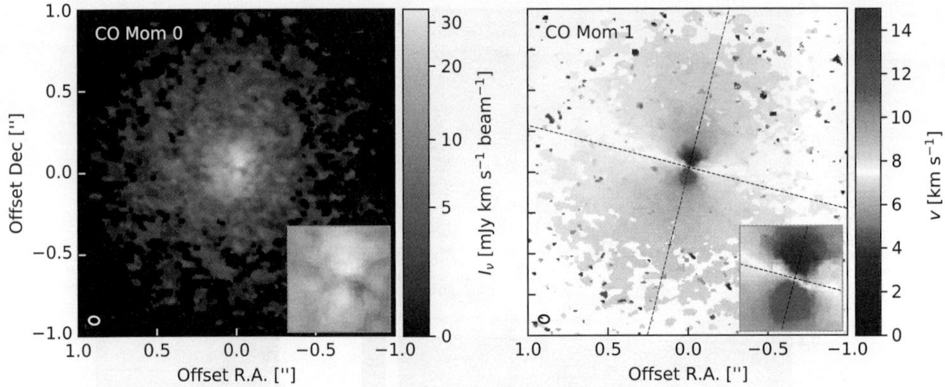

Figure 3. The ALMA view in ^{12}CO gas emission of the disk around HD 143006. The left panel shows the integrated intensity of the ^{12}CO(2-1), also known as moment 0 map. In this particular line tracer there is no clear disk substructure, except for depleted CO emission in the inner disk close to the central star. The right panel shows the velocity field of the CO line, which follows a Keplerian pattern except for the inner disk close to the central star, where the position angle and inclination of the disk appear to change, as revealed by further modeling. Ellipses in the bottom-left corner of each panel indicate the beam size of 66×49 mas (11×8 au). Adapted from Pérez *et al.* (2018).

light imaging cannot probe the inner disk directly, due to the inner working angle of such observations (80 mas for SPHERE) that does not allow for a clear view of the inner disk misalignment.

As part of DSHARP, new ALMA observations of HD 143006 at high angular resolution provided new information about the kinematics of the protoplanetary disk, as well as the substructure in gas and dust (Pérez *et al.* 2018). In the dust continuum emission the disk resolved into three concentric rings, a large arc with three bright peaks outside of the outermost ring, and a bridge-like emission feature detected between the inner and middle ring (see Figure 2). In gas emission, the disk extends to a factor of two larger radii, with no obvious asymmetries or rings detected, although the inner disk appears depleted of emission as in the dust continuum map (see Figure 3). Modeling of the kinematics indicated that a single geometry (inclination, position angle) for the disk cannot reproduce the Keplerian rotation pattern, suggesting a misalignment between the inner disk and the outer disk. Detailed modeling of the dust continuum visibilities obtained by ALMA revealed the actual degree of misalignment: either a $< 10°$ misalignment if the near side of the disk is the same for the inner and outer disk, or a moderate $\sim 40°$ misalignment if the inner and outer disk have opposite near sides. Further hydrodynamical modeling by Zhang *et al.* (2018) suggest that a massive planet of $10 - 20$ M$_{Jup}$ could create a gap between the inner and outer disk and radially segregate the large and small dust grains traced by the ALMA and SPHERE observations.

6. Future

The study of the formation and evolution of protoplanetary disks will likely continue to tackle the topic of substructures and what is their origin. On one hand, future studies that trace the chemistry of disks during planet formation will be a fundamental piece of the puzzle, for which ALMA is perfectly suited. On the other hand, tracing the embedded planets that perhaps are the cause of the observed substructure will be fundamental to understand the origin of the observed features in disks. Future studies with ALMA, VLT, and next-generation instruments like the E-ELT, SKA, and ng-VLA should advance our knowledge on protoplanetary disks and planet formation.

References

ALMA Partnership, Brogan, C. L., Pérez, L. M., *et al.* 2015, *ApJ*, 808, L3
Andrews, S. M., Huang, J., Pérez, L. M., *et al.* 2018, *ApJ*, 869, L41
Andrews, S. M., Wilner, D. J., Espaillat, C., *et al.* 2011, *ApJ*, 732, 42
Ansdell, M., Williams, J. P., van der Marel, N., *et al.* 2016, *ApJ*, 828, 46
Aso, Y., Ohashi, N., Saigo, K., *et al.* 2015, *ApJ*, 812, 27
Barenfeld, S. A., Carpenter, J. M., Ricci, L., *et al.* 2016, *ApJ*, 827, 142
Benisty, M., Juhász, A., Facchini, S., *et al.* 2018, *A&A*, 619, A171
Birnstiel, T., Dullemond, C. P., & Pinilla, P. 2013, *A&A*, 550, L8
Birnstiel, T., Dullemond, C. P., Zhu, Z., *et al.* 2018, *ApJ*, 869, L45
Brown, J. M., Blake, G. A., Qi, C., *et al.* 2009, *ApJ*, 704, 496
Cazzoletti, P., van Dishoeck, E. F., Pinilla, P., *et al.* 2018, *A&A*, 619, A161
Cieza, L. A., Ruíz-Rodríguez, D., Hales, A., *et al.* 2019, *MNRAS*, 482, 698
Dipierro, G., Pinilla, P., Lodato, G., *et al.* 2015, *MNRAS*, 451, 974
Dong, R., Liu, S.-. yuan., Eisner, J., *et al.* 2018, *ApJ*, 860, 124
Dullemond, C. P., Birnstiel, T., Huang, J., *et al.* 2018, *ApJ*, 869, L46
Fukagawa, M., Tsukagoshi, T., Momose, M., *et al.* 2013, *PASJ*, 65, L14
Gaia Collaboration, Brown, A. G. A., Vallenari, A., *et al.* 2018, *A&A*, 616, A1
Geers, V. C., Pontoppidan, K. M., van Dishoeck, E. F., *et al.* 2007, *A&A*, 469, L35
Guzmán, V. V., Huang, J., Andrews, S. M., *et al.* 2018, *ApJ*, 869, L48
Hernández, J., Hartmann, L., Megeath, T., *et al.* 2007, *ApJ*, 662, 1067
Huang, J., Andrews, S. M., Dullemond, C. P., *et al.* 2018, *ApJ*, 869, L42
Huang, J., Andrews, S. M., Pérez, L. M., *et al.* 2018, *ApJ*, 869, L43
Isella, A., Huang, J., Andrews, S. M., *et al.* 2018, *ApJ*, 869, L49
Johansen, A., Youdin, A., & Klahr, H. 2009, *ApJ*, 697, 1269
Kurtovic, N. T., Pérez, L. M., Benisty, M., *et al.* 2018, *ApJ*, 869, L44
Lee, C.-F., Ho, P. T. P., Li, Z.-Y., *et al.* 2017, *Nature Astron.*, 1, 152
Louvet, F., Dougados, C., Cabrit, S., *et al.* 2018, *A&A*, 618, A120
Pascucci, I., Testi, L., Herczeg, G. J., *et al.* 2016, *ApJ*, 831, 125
Pérez, L. M., Isella, A., Carpenter, J. M., *et al.* 2014, *ApJ*, 783, L13
Pérez, L. M., Carpenter, J. M., Andrews, S. M., *et al.* 2016, *Science*, 353, 1519
Pérez, L. M., Benisty, M., Andrews, S. M., *et al.* 2018, *ApJ*, 869, L50
Pinilla, P., van der Marel, N., Pérez, L. M., *et al.* 2015, *A&A*, 584, A16
Plunkett, A. L., Arce, H. G., Mardones, D., *et al.* 2015, *Nature*, 527, 70
Tazzari, M., Testi, L., Ercolano, B., *et al.* 2016, *A&A*, 588, A53
Testi, L., Birnstiel, T., Ricci, L., *et al.* 2014, *Protostars and Planets VI*, 339
Tobin, J. J., Kratter, K. M., Persson, M. V., *et al.* 2016, *Nature*, 538, 483
Tripathi, A., Andrews, S. M., Birnstiel, T., *et al.* 2018, *ApJ*, 861, 64
Vorobyov, E. I., Elbakyan, V. G., Plunkett, A. L., *et al.* 2018, *A&A*, 613, A18
Williams, J. P., & Cieza, L. A. 2011, *Annu. Rev. Astron. Astrophys.*, 49, 67
Wolf, S., & D'Angelo, G. 2005, *ApJ*, 619, 1114
Yusef-Zadeh, F., Wardle, M., Kunneriath, D., *et al.* 2017, *ApJ*, 850, L30
Zhang, S., Zhu, Z., Huang, J., *et al.* 2018, *ApJ*, 869, L47
Zhu, Z., & Stone, J. M. 2014, *ApJ*, 795, 53

Origins: From the Protosun to the First Steps of Life
Proceedings IAU Symposium No. 345, 2018
B. G. Elmegreen, L. Viktor Tóth & M. Güdel, eds.

doi:10.1017/S1743921318008268

Towards Realistic Understandings of Gas Dynamics in Protoplanetary Disks

Xue-Ning Bai

Institute for Advanced Study and Center for Astrophysics, Tsinghua University
Beijing 100084, China
email: xbai@tsinghua.edu.cn

Abstract. The gas dynamics of protoplanetary disks (PPDs) plays a crucial role in almost all stages of planet formation, yet it is far from being well understood largely due to the complex interplay among various microphysical processes. Primarily, PPD gas dynamics is likely governed by magnetic fields, and their coupling with the weakly ionized gas is described by non-ideal magnetohydrodynamic (MHD) effects. Incorporating these effects, I will present the first fully global simulations of PPDs that include the most realistic disk microphysics. Accretion and disk evolution is primarily driven by magnetized disk winds with significant mass loss comparable to accretion rate. The overall disk gas dynamics strongly depends on the polarity of large-scale poloidal magnetic field threading the disk owing to the Hall effect. The flow structure in the disk is highly unconventional with major implications on planet formation.

Keywords. accretion disks, MHD, planetary systems: protoplanetary disks

---

1. Introduction

The gas dynamics of protoplanetary disks (PPDs) is primarily governed by the mechanism that drives angular momentum transport (AMT). This is observationally linked to the typical accretion rate of the order $10^{-8} M_{\odot}$ yr^{-1} (Hartmann *et al.* 1998), as well as outflows that are ubiquitously found from young stellar objects (Hartigan *et al.* 1995, Simon *et al.* 2016). Conventionally, AMT in PPDs is thought to be mediated by turbulence driven by the magnetorotational instability (MRI, Balbus & Hawley 1991, Gammie 1996), which transports angular momentum radially outward. Alternatively, angular momentum can be transported vertically via a magnetized disk wind (Blandford & Payne 1982), which requires a large-scale poloidal field threading the disk.

Both the MRI and disk wind scenarios are sensitive to the level of coupling between gas and magnetic field in the disk, which is tied to the level of ionization. The PPDs largely rely on external sources of non-thermal ionization (except for the innermost region where alkali species can be thermally ionized). These include stellar X-rays, far-UV (FUV), and cosmic-rays, which all have finite penetration depth and only ionize the disk very weakly. They also contribute to substantial heating at the disk surface.

Magnetic fields are no longer frozen in to weakly ionized gas, which is described by non-ideal magnetohydrodynamic (MHD) effects. They are reflected to the induction equation, which describes how magnetic fields evolve (valid in the absence of charged grains):

$$\frac{\partial \boldsymbol{B}}{\partial t} = \nabla \times (\boldsymbol{v} \times \boldsymbol{B}) + \nabla \times \left[\frac{4\pi\eta}{c} \boldsymbol{J} + \frac{\boldsymbol{J} \times \boldsymbol{B}}{e n_e} - \frac{(\boldsymbol{J} \times \boldsymbol{B}) \times \boldsymbol{B}}{c \gamma \rho \rho_i} \right] . \tag{1.1}$$

In the above, $\boldsymbol{J} = c\nabla \times \boldsymbol{B}/4\pi$ is the current density, ρ, ρ_i are the densities of gas and ions, n_e is the electron number density, γ is the coefficient of momentum transfer between ions and neutrals, $\eta \propto 1/n_e$ is the resistivity. The last three terms on the right hand side correspond to Ohmic resistivity (O), the Hall effect (H), and ambipolar diffusion (A). In terms of their strength, they are all inversely proportional to the ionization fraction ($1/n_e$). In the mean time, the relative importance of the three terms is determined by gas density and field strength, with $\eta_O \propto (B/\rho)^0$, $\eta_H \propto (B/\rho)^1$, and $\eta_A \propto (B/\rho)^2$.

Applied to PPDs, it turns out that resistivity is important only in the densest part of the disk, namely, the midplane region of the inner disk (with $R \lesssim 5-10$ AU). Moving towards the surface, as density drops, we transition first to the Hall-dominated regime, followed by ambipolar diffusion dominated region (e.g., Wardle 2007, Bai 2011). In the very upper layer, FUV substantially boosts the ionization fraction, and the gas behaves close to ideal MHD Perez-Becker & Chiang 2011).

The role of resistivity has been extensively studied in the literature (e.g., Fleming & Stone 2003, Fleming & Stone 2003, Ilgner & Nelson 2006, Bai & Goodman 2009), giving the picture of layered accretion where the MRI is suppressed in the Ohmic-dominated midplane region of inner disk, while remain active in the surface layer. This picture has been challenged once the other two non-ideal MHD effects are taken into account: it has been found that in the surface layer of the inner disk, ambipolar diffusion is sufficiently strong that can fully suppress the MRI, and hence disk accretion must be driven by a magnetized disk wind (Bai & Stone 2013, Gressel *et al.* 2015). This means that a large-scale poloidal field threading the disk is essential, which is likely inherited from the star formation process.

In addition, in the presence of background poloidal field, the Hall effect makes the dynamics depend on the polarity of this field, which can be seen from Equation (1.1): the equation remains valid by changing the sign of $\boldsymbol{B}$ except for the Hall term. We call the case with $\boldsymbol{B}\cdot\boldsymbol{\Omega}>0$ the "aligned" case, and $\boldsymbol{B}\cdot\boldsymbol{\Omega}<0$ for "anti-aligned" case. The role of the Hall term has been studied in terms of local shearing-box simulations (e.g., Lesur *et al.* 2014, Bai 2014). Besides certain ambiguities inherited to local simulations, it has been found that the aligned case leads to the Hall-shear instability (HSI, Kunz 2008), which strongly amplifies the horizontal component of the magnetic field, whereas in the anti-aligned case, growth horizontal field is suppressed.

Previous works have largely been conducted using local simulations (aforementioned), or semi-global simulations with truncated vertical/latitudinal boundaries (e.g., Gressel *et al.* 2015, Béthune *et al.* 2017). While they clearly demonstrate wind launching, wind kinematics can not be properly determined due to artificial truncation of the vertical boundary. Moreover, many of these simulations are subject to symmetry issues. It is thus highly desirable to conduct fully global disk simulations with fully covered simulation domain to accommodate wind launching, together with well-resolved disk microphysics.

2. Simulations and Results

In this work, using the Athena++ MHD code (Stone *et al.* in prep), we present the first set of such simulations to date for the inner region ($\sim 1-20$ AU) of PPDs. Since we expect the MRI to be suppressed in this region, our simulations are conducted in 2D in spherical polar coordinates. We set up a thin disk with prescribed surface density profiles $\Sigma = 500 R_{\rm AU}^{-1}$g cm^{-2} as well as temperature profiles in hydrostatic equilibrium. The temperature is set so that disk aspect ratio $\epsilon \sim 0.05-0.1$ in the midplane and increases to $\epsilon = 0.2$ at the atmosphere (to mimic external heating), maintained by a simple cooling prescription. Radial ray tracing is done to track FUV (assumed penetration depth $\Sigma_{\rm FUV} = 0.03$ g cm^{-2}), as well as X-ray and cosmic-ray ionizations with standard prescriptions. All three non-ideal MHD effects are included, with diffusivities interpolated from a pre-computed lookup table. The radial grid to span from $r_{\rm in} = 0.6$ AU to $r_{\rm out} = 60$ AU with logarithmic grid spacing. The θ−grid extends from the midplane all the way to

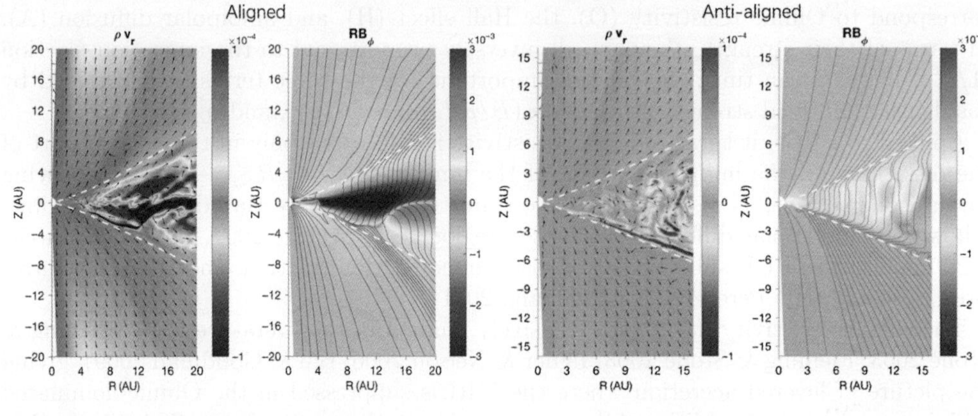

Figure 1. Last snapshot of the fiducial simulation runs. Left/right two panels are for aligned and anti-aligned cases. In each group, the left figure shows the radial mass flux ρv_r (color), with arrows indicating poloidal velocity directions (unit vectors). The right figure shows the magnetic field configurations (color represents B_ϕ, lines indicate poloidal field lines). White dashed lines mark the FUV front. Color bars are in dimensionless code units.

near the poles, and is concentrated around the disk which guarantees adequate resolution. The full grid size is 1152×512 in (r, θ), and we achieve a grid resolution of about $15 - 20$ cells per disk scale height. Poloidal fields are applied to the disks with initial plasma β_0 (ratio of gas to magnetic pressure) at the midplane to be 10^5. The simulations are evolved to $1600 - 1800$ yrs. More details are described in Bai (2017).

Figure 1 shows the magnetic field configurations and mass fluxes for simulations with aligned and anti-aligned field polarities at the end of the simulations. Wind launching are obvious in both cases, which we have found significant mass loss comparable or even in excess of the wind-driven accretion rates. Below we focus on the internal disk dynamics.

For the aligned case, the horizontal field strength is strongly amplified by the HSI. This is most intense in the inner few AU where the Hall effect is the strongest, making B_ϕ of a single sign dominating the entire disk vertical extent, before it flips at the very surface on one side to achieve a physical geometry for wind launching (i.e., to bend radially outward). Towards the outer region ($\gtrsim 10$ AU), as the Hall effect weakens, this flip occurs at the midplane, with a smooth transition. This is directly related to the radial mass flux, which can be shown to be approximately determined by $\rho v_R \approx B_z/(2\pi\Omega) \cdot (\partial B_\phi/\partial z)$. Correspondingly, we see that in the inner few AU, the gas flows inwards and outwards in either half of the disk midplane, plus additional accretion flow at the disk surface where B_ϕ flips. Towards the outer disk, the accretion flow is concentrated at the midplane, with radial mass outflows at the disk upper layers.

For the anti-aligned case, the Hall effect first attempts to reduce horizontal field in the midplane regions towards zero. But this makes the system slightly unstable, eventually leading to symmetry breaking with B_ϕ of one sign dominating the bulk disk, though the field strength is much weaker than that in the aligned case. As a result, the accretion flow is entirely concentrated at the location where B_ϕ flips, with is located at the disk surface and roughly coincides with the location of the FUV front. We find that this accretion flow is supersonic, which makes wind launching from this layer more dynamic.

3. Discussion and Future Directions

The PPD simulations we have conducted so far are the most realistic to date. They demonstrate not only the importance of wind-driven accretion, but also the unusual flow structures in disks that can only be captured when all major disk microphysical effects are properly included. Given that almost all stages of planet formation sensitively depends on disk gas dynamics, this opens up the avenue to explore planet formation under the realistic disk physics, such as grain growth and dust transport, planetesimal formation, pebble accretion, planet migration, etc. Future works should continue to improve upon the thermal physics (e.g., Wang, Bai & Goodman 2019), extend the simulation to three dimensions, and further explore the hot innermost region and the cold outer disk regions.

References

Bai, X.-N. 2011, *ApJ*, 739, 50
Bai, X.-N. 2014, *ApJ*, 791, 137
Bai, X.-N. 2017, *ApJ*, 845, 75
Bai, X.-N., & Goodman, J. 2009, *ApJ*, 701, 737
Bai, X.-N., & Stone, J. M. 2011, *ApJ*, 736, 144
Bai, X.-N., & Stone, J. M. 2013, *ApJ*, 769, 76
Béthune, W., Lesur, G., & Ferreira, J. 2017, *A&A*, 600, A75
Balbus, S. A., & Hawley, J. F. 1991, *ApJ*, 376, 214
Blandford, R. D., & Payne, D. G. 1982, *MNRAS*, 199, 883
Fleming, T., & Stone, J. M. 2003, *ApJ*, 585, 908
Gammie, C. F. 1996, *ApJ*, 457, 355
Gressel, O., Turner, N. J., Nelson, R. P., & McNally, C. P. 2015, *ApJ*, 801, 84
Hartigan, P., Edwards, S., & Ghandour, L. 1995, *ApJ*, 452, 736
Hartmann, L., Calvet, N., Gullbring, E., & D'Alessio, P. 1998, *ApJ*, 495, 385
Ilgner, M., & Nelson, R. P. 2006, *A&A*, 445, 205
Kunz, M. W. 2008, *MNRAS*, 385, 1494
Lesur, G., Kunz, M. W., & Fromang, S. 2014, *A&A*, 566, A56
Perez-Becker, D., & Chiang, E. 2011, *ApJ*, 735, 8
Simon, M. N., Pascucci, I., Edwards, S., *et al.* 2016, *ApJ*, 831, 169
Wang, L., Bai, X.-N., & Goodman, J. 2019, *ApJ*, 874, 90
Wardle, M. 2007, *AP&SS*, 311, 35

Discussion

LINSKY: Do your calculations include heating in the inner disk produced by shearing of the stellar and disk magnetic fields leading to flaring? I think that magnetic reconnection and heating is essential for estimating the ionization in the inner disk.

BAI: I divide the disk into "innermost", "inner", and "outer" regions, where the innermost region (up to a fraction of AU) is the most complex with thermal ionization, dust sublimation, and magnetospheric interactions taking place over a narrow range of radii. This work has focussed on the inner disk (1-20 AU) where the physics is cleaner. Energetic processes in the innermost region generally do not directly impact inner/outer disks, and X-rays/UV from there are jointly incorporated as part of the stellar properties.

LONGMORE: Your simulations focussed on isolated disks, but we heard yesterday that many disks formed next to high mass stars, so sit in a very strong uv-radiation field, and presumably regions of enhanced x-ray and CR fluxes. How would your result change for disks sitting in such environments?

BAI: The simulations we have so far have primarily focussed on the inner disk with $R \sim 1 - 20$ AU, where influence from the host protostar plays the dominant role. Environmental influence is known to be sometimes important in the outer disk (~ 100 pc scale), which we are working to incorporate in future works.

Origins: From the Protosun to the First Steps of Life
Proceedings IAU Symposium No. 345, 2018
B. G. Elmegreen, L. Viktor Tóth & M. Güdel, eds.
doi:10.1017/S1743921319003119

Resolving star and planet formation with ALMA

Per Bjerkeli[1], Daniel Harsono[2], Matthijs H. D. van der Wiel[3], Jon P. Ramsey[4], Lars E. Kristensen[5] and Jes K. Jørgensen[5]

[1]Department of Space, Earth and Environment, Chalmers University of Technology, Sweden
email: per.bjerkeli@chalmers.se

[2]Leiden Observatory, Leiden University, The Netherlands

[3]ASTRON, the Netherlands Institute for Radio Astronomy, The Netherlands

[4]Department of Astronomy, University of Virginia, United States

[5]Centre for Star and Planet Formation, Niels Bohr Institute & Natural History Museum of Denmark, University of Copenhagen, Denmark

Abstract. Disks around young stars are the sites of planet formation. As such, the physical and chemical structure of disks have a direct impact on the formation of planetary bodies. Outflowing winds remove angular momentum and mass and affect the disk structure and therefore potentially planet formation. Until very recently, we have lacked the facilities to provide the necessary observational tools to peer into the wind launching and planet forming regions of the young disks. Within the framework of the *Resolving star formation with ALMA* program, young protostellar systems are targeted with ALMA to resolve the disk formation, outflow launching and planet formation. This contribution presents the first results of the program. The first resolved images of outflow launching from a disk were recently reported towards the Class I source TMC1A (Bjerkeli *et al.* 2016) where we also present early evidence of grain growth (Harsono *et al.* 2018).

Keywords. ISM: jets and outflows, Protoplanetary disks, Stars: winds, outflows, Stars: formation

1. Introduction

The most prominent observational signature of the star formation process are outflows, which remove angular momentum from protostellar systems. Hence, they enhance the ability of stars to grow by accretion (the **why** of outflows). Outflows have in addition been invoked to explain issues such as the low star formation rate, and the low core to star efficiency.

Outflows are closely associated with protostellar disks (Cabrit *et al.* 1990). However, it remains to be understood **where, when** and **how** they are launched. Since the discovery of molecular outflows almost four decades ago (Snell *et al.* 1980), different theories and hypotheses for **how** and **where** they are launched have been presented. These can roughly be divided into three different categories: i) a 'stellar wind' launched from the poles of the protostar (Bouvier *et al.* 2014), ii) An 'X-wind' launching mechanism operating on a fraction of an au from the protostar and where the protostellar and disk magnetic fields interact (Shu *et al.* 1994; Shang *et al.* 2006), iii) the 'disk-wind' scenario where magnetic ejection takes place throughout an extended region of the disk (Blandford & Payne 1982; Königl & Pudritz 2000).

The program *Resolving star formation with ALMA* was initiated in 2015. The first observations were acquired towards the Class I source TMC1A in the Taurus molecular cloud and were published in Bjerkeli *et al.* (2016) and Harsono *et al.* (2018). TMC1A

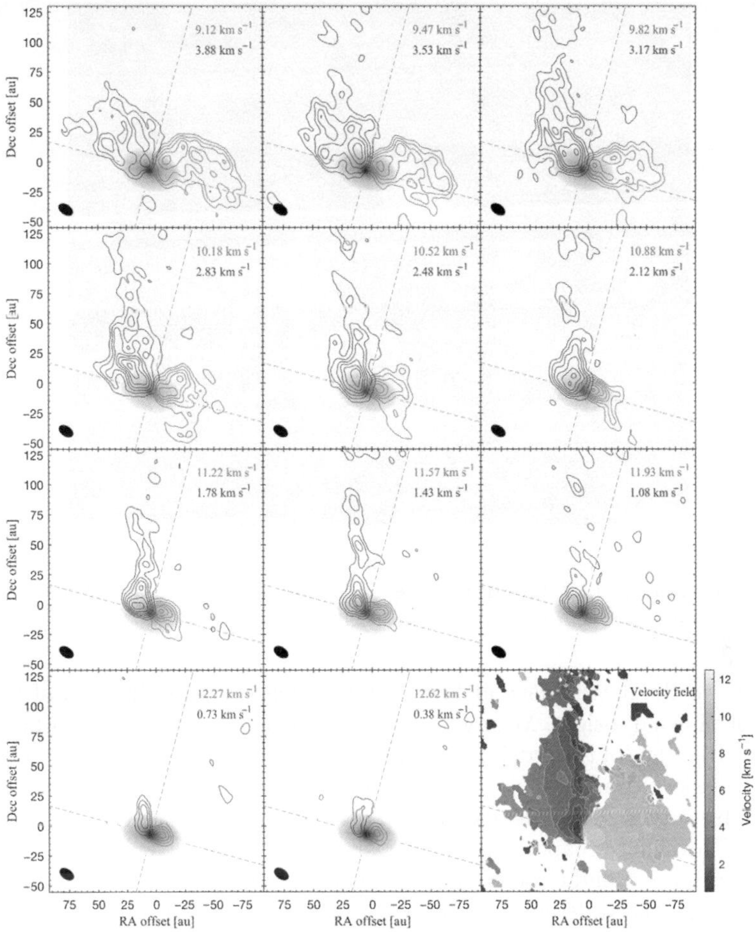

Figure 1. Channelmap of the ^{12}CO emisson in TMC1A Bjerkeli *et al.* (2016). Blue and red-shifted emission is with respect to the source velocity, 6.4 km s^{-1}. The plane of the disk and the outflow axis are indicated with dashed lines. In the lower right panel, the velocity field of the emission is presented.

harbors an outflow that is 6000 au long and a disk that is 200 au in size. Surrounding the system is a ~10 000 au diameter envelope that is collapsing towards the central star and disk. TMC1A is located sufficiently nearby (~140 pc) allowing for resolved observations of both the outflow and disk. ALMA was used for approximately four hours, in its most extended configuration (~6 au spatial resolution), to observe ^{12}CO (2–1), ^{13}CO (2–1), C^{18}O (2–1) and 1.3 mm continuum.

In addition to the TMC1A results presented in this contribution, we are now also including other targets in our study. The objectives of the program are not only to constrain the **when, where** and **how** of outflows that are responsible for mass-loss during early star formation but also to constrain the initial conditions for planet formation.

2. Resolved images of a disk wind in TMC1A

The ^{12}CO channel map (Fig. 1) shows integrated blueshifted and redshifted emission from the outflow in TMC1A, revealing substantial morphological changes with increasing velocity. The northeastern cavity wall is detected above 3σ and at velocities greater than 2 km s^{-1} with respect to $\upsilon_{\rm LSR}$. Outflowing gas is detected out to a distance of more than 100 au from the central source. Already from visual inspection it is clear that there is

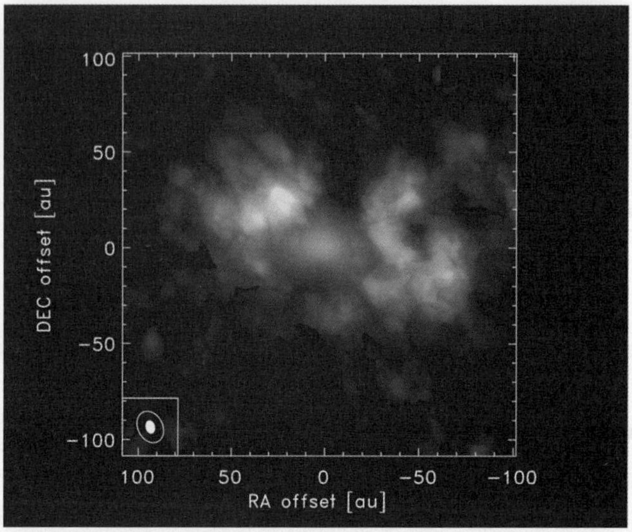

Figure 2. Colours represent integrated ^{13}CO (green) and C^{18}O (blue) emission and dust continuum (red) towards TMC1A (Harsono *et al.* 2018). There is a lack of emission from CO isotopologues and a supression in the continuum emission inside 20 au.

emission close to the disk surface at large radii, which suggests the bulk of the emission can not be of 'X-wind' origin. Meanwhile, only one side of the blueshifted outflow cavity wall and only fractions of the redshifted component to the south are visible. We believe the absence of emission from the north-western and south-eastern cavity walls is a result of the emission from these components coinciding with the velocity of the foreground cloud absorption. The velocity map of the outflow shows an 'onion-layered' pattern, where the highest velocities are found close to the outflow axis. From a careful comparison between the emission from the different CO isotopologues we conclude that the emission from ^{12}CO is from the outflow.

Two different methods are used to determine the outflow launching radii. First, launching radii are fit from the geometry of the emission, yielding launching radii between 5 and 25 au on the disk. The second method is to use magnetohydrodynamic wind theory following Anderson *et al.* (2003). Using this method, we derive launching radii between 1 and 19 au, i.e. consistent with the geometric calculation. Hence we conclude that at least some of the outflow is launched from the disk at radii sufficiently large to be directly resolved in our observations. It is important to point out that this does not rule out the possible existence of a co-eval 'X-wind' operating on smaller scales than probed by these observations.

Since the discovery of the extended disk wind in TMC1A in 2016, similar observations toward other protostellar systems have been presented by other groups. For example HH 212 is particularly interesting since the SiO emission hints that launching takes place very close to the central protostar, i.e. within 0.05 au (Lee *et al.* 2017).

3. Evidence for the start of planet formation in TMC1A

Dust growth is the first step toward planet formation. Such growth has been inferred previously through dust spectral index observations, but only towards systems older than a few million years (Testi *et al.* 2014; Miotello *et al.* 2014). Meanwhile, most observed exoplanets have masses larger than the mass reservoir available in such disks. This alone suggests that grain growth should start very early in protoplanetary disks.

In the case of TMC1A, dust continuum observations do not reveal any gaps or rings, e.g., like the case of HL Tau (ALMA partnership *et al.* 2015). However, a shoulder is clearly visible in the flux density distribution (Supplementary Fig. 1 of Harsono *et al.* 2018). This shoulder was detected already in the initial reduction of the dataset, but the unknown nature of it motivated us to improve the calibration procedure and to include a dataset that was previously omitted due to bad weather. The strength of the shoulder varies slightly in azimuth angle but remains at a fairly constant radius of $\sim$20 au. Inside 20 au, there is, in addition, a striking lack of CO isotopologue emission (Fig. 2). To analyze the lack of emission, we compare the observations to several different disk models. We find that a canonical model with micron sized grains can not reproduce the observed lack of emission, because optical depths would not be high enough. We also conclude that a very massive disk (that could reproduce the hole in emission) is very unlikely because such a disk would be dynamically unstable on short timescales. Deep continuum observations of TMC1A do not show any azimuthal or radial variations consistent with a gravitationally unstable disk. An unusually flat and cold disk structure does produce a lack of emission in the central regions, but this is in contrast with the temperature structure of a disk like TMC1A. We find, that the only way to reproduce the lack of emission is instead to include a population of larger grains in the models that are at least 1 mm in size. This finding is important, since the TMC1A disk is sufficiently young and massive enough to form multiple Jupiter sized planets. This evidence for a substantial reservoir of larger grains in such a young disk therefore implies that planet formation starts very early.

Acknowledgement

P.B. is supported by the Swedish Research Council through contract 2017-04924. D.H. is supported by EU ERC Advanced Grant 291141 CHEMPLAN and by a KNAW professor prize awarded to E. van Dishoeck. The group of J.K.J. is supported by the European Research Council (ERC) under the European Unions Horizon 2020 research and innovation programme (grant agreement no. 646908) through ERC Consolidator Grant S4F.

References

ALMA partnership, *et al.* 2015, *ApJL*, 808, 3
Anderson, J. M, Li, Z. Y., Krasnopolsky, R., *et al.* 2003, *ApJL*, 590, L107
Bjerkeli, P., van der Wiel, M. H. D., Harsono, D., *et al.* 2016, *Nature*, 540, 406
Blandford, R. D. & Payne, D. G 1982, *MNRAS*, 199, 883
Bouvier, J, Matt, S. P, Mohanty, S., *et al.* 2014, *Protostars and Planets VI*, 433
Cabrit, S., Edwards, S., Strom, S. E, & Strom, K. M 1990, *ApJ*, 354, 687
Harsono, D., Bjerkeli, P., van der Wiel, M. H. D., *et al.* 2018, *Nature Astronomy*
Lee, C. F., Ho, P. T. P, Li, Z. Y., *et al.* 2017, *Nature Astronomy*
Königl, A. & Pudritz, R. E 2000, *Protostars and Planets IV*, 759
Miotello, A., Testi, L., Lodato, G, *et al.* 2014, *A&A*, 567, 32
Shang, H., Allen, A., Li, Z. Y., *et al.* 2006, *ApJ*, 649, 845
Shu, F., Najita, J. Ostriker, E., *et al.* 1994, *ApJ*, 429, 781
Snell, R. L, Loren, R. B, & Plambeck, R. L 1980, *ApJL*, 239, L17
Testi, L., Birnstiel, T., Ricci, L., *et al.* 2014, *Protostars and planets VI*, 339

Discussion

PÉREZ: Was the spectral line data processed with continuum subtraction? Because the dust emission is optically thick and bright, if you subtract it from the line it may artificially create a hole.

BJERKELI: We explored different baseline subtraction methods (both in the uv domain and in the image domain) applied to different spectral windows. In all cases we see the $^{13}CO/C^{18}O$ hole, and for ^{12}CO we see the same morphology.

WALLER: What is the status + prospects for observing rotation in the outflows?

BJERKELI: Tentative measurements of rotation has been observed towards outflows for more than a decade. The situation has, however, changed dramatically in recent years, thanks to ALMA. Some recent examples where jet rotation is claimed are: HH212 (Lee *et al.* 2017), IRAS4C (Zhang *et al.* 2018).

WANDEL: The axis of the large scale Cycle 0 observation seems to be different from your small scale outflow. Is this real?

BJERKELI: The position angle of the flow axis is the same in both Aso *et al.* 2015 and Bjerkeli *et al.* 2016. The reason that they look different is that we only see one side of the outflow towards TMC1A, the part of the flow that is rotating towards us.

Origins: From the Protosun to the First Steps of Life
Proceedings IAU Symposium No. 345, 2018
B. G. Elmegreen, L. Viktor Tóth & M. Güdel, eds.

doi:10.1017/S1743921318008372

Unified Picture of Chemical Differentiation in Disk-Forming Regions of Low-Mass Protostellar Sources

Yoko Oya

Department of Physics, The University of Tokyo, 7-3-1, Hongo, Bunkyo-ku, Tokyo, Japan
email: `oya@taurus.phys.s.u-tokyo.ac.jp`

Abstract. Young low-mass protostellar sources are known to show significant chemical diversity in their envelopes at a few 1000s au scale; two distinct cases are hot corino chemistry and warm carbon-chain chemistry (WCCC). It is of great interest how the chemical diversity is inherited to chemistry of disk-forming regions. With the recent ALMA observations, we found that the chemical diversity in envelopes is indeed delivered into the disk-forming regions at a 100 au scale. Moreover, the chemical composition changes drastically from envelopes to disks. We also found sources with the hybrid chemical characteristics; both hot corino chemistry and WCCC occur in spatially separated parts of a single source. This hybrid case may be a common occurrence, while hot corinos and WCCC sources are regarded as distinct cases. This unified view of chemistry in disk-forming regions will be an important clue to tracing the chemical evolution from protostellar cores to protoplanetary disks.

Keywords. astrochemistry, ALMA, ISM: IRAS 16293−2422, L1527, B335, L483

1. Introduction

In order to explore our origin in the Universe, we need to know "what kinds of molecules are delivered from interstellar space into planets". So far, single-dish observations have shown that the envelope gas around the low-mass protostars has significant chemical diversity from source to source. More specifically, two distinct cases have been known (Sakai & Yamamoto 2013); hot corino sources rich in saturated organic molecules (e.g., CH_3OH, CH_3CHO), and WCCC sources rich in unsaturated carbon-chain molecules (e.g., C_4H).

2. Chemical Changes in Young Low-Mass Protostellar Sources

Recently, ALMA allows us to investigate the physical and chemical structures near the protostar. In a prototypical hot corino source IRAS 16293–2422 Source A (Bottinelli *et al.* 2004; Jørgensen *et al.* 2012), it has been reported that the surrounding gas shows rotation motion (Pineda *et al.* 2012; Favre *et al.* 2014). We investigated the chemical structure at a few 10s au scale in this source (Oya *et al.* 2016). We revealed that the gas surrounding the protostar consists of the infalling-rotating envelope and the Keplerian disk inside it. The boundary between them is called as the centrifugal barrier. We found that these three components are traced by different molecules. The OCS line traces only the envelope gas, while the H_2CS line traces both the envelope and the disk. The COM lines highlight the centrifugal barrier. Thus, a drastic chemical change is occurring from the infalling-rotating envelope to the Keplerian disk. IRAS 16293–2422 Source B shows a similar chemical change (Oya *et al.* 2018).

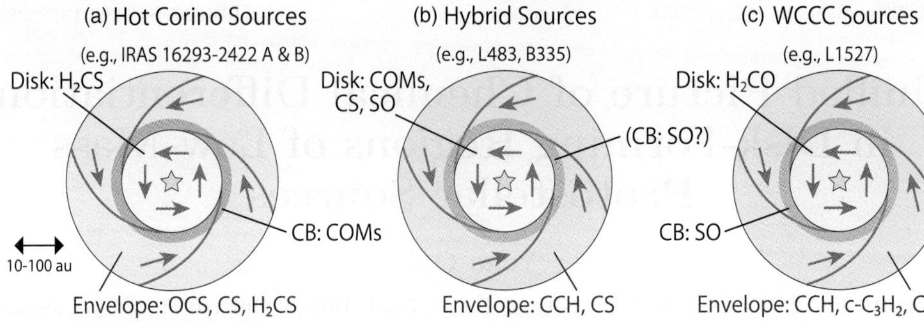

Figure 1. Chemical changes found in the young low-mass protostellar sources.

We found a similar chemical change in a prototypical WCCC source L1527 (Sakai *et al.* 2008, 2014) (Figure 1c). Interestingly, the chemical changes found in IRAS 16293–2422 and L1527 are shown by different molecular species depending on their chemical characteristics. Therefore, the chemical diversity reported for envelopes at 1000s au scale is indeed delivered into disk-forming regions at 10–100s au scale. These findings are important clues to constraining the initial condition of the disk chemistry.

3. Hybrid Chemical Characteristics

The above two sources are the distinct cases of the chemical variation. According to single-dish observations (e.g., Higuchi *et al.* 2018), most of other sources seem to have intermediate chemical characteristics. Then, how do they look like with ALMA?

B335 is a Bok globule (Keene *et al.* 1980) harboring a Class 0 low-mass protostar, and can be a standard template of chemical evolution. In the ALMA observation reported by Imai *et al.* (2016), the COM lines show a compact distribution concentrated around the protostar, while the unsaturated carbon-chain lines show an extended distribution. Figure 2 shows the spectra in two different areas; saturated COMs are prominent at the smaller scale, while unsaturated carbon-chain molecules are prominent at the larger scale. Thus, B335 shows both the chemical characteristics of hot corino chemistry and WCCC at different scales. Also, the Class 0 low-mass protostellar core L483 shows a similar chemical situation (Oya *et al.* 2017); saturated COMs trace the disk around the protostar (~100 au), while unsaturated carbon-chain molecules are prominent in the envelope gas and the bipolar outflow (~1000 au scale). Here, B335 and L483 are interpreted as the third case, the 'hybrid case', for the chemical diversity in disk-forming regions (Figure 1b).

4. Cause for the Chemical Diversity and the Chemical Change

The chemical structures observed in the disk-forming regions are consistent with a chemical model. Aikawa *et al.* (2018) reported a chemical model of a dynamically collapsing core. In their model, CH_3OH resides in the gas-phase near the protostar (~100 au). It sublimates from dust grains in the hot (>100 K) region. On the other hand, CH_4 resides in the gas-phase in the warm (>25 K) region (~1000 au), because CH_4 sublimates at a lower temperature than CH_3OH. Enhancement of CH_4 in the gas-phase triggers efficient production of carbon-chain molecules (WCCC). Thus, hot corino chemistry occurs near the protostar, while WCCC occurs in the outer region. Such a spatial separation is consistent with the observational results in the hybrid sources. This seems to be the basic picture of the chemical change in disk-forming regions (Figure 3b).

This picture will depend on the initial condition of the chemical composition of the grain mantle. If CH_4 is deficient in the grain mantle (Figure 3a), it does not sublimate

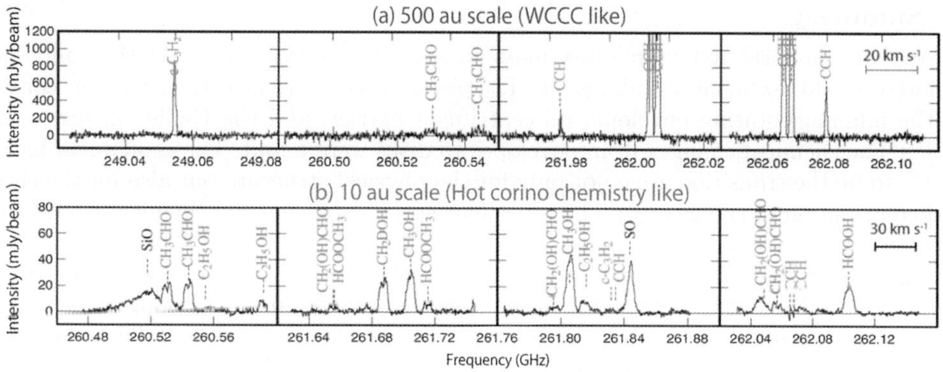

Figure 2. Spectra toward B335 averaged in the circles with the radius of (a) 500 au and (b) 10 au centered at the protostellar position (Imai *et al.* 2016).

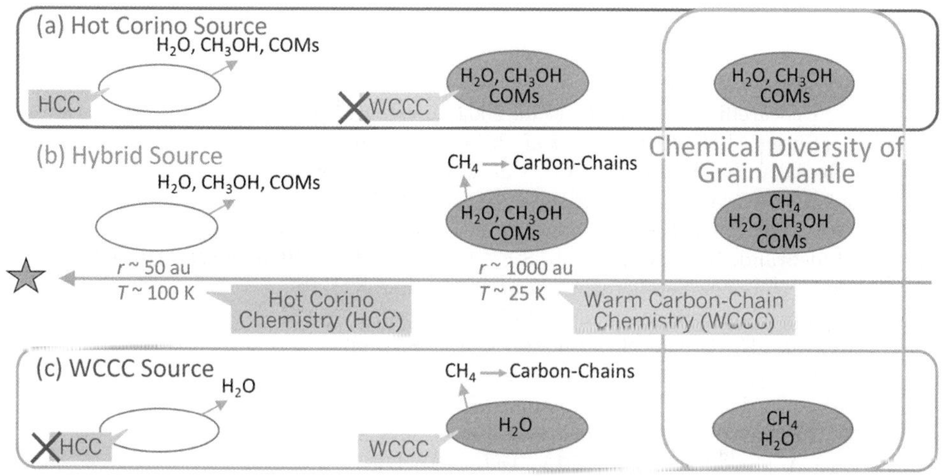

Figure 3. Schematic illustration of the scenario of the chemistry near the protostar. A dust grain falls from the right hand side toward the protostar on the left hand side. Panel (b) shows the basic picture. Panels (a) and (c) show the distinct cases deficient in CH_4 and COMs, respectively.

in the warm region, while water and COMs do in the hot region. This case results in hot corinos. On the other hand, if COMs are deficient (Figure 3c), CH_4 sublimates in the warm region and efficiently produces carbon-chain molecules. This case results in WCCC sources.

Next, why does the chemical diversity of grain mantle happen? It is proposed to depend on the duration time of the starless-core phase after the shielding of UV (Sakai & Yamamoto 2013). If the duration time is long enough compared to the time scale for chemical reactions ($\sim 3 \times 10^5$ yr), C atoms in the gas-phase are well converted to CO and deplete on dust grains. Then, CO will form CH_3OH and other COMs, resulting hot corino chemistry near the protostar. In contrast, if the duration time is too short, atomic carbon can deplete on dust grains as they are. Atomic carbon will be hydrogenated to CH_4 on the grain surface, and it will show WCCC. Of course the duration time can be intermediate; both atomic carbon and CO will exist on dust grains, and this case will result in hybrid sources. Although hot corinos and WCCC sources have extensively been studied so far because of their peculiar chemical characteristics, the intermediate case (i.e. the hybrid case) is most likely the standard case (Figure 3b).

5. Summary

We have investigated young low-mass protostellar sources with ALMA, and have reported the following new findings: (1) The physical structure near the protostar consists of the infalling-rotating envelope, its centrifugal barrier, and the Keplerian disk inside it. (2) The chemical changes from envelopes to disks were found. The centrifugal barrier seems to be the transition zone not only for the physical structure but also for the chemical composition of the gas. (3) The molecular species showing the chemical changes are different among sources.

The above results also raise new questions; (1) what is occurring around the centrifugal barrier ('transition zone')? (2) what is the chemical heritage from envelopes to disks? (3) what is the mechanisms causing the chemical diversity among sources? These questions will be addressed with our new ALMA large program 'FAUST' in the following years.

This work is in collaboration with Imai, M., Sakai, N., López-Sepulcre, A., Ceccarelli, C., Lefloch, B., and Yamamoto, S.

References

Aikawa, Y., Wakelam, V., Garrod, R.T., & Herbst, E. 2008, *ApJ*, 674, 993
Bottinelli, S., Ceccarelli, C., Neri, R., *et al.* 2004, *ApJL*, 617, L69
Favre, C., Jørgensen, J.K., Field, D., *et al.* 2014, *ApJ*, 790, 55
Higuchi, A.E., Sakai, N., Watanabe, Y., *et al.* 2018, *ApJS*, 236, 52
Imai, M., Sakai, N., Oya, Y., *et al.* 2016, *ApJL*, 830, L37
Jørgensen, J.K., Favre, C., Bisshop, S., *et al.* 2012, *ApJL*, 757, L4
Keene, J., Hildebrand, R.H., Whitcomb, S.E., & Harper, D.A. 1980, *ApJL*, 240, L43
Oya, Y., Sakai, N., López-Sepulcre, A., *et al.* 2016, *ApJ*, 824, 88
Oya, Y., Sakai, N., López-Sepulcre, A., *et al.* 2017, *ApJ*, 837, 174
Oya, Y., Moriwaki, K., Onishi, S., *et al.* 2018, *ApJ*, 854, 96
Pineda, J.E., Maury, A.J., Fuller, G.A., *et al.* 2012, *A&A*, 544, L7
Sakai, N., Sakai, T., Hirota, T., & Yamamoto S. 2008, *ApJ*, 672, 371
Sakai, N., & Yamamoto, S. 2013, *Chemical Review*, 113, 8981
Sakai, N., Oya, Y., Sakai, T., *et al.* 2014, *ApJL*, 791, L38

Discussion

LINSKY: Some molecules will be ionized as the accretion from the interstellar medium approaches the inner disk. Ionized and neutral molecules can interact by dissociative recombination reactions. Including this chemistry may explain the different locations of chemical species around protostars.

OYA: It will be interesting to see how ions are distributed in the disk-forming regions. But, so far, our observations focused on neutral species and we have not yet worked on ions. Our new ALMA project 'FAUST' contains some ion lines (e.g., HCO^+, N_2H^+, and their isotopes). We will reveal the distributions of these ions in the near future.

MAURY: If COMs (complex organic molecules) are detected at small disk-like scales in B335, do you detect Keplerian-like motions from COMs kinematics?

OYA: It is unclear whether B335 harbors a Keplerian disk or not at the current stage. Although COMs trace the disk in L483, their detailed emitting region can depend on sources. Observations of the COM lines in B335 at a higher angular-resolution are awaited.

Origins: From the Protosun to the First Steps of Life
Proceedings IAU Symposium No. 345, 2018
B. G. Elmegreen, L. Viktor Tóth & M. Güdel, eds.

doi:10.1017/S174392131900228X

The physical and chemical properties of planet forming disks

Inga Kamp

Kapteyn Astronomical Institute, University of Groningen, PO Box 800, 9700 AV Groningen, The Netherlands
email: kamp@astro.rug.nl

Abstract. VLT instruments and ALMA have revolutionized in the past five years our view and understanding of how disks turn into planetary systems. They provide exquisite insights into non-axisymmetric structures likely closely related to ongoing planet formation processes. The following cannot be a complete review of the physical and chemical properties of disks; instead I focus on a few selected aspects. I will review our current understanding of the physical properties (e.g. solid and gas mass content, snow and ice lines) and chemical composition of planet forming disks at ages of 1-few Myr, especially in the context of the planetary systems that are forming inside them. I will highlight recent advances achieved by means of consistent multi-wavelength studies of gas AND dust in protoplanetary disks.

Keywords. Astrochemistry; planetary systems: protoplanetary disks, formation; solar system: formation, comets, asteroids

1. Introduction

The physical properties have important consequences for the formation of planets such as how much 'solid' surface density is available in the various planet forming regions, the gravitational stability of disks, as well as the total disk gas mass available to keep planet formation ongoing at a few Myr. I will review here the icelines in disks — both from the modeling perspective as well as recent observational insights — and gas masses, mostly summarizing the current caveats and debates on the interpretation of CO rotational line observations and the prospects for using HD instead in future missions.

The chemical properties have implications for the composition of forming planets. The disk composition determines the composition of planetsimals and hence that of growing planetary cores, e.g. the water content as a function of distance from the star and also the C/O ratio of planets and the composition of primary/secondary atmospheres. Two topics that received quite some attention in the past few years are the aspect of heritage versus in-situ for the chemical composition of planet forming disks and the role of (warm) surface chemistry for the 'solid' composition, ice and refractory material among them phyllosilicates; I focus here on these two.

These selected topics all demonstrate the close coupling between gas and dust in planet forming disks and its high relevance for the interpretation of observations. This coupling happens not only thermally through collisions, but also through grain opacities (e.g. non-thermal desorption processes), IR pumping for molecular excitation, gas pressure determining dust grain settling and migration, and varying effective grain surface area for chemical reactions throughout the disk. A proper interpretation of the rich existing and upcoming disk observations requires state-of-the-art disk models to include these coupling terms.

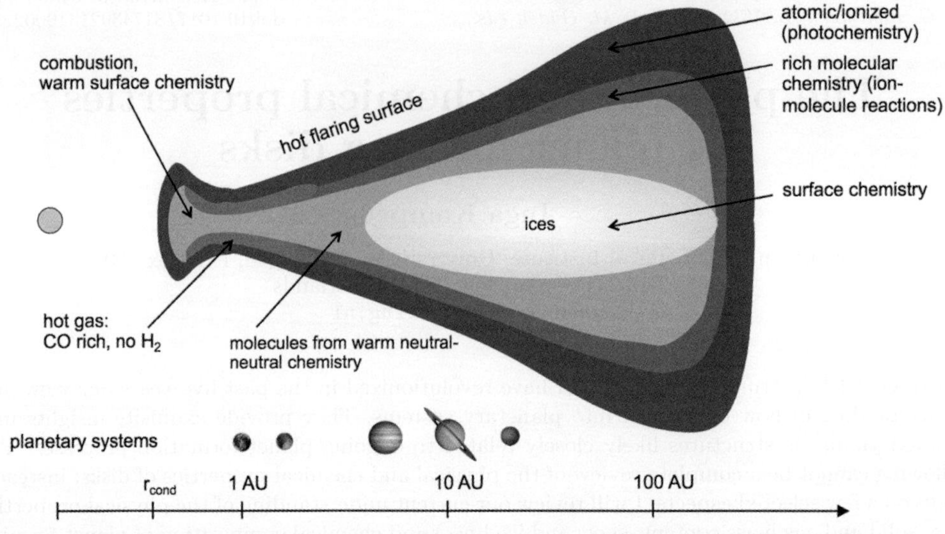

Figure 1. Chemical structure of a planet forming disk outlining the layered structure (atomic/molecular/ice) and the various types of chemistry (photochemistry, ion-molecule chemistry, neutral-neutral chemistry, combustion chemistry) relevant in the different parts of the disk.

2. Disk structure

Our knowledge of disk structure is largely based on seminal papers of steady state accretion disks (Pringle 1981) and irradiated disks (Kenyon & Hartmann 1987; Chiang & Goldreich 1997). Gas and dust temperatures decouple in the disk surface (Kamp & Dullemond 2004; Jonkheid *et al.* 2004). While dust is mainly heated by the bulk luminosity of the central star, the gas thermal balance is driven by far-UV (FUV) and X-rays from the central star, which are absorbed and scattered by dust grains in the disk surface (see Dullemond *et al.* 2007, for a review of disk structure). X-rays mainly interact with the gas that similarly absorbs and scatters them (Nomura *et al.* 2007; Aresu *et al.* 2011); if grains have grown much larger than the ISM, dust opacities become only important for high energy X-rays ($E \gtrsim 5 - 10$ eV, Rab *et al.* 2018).

Cosmic rays (CR) are less relevant for heating the gas, but they drive chemical processes close to the disk midplane such as non-thermal desorption and cosmic ray ionization. Disks are layered structures (e.g. Aikawa *et al.* 2002; Kamp & Dullemond 2004, see Fig. 1) that can be divided into

(a) a hot ionized/atomic layer in which photochemistry dominates
(b) a warm molecular layer in which ion-molecule chemistry dominates
(c) a cold icy midplane in which CR ionization and surface chemistry dominate.

Surface chemistry encompasses a wide range of non-thermal desorption processes and has been reviewed recently by Cuppen *et al.* (2017). In addition to these three disk layers, Lee *et al.* (2010) discussed implications of combustion chemistry for the innermost parts of protoplanetary disks. A more general review of chemical evolution in disks including observations can be found in Bergin *et al.* (2007) and Dutrey *et al.* (2014).

Planet forming disks are shaped by dynamical processes such as accretion onto the central star, winds (radiation or magnetically driven), gas turbulence, vertical settling, collisional growth and radial migration of dust grains, as well as disk-planet interaction (e.g. Weidenschilling 1977; Miyake & Nakagawa 1995; Dullemond & Dominik 2004). Alexander (2008), Armitage (2011) and Birnstiel *et al.* (2016) provide recent reviews of the various processes.

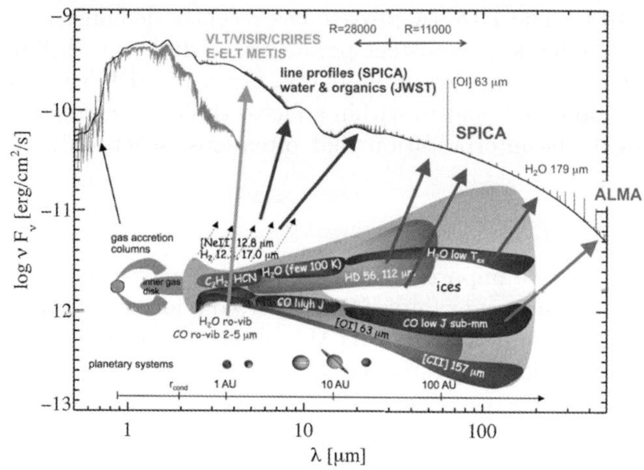

Figure 2. An overview of line emission from near-IR to submm wavelengths from planet forming disks and existing and upcoming instrumentation to detect it. The model SED is that of a typical T Tauri disk using two different spectral resolutions ($R = 28\,000$ and $R = 11\,000$ relevant for the planned SMI and SAFARI instruments onboard the proposed SPICA mission).

Over the past decade, there has been increasing observational evidence that confirms the general picture described above. Avenhaus *et al.* (2018) found in their VLT/SPHERE survey of T Tauri disks that the scattered light surface displays a typical flaring structure; previously this had been traced with resolved PAH emission for the Herbig disk around HD 97048 (Lagage *et al.* 2006). In addition, dust settling has now been directly seen with ALMA. Mulders & Dominik (2012) show that in a self-consistent settling model, mm-sized dust grains are concentrated at heights of $z/r \lesssim 0.02$. Modeling the high spatial resolution HL Tau ALMA data (ALMA Partnership *et al.* 2015), Pinte *et al.* (2016) demonstrate that the similarity in gap width between major- and minor-axis requires a typical z/r of $\lesssim 0.01$, consistent with a very low turbulence, $\alpha = 3 \times 10^{-4}$.

Understanding the physical conditions of the gas in these planet forming disks is challenging because gas lines emit according to their respective excitation conditions which depend to first order on the gas temperature, the density and the optical depth. Figure 2 illustrates how emission from hot gas in the inner disk occurs preferentially at near- to mid-IR wavelengths. For these lines, the dust optical depth at the corresponding wavelength determines how deep we can peer into these disks. Molecules such as CO and water, which possess a wide range of lines across the near- to far-IR wavelength range, can scan the surface and also probe successively deeper into the disk as the dust becomes more optically thin towards longer wavelengths. Since each line probes only a small region in such a planet forming disk, we need a suite of instruments to trace the gas across the entire disk such as VLT/VISIR and CRIRES (ELT/METIS in the future) for the hot (few 100 - few 1000 K) gas at short wavelengths, JWST/NIRSPEC and MIRI in the near- to mid-IR, the proposed SPICA mission ranging from mid- to far-IR and bridging the gap to detecting cold (few 10 K) gas with ALMA at submm wavelengths.

We are now in an era where we have tens of disks with very rich multi-wavelengths and multi-kind observational data: X-ray to mm wavelengths and photometric, spectroscopic, spectro-astrometric, imaging and interferometric datasets for both gas and dust. Disk modeling has caught up with the challenge to interpret such complex datasets in the form of various radiation thermo-chemical models such as ProDiMo (Woitke *et al.* 2009; Kamp *et al.* 2010; Thi *et al.* 2011), DALI (Bruderer *et al.* 2012), and ANDES

(Akimkin *et al.* 2013). The DIANA project has recently demonstrated for a sample of 14 objects that individual self-consistent parametrized dust+gas disk models are able to match the entirety of such complex datasets (Woitke *et al.* 2019, e.g. line fluxes across all wavelengths match observed ones to within a factor three). Such modeling can break the large degeneracies in the interpretation that often exist when only individual datasets are analysed.

3. Disk physical properties

These properties provide important constraints for planet formation models. The following paragraphs present two examples that received a lot of attention over the last few years, especially in the context of new ALMA observations: The location of icelines in disks and the determination of gas masses.

3.1. *Icelines*

Icelines in disks are not sharp transitions. Different binding energies for water on bare and icy grains as well as turbulent mixing lead to so-called 'snow borders' in which bare grains and icy grains coexist. Marseille & Cazaux (2011) explore in their models the radial width of such 'snow borders' and find typical values of $0.03 - 0.2$ au. Using a completely different modeling approach, Krijt *et al.* (2016) explore the vertical distribution of water vapor and ice in the presence of dust grain growth, settling and turbulent mixing. At low turbulence levels ($\alpha \leqslant 10^{-3}$), the snowline resides about one pressure scale hight deeper in the disk compared to simple equilibrium models. Ciesla & Cuzzi (2006) and Furuya & Aikawa (2014) showed that — besides relocating snowlines — turbulent mixing and transport processes have large effects on the chemical composition of disks, especially the local abundances of water, ices and complex organic molecules. It is however notable that for example vertical column densities of gaseous CO in the outer disk ($\gtrsim 50$ au) are hardly affected by turbulent mixing; this confirms earlier results by Willacy *et al.* (2006).

At lower temperatures (< 100 K), in the outer disk, photodesorption plays an important role in setting the height of the icelines. Ceccarelli & Dominik (2005) postulated this effect to explain the high level of deuterated cold water emission in DM Tau; later Herschel/HIFI observations detected this "photodesorbed" water reservoir through the ground state lines of the main isotopologue H_2O (Hogerheijde *et al.* 2011; Podio *et al.* 2013). Laboratory and thermo-chemical disk modeling efforts subsequently quantified the process (e.g. Fraser *et al.* 2001; Öberg *et al.* 2009; Kamp *et al.* 2013).

Very recently, Schwarz *et al.* (2016) and Pinte *et al.* (2018) directly observed the effect of photodesorption using ALMA CO isotopologue observations in the disks of TW Hya and IM Lup respectively. The CO temperature plateaus at 21 K at intermediate radii in both datasets; this is consistent with laboratory measurements of CO binding energies. In addition, the radial extent and location of CO as derived from the high spatial resolution channel maps can only be reproduced with models that take into account CO photodesorption (Pinte *et al.* 2018). The depth to which photons can penetrate these disks and drive photodesorption depends strongly on the local dust properties, demonstrating once more the high relevance of consistent dust and gas models in interpreting the new data.

3.2. *Gas masses*

For decades, many continuum surveys measured disk masses as a key input for the "planet forming potential" of these disks (e.g. Beckwith *et al.* 1990; Andre & Montmerle 1994; Andrews & Williams 2005; Andrews & Williams 2007). Photometric measurements at long wavelength, where the emission is thought to be optically thin, have been used together with a representative dust emissivity (opacity) to quantify disk dust masses†. To probe how disks evolve into planetary systems, the gas-to-dust mass ratio was often taken as an indicator, primordial disks being rich in gas with a gas-to-dust ratio of ~ 100 and debris disks being gas-poor. Several gas surveys (e.g. FEPS, GASPS) aimed at deriving this key quantity for various star forming regions using tracers such as mid-IR line emission (warm gas) and CO submm lines (cold gas) and the [O I] 63 μm fine structure line (e.g. Pascucci *et al.* 2006; Dent *et al.* 2013).

With the advent of more sensitive submm interferometers such as SMA and ALMA, Williams & Best (2014) suggested to use the two rarer isotopologues ^{13}CO and C^{18}O to determine cold gas masses in disks (diagnostic CO isotopologue mass diagram) and found gas-to-dust ratios below 20 in several Taurus disks. The work of Willacy *et al.* (2006) and Furuya & Aikawa (2014) had shown that low levels of turbulence as detected in several planet forming disks (e.g. Flaherty *et al.* 2018, for TW Hya) should not affect the CO abundance in the outer disk. Miotello *et al.* (2016) show that isotope selective photodissociation can affect the gas measurements by an order of magnitude. Still ALMA surveys report gas-to-dust mass ratios well below 100 in 1-3 Myr old star forming regions (e.g. Miotello *et al.* 2017, for Lupus). Recently, Woitke *et al.* (2016) and Kamp *et al.* (2018) quantify the effects of changing dust properties and disk outer edges (tapering off) for the CO isotopologue ratio and the diagnostic CO isotopologue mass diagram. The DIANA multi-wavelength disk modeling project has shown that a sample of ten well known disks span over two orders of magnitude in CO isotopologue line fluxes, while the gas-to-dust mass ratio in the thermo-chemical dust+gas models, that are consistent with an even wider observational dataset, span gas-to-dust mass ratios between 50 and 450 (Kamp *et al.* 2018). It is important to note that the "main sequence" of CO isotopologue lines is consistently found in all disk modeling works (Williams & Best 2014; Miotello *et al.* 2016; Kamp *et al.* 2018), but disk mass is not the only parameter shifting line fluxes up and down this main sequence; for example the maximum grain size or the power law index of the grain size distribution also play a key role. Using the now available multi-band medium resolution ALMA surveys of several star forming regions, the underlying dust properties can be constrained observationally and consistent gas+dust modeling should be employed to derive better estimates of gas masses from CO isotopologue data. However, the statistics are often limited since many of the surveys are fairly shallow and the number of sources detected in both ^{13}CO and C^{18}O remains small compared to the total number of sources detected in the dust continuum.

Since the HD abundance depends only on the primordial D abundance, an alternative to determine warm gas masses is the use of the two lowest excitation HD lines at 112 and 56 μm ($T_{\rm ex} = 128.5$ and 384.3 K respectively). Bergin *et al.* (2013) and later McClure *et al.* (2016) report HD masses or upper limits for seven T Tauri disks. Interestingly, the gas masses derived from the warm HD lines via disk modeling are often larger than those derived via the CO lines. Trapman *et al.* (2017) use thermo-chemical disk models to explore the predictive power of these HD lines for disk gas masses and show that deep HD surveys with the proposed JAXA/ESA SPICA mission can detect disk gas masses down to 10^{-5} M$_\odot$ for typical T Tauri disks in nearby low mass star forming regions.

4. Disk chemical properties

These properties provide constraints for the composition of the planets forming within these disks. The following paragraphs review two controversial issues relating to these

† Note that dust mass refers here to the solid mass in small dust grains up to mm-sizes; the solid mass in larger planetesimals and even protoplanets does not contribute significantly to the mm fluxes of disks and can hence not be measured this way.

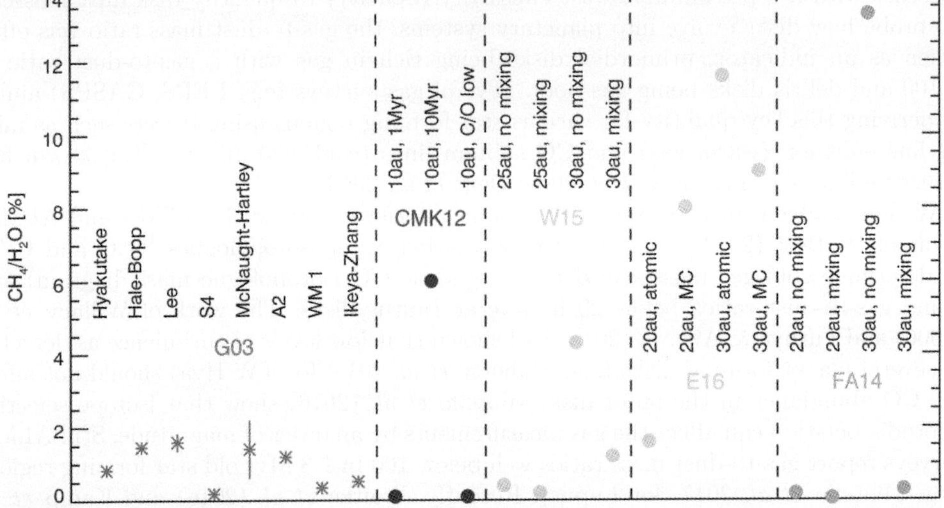

Figure 3. Comparsion between a collection of methane/water ice abundance ratios (with error bars) in Oort cloud comets (Gibb *et al.* 2003) and a sample of disk models: Chaparro *et al.* (2012 blue, C/O= 0.45, low C/O= 0.16), Willacy *et al.* (2015, cyan), Eistrup *et al.* (2016, green), and Furuya & Aikawa (2014, orange). Figure updated from a previous version by E. Gibb (private communication).

chemical properties: The question of heritage versus in-situ for the chemical composition of disks and the relevance of surface chemistry in disks.

4.1. *Heritage versus in-situ*

The thermal history of the material forming a disk is strongly affected by processes such as stellar outbursts, and grain growth during the collapse/disk formation phase. The collapse of a dense core is dynamically complex involving magnetic fields and possibly instabilities. A showcase example of this complexity is the ALMA high spatial resolution imaging of the binary protostar system IRAS 16293-2422 (PILS, Jørgensen *et al.* 2016).

Several works studied the ice processing during the disk formation phase (e.g. Chick & Cassen 1997 1997; Visser *et al.* 2009; Visser *et al.* 2011; Furuya *et al.* 2012; Drozdovskaya *et al.* 2014). The comparison of comets and outer disks composition can provide important clues as to the amount of additional processing within the disk and the thermal history of ices. Willacy *et al.* (2015) show that most $(D/H)_{water}$ measurements are only compatible with disk models that include mixing processes. Cleeves *et al.* (2014) go one step further using a comprehensive range of disk ionisation sources and a detailed chemical model of the protosolar disk to claim that a significant fraction of HDO in the disk stems from accretion and inheritance of interstellar ices. Figure 3 shows a comparison between the methane/water abundance ratio in cometary ices and the same ratio in a series of disk models. It becomes clear that this ice ratio depends strongly on the initial condition (atomic versus molecular), the elemental abundances (not only C and O, but also other metals as electron sources), mixing, as well as the chemical age of the disk. The no-mixing models and the young chemical ages give a better agreement with the observed CH_4/H_2O ratio than mixing/old models (contrary to D/H). The initial composition (atomic or molecular cloud values) as well as the element abundances (C/O) play a large role in setting the ice abundances in the outer disk (beyond 10 au). However, we are only at the beginning since many other effects also impact the ice composition, especially also the relative abundances of ices, such as layering and diffusion within ices, grain growth and migration, and the role of irradiation within ices.

4.2. *(Warm) Surface chemistry*

Surface chemistry itself is less well understood compared to gas-phase chemistry and the status and current limitations are comprehensively reviewed by Cuppen *et al.* (2017). Surface chemistry is often invoked to explain the formation of COMs (complex organic molecules with six or more atoms, at least two of which are C, N, O). Ciesla & Sandford (2012) form COMs via mixing of icy grains in disks, subsequently exposing those icy grains to different levels of UV dosage. Drozdovskaya *et al.* (2016) propose that the abundances of COMs are strongly affected by the dynamical history of the disk build-up evolution (pure infall versus infall+viscous spreading). Walsh *et al.* (2014) show that even without taking details of disk build-up into account, significant abundances of COMs can be synthesized in disks through grain surface chemistry.

But also warm surface chemistry has recently received more attention in studying the composition of the inner few au as well and the link to terrestrial planets and asteroids. Combustion chemistry can burn carbonaceous dust inside ~ 1 au (e.g. Lee *et al.* 2010; Anderson *et al.* 2017), although the level of carbon depletion may be limited by the efficiency of radial and vertical mixing processes, which replenish carbon rich material faster than burning can proceed (Klarman *et al.* 2018).

A second process driven by warm surface chemistry is the transformation of silicates into phyllosilicates (sheet silicates with intercalated water). Early work by Fegley & Prinn (1989) dismissed the relevance of this process based on the long timescales it required compared to the lifetime of disks. More recent works by Stimp *et al.* (2006) and Muralidharan *et al.* (2008) using realistic silicate surfaces and adsorption energies find higher efficiencies in the solar nebula around the region where Earth has formed. We recently revisited these processes using both Monte Carlo simulations to assess the efficiency of water adsorption under high temperature/pressure conditions inside ~ 1 au (D'Angelo *et al.* 2019) and the subsequent surface chemistry that transforms part of the silicate grains into phyllosilcate material (Thi *et al.* 2018). Inside the water snowline, this process can proceed within 1 Myr to produce grains that contain a significant fraction of water (up to a few %). If the process occurs before grains have substantially grown, this is an alternative pathway to incorporate water into the proto-Earth, i.e. contribute to the water content within the Earth mantle.

5. Outlook

The future has great prospects to employ disk observations in conjunction with detailed disk models (radiation, physics, chemistry, dynamics) to achieve a more comprehensive view on planet formation. ALMA and VLT (now and in the future) will provide molecular channel maps to confront thermo-chemical disk models (gas+dust), to characterize rings and non-axisymmetric gas structures (embedded protoplanets, vortices), to detect circumplanetary disks, and study inner disk dispersal processes. In the near future, SOFIA/HIRMES (~ 2019) will detect thermal ice features and HD in a about a dozen disks showcasing the potential of such data. From 2021 onwards, JWST will provide data to study the warm gas+dust composition/evolution in the inner ~ 10 au. ELT/METIS (beyond ~ 2025) will provide detailed kinematics (channel maps) in the inner ~ 10 au of disks, and study the formation and properties of circumplanetary disks. Further ahead beyond 2030, the proposed SPICA mission can study the trail of water, ice and solids during the entire planet formation phase through large unbiased surveys of water, HD, ices, and dust mineralogy well into the debris disk stage.

References

Aikawa, Y., van Zadelhoff, G. J., van Dishoeck, E. F., & Herbst, E. 2002, *A&A*, 386, 622

Akimkin, V., Zhukovska, S., Wiebe, D., *et al.* 2013, *ApJ*, 766, 8

Alexander, R. 2008, *NewAR*, 52, 60

ALMA Partnership, Brogan, C. L., Pérez, L. M., *et al.* 2015, *ApJL*, 808, L3

Anderson, D. E., Bergin, E. A., Blake, G. A., *et al.* 2017, *ApJ*, 845, 13

Andre, P. & Montmerle, T. 1994, *ApJ*, 420, 837

Andrews, S. M. & Williams, J. P. 2005, *ApJ*, 631, 1134

Andrews, S. M. & Williams, J. P. 2007, *ApJ*, 671, 1800

Aresu, G., Kamp, I., Meijerink, R., *et al.* 2011, *A&A*, 526, A163

Armitage, P. J. 2011, *Annu. Rev. Astron. Astrophys.*, 49, 195

Avenhaus, H., Quanz, S. P., Garufi, A., *et al.* 2018, *ApJ*, 863, 44

Beckwith, S. V. W., Sargent, A. I., Chini, R. S., & Guesten, R. 1990, *AJ*, 99, 924

Bergin, E. A., Aikawa, Y., Blake, G. A., & van Dishoeck, E. F. 2007, *Protostars and Planets V*, 751

Bergin, E. A., Cleeves, L. I., Gorti, U., *et al.* 2013, *Nature*, 493, 644

Birnstiel, T., Fang, M., & Johansen, A. 2016, *SSRv*, 205, 41

Bruderer, S., van Dishoeck, E. F., Doty, S. D., & Herczeg, G. J. 2012, *A&A*, 541, A91

Ceccarelli, C. & Dominik, C. 2005, *A&A*, 440, 583

Chaparro Molano, G. & Kamp, I. 2012, *A&A*, 537, A138

Chiang, E. I. & Goldreich, P. 1997, *ApJ*, 490, 368

Chick, K. M. & Cassen, P. 1997, *ApJ*, 477, 398

Ciesla, F. J. & Cuzzi, J. N. 2006, *Icarus*, 181, 178

Ciesla, F. J. & Sandford, S. A. 2012, *Science*, 336, 452

Cleeves, L. I., Bergin, E. A., Alexander, C. M. O., *et al.* 2014, *Science*, 345, 1590

Cuppen, H. M., Walsh, C., Lamberts, T., *et al.* 2017, *SSRv*, 212, 1

D'Angelo, M., Cazaux, S., Kamp, I., Thi, W.-F., & Woitke, P. 2019, *Astronomy & Astrophysics*, 622, A208

Dent, W. R. F., Thi, W. F., Kamp, I., *et al.* 2013, *PASP*, 125, 477

Drozdovskaya, M. N., Walsh, C., van Dishoeck, E. F., *et al.* 2016, *MNRAS*, 462, 977

Drozdovskaya, M. N., Walsh, C., Visser, R., Harsono, D., & van Dishoeck, E. F. 2014, *MNRAS*, 445, 913

Dullemond, C. P. & Dominik, C. 2004, *A&A*, 421, 1075

Dullemond, C. P., Hollenbach, D., Kamp, I., & D'Alessio, P. 2007, *Protostars and Planets V*, 555

Dutrey, A., Semenov, D., Chapillon, E., *et al.* 2014, *Protostars and Planets VI*, 317

Eistrup, C., Walsh, C., & van Dishoeck, E. F. 2016, *A&A*, 595, A83

Fegley, Jr., B. & Prinn, R. G. 1989, in The Formation and Evolution of Planetary Systems, ed. H. A. Weaver & L. Danly, 171-205

Flaherty, K. M., Hughes, A. M., Teague, R., *et al.* 2018, *ApJ*, 856, 117

Fraser, H. J., Collings, M. P., McCoustra, M. R. S., & Williams, D. A. 2001, *MNRAS*, 327, 1165

Furuya, K. & Aikawa, Y. 2014, *ApJ*, 790, 97

Furuya, K., Aikawa, Y., Tomida, K., *et al.* 2012, *ApJ*, 758, 86

Gibb, E. L., Mumma, M. J., Dello Russo, N., DiSanti, M. A., & Magee-Sauer, K. 2003, *Icarus*, 165, 391

Hogerheijde, M. R., Bergin, E. A., Brinch, C., *et al.* 2011, *Science*, 334, 338

Jonkheid, B., Faas, F. G. A., van Zadelhoff, G.-J., & van Dishoeck, E. F. 2004, *A&A*, 428, 511

Jørgensen, J. K., van der Wiel, M. H. D., Coutens, A., *et al.* 2016, *A&A*, 595, A117

Kamp, I., Antonellini, S., Carmona, A., Ilee, J., & Rab, C. 2018, in The Cosmic Wheel and the Legacy of the AKARI Archive: From Galaxies and Stars to Planets and Life, ed. T. Ootsubo, I. Yamamura, K. Murata, & T. Onaka, 89-96

Kamp, I. & Dullemond, C. P. 2004, *ApJ*, 615, 991

Kamp, I., Thi, W.-F., Meeus, G., *et al.* 2013, *A&A*, 559, A24

Kamp, I., Tilling, I., Woitke, P., Thi, W., & Hogerheijde, M. 2010, *A&A*, 510, A260000

Kenyon, S. J. & Hartmann, L. 1987, *ApJ*, 323, 714
Klarmann, L., Ormel, C. W., & Dominik, C. 2018, *A&A*, 618, L1
Krijt, S., Ciesla, F. J., & Bergin, E. A. 2016, *ApJ*, 833, 285
Lagage, P.-O., Doucet, C., Pantin, E., *et al.* 2006, *Science*, 314, 621
Lee, J.-E., Bergin, E. A., & Nomura, H. 2010, *ApJL*, 710, L21
Marseille, M. G. & Cazaux, S. 2011, *A&A*, 532, A60
McClure, M. K., Bergin, E. A., Cleeves, L. I., *et al.* 2016, *ApJ*, 831, 167
Miotello, A., van Dishoeck, E. F., Kama, M., & Bruderer, S. 2016, *A&A*, 594, A85
Miotello, A., van Dishoeck, E. F., Williams, J. P., *et al.* 2017, *A&A*, 599, A113
Miyake, K. & Nakagawa, Y. 1995, *ApJ*, 441, 361
Mulders, G. D. & Dominik, C. 2012, *A&A*, 539, A9
Muralidharan, K., Deymier, P., Stimp, M., de Leeuw, N. H., & Drake, M. J. 2008, *Icarus*, 198, 400
Nomura, H., Aikawa, Y., Tsujimoto, M., Nakagawa, Y., & Millar, T. J. 2007, *ApJ*, 661, 334
Öberg, K. I., Linnartz, H., Visser, R., & van Dishoeck, E. F. 2009, *ApJ*, 693, 1209
Pascucci, I., Gorti, U., Hollenbach, D., *et al.* 2006, *ApJ*, 651, 1177
Pinte, C., Dent, W. R. F., Ménard, F., *et al.* 2016, *ApJ*, 816, 25
Pinte, C., Ménard, F., Duchêne, G., *et al.* 2018, *A&A*, 609, A47
Podio, L., Kamp, I., Codella, C., *et al.* 2013, *ApJL*, 766, L5
Pringle, J. E. 1981, *Annu. Rev. Astron. Astrophys.*, 19, 137
Rab, C., Güdel, M., Woitke, P., *et al.* 2018, *A&A*, 609, A91
Schwarz, K. R., Bergin, E. A., Cleeves, L. I., *et al.* 2016, *ApJ*, 823, 91
Stimp, M., Walker, A., Drake, M., de Leeuw, N., & Deymier, P. 2006, *Journal of Crystal Growth*, 294, 83
Thi, W. F., Hocuk, S., Kamp, I., *et al.* 2018, arXiv e-prints
Thi, W.-F., Woitke, P., & Kamp, I. 2011, *MNRAS*, 412, 711
Trapman, L., Miotello, A., Kama, M., van Dishoeck, E. F., & Bruderer, S. 2017, *A&A*, 605, A69
Visser, R., Doty, S. D., & van Dishoeck, E. F. 2011, *A&A*, 534, A132
Visser, R., van Dishoeck, E. F., & Black, J. H. 2009, *A&A*, 503, 323
Walsh, C., Millar, T. J., Nomura, H., *et al.* 2014, *A&A*, 563, A33
Weidenschilling, S. J. 1977, *MNRAS*, 180, 57
Willacy, K., Alexander, C., Ali-Dib, M., *et al.* 2015, *SSRv*, 197, 151
Willacy, K., Langer, W., Allen, M., & Bryden, G. 2006, *ApJ*, 644, 1202
Williams, J. P. & Best, W. M. J. 2014, *ApJ*, 788, 59
Woitke, P., Kamp, I., Antonellini, S., *et al.* 2019, *Publications of the Astronomical Society of the Pacific*, 131, 064301
Woitke, P., Kamp, I., & Thi, W.-F. 2009, *A&A*, 501, 383
Woitke, P., Min, M., Pinte, C., *et al.* 2016, *A&A*, 586, A103

Discussion

MAROV: What mechanism of dust particles integration do you postulate in your model — whether you believe it could be by integration per se through a direct collision mechanism, or rather some interaction process involving, say, original dust cluster formation is to be invoked? In my mind the latter would significantly ease the integration and particle growth process, which is supported by the results of my team computer modeling.

KAMP: First of all, I myself have no model of dust particle growth. Growth is now already detected at very early stages of star formation, e.g. coreshine. The very first condensation of dust is indeed thought to proceed via clusters that form e.g. around AGB stars. However, similar processes can occur if refractory material sublimates in disks close to the star due to e.g. outbursts and then recondenses.

Origins: From the Protosun to the First Steps of Life
Proceedings IAU Symposium No. 345, 2018
B. G. Elmegreen, L. Viktor Tóth & M. Güdel, eds.

doi:10.1017/S1743921319001923

3D radiation MHD simulations of gas and dust in protoplanetary disks

Mario Flock

Max-Planck Institut for Astronomy
Königstuhl 17, 69117, Heidelberg, Germany
email: flock@mpia.de

Abstract. 3D global radiation MHD simulations of gas and dust in protoplanetary disks allow us to understand the dynamical and thermal evolution of protoplanetary disks. At the same time, recent observations in the mm-dust emission by the Atacama Large Millimeter Array (ALMA) allow us to resolve structures at scales of the disk scale height.

From our recent simulation results by Flock *et al.* (2015) and Flock *et al.* (2017) we are able to directly compare for the first time detailed observational constraints from high-resolution observations by ALMA with the gas and dust dynamics obtain in 3D state-of-art simulations of protoplanetary disks. Especially measurements of the dust scale height obtained from the disk around the young system HL Tau allow us to compare for different gas disk instability models. Further we use Monte Carlo radiation transfer models of the dusty disk to compare our results of the dust scale height in 3D radiation HD and MHD simulations. Our findings are that magnetized models fit perfectly the observational constraints, showing a strongly settled disk, while hydrodynamical turbulence leads to a dust uplifting which is larger than expected. These results open a new window to compare future multi-wavelength observations to simulations.

Keywords. MHD simulations, accretion disks, protoplanetary disks, mm observations

---

1. Introduction

Planet formation is one of the key research topics in modern astrophysics. Especially the discovery of thousand new exoplanetary systems (Dressing & Charbonneau 2015) have raised the question about their formation. For this we have to look into young circumstellar disks, often called protoplanetary disks which are typically less then 10 million years old. The advancement in telescopes and observations allow us now to get a direct constrain on the physical structure of the disk. Dust grains around micron size and smaller are coupled to the gas structure and motion in protoplanetary disks. Observations of the near infrared traces the scattered light from tiny dust grains in the disk atmosphere and they allow us to learn about the gas structure as well. Recently this was done for the disk system around IM Lup by Avenhaus *et al.* (2018) showing the flared disk structure or for the disk around HD 169142 by Bertrang *et al.* (2018) showing a more complex ring and gap structure. Larger grains of millimeter and centimeter sizes start to decouple from the gas motion. They start to settle to the disk midplane and they drift to disk regions which have a local pressure maximum (Weidenschilling 1977). The structure of these large grains were observed in the disk around HL Tau (ALMA Partnership *et al.* 2015) with a so far surpassing resolution. These observations alow to determine the dust scale height in the disk. Pinte *et al.* (2016) concluded from best fit radiation transfer models that a dust scale height of $H_{dust}/R = 0.007$ fits best the scale height of mm sized grains. In the following we will investigate the dust scale height from hydrodynamical and magnetohydrodynamical models.

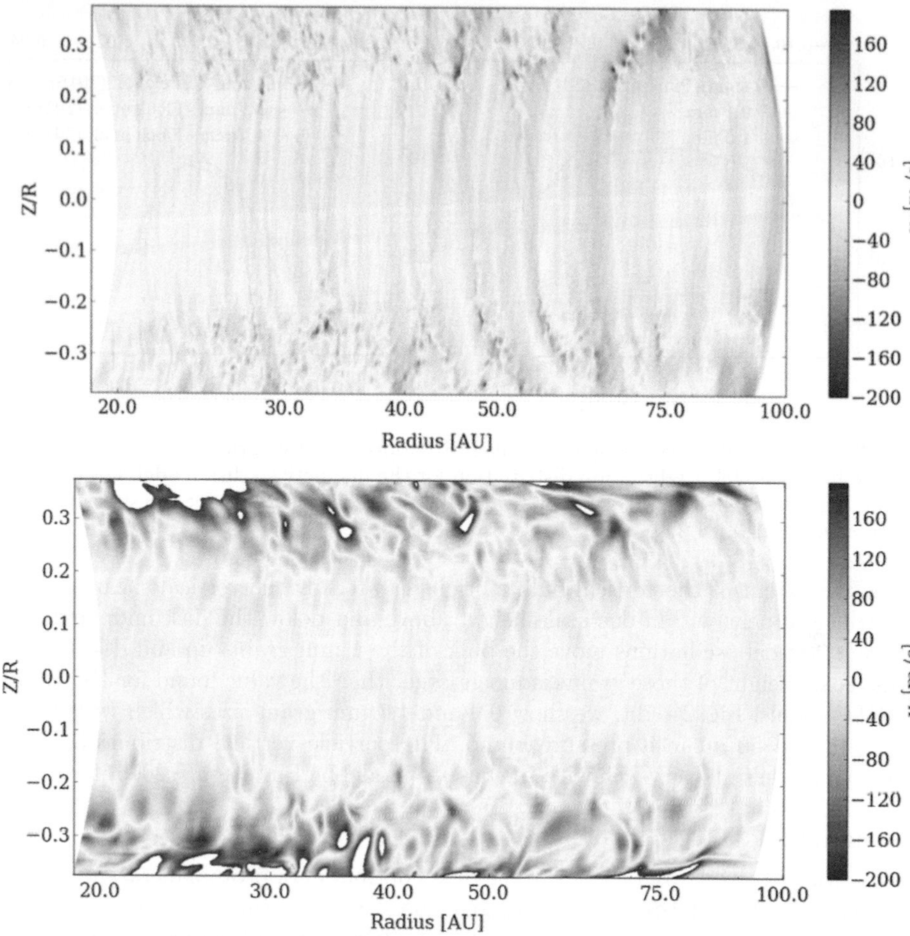

Figure 1. Top: Snapshot of the vertical velocity component for the 3D radiation hydrodynamical model Flock *et al.* (2017). Bottom: Snapshot of the vertical velocity component for the magnetized disk model by Flock *et al.* (2015). Both are in the R - Z/R plane for a given azimuthal slice.

2. Models and results

The 3D models are based on 3D radiation HD and MHD simulations with embedded grains of different sizes. Both models assume the same stellar and disk parameter for the initial conditions. For details on the models we refer to Flock *et al.* (2017), Ruge *et al.* (2016) and Flock *et al.* (2015). Fig. 1 presents a snapshot of the vertical velocity in the R - Z/R plane for a given azimuthal slice. The hydro model, Fig. 1, top, shows characteristic vertical motions of the disk bulk material of around 50 m/s at the disk midplane. Those motions easily overshoot the midplane region which emphasizes the need of fully stratified models of both hemispheres. In contrast, the corresponding magnetized model by Flock *et al.* (2015), Fig. 1 bottom, shows turbulent motions with smaller amplitude at the midplane compared to the hydro model.

Fig. 2 summarizes the results of two grain sizes which are embedded in the simulation by including individual particle motion with a drag term. The hydro model, Fig. 2 left, shows 100 μm grains well mixed. The 1 mm grains are mixed less efficiently. Their profile shows a quasi-plateau with a dropoff above one gas scale height. This extent of the plateau

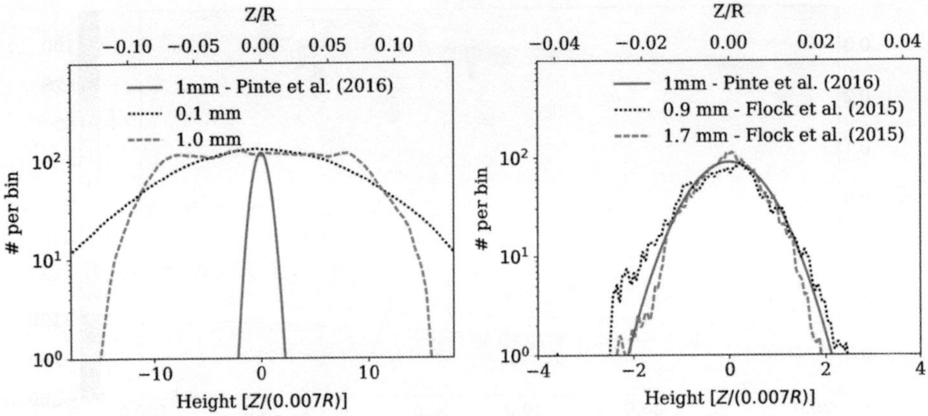

Figure 2. Time- and space-averaged vertical distribution of two grain sizes. Left: grain data set for the hydro model. Right: grain distribution for the magnetized disk model. For both plots, the green line shows the best-fit value of $H/R = 0.007$ which was found for the systems HL Tau by Pinte *et al.* (2016).

matches the extent of the vertical bulk motions created by the vertical shear instability (VSI), which appear within one scale height above and below the disk midplane. These large-scale vertical oscillations move the bulk of the 1 mm grains up and down. Clearly the dust scale height of those grains is much larger than the value found for HL Tau. For the MHD models, Fig. 2 right, we show 0.9 and 1.7 mm grain sizes which were trapped between 40 and 50 au in a pressure bump. The particle vertical distribution of those magnetized models fits very well with the best-fit model by Pinte *et al.* (2016) for HL Tau.

3. Conclusion

New spatial resolved observations of the dust continuum of protoplanetary disks in the mm emission enable for the first time to give constraints on the dust scale height. Such measurements are a promising way to learn also about the underlying disk physics, e.g. the strength of gas turbulence and mixing. By comparing 3D radiation non-ideal HD and MHD simulations with gas and dust we found that magnetized disk models fit best the observational constraints for the young protoplanetary disk HL Tau. Future simulations and observations should target the determination of the dust scale height. Especially highly-inclined disks should be suited to determine the dust scale height by using the thermal emission for different wavelengths. At the same time future simulations of gas and dust are needed to learn more about their interaction and to finally understand the relevant physical processes in these young disks.

References

Avenhaus H., Quanz S., Garufi A., Perez S., Casassus S., Pinte C., Bertrang G., Caceres C., Benisty M., & Dominik C. 2018, *AJ*, 863, 44

Bertrang, G. H. -M., Avenhaus, H., Casassus, S., Montesinos, M., Kirchschlager, F., Perez, S., Cieza, L., & Wolf, S., 2018 *MNRAS*, 474, 5105

Dressing, Courtney D., & Charbonneau, David 2015, *ApJ*, 807, 45

Flock, M., Ruge, J. P., Dzyurkevich, N., Henning, Th., Klahr, H., & Wolf, S. 2015, *A&A*, 574, 68

Flock, M., Nelson, R. P., Turner, N. J., Bertrang, G. H. -M., Carrasco-González, C., Henning, Th., Lyra, W., & Teague, R., 2017 *ApJ*, 850, 131

Ruge, J. P., Flock, M., Wolf, S., Dzyurkevich, N., Fromang, S., Henning, Th., Klahr, H., & Meheut, H., 2016 *A&A*, 590, 17

ALMA Partnership *et al.* 2015, *ApJ* 8083 (https://ui.adsabs.harvard.edu/abs/2015ApJ...808L...3A)

Pinte, C., Dent, W. R. F., Ménard, F., Hales, A., Hill, T., Cortes, P., & de Gregorio-Monsalvo, I. 2016, *ApJ*, 816, 25

Weidenschilling, S. J., 1977 *MNRAS*, 180, 57

Discussion

KHAIBRAKHMANOV: What value of plasma β did you observe in your simulations? Can the radial pressure of the magnetic field influence the position of pressure bumps in the disk?

FLOCK: The initial plasma β is 10^4 at the midplane. It is a vertical magnetic field. During the simulation the plasma β changes locally, being high in the pressure bump and low inside the gap.

KAMP: Can you explain if there are any limitations arising from the fact that you consider trace dust of single size, so neglecting dust colliding with itself (maybe leading to growth or destruction)?

FLOCK: The dust growth becomes important on timescales of roughly 100 orbits. For the simulation I presented, this is mostly not important. However, locally, the dust could concentrate which would decrease the timescale.

KAMP: Do your 1μm particles agree with SPHERE data spatial distribution?

FLOCK: We are going to investigate this soon.

Origins: From the Protosun to the First Steps of Life
Proceedings IAU Symposium No. 345, 2018
B. G. Elmegreen, L. Viktor Tóth & M. Güdel, eds.

doi:10.1017/S1743921319001625

A mid-infrared interferometric survey of the planet-forming region around young Sun-like stars

József Varga[1,2], Péter Ábrahám[1], Lei Chen[1], Thorsten Ratzka[3], K. É. Gabányi[1,4] and Á. Kóspál[1,5]

[1]Konkoly Observatory, Research Centre for Astronomy and Earth Sciences, Hungarian Academy of Sciences, Konkoly Thege Miklós út 15-17., H-1121 Budapest, Hungary
email: varga.jozsef@csfk.mta.hu

[2]Leiden Observatory, Leiden University, PO Box 9513, NL2300, RA Leiden, The Netherlands

[3]Institute for Physics/IGAM, University of Graz, Universitätsplatz 5/II, 8010, Graz, Austria

[4]MTA-ELTE Extragalactic Astrophysics Research Group, Pázmány Péter sétány 1/A H-1117 Budapest Hungary

[5]Max Planck Institute for Astronomy, Königstuhl 17, 69117 Heidelberg, Germany

Abstract. We present our results from a mid-infrared interferometric survey targeted at the planet-forming region in the circumstellar disks around low- and intermediate-mass young stars. Our sample consists of 82 objects, including T Tauri stars, Herbig Ae stars, and young eruptive stars. Our main results are: 1) Disks around T Tauri stars are similar to those around Herbig Ae stars, but are relatively more extended once we account for stellar luminosity. 2) From the distribution of the sizes of the mid-infrared emitting region we find that inner dusty disk holes may be present in roughly half of the sample. 3) Our analysis of the silicate spectral feature reveals that the dust in the inner ∼1 au region of disks is generally more processed than that in the outer regions. 4) The dust in the disks of T Tauri stars typically show weaker silicate emission in the N band spectrum, compared to Herbig Ae stars, which may indicate a general difference in the disk structure. Our data products are available at VizieR, and at the following web page: http://konkoly.hu/MIDI_atlas.

Keywords. stars: pre–main-sequence, planetary systems: protoplanetary disks, techniques: interferometric, infrared: stars

1. Introduction

Circumstellar disks around young stellar objects (YSOs) are the places where planetary systems are born. These disks have a lifetime of about 10 million years. During this relatively short time disk material is accreted by the central star. Dust grains grow and stick together, gradually building planets. The protoplanetary disk eventually disperses, leaving a debris disk from the remaining material. Important aspects of disk and planetary evolution, e.g., dust evolution or the clearing of the dusty disk, are still not well understood, since they can only be tackled with high angular resolution observations.

Long-baseline interferometers at mid-infrared (mid-IR) wavelengths can reach angular resolution around ∼10 mas. With mid-IR interferometry we can probe the structure of the warm dust, located at typically $1-10$ au from the star (in the planet forming region). Additionally, spectrally resolved mid-IR interferometric observations provide an insight to the composition and spatial distribution of warm dust throughout the disk.

Here we present an overview of our mid-IR interferometric survey of young Sun-like stars, and highlight some results from the statistical analysis of our sample.

2. Overview

The now decommissioned Mid-infrared Interferometric Instrument (MIDI, Leinert *et al.* 2003) at the Very Large Telescope Interferometer (VLTI) left a rich heritage of high spatial resolution observations of planet-forming disks. It provided spectrally resolved interferometric observations in the N band (7.5 – 13 μm), covering the silicate spectral feature. However, large part of the data remained unpublished, partly because the reduction and the interpretation of IR interferometric data is not always straightforward. In Varga *et al.* (2018) we published significant part of these data, focusing on low-mass YSOs (T Tauri stars), the younger siblings of the Sun. We collected MIDI observations on 45 T Tauri stars, but for comparison we also included in our sample 26 Herbig Ae stars, and 11 young eruptive stars. In total we selected 82 objects with ~600 interferometric measurements over ~200 nights between 2003 and 2015. The data were reduced homogeneously using the standard MIDI pipeline, EWS (e.g., Chesneau 2007), with the help of the routines of Menu *et al.* (2015). The resulting total spectra (representing the whole disk's emission), correlated spectra (the emission from the inner disk region, where the size of this region depends on the baseline length), spectrally resolved visibilities, and differential phases are published online in the form of an atlas at http://konkoly.hu/MIDI_atlas, and also at VizieR (http://cdsarc.u-strasbg.fr/viz-bin/cat/J/A+A/617/A83).

To estimate the extent of the warm dusty part of the disks, for each object we fitted a simple geometric disk model to the 10.7 μm correlated fluxes. The result of the modelling is the half-light radius of the mid-IR emitting region, defined as the radius at which the integrated intensity is half of the total flux density.

3. Implications

Size of the mid-IR emitting region. Disks around T Tauri and Herbig Ae/Be stars are illuminated by the central star, and re-emit this radiation at IR wavelengths. Thus, there is a correlation between the IR appearance of the disks and the luminosity of the central star. In the near-IR this correlation is quite strict (Eisner *et al.* 2007), indicating that the near-IR emitting region is usually confined to a region near the dust sublimation radius, which is almost entirely determined by the star (as the place where the temperature is ~1500 K). In Fig. 1 we plot the size-luminosity relation from our data. The correlation between the mid-IR emitting region and the stellar luminosity is much weaker compared to the near-IR. This reflects the structural diversity of the mid-IR emitting region. The radii that correspond to temperatures of 250 K and 900 K for optically thin gray dust are also plotted. Thus, the location of points in the figure can be connected to the typical temperature of the dust. Points located above the 250 K line indicate disks that are relatively cold and oversized. This can be caused by the inner clearing of the disk, and it is usually observed among transitional disks (e.g., TW Hya, GM Aur, Haro 6-10N, HD 169142). Several Herbig Ae stars are located close to the 900 K line, indicating relatively compact disks. In contrast, T Tauri stars preferentially lie closer to the 250 K line, suggesting that disks around T Tauri stars are relatively more extended and colder than those around Herbig Ae stars, once we account for stellar luminosity. Inner disk clearing can be caused by e.g., photoevaporation, grain growth, and planet formation. We simulated a population of disk models with radiative transfer modeling. The model population had two groups: in the first group the dusty disk begins at the sublimation radius, while in the other it starts further out due to an inner hole. By comparing the

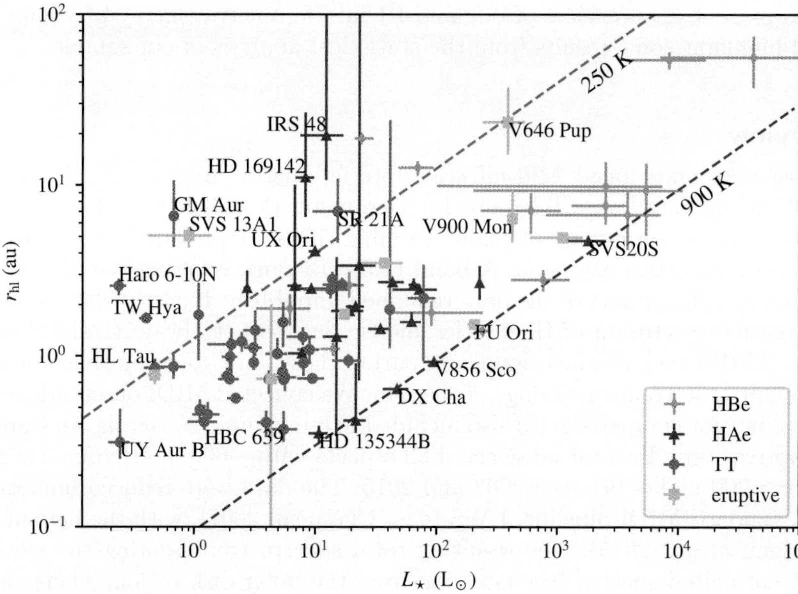

Figure 1. Size (half-light radius) of the mid-IR emitting region of circumstellar disks as a function of the stellar luminosity. Points for T Tauri (red circle), Herbig Ae (blue triangle), and young eruptive (orange square) stars are from our work. Data for Herbig Be stars (grey diamonds) are from Menu *et al.* (2015). The radii that correspond to temperatures of 250 K and 900 K for optically thin gray dust are plotted with red and blue dashed lines, respectively.

size distribution of our objects with the disk models, we found that roughly half of our objects can have an inner dusty disk hole.

Dust mineralogy. The first important step towards the formation of planetesimals is dust processing. It is known from mid-IR spectroscopy of the N band silicate feature that the original small amorphous dust grains interact with each other, grow and/or partly transform into crystalline species (van Boekel *et al.* 2003; Przygodda *et al.* 2003). van Boekel *et al.* (2004) studied MIDI spectra of Herbig Ae/Be stars, and revealed that primordial dust is mainly found in the outer disk, while more evolved dust is located closer to the star. Here we study dust processing both in T Tauri and Herbig Ae stars from our sample. The shape of the N band silicate feature reflects the composition and size of the dust grains. For small (∼0.1 μm) amorphous grains the feature has a triangular shape peaking at 9.8 μm, while for larger (∼$1-2$ μm) grains the spectrum will show a plateau between 10 and 11 μm. Crystalline silicates have smaller sharp features at several wavelengths (e.g., at 9.2, 10.5 and 11.3 μm). In Fig. 2 we show two correlated spectra of the T Tauri star GW Ori as an example. The spectra represent the emission from the disk inside $r=3.9$ au (9.8 mas, dashed blue line), and inside $r=2.5$ au (6.3 mas, solid red line), respectively. The $r<2.5$ au spectrum show a plateau characteristic of large grains, while in the $r<3.9$ au spectrum a triangular peak centered at 9.8 μm appears. It clearly indicates dust processing, with larger grains located closer to the star, in agreement with the findings of van Boekel *et al.* (2004). We found a similar trend for most objects in our sample. The different radial distribution of small and large grains indicate that grain processing mostly happens in the inner ∼1 au region of planet-forming disks, where the temperature and density are higher, and the dynamical timescales are smaller. We also found that the objects showing the strongest silicate emission in their spectra are all Herbig Ae stars. T Tauri stars typically have weaker features, indicating a possible

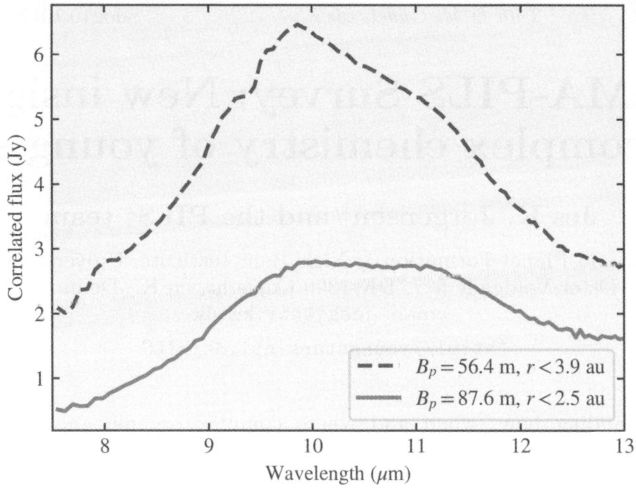

Figure 2. VLTI/MIDI correlated spectra of GW Ori showing the N band silicate spectral feature. The blue dashed line represents the emission from the inner 3.9 au, while the red solid line represents the emission from the inner 2.5 au.

difference in the disk structure. As the silicate emission feature originates from the disk surface, disks with larger surface area in the mid-IR emitting region are expected to show a stronger silicate feature. The size of the surface heavily depends on the disk flaring. Thus, we propose that T Tauri disks may be less flared than Herbig Ae disks.

Acknowledgements

This project has received funding from the European Research Council (ERC) under the European Union's Horizon 2020 Research and Innovation programme under grant agreement No. 716155 (SACCRED).

References

Chesneau, O. 2007, *New Astron. Revs*, 51, 666
Eisner, J. A., Hillenbrand, L. A., White, R. J., *et al.* 2007, *ApJ*, 669, 1072
Leinert, C., Graser, U., Richichi, A., *et al.* 2003, *The Messenger*, 112, 13
Menu, J., van Boekel, R., Henning, T., *et al.* 2015, *A&A*, 581, A107
Przygodda, F., van Boekel, R., Ábrahám, P., *et al.* 2003, *A&A*, 412, L43
van Boekel, R., Waters, L. B. F. M., Dominik, C., *et al.* 2003, *A&A*, 400, L21
van Boekel, R., Min, M., Leinert, C., *et al.* 2004, *Nature*, 432, 479
Varga, J., Ábrahám, P., Chen, L., *et al.* 2018, *A&A*, 617, A83

Discussion

JOHNSTONE: Beautiful observations. How much degeneracy remains in the interpretation? You provide very nice results but what else is needed to really confirm that these observations really lead to the physical trends suggested?

VARGA: Due to sparse baseline coverage, many different disk models can fit the interferometric data. By using simple models we can get robust estimates, like the half-light radius. An other ambiguity lies in the interpretation of the silicate spectral feature, as both large amorphous grains and crystalline grains can produce a similar feature shape. To tackle these issues, better baseline coverage and higher spectral resolution is needed.

Origins: From the Protosun to the First Steps of Life
Proceedings IAU Symposium No. 345, 2018
B. G. Elmegreen, L. Viktor Tóth & M. Güdel, eds.

doi:10.1017/S1743921319002849

The ALMA-PILS Survey: New insights into the complex chemistry of young stars

Jes K. Jørgensen[1] and the PILS[2] team

[1]Centre for Star and Planet Formation & Niels Bohr Institute, University of Copenhagen, Øster Voldgade 5–7, DK-1350 Copenhagen K., Denmark
email: `jeskj@nbi.ku.dk`

[2]`http://youngstars.nbi.dk/PILS`

Abstract. Understanding how, when and where complex organic and potentially prebiotic molecules are formed is a fundamental goal of astrochemistry. Since its beginning the Atacama Large Millimeter/submillimeter Array (ALMA) has demonstrated its capabilities for studies of the chemistry of solar-type stars. Its high sensitivity and fine spectral and angular resolution makes it possible to study the chemistry of young stars on Solar System scales. We here present an unbiased spectral survey, *Protostellar Interferometric Line Survey (PILS)*, of the astrochemical template source and Class 0 protostellar binary IRAS 16293-2422 using ALMA. The high quality ALMA data have allowed us to detect a wealth of species previously undetected toward solar-type protostars as well as the interstellar medium in general. Also, the data show the presence of numerous rare isotopologues of complex organic molecules and other species: the exact measurements of the abundances of the complex organic molecules and their isotopologues shed new light onto the formation of these species and provide a chemical link between the embedded protostellar stages and the early Solar System.

Keywords. astrochemistry, stars: formation, stars: protostars, ISM: molecules, ISM: individual (IRAS 16293−2422), Submillimeter: ISM

---

1. Introduction: astrochemistry in the ALMA era

One of the important tasks of astrochemistry is to address how, when and where complex organic and potentially prebiotic molecules are formed in the environments of young stars. A particularly interesting question in this context is what is the chemical composition of material as it is accreted onto the circumstellar disks and, in particular, whether the rich chemistry of the early stages can be transferred to the later planet-forming stages. The Atacama Large Millimeter/submillimeter Array (ALMA) has already significantly changed the field of astrochemistry – in particular, through its capabilities for studies of the organic content in star forming regions. The key features of ALMA for this include its:

Large frequency coverage: allowing secure assignments of complex and rare species through identifications of a large number of lines.
High spectral resolution: which limits confusion and makes it possible to push deeper toward sources with narrow lines.
High sensitivity: making (fast) unbiased surveys of large spectral ranges possible.
Excellent calibration: providing consistent relative line intensities over large frequency ranges, critical for extracting physical conditions from the spectra.
High spatial resolution: making it possible to zoom in on Solar System scales toward young stars in nearby star forming regions and study the spatial coincidence of different

lines and species, providing important information about the links between the physics and chemistry there.

2. The ALMA Protostellar Interferometric Line Survey (PILS)

One of the prime targets for studies of the astrochemistry of solar-type protostars is the nearby protostellar binary IRAS 16293-2422 (IRAS16293 in the following). IRAS16293 has a rich spectrum with bright line emission on different scales (e.g., Blake *et al.* 1994, Bisschop *et al.* 2008, Caux *et al.* 2011, Jørgensen *et al.* 2011) and is the place where most of the first detections of complex organics toward solar-type protostars have been made (e.g., van Dishoeck *et al.* 1995, Cazaux *et al.* 2003, Bottinelli *et al.* 2004, Kuan *et al.* 2004, Jørgensen *et al.* 2012, Kahane *et al.* 2013). IRAS16293 was therefore also a natural target for early Science Verification (SV) observations with ALMA through which the first discovery of the prebiotic molecule glycolaldehyde ("a simple sugar-like molecule") toward a low-mass protostar was made (Jørgensen *et al.* 2012).

Building on that discovery we initiated a large unbiased survey of IRAS16293 with ALMA, the *"Protostellar Interferometric Line Survey (PILS)"*, covering one of the important atmospheric windows at submillimeter wavelengths (Jørgensen *et al.* 2016). The observations cover the spectral range from 329 to 363 GHz (ALMA's Band 7) with 0.2 km s^{-1} spectral resolution and 0.5″ angular resolution ($\approx$ 70 AU). The survey is 1–2 orders of magnitude deeper than previous single-dish surveys and furthermore produce resolved maps of the line emission across the entire structure of the protostellar binary.

3. Selected results concerning the origin of complex organics

Fig. 1 illustrates the richness of the data with spectra extracted toward each of the two components of the binary: IRAS16293A, which appears elongated in the dust continuum maps and with wide lines corresponding to an "edge-on structure", and IRAS16293B, which appears more circular in the dust continuum maps and show narrow lines (FWHM $\approx$ 1 km s^{-1}) consistent with a more face-on morphology. Toward IRAS16293B more than 10,000 lines can be seen corresponding to approximately one line per 3 km s^{-1}, i.e. close to the line confusion limit across significant parts of the spectrum (Jørgensen *et al.* 2016).

Through LTE modeling and careful evaluation of the spectroscopy, so-far more than 100 species (including isotopologues) have been assigned toward IRAS16293B of which 27 species have not previously been seen toward solar-type protostars and 16 species represent new detections in the interstellar medium altogether.

One example of a first detection toward a solar-type protostar with PILS was the of identification of methyl isocyanate, CH_3NCO, reported by Ligterink *et al.* (2017) and independently by Martín-Doménech *et al.* (2017). Methyl isocyanate is an interesting organic molecule with a peptide-bond characteristic of amino-acids. Its origin has been debated but through UHV experiments with irradiation of CH_4:HNCO ice mixtures, Ligterink *et al.* demonstrated that it could be formed via CH_3 and (H)NCO recombinations in the solid-state (see also Chuang *et al.* 2017, Ligterink *et al.* 2018).

Another example was the first discovery of an organohalogen methyl chloride or chloromethane, CH_3Cl, in the interstellar medium reported in Fayolle *et al.* (2017). The identification of CH_3Cl in the PILS data triggered a search for this species in data from the ROSINA instrument on the Rosetta mission toward Comet 67P/C-G also reported in Fayolle *et al.* (2017). The main implication of this discovery is the direct link between protostellar chemistry and the constituents observed in icy bodies in our own Solar System prompting further comparisons between the protostellar and cometary inventories (Drozdovskaya *et al.* 2018 and M. Drozdovskaya *et al.* in prep.).

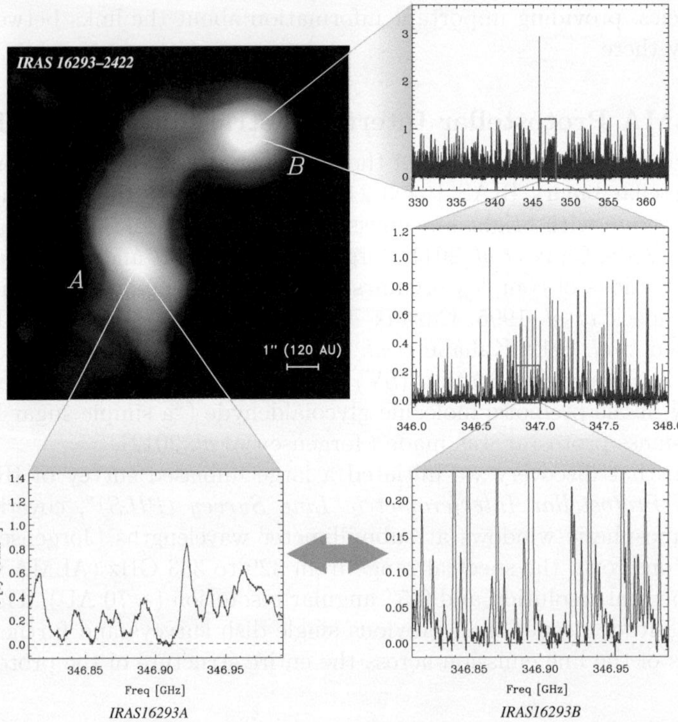

Figure 1. Example of the ALMA-PILS data for IRAS 16293-2422: the panel in the top left corner shows a false color image of the dust continuum radiation (blue: 0.8 mm; green: 1.3 mm and red 3.0 mm) toward the protostellar binary. The panels in the righthand column shows the spectra toward IRAS16293B zooming-in from the full $\approx$ 33 GHz spectral range to a small (180 MHz) window illustrating the richness of the data. The panel in the lower left shows the same zoomed-in spectrum toward IRAS16293A where the significantly broader line are visible.

The high sensitivity of the PILS data makes it possible to further explore the degree of complexity that arises around solar-type protostars. Examples of that are the first detections of the complex organics acetone and propanal each containing three carbon atoms, previously only seen toward the Galactic Center and selected hot cores (see Lykke *et al.* 2017, and references therein). Besides serving as important tests of astrochemical models, these discoveries raise the interesting question whether there are any limitations to the "size" of these complex organics, e.g., in terms of the number of carbon-atoms.

The high sensitivity also allow more systematic characterisations of the isotopic compositions of complex organics toward IRAS16293B. The survey has yielded detections of numerous isotopologues of complex organics toward IRAS1629B, including the first detections of deuterated forms of formamide (NH_2CHO; Coutens *et al.* 2016), deuterated and ^{13}C-isotopologues of glycolaldehyde ($CH_2(OH)CHO$; Jørgensen *et al.* 2016), and other oxygen-bearing organics (Jørgensen *et al.* 2018) as well as cyanamide (NH_2CN; Coutens *et al.* 2018). In particular, the survey provides systematic estimates of the D/H and $^{12}C/^{13}C$ ratios for full sets of O- and N-bearing organics (Jørgensen *et al.* 2018, Calcutt *et al.* 2018) as well as S-species (Drozdovskaya *et al.* 2018). Interestingly, no systematic differences are found between between the D/H ratios for molecules where the deuterium is substituted in different functional groups for given molecules – but the D/H ratios do vary between, e.g., "simpler" and more "complex" organics (e.g., Jørgensen *et al.* 2018).

The data also show the detection of a number of species where more than one hydrogen atom is substituted by deuterium, for example formaldehyde (H_2CO; Persson *et al.* 2018) and methyl formate (CH_3OCHO; Manigand *et al.* 2018). Similar to the case for water (e.g., Coutens *et al.* 2014) it is found that the doubly-substituted species are more abundant relative to singly-substituted variants than the singly-substituted to non-substituted species. Both the systematic differences in the deuteration from species-to-species as well as the relation between the differently substituted molecules appear to be related to the formation times of the individual variants in the icy mantles of dust grains (e.g., Jørgensen *et al.* 2018, Persson *et al.* 2018, Manigand *et al.* 2018).

4. Summary

The results presented here are a direct illustration of how ALMA is moving studies of the chemistry of the earliest stages of star and planet formation into a new era – even though the current studies are only scratching the surface of what is achievable in terms of angular resolution, much broader surveys etc. It should be emphasised that this effort is not just purely observational, but its success depends on a tight collaboration between astronomers, spectroscopists, modelers, laboratory experimentalists and more.

References

Bisschop, S. E., Jørgensen, J. K., Bourke, T. L., *et al.* 2008, *A&A*, 488, 959
Blake, G. A., van Dishoek, E. F., Jansen, D. J., *et al.* 1994, *ApJ*, 428, 680
Bottinelli, S., Ceccarelli, C., Neri, R., *et al.* 2004, *ApJL*, 617, L69
Calcutt, H., Jørgensen, J. K., Müller, H. S. P., *et al.* 2018, *A&A*, 616, A90
Caux, E., Kahane, C., Castets, A., *et al.* 2011, *A&A*, 532, A23
Cazaux, S., Tielens, A. G. G. M., Ceccarelli, C., *et al.* 2003, *ApJL*, 593, L51
Chuang, K.-J., Fedoseev, G., Qasim, D., *et al.* 2017, *MNRAS*, 467, 2552
Coutens, A., Jørgensen, J. K., Persson, M. V., *et al.* 2014, *ApJL*, 792, L5
Coutens, A., Jørgensen, J. K., van der Wiel, M. H. D., *et al.* 2016, *A&A*, 590, L6
Coutens, A., Willis, E. R., Garrod, R. T., *et al.* 2018, *A&A*, 612, A107
Drozdovskaya, M. N., van Dishoeck, E. F., Jørgensen, J. K., *et al.* 2018, *MNRAS*, 476, 4949
Fayolle, E. C., Öberg, K. I., Jørgensen, J. K., *et al.* 2017, *Nature Astronomy*, 1, 703
Jørgensen, J. K., Bourke, T. L., Nguyen Luong, Q., & Takakuwa, S. 2011, *A&A*, 534, A100
Jørgensen, J. K., Favre, C., Bisschop, S. E., *et al.* 2012, *ApJL*, 757, L4
Jørgensen, J. K., Müller, H. S. P., Calcutt, H., *et al.* 2018, *A&A*, 620, A170
Jørgensen, J. K., van der Wiel, M. H. D., Coutens, A., *et al.* 2016, *A&A*, 595, A117
Kahane, C., Ceccarelli, C., Faure, A., & Caux, E. 2013, *ApJL*, 763, L38
Kuan, Y., Huang, H., Charnley, S. B., *et al.* 2004, *ApJL*, 616, L27
Ligterink, N. F. W., Coutens, A., Kofman, V., *et al.* 2017, *MNRAS*, 469, 2219
Ligterink, N. F. W., Terwisscha van Scheltinga, J., Taquet, V., *et al.* 2018, *MNRAS*, 480, 3628
Lykke, J. M., Coutens, A., Jørgensen, J. K., *et al.* 2017, *A&A*, 597, A53
Manigand, S., Calcutt, H., Jørgensen, J. K., *et al.* 2018, *A&A* in press. (arXiv:1811.09102)
Martín-Doménech, R., Rivilla, V. M., Jiménez-Serra, I., *et al.* 2017, *MNRAS*, 469, 2230
Persson, M. V., Jørgensen, J. K., Müller, H. S. P., *et al.* 2018, *A&A*, 610, A54
van Dishoeck, E. F., Blake, G. A., Jansen, D. J., & Groesbeck, T. D. 1995, *ApJ*, 447, 760

Acknowledgements

JKJ acknowledges support by the European Research Council (ERC) under the European Union's Horizon 2020 research and innovation programme through ERC Consolidator Grant "S4F" (grant agreement No 646908) and the Danish National Research Foundation through Centre for Star and Planet Formation.

Discussion

WANDEL: What is the forecast for the detection of complex molecules with higher molecular weight using ALMA in the near future?

JØRGENSEN: The challenge with going for larger/more complex molecules is mainly matching the available laboratory spectroscopy with the extracted spectra from the ALMA data. We can do a reasonably good job on systematically identifying molecules with two to three carbon atoms. Going much beyond that becomes challenging because of lacking laboratory spectroscopy and line confusion.

LISEAU: The source is in the western part of the active star-forming cloud ρ Oph, where nowhere else the spectral line density is observed to be as high and diverse. Why is your source so special?

JØRGENSEN: IRAS16293 is surrounded by a massive envelope and has a high luminosity compared to other protostars in Ophiuchus. The region where the molecules sublimate is therefore relatively extended and represents a high column density. Thus, I think it is largely a sensitivity issue. In fact, when we are integrating deeper on other protostars with ALMA, many start showing lines from similar organics as IRAS16293.

Origins: From the Protosun to the First Steps of Life
Proceedings IAU Symposium No. 345, 2018
B. G. Elmegreen, L. Viktor Tóth & M. Güdel, eds.

doi:10.1017/S1743921318008311

The first 200 kyr of the Solar System: making the planetary material diversity

Francesco C. Pignatale[1,2], Sébastien Charnoz[2], Marc Chaussidon[2] and Emmanuel Jacquet[1]

[1]Muséum national d'Histoire naturelle, UMR 7590, CP52, 57 rue Cuvier, 75005, Paris, France
email: pignatale@ipgp.fr

[2]Institut de Physique du Globe de Paris (IPGP)
1 rue Jussieu, 75005, Paris, France

Abstract. Chondrites are made of a mixture of solids formed at high and low temperatures. This heterogeneity was thought to be produced by large scale transport processes in the Sun's isolated accretion disk. However, mounting evidences suggest that refractory inclusions in chondrites were produced together with the disk formation.

We present numerical simulations of the formation and transport of rocky materials during the collapse of the Solar Nebula's parent cloud and the consequent disk assembling.

We find that the interplay between the cloud collapse, the dynamics of gas and dust and thermal processing of different species in the disk, results in a local mixing of solids with different thermal histories. Our simulations return an heterogeneous distribution of refractory material with higher concentration in the outer disk. This refractory material has a short formation timescales, during the first tens of kyr of the Sun (class 0-I). Our results open new frontiers into the origin of the compositional diversity of chondrites.

Keywords. solar system: formation, meteors, meteoroids, stars: formation, (stars:) planetary systems: protoplanetary disks

1. Introduction

Chondrites are undifferentiated meteorites formed within the first 4 Myr of the Solar System history (with the age determined by the oldest compoments, Ca-Al-rich inclusions, CAIs). They are a mixture of components formed at high and low temperature (see Scott & Krot (2003)). This peculiar mixture is one important ingredient in the composition of terrestrial planets and its formation processes are still unknown. A paradox is that carbonaceous chondrites, which are thought to have accreted in the outer Solar System at temperatures lower than 150 K (see Warren (2011)) are enriched in CAIs (formed at $T > 1600$ K) and close to the Sun) when compared with other chondrites (see Scott & Krot (2003)).

The chemical composition of CAIs links them with condensate precursors (see Richter *et al.* (2006)) formed during a short interval of time, around $t \sim 160$ kyr (see Connelly *et al.* (2012)). This timescale is shorter than the formation time of a protoplanetary disk from its parent collapsing cloud, and suggests that the formation of the first solids occurred simultaneously with the building of our Solar Nebula. This is supported by observational evidence of young sellar objects that show crystalline dust in their infrared spectra as soon as the star begins to form, as described by Williams & Cieza (2011) and Chiang *et al.* (2012).

In order to understand the formation and accretion processes of chondrites components, we developed a 1D simulation to track the thermal evolution of dust with

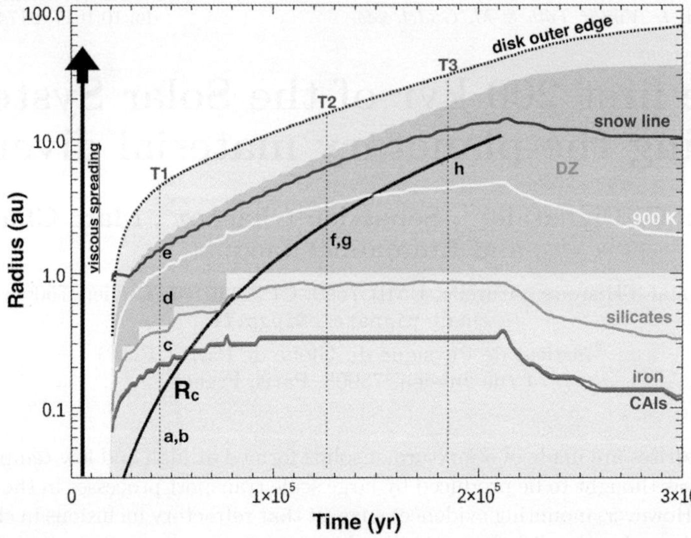

Figure 1. Time evolution of the centrifugal radius (black solid line) and the locations of disk outer edge, different condensation fronts, and dead zone (dark grey). Figure adapted from Pignatale *et al.* (2018) and reproduced by permission of the AAS.

different composition during the collapse of the parent cloud and the formation of the protoplanetary disk.

2. Method

We model a classical alpha-disk starting from Hueso & Guillot (2005) that considers a collapsing cloud with radiative and viscous heating. Our 1D disk is described by a logarithmic grid (100 cells, between 0.01 and 1000 au). We include the modellisation of a dead zone (DZ) in the form of a parameterized description. We consider a layered accretion model with a low-α midplane and high-α surface layers, dust transport through advection and diffusion, and growth/fragmentation. We also include gravitational instability. The opacity is calculated using an opacity table. A detailed description of the code, methods and references is reported in Pignatale *et al.* (2018) and in Charnoz *et al.* (2019).

Improving on the work of Yang & Ciesla (2012), we now include multiple chemical species allowing material condensation, sublimation and thermal processing using a simple chemical model. We then track the formation and transport in the disk of gaseous species and different solid species with their vaporisation temperatures: (i) condensed refractory phases, that we consider potential precursors of CAIs, (ii) condensed metal (iii) condensed silicates, (iv) thermally processed and unprocessed silicates, (v) thermally processed and unprocessed iron, (vi) water ice, and (vii) CO ice. The main gas species is molecular hydrogen.

3. Results

Figure 1 shows the time evolution of the centrifugal radius ($R_c(t)$), the locations of disk outer edge, different condensation fronts, and dead zone (dark grey). As the dust is injected into the forming disk, its composition and thermal structure change because of heating of species with different vaporisation temperatures. Because of angular momentum constraints, R_c(t) is initially close to the forming Sun fed from the parent cloud's innermost regions (see Kimura *et al.* (2016)). Then R_c(t) increases with time due to the arrival of originally more distant cloud materials with higher specific angular momentum.

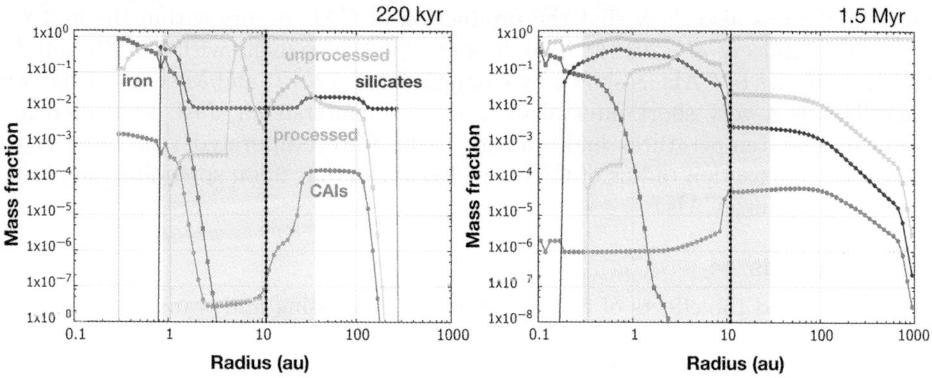

Figure 2. Mass fraction of various types of rocky material in the disk as a function of radius and time. The different types of solids considered are CAIs (purple), metallic iron (red), condensed silicates (brown), all processed dust (green), all unprocessed dust (cyan). Figure adapted from Pignatale *et al.* (2018) and reproduced by permission of the AAS.

At $t \sim 50$ kyr (T1), $R_c(t)$ is within the condensation fronts of the most refractory species (CAIs and iron) so that all the dust injected with the gas (a) is vaporised (b). At the same time, further out in the disk, condensation of CAIs and iron (c) silicates (d) takes place from a gas which is spreading outwards. Iron condensed and transported with the spreading gas can be processed further out (e). At $t \sim 120$ kyr (T2), R_c(t) has reached a region in the disk where the temperature is below 1500 K. Presolar refractory material and silicates can be injected into the disk without being vaporised (f) but they can experience thermal processing (g). At $t \sim 180$ kyr (T3), parent cloud material is injected in the disk likely retaining its primordial texture as the temperature is too low for thermal processing (h).

Figure 2 shows two time snapshots of the mass fraction of the considered rocky material as a function of the distance from the central star. The light grey zone shows the extension of the dead zone. The black dotted line shows the location of the centrifugal radius $R_c(t)$. It can be seen that, at the end of collapse, $t \sim 220$ kyr, $R_c(t)$ penetrates the DZ where dust and gas can be stored because of very low turbulence. This lowers dramatically the concentration of refractory condensates within the DZ. This leads to a DZ poor in refractory condensates and a CAIs-enriched outer disk beyond the DZ. This dust distribution will qualitatively preserve its heterogeneity over the lifetime of the disk ($t \sim 1.5$ Myr). Inward the DZ inner edge ($R < 0.3$ AU), the disk becomes CAIs-poor because of strong accretion of gas that removes solids.

The resulting distribution for the various dust components in the disk ($t \sim 1.5$ Myr) shows an overall agreement with the composition of enstatite, ordinary, carbonaceous chondrites as defined by their fractions of CAIs, metal, chondrules and matrix (see Scott & Krot (2003)). We consider the unprocessed dust as a proxy for the fine-grained matrix containing different kinds of circumstellar grains which are likely surviving interstellar grains. In the inner disk region ($1 < R < 10$ au), dust is dominated by disk-born silicates and iron (processed or condensed) and poor in CAIs and pristine interstellar dust. This reminds of (matrix-poor, chondrule-rich) non-carbonaceous chondrites (ordinary chondrites and enstatite). At larger distances from the central star, the dust is dominated by unprocessed dust that was injected late when $R_c(t) > 5$ au, processed dust and condensed silicates. This region, although beyond the snow-line and having water ice-bearing dust, has the highest fraction of CAIs. This reminds the carbonaceous chondrites.

Our simulations also show that the production of CAIs occurs within the first $t \sim 80$ kyr (see Pignatale *et al.* (2018)). This in very good agreement with Pb–Pb and ^{26}Al–^{26}Mg chronologies of CAIs as shown by Connelly *et al.* (2012) and Mishra & Chaussidon (2014). This is a very short interval of time, when interstellar dust is injected in the forming disk at temperatures high enough to be fully evaporated (i.e. when $R_c(t)$ is inward the condensation radius of CAIs, see Fig.1), so that upon spreading outward, the gas re-condense into CAIs.

4. Conclusions

We investigated the effects of coupling thermal processing and transport of dust during the collapse of a parent cloud and the consequent disk formation. This naturally results in a local mixing of materials with different thermal histories. Our resulting dust radial distribution shows qualitative agreement with enstatite, ordinary and carbonaceous chondrites. Our model reconciles with the overabundance of refractory materials at large disk radii and their formation timescales, placing the time 0 of the Solar System during the class 0 to I of the Sun's formation.

References

Charnoz, S., Pignatale, F.C., Hyodo, R., *et al.* 2019, *A&A*, Astronomy and Astrophysics, 627, A50.

Chiang, H.-F., Looney, L. W., & Tobin, J. J. 2012, *ApJ*, The Astrophysical Journal, 756, 168

Connelly, J. N., Bizzarro, M., Krot, A. N., *et al.* 2012, Science, 338, 651

Hueso, R., & Guillot, T. 2005, *A&A*, Astronomy and Astrophysics, 442, 703

Kimura, S. S., Kunitomo, M., & Takahashi, S. Z. 2016, *MNRAS*, Monthly Notices of the Royal Astronomical Society, 461, 2257

Mishra, R. K., & Chaussidon, M. 2014, Earth and Planetary Science Letters, 390, 318

Pignatale, F. C., Charnoz, S., Chaussidon, M., & Jacquet, E. 2018, *ApJL*, 867, L23

Richter, F. M., Mendybaev, R. A., & Davis, A. M. 2006, Meteoritics and Planetary Science, 41, 83

Scott, E. R. D., & Krot, A. N. 2003, Treatise on Geochemistry, 1, 711

Yang, L., & Ciesla, F. J. 2012, Meteoritics and Planetary Science, 47, 99

Warren, P. H. 2011, Geo Cos Acta, 75, 6912

Williams, J. P., & Cieza, L. A. 2011, *ARAA* Annual Review of Astronomy and Astrophysics, 49, 67

Discussion

LISSAUER: What causes the differences you find between the distribution of condensed iron and silicates which have similar condensation temperature in standard model?

PIGNATALE: This depends on the steepness of the resulting temperature gradient as a function of the distance from the central star. And also on the fiducial condensation temperatures chosen for the selected species.

LINSKY: Thank you for the excellent presentation of a thermal steady state simulation. However star flares and interaction between the stellar and disk magnetic field can produce heat, and episodic UV and X-ray emission that can lead to different results from your calculations.

PIGNATALE: That is right. However this will not affect how the material will distribute during the disk building, and temperature in the inner disk is dominated by the dissipation of turbulence. Nevertheless, the events you mentioned can actually make the production, alteration and transport of refractory material easier.

Origins: From the Protosun to the First Steps of Life
Proceedings IAU Symposium No. 345, 2018
B. G. Elmegreen, L. Viktor Tóth & M. Güdel, eds.

doi:10.1017/S1743921319002308

Formation of terrestrial planets

Eiichiro Kokubo

National Astronomical Observatory of Japan,
2-21-1, Osawa, Mitaka, Tokyo 181-8588 Japan
email: kokubo.eiichiro@nao.ac.jp

Abstract. In the standard formation scenario of planetary systems, planets form from a protoplanetary disk that consists of gas and dust. The scenario can be divided into three stages: (1) formation of planetesimals from dust, (2) formation of protoplanets from planetesimals, and (3) formation of planets from protoplanets. In stage (1), planetesimals form from dust through coagulation of dust grains and/or some instability of a dust layer. Planetesimals grow by mutual collisions to protoplanets or planetary embryos through runaway and oligarchic growth in stage (2). The final stage (3) of terrestrial planet formation is giant impacts among protoplanets while sweeping residual planetesimals. In the present paper, we review the elementary processes of terrestrial planet formation and discuss the extension of the standard scenario.

Keywords. solar system: formation, planets and satellites: formation

1. Introduction

Needless to say, the formation of terrestrial planets is one of the most important problems in modern planetary astrophysics. At the beginning we overview the important properties of the terrestrial planets in the solar system when we consider their origin. The terrestrial planets have a bimodal mass distribution, namely, large planets with $\sim M_\oplus$ (Venus and Earth) and small planets with $\sim 0.1 M_\oplus$ (Mercury and Mars), where $M_\oplus$ is the Earth mass (Fig. 1). Their orbits have the semimajor axis of $\simeq$0.4-1.5 au in which the large planets reside inside and the small ones outside. The large planets have smaller eccentricities and inclinations ($e, i \sim 0.01$) than the small ones. The obliquity of the planets ranges from 0° to about 180° and the spin periods are from ~ 1 d to ~ 100 d. The effective spin period for the Earth-Moon system is $\simeq 4$ hr. The core (iron) fraction of the planets ranges from 20% to 70%. Note that all these properties affect the habitability of planets.

In the standard scenario for solar system formation, planets form from a protosolar or protoplanetary disk that consists of gas and dust (e.g., Hayashi *et al.* 1985, Raymond *et al.* 2014). The formation scenario can be divided into three stages: (1) formation of planetesimals from dust, (2) formation of protoplanets or planetary embryos from planetesimals, and (3) formation of planets from protoplanets (Fig. 2). In stage (1), planetesimals form from dust through coagulation of dust grains and/or some instability of a dust layer. Planetesimals are small building blocks of solid planets. Planetesimals grow by mutual collisions to protoplanets through runaway and oligarchic growth in stage (2). The final stage (3) depends on the type of planets. The final stage of terrestrial planet formation is giant impacts among protoplanets while sweeping residual planetesimals. This scenario gives a general framework of terrestrial planet formation though it still has many problems.

In the present paper, we review the basic elementary processes of terrestrial planet formation, showing some recent simulations (e.g., Kokubo & Ida 2012). We also discuss the extension of the standard scenario and the origin of the diversity of terrestrial planets.

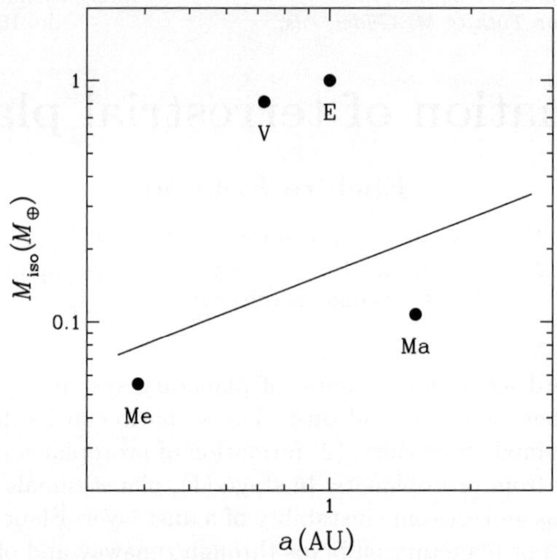

Figure 1. Mass of the terrestrial planets, Mercury, Venus, Earth, and Mars against the semimajor axis. The solid line shows the isolation mass of protoplanets for the standard protoplanetary disk model.

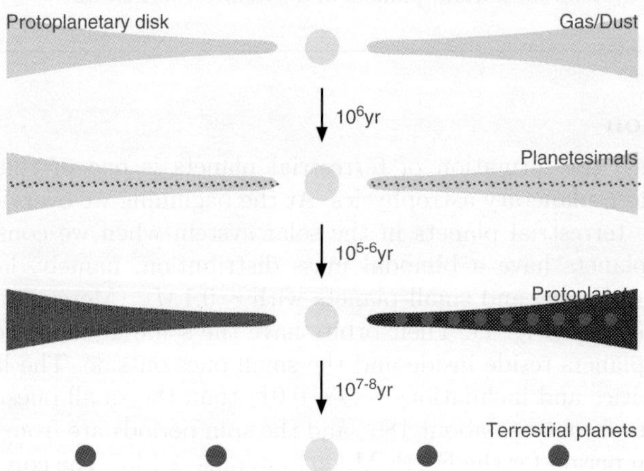

Figure 2. Schematic illustration of the standard scenario of terrestrial planet formation.

2. Planetesimal Formation

The formation of planetesimals from dust is one of the most important unsolved problems in planet formation theory. Since planetesimals are building blocks of planets, understanding their formation is a key to understand the structure and evolution of protoplanetary systems.

There are two classical models, the pairwise coagulation of (compact) dust grains (e.g., Weidenschilling & Cuzzi 1993) and the (dynamical) gravitational instability of a dust layer (e.g., Goldreich & Ward 1973). The coagulation model has serious difficulties known as fragmentation, bouncing and radial drift barriers, while the protoplanetary gas disk seems too turbulent for the gravitational instability model to operate (for a review, see e.g., Chiang & Youdin 2010).

A few new models that can possibly overcome these difficulties have been proposed recently. In the streaming instability model, due to clogging of dust grains through radial drift, dust clumps develop to form planetesimals (e.g., Youdin & Goodman 2005). In the

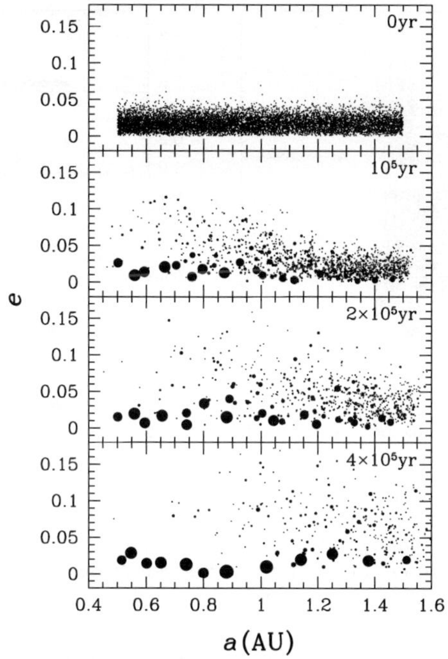

Figure 3. Snapshots of a planetesimal system on the semimajor axis-eccentricity plane at $t = 0, 10^5, 2\times10^5, 2\times10^5$, and 4×10^5 years (Kokubo & Ida 2002). The size of circles is proportional to the radius of planetesimals.

secular gravitational instability model, the gravitational instability of a dust layer driven by gas drag can take place on the longer timescale than the classical dynamical instability (e.g., Michikoshi *et al.* 2012). The porous dust model assumes a pairwise coagulation of low density dust aggregates. In this model, porous dust aggregates with fractal dimension of about 2 grow by pairwise coagulation. They can grow beyond the three barriers because of their low-density (e.g., Kataoka *et al.* 2013). As the porous aggregates grow their layer becomes gravitationally unstable to fragment to form planetesimals (e.g., Michikoshi & Kokubo 2016).

The planetesimal formation process depends on the condition of the protoplanetary disk. The high-resolution observation of the early stage of protoplanetary disks is necessary to understand the realistic formation process.

3. Protoplanet Formation

The formation of protoplanets from planetesimals is relatively well understood, compared with planetesimal formation (Fig. 3). For the details of the basic dynamical models, see Kokubo & Ida (2012).

We assume that the relative growth rate of planetesimals with mass M by collisions among themselves $(1/M)(\mathrm{d}M/\mathrm{d}t)$ is proportional to M^p. There are two modes of planetesimal growth, namely orderly $(p<0)$ and runaway $(p>0)$ (e.g., Kokubo & Ida 1996). In the orderly growth planetesimals tend to grow equally, while larger planetesimals grow faster than smaller ones in the runaway growth.

In the course of planetesimal growth $(1/M)(\mathrm{d}M/\mathrm{d}t)$ can be approximated as

$$\frac{1}{M}\frac{\mathrm{d}M}{\mathrm{d}t} \propto M^{1/3} v_{\mathrm{ran}}^{-2}, \tag{3.1}$$

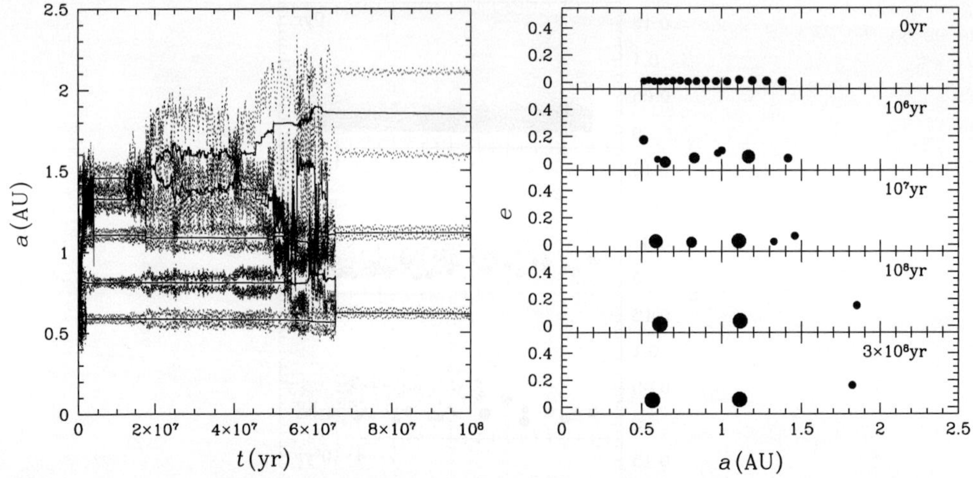

Figure 4. Evolution of a protoplanet system on the time-semimajor axis plane (left) and the semimajor axis-eccentricity plane at $t = 0, 10^6, 10^7, 10^8$, and 3×10^8 years (right) (Kokubo *et al.* 2006). The dashed curves present apocenter and pericenter distances (left) and the size of circles are proportional to the radius of planets (right).

where $v_{\rm ran}$ is the random velocity of planetesimals, which is a deviation velocity from a non-inclined circular orbit approximately given by $(e^2 + i^2)^{1/2} v_{\rm K}$, where e and i are the eccentricity and inclination of planetesimals and $v_{\rm K}$ is the Kepler circular velocity. Thus it is the random velocity of planetesimals that controls the growth mode and timescale. There exists the equilibrium random velocity at which the effects of gravitational scattering among planetesimals and drag form the disk gas balance. At least in the initial stage of planetesimal growth, the growth mode is runaway, where the random velocity is independent of the mass of a growing planetesimal. In this case $(1/M)({\rm d}M/{\rm d}t) \propto M^{1/3}$, which shows the runaway growth. We call a runaway-growing planetesimal as a protoplanet.

Once the mass of a protoplanet exceeds about a hundred times the mass of surrounding planetesimals, the runaway growth slows down to shift to the oligarchic growth where the random velocity of planetesimals becomes proportional to the Hill radius of the protoplanet ($r_{\rm H} \propto M^{1/3}$). In this case $(1/M)({\rm d}M/{\rm d}t) \propto M^{-1/3}$, which means the orderly growth. Among protoplanets orbital repulsion keeps their orbital separation $b \simeq 10 r_{\rm H}$. In summary, on the oligarchic growth stage, protoplanets dominantly grow in the orderly way with certain orbital separations (e.g., Kokubo & Ida 1998). Given the orbital separation b of adjacent pairs we can estimate the isolation or final mass of protoplanets formed by the oligarchic growth. For the standard disk with the planetesimal surface density $\Sigma_{\rm d} = 10(a/1\,{\rm au})^{-3/2}\,{\rm g\,cm^{-2}}$, we obtain

$$M_{\rm iso} \simeq 2\pi a b \Sigma_{\rm d} = 0.16 \left(\frac{b}{10 r_{\rm H}}\right)^{3/2} \left(\frac{a}{1\,{\rm au}}\right)^{3/4} M_\oplus \tag{3.2}$$

(Kokubo & Ida 2002). The mass of protoplanets around the terrestrial planet region is on the order of 0.1 $M_\oplus$, in other words, about Mercury to Mars mass (Fig. 1).

4. Giant Impacts

In order to complete the large terrestrial planets such as Venus and Earth, further accretion among protoplanets is necessary. While protoplanets are in the disk gas, their orbits are stable since their eccentricities are damped by the gravitational drag from gas (e.g., Tanaka & Ward 2004). However, from the observations, the lifetime of disk gas is estimated to $\lesssim 10^7$ years. After the disk gas dispersal, protoplanets gradually build up their eccentricities by gravitational interaction. It is empirically known from the numerical experiments that for an equal-mass protoplanet system with equal orbital separations in the mutual Hill radius the instability timescale of the systems is given as

$$\log t_{\rm inst} \simeq c_1(b/r_{\rm H}) + c_2, \tag{4.3}$$

where c_1 and c_2 are the functions of e and i (e.g., Chambers *et al.* 1996; Yoshinaga *et al.* 1999). In this timescale close encounters between protoplanets occur and the giant impact stage begins, where protoplanets collide with one another.

The giant impacts continue until the system becomes orbitally stable where all orbits are nearly circular and widely separated. In other words, the mass and orbital properties of planets are determined by the system stability. The timescale of this stage is on the order of $\sim 10^8$ years (e.g., Chambers & Wetherill 1998). A typical outcome for the standard disk is two Earth-sized planets and one or two leftover protoplanets (Fig. 4) (e.g., Kokubo *et al.* 2006). Their eccentricities and inclinations are $\simeq 0.1$ that are higher than those of Venus and Earth. Therefore some damping is necessary after formation, which may be dynamical friction from residual planetesimals (collisional debris) and/or disk gas.

The spin parameters are also determined on this stage (Kokubo & Genda 2010). The typical spin angular velocity is about 70% of the critical spin angular velocity for rotational breakup

$$\omega_{\rm cr} = 3.3 \left(\frac{\rho}{3\,{\rm gcm}^{-3}}\right)^{1/2} {\rm hr}^{-1}, \tag{4.4}$$

where ρ is the internal density of protoplanets, and is independent of the planet mass. The obliquity distribution obeys an isotropic distribution

$$n d\varepsilon = \frac{1}{2} \sin \varepsilon d\varepsilon, \tag{4.5}$$

which means that the typical obliquity is around 90°. This isotropic distribution is an outcome of isotropic collisions of protoplanets due to much larger scale-height of the protoplanet system than the physical size of planets.

5. Extension of the Standard Scenario

The basic dynamical model of terrestrial planet formation gives us a general framework. However we may need to modify it when we apply it to individual systems.

For simplicity, the standard scenario assumes a continuous power-law disk except for the snow-line, in-situ formation, and km-sized planetesimals as building blocks. Of course, these assumptions may not be always justified in a realistic system. Now many authors are extending the standard scenario by including the effects of an evolving disk, radial migration of planet-building materials, and larger planetesimals and small pebbles. Here we focus on the localization of terrestrial planets in the solar system and introduce two models to explain it.

In the solar system, terrestrial planets reside in the limited radial range $0.4\,{\rm au} \lesssim a \lesssim 1.5\,{\rm au}$ and the two planets, Mercury and Mars, near the inner and outer edges are smaller than those in the middle, Venus and Earth. Explaining this property of the mass distribution is often referred to as the small Mars problem. Hansen (2009) showed that if planets are accreted from a narrow ring of protoplanets and planetesimals with $0.7\,\mathrm{au} \lesssim a \lesssim 1.0\,\mathrm{au}$, the mass distribution of the planets is easily reproduced. Then the question is how to form such a narrow ring. Here the so called grand-tack model comes in (Walsh *et al.* 2011). In the model first Jupiter migrates inward to cut off the planetesimal disk around 1 au and then tacks to migrate outward with Saturn. This model succeeds in forming a narrow ring, depletion of the asteroid belt, and water delivery to the Earth. Note that the grand-tack model does not explain the inner cut-off of the planetesimal disk.

Another model is the disk bump model. The evolution of a protoplanetary disk with viscous accretion and disk winds naturally leads to formation of a global radial pressure bump in the disk (Suzuki *et al.* 2016). Around the pressure bump planetesimals accumulate due to gas drag, which forms a narrow planetesimal ring. In the evolved gas disk planetesimals and protoplanets barely migrate radially due to an almost flat surface density distribution of gas. The accretion from such a disk can reproduce the localization of terrestrial planets (Ogihara *et al.* 2018).

6. Summary

We have reviewed the standard dynamical model of terrestrial planet formation in which rocky dust grains grow through planetesimals and protoplanets to terrestrial planets on the timescale of 10^8 years. At the present after the discovery of diverse terrestrial planets in exoplanetary systems, the formation model is being modified and extended. This extension includes a discrete distribution of planetesimals, formation with migration, collisional disruption, and pebbles as building blocks. We are aiming to construct a general comprehensive formation model that explains not only the solar system terrestrial planets but also the diverse terrestrial planets in exoplanetary systems.

References

Chambers, J. E., Wetherill, G. W., & Boss, A. P. 1996, *Icarus*, 119, 261

Chiang, E., & Youdin, A. N. 2010, *AREPS*, 38, 493

Goldreich, P., & Ward, W. R. 1973, *ApJ*, 183, 1051

Hansen, B. M. S. 2009, *ApJ*, 703, 1131

Hayashi, C., Nakazawa, K., & Nakagawa, Y. 1985, *Protostars and Planets II*, 1100

Kataoka, A., Tanaka, H., Okuzumi, S., & Wada, K. 2013, *A&A*, 557, L4

Kokubo, E., & Genda, H. 2010, *ApJL*, 714, L21

Kokubo, E., & Ida, S. 1996, *Icarus*, 123, 180

Kokubo, E., & Ida, S. 1998, *Icarus*, 131, 171

Kokubo, E., & Ida, S. 2002, *ApJ*, 581, 666

Kokubo, E., & Ida, S. 2012, *Progress of Theoretical and Experimental Physics*, 01A308

Kokubo, E., Kominami, J., & Ida, S. 2006, *ApJ*, 642, 1131

Michikoshi, S., Kokubo, E., & Inutsuka, S. 2012, *ApJ*, 746, 35

Michikoshi, S., & Kokubo, E. 2016, *ApJL*, 825, L28

Ogihara, M., Kokubo, E., Suzuki, T. K., & Morbidelli, A. 2018, *A&A*, 612, L5

Raymond, S. N., Kokubo, E., Morbidelli, A., Morishima, R., & Walsh, K. J. 2014, *Protostars and Planets VI*, 595

Suzuki, T. K., Ogihara, M., Morbidelli, A., Crida, A., & Guillot, T. 2016, *A&A*, 596, A74

Tanaka, H., & Ward, W. R. 2004, *ApJ*, 602, 388

Walsh, K. J., Morbidelli, A., Raymond, S. N., O'Brien, D. P., & Mandell, A. M. 2011, *Nature*, 475, 206

Weidenschilling, S. J., & Cuzzi, J. N. 1993, *Protostars and Planets III*, 1031

Yoshinaga, K., Kokubo, E., & Makino, J. 1999, *Icarus*, 139, 328

Youdin, A. N., & Goodman, J. 2005, *ApJ*, 620, 459

Discussion

LINSKY: Please comment on the role of resonance in determining the localization of planets. TRAPPIST-1 has 7 planets all in resonances.

KOKUBO: Capturing into resonances is possible when planets migrate in a gas disk. Since the giant impact stage starts after disk gas dispersal, planets formed by giant impacts are not in resonances.

MAROV: Did you try to figure out in the collisional processes assembly/repulsion ratios depending on size particles specifically beyond the size barrier, and how it depends on the assumed fractal dimension?

KOKUBO: We investigated the accretionary conditions of planetesimals and protoplanets and used them in the N-body simulations.

Origins: From the Protosun to the First Steps of Life
Proceedings IAU Symposium No. 345, 2018
B. G. Elmegreen, L. Viktor Tóth & M. Güdel, eds.

doi:10.1017/S1743921319001455

Formation of the Earth-Moon system

Sergei I. Ipatov

Vernadsky Institute of Geochemistry and Analytical Chemistry of RAS,
119991, 19 Kosygin st., Moscow, Russia
email: siipatov@hotmail.com

Abstract. Trans-Neptunian satellite systems and embryos of the Earth-Moon system could be formed as a result of contraction of rarefied condensations. The angular momenta of rarefied condensations needed for such formation could be acquired at collisions of condensations. The angular momentum of the present Earth-Moon system could be acquired at a collision of two rarefied condensations with a total mass not smaller than $0.1M_E$, where M_E is the mass of the Earth. The mass of the condensation that was a parent for the embryos of the Earth and the Moon could be about $0.01M_E$, if we take into account the growth of the angular momentum of the embryos with growth of their masses. The Moon embryo could get by an order of magnitude more material ejected from the Earth embryo than that fell directly onto the Moon embryo.

Keywords. methods: n-body simulations, Earth, Moon, solar system: formation

1. Introduction

Galimov & Krivtsov (2012) presented arguments that the popular giant impact concept (e.g., Canup *et al.* 2013) of the formation of the Moon has several weaknesses. They suggested that solid embryos of the Earth and the Moon have been formed from the same rarefied dust condensation. Formation of massive (up to $0.1M_E - 0.6M_E$, where M_E is the mass of the Earth) condensations was studied by several scientists (e.g. Lyra *et al.* 2008). Surville *et al.* (2016) considered a possible formation of a narrow dust disk with a mass up to $0.6M_E$ and the width of $(2-3)10^{-3}$ of the distance from the Sun. Ipatov (2010, 2014, 2017a,b) and Nesvorny *et al.* (2010) studied formation of trans-Neptunian satellite systems as a result of contraction of rarefied condensations. In my opinion, the model of formation of trans-Neptunian satellite systems and that of formation of the Earth-Moon system can be similar.

2. Formation of solid embryos of the Earth-Moon system at the stage of rarefied condensations

Ipatov (2014, 2017a) showed that the angular momentum of a condensation needed in calculations by Nesvorny *et al.* (2010) for formation of a satellite system could be acquired at a collision of two condensations moved around the Sun in circular orbits. The radii of the condensations did not differ much from their Hill radii. The angular momentum K_{sEM} of the present Earth-Moon system could be acquired at a collision of two rarefied condensations with radii about their Hill radii and with a total mass not smaller than $0.1M_E$ (Ipatov 2015, 2018a). The mass of the condensation that was a parent for the embryos of the Earth and the Moon could be relatively small ($0.01M_E$ or even less), if we take into account the growth of the angular momentum of the embryos with growth of their masses. The angular momentum K_s of the condensation used by Galimov & Krivtsov (2012) in their computer simulations of the formation of the embryos of the

Earth-Moon system as a result of contraction of a rarefied condensation could not be acquired during formation of the condensation from a protoplanetary disk.

Some fraction of the angular momentum of the parental condensation could be acquired by its accumulation of small objects. If we consider accumulation only of small objects, then $K_s = K_{sEM}$ for a parental condensation with mass $m > 0.2M_E$. However, for such accumulation of only small objects, other terrestrial planets could have large satellites because their parental condensations also could get the angular momenta needed for formation of satellites. On the other hand, large concentration of matter could be only at 1 AU. Probably, the condensations that contracted and formed the embryos of the terrestrial planets other than the Earth did not collide with similar massive condensations, and therefore they did not get a large enough angular momenta needed for formation of massive satellites. There could be also the second main collision (or a series of similar collisions) of condensations or solid bodies that changed the tilt of the Earth. For the second main collision of condensations, the radius of the Earth embryo condensation had to be smaller than the semi-major axis of the orbit of the Moon embryo condensation. Note that for two condensations with masses equal to $0.01M_E$ moved in circular close orbits with semimajor axes about 1 AU, their typical collision would be in about 100 yrs. During this time the condensations could approximately keep their original sizes.

3. Growth of solid embryos of the Earth-Moon system

First let us consider the model for which the solid embryos of the Earth-Moon system that formed from the parental condensation both grew only by accumulation of smaller objects from the same source. In the case of small relative velocities of planetesimals, effective radii r_{ef} of the embryos are proportional to r^2, where r is the radius of a considered embryo. In this case,

$$m_{Mo}^{-1/3} = m_M^{-1/3} + k_2 m_{Eo}^{-1/3} - k_2 m_E^{-1/3}, \tag{1}$$

where $k_2 = k_d^{1/3}$, k_d is the ratio of the density of the growing Earth of mass m_E to that of the growing Moon of mass m_M ($k_d \approx 1.65$ for the present Earth and Moon), m_{Mo} and m_{Eo} are initial values of m_M and m_E. From equation (1) one can obtain that the mass of the Moon embryo grew by a factor of 2 - 2.3 while the Earth embryo grew by a factor of 10. At $r_{ef} \propto r^2$, the embryo of the Earth grew faster than that of the Moon. For large enough eccentricities of planetesimals, the effective radii of proto-Earth and proto-Moon were proportional to r. In this case

$$m_{Mo}^{1/3} = m_M^{1/3} + k_1 m_{Eo}^{1/3} - k_1 m_E^{1/3} \tag{2}$$

(where $k_1 = k_d^{2/3}$) and the increase of m_M/m_{Mo} is greater than that of m_E/m_{Eo}.

Galimov & Krivtsov (2012) supposed that initial embryos of the Earth and the Moon were depleted in iron, and the Earth got a larger fraction of iron than the Moon because it grew faster by accumulation of dust. To estimate the maximum growth of m_M, let us consider the following simple model: The fraction of iron in the initial embryos is denoted by β, and the incoming material contained 33% of iron. The fraction of iron in the Moon would be equal to $0.33 + (\beta - 0.33)m_{rMo}$, where m_{rMo} is the ratio of the initial mass of the Moon embryo to the present mass of the Moon. Taking the present fraction of iron in the Moon to be equal to 8% and solving $0.33 + (\beta - 0.33)m_{rMo}$)=0.08, we get $m_{rMo} \geqslant 0.76$ and the growth of the Moon embryo mass by a factor of not more than 1.3. As $m_{rMo} \leqslant 1$, then this equation can be solved only for $\beta \geqslant 0.08$. At r_{ef} proportional to r^2 from formula (1), one can obtain that for the growth of the mass of the Moon embryo from $M_M/1.3$ to the present mass M_M of the Moon, the mass of the Earth embryo grew (to M_E) by a factor of 2.4 or 2.7 at k_d equal to 1.65 and 1, respectively. In this case for the above simple model, the fraction of iron in the Earth did not exceed 0.33(1-1/2.7)≈0.21 and was less than the present value equaled to 0.32. In order to obtain the current iron content in the Earth and the Moon, the increment dm of the embryo mass should be proportional to m^p, where $p \geqslant 2$, as it was considered by Galimov & Krivtsov (2012). For the motion of solid bodies in a gas-free medium, parameter p does not exceed 4/3. In my opinion, if the embryos are assumed to have grown only by accumulation of solid planetesimals (without the ejection of matter from the embryos), it is hard to reproduce the current lunar and terrestrial iron abundances at any initial abundance in the embryos.

Besides direct collisions with planetesimals, the Moon embryo should also grew by accumulation of iron-depleted material ejected from the Earth embryo at impacts of planetesimals with the Earth embryo. If the iron abundance in the initial Moon embryo and in planetesimals was 0.33 and the iron abundance in the crust of the Earth and on the Moon was 0.05 and 0.08, respectively, than the fraction k_E of matter of the Earth crust in the Moon should be about 0.9 (this follows from the relation $0.05k_E + 0.33(1 - k_E) = 0.08$). The greater fraction of matter incorporated into the Moon embryo could be ejected from the Earth in its multiple collisions with planetesimals (and smaller bodies). This model differs from the known multiple impact models (e.g., Rufu & Aharonson 2017) by that the initial embryo of the Moon in my model was formed from the same rarefied condensation, as the Earth embryo, but not from a disk of material ejected from the Earth embryo. The previous multiple impact models do not explain clearly why only the Earth, but not other terrestrial planets, formed with a large satellite. In my model it is not needed to form the initial Moon embryo from material ejected from the Earth. The model of the formation of a solid planet with a large satellite can also work for some exoplanet.

Based on our runs of migration of planetesimals from the feeding zone of Jupiter and Saturn (Marov & Ipatov 2018) and from different distances from the Sun (from 5 to 40 AU) (Ipatov 2018b) and of migration of Jupiter-crossing objects (Ipatov & Mather 2007), I calculated probabilities of collisions of such planetesimals and objects with the Moon. Such probabilities were typically smaller than the probabilities of collisions with the Earth by a factor of 16 or 17 for the planetesimals and many Jupiter-family comets. For some Jupiter-family comets, the factor was up to 25. Based on the above results, we can conclude that the amount of the material, including water, collided with the Moon and migrated from outside of the Jupiter's orbit could be smaller by about a factor of 20 than that delivered to the Earth from this region. However, a greater fraction of water could be evaporated at such collisions for the Moon than for the Earth.

I also studied the migration of planetesimals from different regions of the feeding zone of the terrestrial planets and the probabilities of their collisions with these planets or their embryos. When the embryos were small, they accumulated planetesimals from their neighbourhood. For masses of the embryos close to masses of the terrestrial planets, for different source regions of the planetesimals, the fraction of planetesimals collided with Earth differed by less than a factor of 2 from that with Venus. The ratio of planetesimals collided with the Earth to those collided with the Moon was about 35 for planetesimals initially moved in low eccentric orbits near the Earth's orbit and was about 25 for more distant regions of the feeding zone of the terrestrial planets. In calculations with masses and orbits of planets close to their present values, most of planetesimals collided with the Earth and the Moon mainly in 20 Myr both for planetesimals from the feeding zone of the terrestrial planets and from initial distances between 5 and 30 AU. However, some planetesimals from the feeding zones of Uranus and Neptune could collide the Earth and the Moon after hundreds of Myrs.

4. Conclusions

The embryos of the Earth and the Moon could form as a result of contraction of the same parental rarefied condensation. A considerable fraction of the angular momentum of such condensation could be acquired at a collision of two rarefied condensations. The present angular momentum of the Earth-Moon system could be acquired at the collision of two identical rarefied condensations with sizes of their Hill spheres, which total mass was about 0.1 of the mass M_E of the Earth and which heliocentric orbits were circular. The initial mass of the rarefied condensation that was a parent for the embryos of the Earth and the Moon could be relatively small ($0.01M_E$ or even less) if we take into account the growth of the angular momentum of the embryos during growth of masses of the embryos. The Moon embryo could get by an order of magnitude more material ejected from the Earth embryo than that fell directly onto the Moon embryo.

The work was supported by the grant of Russian Science Foundation N 17-17-01279.

References

Canup, R. M., Barr, A. C., & Crawford, D. A. 2013, *Icarus*, 222, 200

Galimov, E. M., & Krivtsov, A. M. 2012, *Origin of the Moon. New concept* (Berlin: De Gruyter), 168 p

Ipatov, S. I. 2010, *MNRAS*, 403, 405

Ipatov, S. I. 2014, in N. Haghighipour (ed.), *Formation, detection, and characterization of extrasolar habitable planets*, Proc. of the IAU, vol. 8., Symp. S293 (Cambridge: Cambridge Univ. Press), p. 285, http://arxiv.org/abs/1412.8445

Ipatov, S. I. 2015, *SPACEKAZAN-IAPS-2015* (Kazan: Kazan University), p. 97, http://arxiv.org/abs/1607.07037

Ipatov, S. I. 2017a, *Solar System Research*, 51, 294, https://arxiv.org/abs/1801.05217

Ipatov, S. I. 2017b, *Solar System Research*, 51, 409, https://arxiv.org/abs/1801.05254

Ipatov, S. I. 2018a, *Solar System Research*, 52, 401

Ipatov, S. I. 2018b, *The Ninth Moscow Solar System Symposium 9M-S3* (Space Research Institute, Moscow, Russia, October 8-12, 2018), 9MS3-SB-11, p. 104, https://ms2018.cosmos.ru/

Ipatov, S. I. & Mather, J. C. 2006, *Advances in Space Research* 37, 126, http://arxiv.org/format/astro-ph/0411004

Ipatov, S. I. & Mather, J. C. 2007, in: G. B. Valsecchi, D. Vokrouhlicky, & A. Milani (eds.), *Near-Earth Objects, Our Celestial Neighbors: Opportunity and Risk*, Proc. IAU Symp. No. 236 (Cambridge: Cambridge Univ. Press), p. 55, http://arXiv.org/format/astro-ph/0609721

Lyra, W., Johansen, A., Klahr, H., & Piskunov, N. 2008, *A&A*, 491, L41

Marov, M. Ya. & Ipatov, S. I. 2018, *Solar System Research*, 52, 392

Nesvorny, D., Youdin, A. N., & Richardson, D. C. 2010, *AJ*, 140, 785

Rufu, R. & Aharonson, O. 2017, *Nature Geoscience*, 10, 89

Surville, C., Mayer, L., & Lin, D. N. 2016, *ApJ*, 831, A82

Discussion

GÜDEL: You mentioned that planetary tilt may be due to early collisions. But wouldn't tilt vary in later evolution due to perturbations?

IPATOV: Two condensations that formed in the parental condensation could move one above another (not in one plane). The angular momentum of the formed condensation could be not perpendicular to ecliptic. During accumulation of planetesimals, the Earth embryo's angular momentum became more perpendicular to the ecliptic, but some tilt remained.

Origins: From the Protosun to the First Steps of Life
Proceedings IAU Symposium No. 345, 2018
B. G. Elmegreen, L. Viktor Tóth & M. Güdel, eds.

doi:10.1017/S1743921319001662

Transitional disk archeology from exoplanet population synthesis

Germán Chaparro Molano[1], Frank Bautista[2] and Yamila Miguel[3]

[1]Vicerrectoría de Investigación, Universidad ECCI
Bogotá, Colombia
email: gchaparrom@ecci.edu.co

[2]Departamento de Física, Universidad Nacional de Colombia
Bogotá, Colombia
email: fjbautistas@unal.edu.co

[3]Sterrewacht Leiden, Leiden University
Leiden, The Netherlands

Abstract. Increasingly better observations of resolved protoplanetary disks show a wide range of conditions in which planets can be formed. Many transitional disks show gaps in their radial density structure, which are usually interpreted as signatures of planets. It has also been suggested that observed inhomogeneities in transitional disks are indicative of dust traps which may help the process of planet formation. However, it is yet to be seen if the configuration of fully evolved exoplanetary systems can yield information about the later stages of their primordial disks. We use synthetic exoplanet population data from Monte Carlo simulations of systems forming under different density perturbation conditions, which are based on current observations of transitional disks. The simulations use a core instability, oligarchic growth, dust trap analytical model that has been benchmarked against exoplanetary populations.

Keywords. planetary systems: formation, protoplanetary disks

---

1. Introduction

Planet-forming disks can have either smooth or density perturbed profiles (van der Marel *et al.* 2015). Gas-depleted, density perturbed disks are often called transitional disks. Gaps, cavities, or radial perturbations in such disks sometimes show dust traps in which planet formation may be taking place. Such dust traps may be caused by pressure bumps, which are often theorized as a consequence of the presence of an already-formed planet in the disk (Pinilla *et al.* 2011). In this work, we are interested in the impact of disk density perturbations in planetary populations while making no assumptions about the exact mechanism that causes such inhomogeneities in the disk. In order to explore the link between transitional disks and exoplanetary systems, we intend to compare synthetic and observed populations of exoplanetary systems. However, instead of looking at individual cases (Raymond *et al.* 2018) we take a Bayesian inference approach to reconstruct probability distributions of general properties of 3000+ simulated synthetic planetary systems. We thus study the effect of radial density perturbations in the disk structure on the formation of exoplanetary systems. For many transitional disks, a radial density perturbation can be described as a succession of over-dense and under-dense regions which appear as the radial distance r changes (Pinilla *et al.* 2011),

$$\Sigma_p(r) = \Sigma(r)\left(1 + A\cos\left(2\pi\frac{r}{fH(r)}\right)\right). \tag{1.1}$$

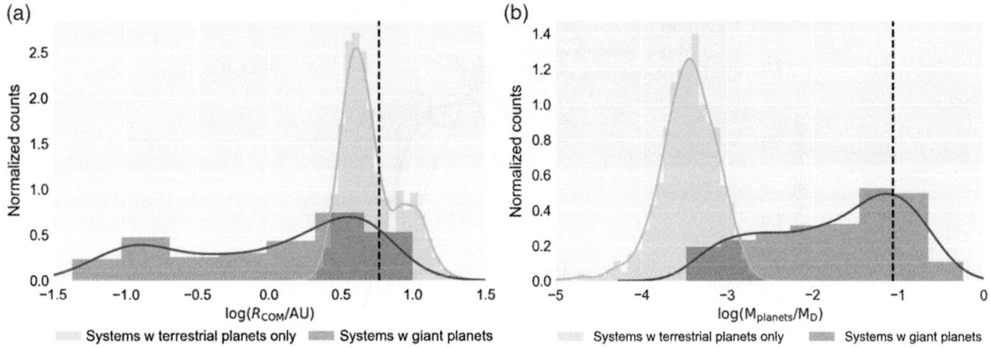

Figure 1. Marginalized distributions of center of mass and total planet mass/disk mass ratio for systems with giant planets (green) and with only terrestrial planets (blue). The distributions are approximated by a Kernel Density Estimation (solid lines).

Here $\Sigma(r)$ is the surface density distribution (Miguel *et al.* 2011), A is the amplitude of the perturbation, f is the length scale of the perturbation and $H(r)$ is the scale height of the disk. Here we consider $A=0$ for smooth disks, and $A=0.3$ for transitional disks.

2. Synthetic planetary systems

Population synthesis models are often used for modeling individual exoplanetary systems (Raymond *et al.* 2018) in order to estimate properties of individual planets in the system. We extend this method for 3000 synthetic planetary systems formed in smooth and transitional disks (following the perturbation recipe described above). Each system has initial conditions drawn from prior probability distributions on stellar mass, disk mass and radial extent, stability, metallicity, and gas dissipation timescale from Miguel *et al.* (2011). This planet population synthesis framework is also described as part of the review in Benz *et al.* (2014). The result of each simulation is a system with orbital data like planetary semi-major axis and mass (solid+gas). We calculate consolidated quantities that summarize general properties of each simulated system. Thus, we use quantities like the number of terrestrial and giant planets, total terrestrial planetary mass, average terrestrial planet mass, total planetary mass/disk mass ratio, and what we refer here to as center of mass. This center of mass is not the barycenter of the system but rather the first moment of the mass distribution of planets with respect to their semi-major axes. With the consolidated results per system, we can reconstruct the posterior probability distribution from the simulated quantities mentioned above. This reconstructed posterior can be used to make educated predictions for existing exoplanetary systems based on known properties like stellar mass, metallicity, planetary masses and semi-major axes distributions, etc.

3. Results

As an initial benchmark, we looked at the resulting distributions of the location of the center of mass and the total planetary mass/disk mass ratio for systems formed in smooth disks. Figure 1(a) shows that systems with giant planets are more spread inwards (closer to the star) than systems that only formed terrestrial planets, due to giant planet migration. On the other hand, more of the original disk mass goes to form planets in systems with giant planets than in terrestrial planet-only systems (Figure 1(b)). Both of these results are expected from exoplanet and planet-forming disk observations. The resulting synthetic planetary systems were divided among systems with giant planets (20-25% of all systems), and systems with terrestrial planets only. We benchmarked our results of the distribution of planetary systems with respect to metallicity. Figure 2),

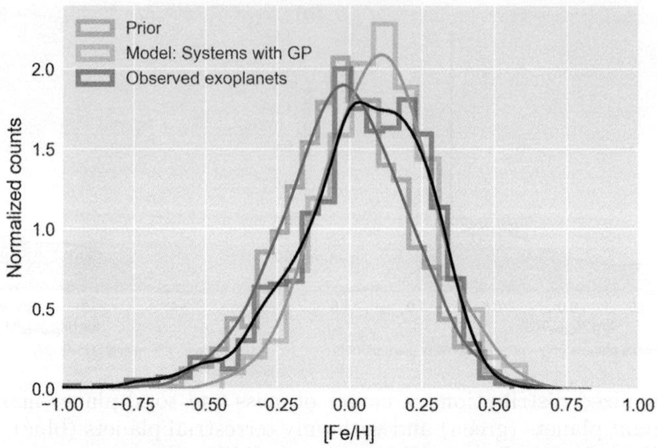

Figure 2. Marginalized distribution of exoplanetary systems metallicities for synthetic systems with giant planets (green), observed systems with giant planets (red) and from the prior used in the simulation (blue) from Mayor *et al.* (2011). The distributions are approximated by a Kernel Density Estimation (solid lines).

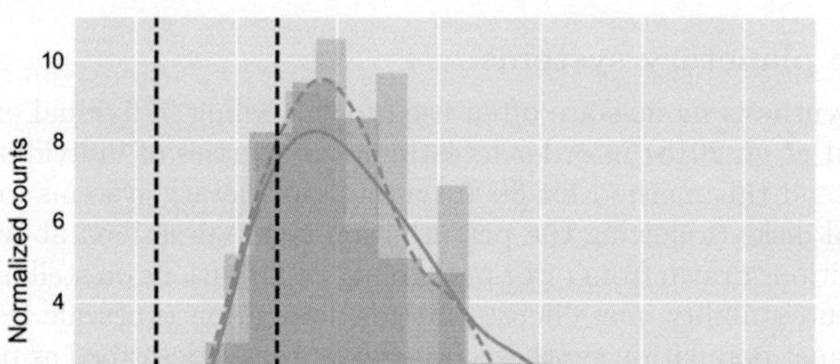

Figure 3. Marginalized posterior distribution of parent disk masses for synthetic systems with giant planets for smooth and transitional disks. The dashed black lines show lower and upper estimations for the minimum mass solar nebula. The distributions are approximated by a Kernel Density Estimation (solid lines).

shows that systems with giant planets tend to be formed in metal-rich systems, both in our simulations and in observations from `exoplanet.eu`. Comparatively, most exoplanetary systems discovered to date tend to be distributed uniformly around the solar value in the HARPS 2011 catalog (Mayor *et al.* 2011), which we used as a prior for our simulations. For the same initial disk mass, transitional disks form more systems with giant planets than smooth disks (Figure 3). This is due to transitional disks having over-dense regions that favor planet growth. Figure 4 shows the distribution of stellar mass vs. center of mass for synthetic and observed planetary systems. The 1σ and 2σ contours show that there is a significant overlap in parameter space between synthetic and observed planetary systems. For exoplanetary systems with a low center of mass the overlap breaks down, which is likely a result of observational bias.

4. Conclusions

Our comparison between smooth and transitional disks shows that transitional disks favor the formation of giant planets at lower disk masses than smooth disks.

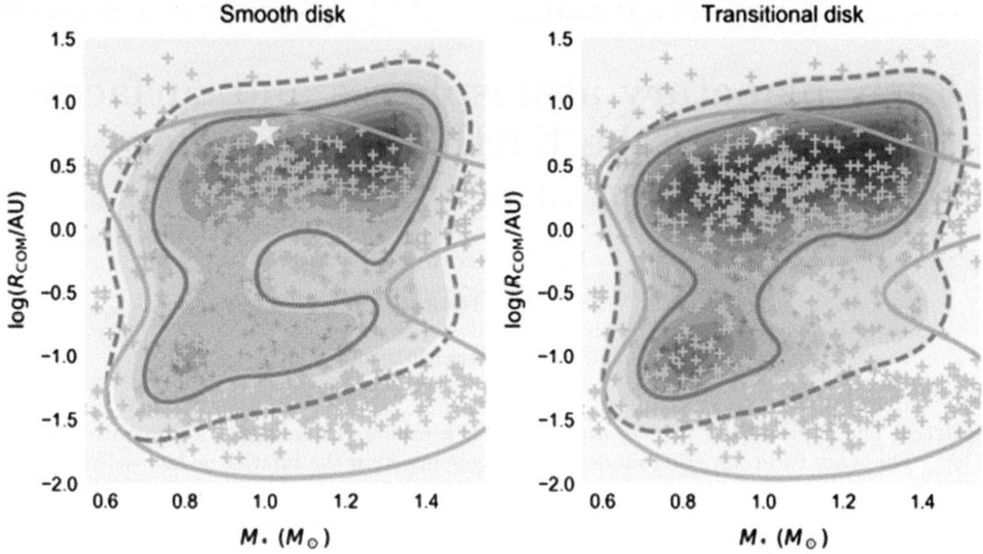

Figure 4. Center of mass vs. stellar mass for observed systems (red) and synthetic systems (blue) formed in smooth disks (left) and in transitional disks (right). The 1σ (solid line) amd 2σ (dashed line) contours were obtained using a Kernel Density Estimation for each set of data points.

Benchmarking of the results of our simulations show a very good parameter space overlap between synthetic and observed exoplanetary systems. The method of approximating a posterior probability distribution for exoplanetary parameters from our simulations can be used to make educated predictions that direct future surveys of observed exoplanetary systems.

References

Benz, W., Ida, S., Alibert, Y., Lin D. N. C., & Mordasini, C. 2014, in H. Beuther, R. S. Klessen, C. P. Dullemond, & T. Henning (eds.) *Protostars and Planets VI*, Univ. of Arizona Press, 944

Hughes, A. M., Duchêne, G., & Matthews B. C. 2018, *Annu. Rev. Astron. Astrophys.*, 56, 541

Mayor G. M., Marmier M., Lovis, C., *et al.* 2011, arXiv;1109.2497

Miguel, Y., Guilera, M., & Brunini, A. 2011, *MNRAS*, 417, 314

Pinilla, P., Birnstiel, T., Ricci, L., Dullemond, C. P., Uribe, A. L., Testi, L., & Natta, A. 2012, *A&A*, 538, A114

Pinilla, P., Tazzari, M., Pascucci, I., *et al.* 2018, *ApJ*, 859, 32

Pinilla, P., van der Marel, N., Pérez, L. M., van Dishoeck, E. F., Andrews, S., Birnstiel, T., Herczeg, G., Pontoppidan, K. M., & van Kempen, T. 2015, *A&A*, 584, A16

Raymond, S. N., Boulet, T., Izidoro, A., Esteves, L., & Bitsch, B. 2018, *MNRAS*, 479, L81

van der Marel, N., van Dishoeck, E. F., Bruderer, S., Andrews, S. M., Pontoppidan, K. M., Herczeg, G. J., van Kempen, T. & Miotello, A. 2016, *A&A*, 585, A58

van der Marel N., van Dishoeck E. F., Bruderer, S., Pérez, L., & Isella, A. 2015, *A&A*, 579, A106

Discussion

ALFARO: What kind of priors are you using in the Bayesian inference methods?

CHAPARRO: The functional form of priors for each input parameter are taken from the literature, which are summarized in the 2011 work by Miguel *et al.*

Origins: From the Protosun to the First Steps of Life
Proceedings IAU Symposium No. 345, 2018
B. G. Elmegreen, L. Viktor Tóth & M. Güdel, eds.

doi:10.1017/S1743921319001807

Early planetary atmospheres and surfaces: Origin of the Earth's water, crust and atmosphere

Doris Breuer

German Aerospace Center (DLR), Institute of Planetary Research,
Rutherfordstr. 2, 12489 Berlin, Germany
email: doris.breuer@dlr.de

Abstract. The origin of the planets atmosphere is a profound question of comparative planetology. There are two competing models, i.e. outgassing from the interior or late delivery from comets or volatiles-rich asteroids after most of the planet has been formed, of which the former is currently preferred. Meteorite compositions as well as radial mixing during accretion derived from accretion models suggest that the building blocks of the terrestrial planets contained some volatiles. Processes like dehydration by hydrous melting, oxidation, impact devolatilization, and in particular degassing during magma ocean solidification will then lead to a significant volatile loss of the interior and to the formation of a dense atmosphere during the early stages of planetary evolution. These processes are also responsible for the oxidation state of this early atmosphere, i.e. whether it was more reduced or oxidized. Although this early volatile loss was very efficient, the interior probably retained some water. This was distributed in the subsequent evolution between interior and atmosphere, as well as on the surface as liquid water in case of favorable temperature and pressure conditions. The main processes responsible for the water distribution are volcanic outgassing driven by partial melting of the silicate mantle and formation of the crust and recycling of water-rich crustal material. Here, an important difference between the terrestrial planets is the tectonic style prevailing on the planet. For the Earth with its plate tectonics, recycling of water is very efficient and can even balance the outgassing. For terrestrial planets in the stagnant lid regime of mantle convection such as Mars, the exchange of water between the interior and the surface/atmosphere is mainly in one direction and results in a continuous depletion of the interior. In this talk, I will briefly review our current knowledge on these interactions between interior and atmosphere and on the problem we are facing to better understand the influence of the interior on the habitability of a planet.

Keywords. Earth, planets and satellites: formation, convection, astrobiology

---

1. Introduction

The initial volatile inventory of a planet is acquired from colliding planetesimals, impacting asteroids and comets (e.g., Lunine *et al.* 2003, Brasser 2013). Based on the knowledge of the composition of these building blocks and the accretion scenarios, the initial volatile content for terrestrial planets can be estimated. Water and CO_2 are among the most important greenhouse gases in the atmospheres of terrestrial planets. We will focus in the following on the origin and outgassing of water, although some of the described processes and effects also apply to CO_2 but some are different.

The assumption about the amount of water that accreted on the terrestrial planets varies from almost 0 wt to 5 wt % (e.g., Lunine *et al.* 2003, Raymond *et al.* 2006). Three possible scenarios for the terrestrial planets are shown in Fig. 1. In the first scenario (upper row), the accreting material originates from the region of the solar nebula in

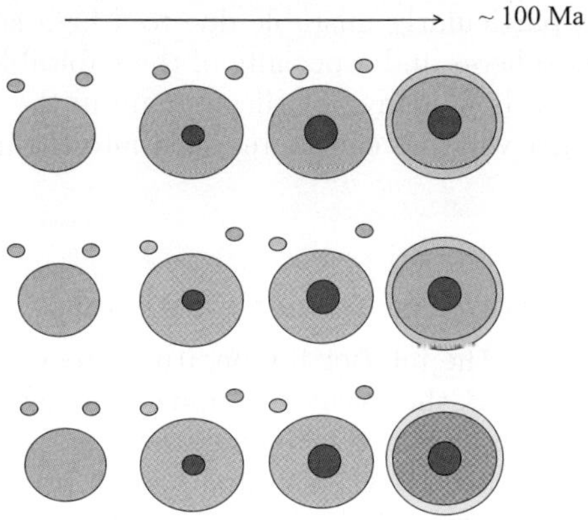

Figure 1. Accretion scenarios which assume a different time of accumulation of volatile-rich material and show the distribution of volatiles between interior and atmosphere. The brownish color represents dry material, bluish color is wet material, red is iron-rich core material, the light blue represents the atmosphere. Further description see in the text.

which the planet is forming. In this region, the inner solar system, the planetsimals are assumed to be dry and form a dry proto-planet. Only after the end of the main accretion phase and core formation, water is brought into the inner solar system by volatile-rich impacting bodies such as carbonaceous chondrites or comets. However, the analysis of the deuterium/hydrogen (D/H) ratio on comets, i.e., water-rich bodies from the Kuiper belt or the Oort cloud, suggests that these objects probably didn't bring water to Earth. Instead, asteroids as the origin for ocean water seem to be more likely (Genda & Ikoma 2008). The late accretion of volatiles after core formation, is also known as the Late Veneer model. The interior of the planets remains mainly 'dry' and an early primordial atmosphere is formed. In the second scenario (middle and lower row), volatile-rich bodies accrete much earlier, i.e., while the terrestrial planets are still growing and core formation is not completed. The volatile-rich bodies originate from regions beyond the planets feeding zone, i.e., beyond the snow line where it is cold enough for volatile compounds such as water and carbon dioxide to condense into solid ice grains. Planet formation models show that the exchange between inner and outer solar system is likely but the timing and amount is not well constrained (e.g., Morbidelli *et al.* 2000).

Even if the water content and the origin of the building blocks were known, the open question remains of how much is lost from the interior due to oxidation, impact devolatilization, and degassing during magma ocean solidification. The second scenario can therefore be subdivided further which differ in the initial amount of volatiles in the planet's interior after planet formation and magma ocean solidification (Fig. 1, middle and lower row). All these early processes tend to strongly dehydrate the planet, leaving the interior dry and producing an early atmosphere - possibly similar to the one of the first scenario but the atmosphere is already formed during accretion not shown in Fig. 1. Please note that this early atmosphere is not a primary atmosphere defined as gas captured gravitationally from the solar nebula. Such a primary atmosphere is mainly made of hydrogen. In principle small planets like Earth could have captured significant primary atmospheres, depending on how long the nebula lasted (Hayashi *et al.* 1979). Alternatively to efficient and almost complete early outgassing (Fig. 1, middle row), some volatiles remained in the interior and are released during secondary volcanism (Fig. 1, lower row). The timing of volatile loss from the interior in these first ~ 100 Ma (assumed time for accretion and magma ocean solidification in the case of Earth), however, determines the loss of the volatiles from the atmosphere into space because an early atmosphere was particularly unstable due to a high activity of the young sun. The evolution of the atmospheres and especially of the habitable surface conditions may therefore vary depending on how many volatiles remain in the early planet, and/or can degas later but also whether volatiles can be recycled into the interior depending on the tectonic style of a planet.

2. Early volatile loss from the planet's interior

During planetary accretion, the interior temperature progressively increases as a consequence of impact heating and the decay of radioactive elements. Before accretion is completed and core formation starts, the temperature can reach the stability fields of the hydrous silicate phases and the H_2O saturated solidus of mantle material. The latter is significantly reduced by several hundred degrees with respect to dry mantle material (Medard & Grove 2006). Thus, before complete accretion, hydrous phases can be dehydrated and mantle material starts to melt resulting in efficient H_2O outgassing. A primordial water-rich atmosphere is formed and as water acts as an efficient greenhouse gas, the surface temperature increases (Abe & Matsui 1986). The latter makes it even more difficult for hydrous phases in the accreting material to retain their water. However, as the temperature of an accreting body increases toward the surface, it has been suggested that H_2O can be preserved in the interior of a planet (Medard & Grove 2006): Material that accreted before the planet reached 55% its final radius does not dehydrate, assuming the temperature distribution derived from the accretion model by Senshu *et al.* (2002). Water in this deeper region will become unstable with further heating and hydrous phases may transform to nominally anhydrous minerals. Anhydrous minerals in the transition zone of the Earth and in the deep Martian mantle are typically wadsleyite and ringwoodite (Bertka & Fei 1997) that can store up to 2 wt.% water while in the deeper mantle of the Earth, the water storage is assumed to be much smaller (Bolfan-Casanova 2005).

In addition to early dehydration and hydrous melting, oxidation can be very efficient as long as free iron (and other metallic phases) and water is available, i.e., before complete formation of the core. Water delivered after core formation will not bind with free iron and can be sequestered in the silicate portion of the planet. Dreibus & Wänke (1987) argue that for homogeneous accretion, i.e., the dry and the volatilerich components are delivered almost at the same time and before core formation, almost all of the H_2O added during accretion is converted on reaction with metallic Fe to FeO and H_2, of which the former remained in the mantle and the latter escaped. The possibility of significant oxidation has been questioned by Rubie *et al.* (2014) who demonstrate that at high temperatures and pressures in terrestrial magma oceans, iron moves into a metallic phase preferentially to the oxidized phase. Thus, in deep magma oceans metallic iron could be stable and water may remain in the liquid magma ocean. Furthermore, iron may sink quickly through any potentially oxidizing magma ocean and aggregate into a core before oxidation can occur.

Water can also be lost by impact devolatilization during accretion. The efficiency of these processes to dehydrate an accreting planetary body increases with the amount of accretional heating, which on the other hand is related to the size of the growing body and the size of the impactor. As a planet grows, the impact velocities increase owing to the increase in radius until first partial and then, at a larger radius, complete devolatilization occurs. Volatiles in minerals are released because of the resulting high temperatures. At what planetary radius devolatilization becomes effective depends also on the hydrated minerals and estimates for complete volatilization range from 1300 km to 2500 km (e.g., Tyburczy *et al.* 2001). However, recent experiments have shown that up to 30% volatiles can be stored in impact melt and in projectile survivors (Daily & Schulz 2018).

With growing size of the planet and of the impactor in the last stage of accretion, the amount of melting increases. In case of large scale melting and the formation of a magma ocean, a substantial part of the water inventory can be outgassed as a consequence of magma ocean solidification (Elkins-Tanton 2008, Lebrun *et al.* 2013). Water enters solidifying minerals in only small quantities and is enriched in magma ocean liquids as solidification proceeds. Water in excess of the saturation capacity of the magma will then degas into the atmosphere when convection transports liquid close enough to the surface as the consequence of the pressure dependent H_2O solubility in silicate melts (e.g., Burnham 1994). Therefore, the efficiency of outgassing depends on the process of magma ocean crystallization, i.e. how much of the silicate melt was effectively transported toward the surface, how much of the melt was retained in the solidified mantle and when and whether a floating or stable conductive lid formed on top the magma ocean preventing melt to rises to the surface. If magma is transported close to the surface, water will then partition between the atmosphere and the magma ocean liquids according to their equilibrium partial pressures.

A magma ocean would freeze from the bottom (i.e., from the core-mantle boundary, if it comprises the entire mantle) to the surface because of the steeper slope of the mantle adiabat compared to the slope of the solidus (e.g.; Solomatov 2000, Elkins-Tanton *et al.* 2003). For larger planets like Earth, the situation can be different and crystallization begins in the middle of the mantle, which can also lead to a basal magma ocean that may still exist today (e.g., Labrosse *et al.* 2007). Typically, the forming crystals are denser than the liquid and sink to the bottom of the magma ocean, i.e. fractional crystallization. Water and CO_2 as incompatible elements are continuously enriched in the liquid and degases into the atmosphere. Whether the water contained in the melts is outgassed into the atmosphere or retained in the solidifying melt depends on its solubility in surface lavas at the evolving pressure and temperature conditions of the atmosphere (Gaillard & Scaillet 2014), whereby the effect of pressure dominates. The solubility of H_2O is much larger than that of CO_2 (by more than two orders of magnitude at atmosphere-relevant pressures below 100 bar). Therefore, the outgassing of water can be significantly limited by its high solubility in the melts, while all extracted CO_2 is easily released into the atmosphere. As a result, CO_2 is first outgassed and H_2O only at the end of magma ocean crystallization when the melt is increasingly enriched. It should be noted though that for a low oxygen fugacity (IW4 to IW+1) in the magma ocean, the equilibria of C-O-H fluids predicts degassing of CH_4, CO, and H_2 rather than CO_2 and H_2O (e.g., Holloway & Jakobsson 1986).

In any case, degassing is efficient and most volatiles got lost from the interior (e.g., Elkins-Tanton 2008, Lebrun *et al.* 2013). Although the formation of an insulating atmosphere can increase substantially the lifetime of a magma ocean (Abe 1997), the entire process of solidification is rapid (Solomatov 2000) and may occur within a few million years (Elkins-Tanton 2008). However, this degassing efficiency may limit the assumptions of the initial magma ocean water inventory. It has been speculated that 0.1 wt% was unlikely because this would imply initial atmosphere pressures of thousands of bars and times related to cooling of clement surface conditions may have lasted hundreds of Myr (Elkins-Tanton 2008). But old zircons in Earth rocks provide compelling evidence that liquid water was in place before 4.2 Ga and probably before 4.3 Ga (Valley *et al.* 2005). In addition to interior outgassing, magma ocean solidification influences strongly the initial distribution of water and the initial density stratification. As a certain amount is partitioned into the solidifying minerals, an exponential increase of water toward the surface can be expected. In the process of fractional crystallization, the evolving magma ocean liquid is also continuously enriched in iron and therefore Mg-rich cumulates can be found at the CMB and Fe-rich cumulates close to the surface (Elkins-Tanton *et al.*

2003). This leads to a non-monotonic density increase toward the surface, resulting in a gravitationally unstable configuration and initiating early mantle convection.

As described above, fractional crystallization suggests efficient degassing. A depletion of more than 90% of the initial amount of volatiles has been assumed and even small initial volatile contents (0.05 wt.% H_2O, 0.01 wt.% CO_2) can produce atmospheres in excess of 100 bars (Elkins-Tanton 2008). However, fractional crystallization and degassing is disturbed among others by inefficient crystal-melt segregation, i.e., melt in which incompatible elements are enriched can be sequestered in the solid cumulates. The faster the cooling and the slower the compaction, the less efficient is the crystal-melt separation (e.g., Solomatov 2000, Hier-Majumder & Hirschmann 2017). The melt fractions that are trapped in the cumulate pile have been estimated to values of between 1 and 10% (Elkins-Tanton 2008, Hier-Majumder & Hirschmann 2017). Furthermore, if the magma ocean did not comprise the entire mantle, the lower mantle remains primordial and thus more water-rich in comparison to the dehydrated upper mantle due to magma ocean solidification. This means that a substantial amount of volatiles may still be present within planets after this early accretion and differentiation process.

3. Secondary volatile loss from the planet's interior

In the subsequent evolution after accretion, core formation and magma ocean crystallization, the amount and distribution of water in the terrestrial mantle is significantly influenced by mantle convection and partial melting of the mantle. Vice versa, water influences the mantle dynamics and melting as the rheology (viscosity) and solidus of mantle rocks strongly depend on water content.

Effect of water on solidus and viscosity The viscosity of mantle rocks is strongly temperature, pressure and water dependent and it is usually assumed that the flow behavior in terrestrial mantles is dominated by olivine. Thus, typically values of activation energy, activation volume and the stress exponent for olivine (e.g., Karato & Wu 1993, Hirth & Kohlstedt 1995) are used to calculate the viscosity. Olivine like the other mantle minerals is a nominally anhydrous minerals and can not store a lot of water. However, the critical water content above which viscosity is significantly reduced is only a few ppm wt.% (Mei & Kohlstedt 2000) and for a water content close to 100 ppm, the viscosity is reduced by a factor of 100 in comparison to a completely dry mantle. Therefore, a few tens of ppm water can be rheologically significant, but the exact value depends on the operating deformation mechanism (e.g., Mei & Kohlstedt 2000). As limiting cases, viscosities of 10^{19} and 10^{21} Pa s at a temperature of 1600 K and a pressure of 3 GPa are generally assumed to represent wet and dry mantle rheologies, respectively. In general, low viscosity suggests vigorous convection and a strong planetary cooling while a high viscosity suggests a more sluggish convection and a slower cooling.

In addition to the influence of water on the rheology, water can significantly reduce the solidus of mantle materials (e.g., Hirschmann 2006). The solidus depression is approximately linear with dissolved water and for instance a bulk water content of 250 ppm (0.025 wt.%) can lower the peridotite solidus by 100 K. It is important to note that in contrast to the effect of water on rheology for which a few tens of ppm water can significantly reduce the viscosity, a water content above 100 ppm is required to notably reduce the mantle solidus (e.g., Green *et al.* 2014).

Interaction between volatiles, partial melting and convection To understand the evolution of volatiles in the planetary interior and its relation to the thermo-chemical and atmospheric evolution of a planet, it is therefore important to study the interaction between volatiles, partial melting, convection and outgassing. It should be noted, however, that in addition to the unknown initial water content, the present content of water in the interior of the planets is also not well known although this knowledge may help constraining potential evolution scenarios. For the Earth, the surface ocean corresponds to ~ 300 ppm water in the interior but the estimates of the present interior water vary between 0.5 and 10 ocean masses (i.e., 150 ppm - 3000 ppm water). We have even less evidence about the amount and distribution of water in other terrestrial planets although there's more and more evidence that planets like Mars and even Mercury aren't completely dry.

Volatiles present in the mantle minerals are transported and mixed by mantle convection. During melting, volatiles are preferentially enriched in the liquid phase. The enrichment depends on the partition coefficient, which is for water about 0.01. Modelling of CO_2 extraction is more complicated by the fact that carbon is not directly soluble in silicate minerals, but occurs in separate phases depending on pressure, temperature, and oxygen fugacity (e.g. Dasgupta & Hirschmann 2010). The partial melt, typically having a lower density than the remaining solid mantle, rises toward the surface and redistributes the volatiles: Erupting material will release essentially all dissolved volatiles into the atmosphere as the solubility of volatiles in magmas at surface pressure is low (Burnham 1994). Volatiles in intrusive material will mostly remain in the crust, although dissolved volatiles may outgas and reach the atmosphere when some crustal porosity is present at the depth of the intrusion (Hirschmann & Withers 2008). In any case, the mantle water concentration decreases due to dehydration by melting, resulting in an increase of the mantle viscosity by several orders of magnitude (e.g., Hirth & Kohlstedt 1995). This effect of dehydration stiffening tends to thicken the upper thermal boundary layer, which then limits melt production, crustal growth, and mantle degassing (Morschhauser *et al.* 2011).

In general, the water concentration in the interior decreases and the atmospheric pressure increases with time - neglecting volatile loss of the atmosphere. This effect is strongest in the early evolution of stagnant lid planets like Mars, Mercury and possibly Venus, while volcanism is strongest. Furthermore, this effect is strengthen with an increasing water content due to lower melting temperatures and a lower mantle viscosity. The situation is different for Earth with its plate tectonics because crust formation and degassing is longstanding until present at the diverging plate boundaries (mid-ocean ridges) and, most importantly, volatiles can be efficiently recycled into the mantle. Water and CO_2 stored in the subducting plates either in the sediments or in the basaltic oceanic crust can be transported back into the mantle. It has been even speculated that the interior has become volatile-rich over time due to recycling by plate tectonics (Korenaga *et al.* 2017) - in contrast to the stagnant lid planets. However, this depends on the time of the onset of plate tectonics, which is controversially discussed. It may have started just after magma ocean crystallization or only 2.5 Ga ago (Korenaga 2013). For a late onset of plate tectonics and with an initially dry mantle, the interior would have remained dry for a longer time of the planet's evolution while for an initially wet mantle, the interior would have been dehydrated first. The implication for the early atmosphere and surface conditions is however unclear. Thus, the initial amount of interior water and the onset of plate tectonics seem to be crucial in our understanding of the habitable condition on Earth.

Recycling of volatiles also requires low enough surface temperatures for precipitation (fluid water) and weathering to occur. The evidence for liquid H_2O on Earth's surface from zircons indicates that these conditions occurred already during the fist few hundred million years (Valley *et al.* 2005). Precipitation will remove CO_2 from the atmosphere, which is then stored due to weathering into carbonates. The interplay between weathering and plate tectonics, known as silicate-carbon cycle, allows stable clement surface conditions in the long-term evolution. This is particularly important in view of the increase in solar flux over time (faint young sun problem) (e.g., Kasting & Catling 2003). For a late start of plate tectonics, stable surface temperatures must be ensured without recycling of the volatiles.

4. Summary and conclusions

According to the discussion above, volatile loss from the interior and earlyatmosphere formation was efficient and took place in several phases. Nevertheless, it is to be expected that terrestrial planets likely started with some amount of water in its interior after its accretion, core differentiation and magma ocean crystallization - although the amount is unknown and can vary from dry to a few hundred ppm of water, it is crucial for the subsequent evolution. The efficient early degassing created a dense primordial, H_2O and CO_2 rich or a more reducing CH_4, CO, and H_2 atmosphere. This should not to be confused with a nebular-based proto-atmosphere, which, if existent, has likely been removed by a high EUV flux of the young sun. But also the primordial atmosphere could be lost to space. How much got lost depends on the early sun activity and the planetary mass, i.e. smaller planets such as Mars lost most of not all of their early atmosphere while Earth and Venus may have kept a substantial part (e.g., Tian *et al.* 2009, Lammer *et al.* 2013). Efficient precipitation and weathering has significantly changed the Earth's primordial atmosphere in contrast to Venus with its higher surface temperatures due to its closer proximity to the sun.

Degassing by secondary volcanism helps then to form (Mars) or maintain (Earth) the atmosphere. The atmosphere would otherwise disappear slowly into space due to continuous loss processes by solar activity or by silicate weathering. It is to be speculated that the less efficient the early degassing (i.e. in case of wet initial conditions in the interior) and the more pronounced the secondary volcanism and degassing, the better the habitable conditions of a planet. This is in particular true for a small planet like Mars that loses rapidly its early atmosphere and requires a sufficiently thick atmosphere for clement surface conditions. This probably not the case for planets with a dense primordial atmosphere like Venus' that did not get lost and for which surface temperatures are too high for precipitation due to the greenhouse effect. In the case of the Earth, however, the question arises as to whether an initially volatile-rich mantle was necessary for its habitability, since plate tectonics could have provided the interior with water for later volcanic degassing. But is it possible to initiate plate tectonics with a dry interior or is a wet interior necessary? What if plate tectonics started much later (Korenaga 2013)? Can the mantle have stayed 'dry' for so long and thus contributed little to the atmosphere and surface water and is this consistent with the observations of clement conditions? If this is not the case, it would support the assumption that secondary volcanism and degassing from an initially more volatile-rich interior is important for habitability. Thus, processes that influence the distribution of volatiles between interior and atmosphere - especially in the early planetary phase of accretion and differentiation - need to be investigated more in the future.

References

Abe, Y. 1997, *PEPI*, 100, 27

Abe, Y. & Matsui, T. 1986, *J. Geophys. Res.*, 91, 291

Amari, S., Hoppe, P., Zinner, E., & Lewis R.S. 1995, *Meteoritics*, 30, 490

Bertka, C. M. & Fei, Y. 1997, *J. Geophys. Res.R*, 102, 5251

Bolfan-Casanova, N. 2005, *Mineralogical Magazine*, 69, 229

Brasser, R. 2013, *Space Sci. Revs*, 174, 11

Burnham, C. W. 1994, *Rev. Mineral.*, 30, 123

Daly, R. T., & Schultz, P. H. 2018, *Sci. Advances*, 4

Dreibus, D. & Wänke, H. 1987, *Icarus*, 71, 225

Elkins-Tanton, L. T. 2008, *Earth Planet. Sci. Lett.*, 271, 181

Elkins-Tanton, L.T., Parmentier, E.M., & Hess, P.C. 2003, *J. Geophys. Res.*, 271, 181

Gaillard, F., & Scaillet, B. 2014, *Earth Planet. Sci. Lett.*, 403, 307

Genda, H., & Ikoma, M. 2008, *Icarus*, 194, 42

Green, D. H., Hibberson, W. O., Rosenthal, A., Kovcs, I., Yaxley, G. M., Falloon, T. J., & Brink, F. 2008, *J. Petrology*, 55, 2067

Hayashi, C., Nakazawa, K., & Mizuno, H. 1979, *Earth Planet. Sci. Lett.*, 43, 22

Hier-Majumder. S., & Hirschmann, M. M. 2014, *Geochemistry, Geophysics, Geosystems*, 18, 3078

Hirschmann, M. M., & Withers, A. C. 2008, *Earth Planet. Sci. Lett.*, 270, 147

Hirth, G., & Kohlstedt, D.L. 1995, *Earth Planet. Sci. Lett.*, 144, 93

Hirschmann, M. M. 2006, *Annu. Rev. Earth Planet. Sci.*, 34, 629

Holloway, J. R. & Jakobsson, S. 1986, *J. Geophys. Res.*, 91, 505

Karato. S., & Wu, P. 1993, *Science*, 260, 771

Kasting, J.F., & Catling, D. 1993, *Annu. Rev. Astron. Astrophys.*, 41, 429

Korenaga, J. 2013, *Annu. Rev. Earth Planet. Sci.*, 41, 117

Korenaga, J., Planavsky, N. J., & Evans, D. A. 2017, *Phil. Trans. R. Soc. A*, 375, 20150393

Labrosse, S., Hernlund, J. W., & Coltice, N. 2007, *Nature*, 450, 866

Lammer, H., Chassefire, E., Karatekin, , Morschhauser, A., Niles, P. B., Mousis, O., Odert, P., Mstl, U. V., Breuer, D., Dehant, V., Grott, M., Grller, H., Hauber, E. & L Binh San Pham 2012, *Space Sci. Revs*, 1744, 13

Lebrun, T., Massol, H., Chassefire, E., Davaille, A., Marcq, E., Sarda, P., & Brandeis, G. 2013, *J. Geophys. Res.*, 118, 1155

Lunine, J. I., Chambers, J., Morbidelli, A., & Leshin, L. A. 2003, *Icarus*, 165, 1

Mdard, E., & Grove, T. L. 2006, *J. Geophys. Res.*, 111

Mei, S. & Kohlstedt, D.L. 2000, *J. Geophys. Res.*, 105, 21457

Morbidelli, A., Chambers, J., Lunine, J. I., Petit, J. M., Robert, F., Valsecchi, G. B., & Cyr, K. E. 2000, *MAPS*, 35, 1309

Morschhauser, A., Grott, M., & Breuer, D. 2000, *Icarus*, 212, 541

Raymond, S. N., Quinn, T., & Lunine, J. I. 2006, *Icarus*, 183, 265

Rubie, D. C., Jacobson, S. A., Morbidelli, A., OBrien, D. P., Young, E. D., de Vries, J., & Frost, D. J. 2014, *Icarus*, 248, 89

Senshu, H., Kuramoto, K., & Matsui, T. 2002, *J. Geophys. Res.*, 107, 1

Solomatov V. S. 2000, *Origin of the Earth and Moon*, 1, 323

Tian, F., Kasting, J. F., & Solomon, S. C. 2009, *Geophys. Res. Lett.*, 36

Tyburczy, J. A., Xu, X., Ahrens, T. J., & Epstein S. 2009, *Earth Planet. Sci. Lett.*, 192, 23

Valley, J. W., Lackey, J. S., Cavosie, A. J., Clechenko, C. C., Spicuzza, M. J., Basei, M. A. S., & Peck, W. H. 2009, *Contributions to Mineralogy and Petrology*, 150, 561

Origins: From the Protosun to the First Steps of Life
Proceedings IAU Symposium No. 345, 2018
B. G. Elmegreen, L. Viktor Tóth & M. Güdel, eds.

doi:10.1017/S1743921319001479

Water inventory from beyond the Jupiter's orbit to the terrestrial planets and the Moon

Mikhail Ya. Marov and Sergei I. Ipatov

Vernadsky Institute of Geochemistry and Analytical Chemistry of RAS,
119991, 19 Kosygin st., Moscow, Russia
email: marovmail@yandex.ru, email: siipatov@hotmail.com

Abstract. Computer simulations of migration of planetesimals from beyond the Jupiter's orbit to the terrestrial planets have been made. Based on obtained arrays of orbital elements of planetesimals and planets during the dynamical lifetimes of planetesimals, we calculated the probabilities of collisions of planetesimals with planets, the Moon, and their embryos. The results of calculations showed that for the total mass of planetesimals of about 200 Earth masses, the mass of water delivered to the Earth from beyond the orbit of Jupiter could be about the mass of the terrestrial oceans. For the growth of the mass of the Earth embryo up to a half of the present mass of the Earth, the mass of water delivered to the embryo could be up to 30% of all water delivered to the Earth from the zone of Jupiter and Saturn. The water of the terrestrial oceans and its D/H ratio could be the result of mixing of water from several exogenic and endogenic sources with large and low D/H ratios. The ratio of the mass of water delivered from beyond the orbit of Jupiter to a planet to the mass of the planet for Venus, Mars, and Mercury was not smaller than that for the Earth. The mass of water in planetesimals that collided the Moon and migrated from beyond the Jupiter's orbit could be not more than 20 times smaller than that for the Earth.

Keywords. methods: n-body simulations, Earth, Moon, Solar System: formation

---

1. Introduction

A source of water storage on the Earth and the terrestrial planets is one of the key problems of planetary sciences (Marov 2018). Basically, two main mechanisms are discussed: Endogenous sources (Water adsorption by grains before gas dissipation from the inner Solar System. Direct adsorption of hydrogen from the nebular gas into magma oceans through the reaction of H_2 with FeO. Water coupling in minerals of the early silicate mantle with a potential storage up to a few Earth's oceans) and Exogenous sources (Small bodies coming from the outer asteroid belt and volatile-rich planetesimals and asteroids/comets coming from beyond the Jupiter's orbit). Both endogenous and exogenous models have some constraints. Endogenous model is constrained by the facts that: (1) Temperature in the regions of protoplanetary disk where the inner planets formed exceeded 500-700 K and water/volatiles could not be retained. (2) Efficiency and timescale of the mechanism of water transport from upper mantle to the surface is not clear. Exogenous model is constrained by some geochemical evidence: (1) Os isotopic composition of the Earth's primitive upper mantle matches anhydrous ordinary chondrites rather than hydrous carbonaceous chondrites pertinent to the outer asteroid belt. (2) D/H ratio in water of the majority of known comets is greater than that in the Earth's oceans. We favor exogenous model focusing on the computer simulation of planetesimals and comets coming from beyond the Jupiter's orbit, with the follow up estimates of water delivery.

2. Computer simulations of migration of planetesimals from the feeding zone of Jupiter and Saturn

We made several series of calculations of migration of bodies under the gravitational influence of planets. In most calculations, all planets exclusive of Mercury were considered in their present orbits. In some runs, Uranus and Neptune were excluded. For some integrations, we used the symplectic code. For other runs, we used the Bulirsh-Stoer code. The main results were similar for both methods. Based on arrays of the orbital elements of bodies during their dynamical evolution, we calculated probabilities of collisions of the bodies with planets and the Moon (or with their embryos). Integrations were usually made until all bodies reached 2000 AU from the Sun or collided with the Sun. In a few runs considered time intervals equaled to the age of the Solar System. Earlier (Ipatov & Mather (2004, 2006, 2007) studied the orbital evolution of >30,000 bodies with initial orbits close to those of Jupiter-family comets (JFCs), Halley-type comets, long-period comets, and asteroids in the resonances 3/1 and 5/2 with Jupiter.

Marov & Ipatov (2018) made computer simulations of migration of 10^4 planetesimals from the feeding zone of Jupiter and Saturn to forming terrestrial planets. In series JN, all planets were assumed as having their present orbits and masses. In series JS, Uranus and Neptune were excluded. Initial eccentricities and inclinations of planetesimals were 0.3 and 0.15 rad, respectively. The initial semi-major axes of the planetesimals were between 4.5 and 12 AU. Masses of planets moving in the orbits of the terrestrial planets were equal to present masses of the planets in series JS and JN. In series JS_{01} and JN_{01}, they were smaller by a factor of 10 than the present masses. We also made calculations for which the giant planets of present masses initially were located more close to each other than the present giant planets. For such runs, at least one giant planet (not Jupiter) was ejected into a hyperbolic orbit during evolution. The values of the probability p_E of a collision of a planetesimal with the Earth for such runs were usually not smaller than those for series JS, JN, JS_{01}, and JN_{01}.

In Ipatov (2018), initial semi-major axes a varied from a_{om} to a_{om}+2.5 AU with a number of initial planetesimals proportional to $a^{1/2}$. For different runs, a_{om} varied from 2.5 to 40 AU with a step equaled to 2.5 AU. Initial eccentricities e_o equaled to 0.3 or 0.05, and initial inclinations $i_o = e_o/2$ rad. Eccentricities e_o=0.3 could be reached due to mutual gravitational influence of planetesimals during evolution of a disk of planetesimals in the feeding zone of the giant planets (Ipatov 1993).

3. Delivery of water and volatiles to the terrestrial planets and the Moon

Water and volatiles could be delivered to the terrestrial planets and the Moon from different distances from the Sun. It is often supposed (Petit *et al.* 1997) that the outer asteroid belt was the main source of the delivery of water to the terrestrial planets. Drake & Campins (2006) argued against an asteroids' source of Earth's water because oxygen isotopic composition in its primitive upper mantle matches that of anhydrous ordinary chondrites, rather than hydrous carbonaceous chondrites. Hallis *et al.* (2015) noted that the deep mantle water has a low D/H ratio and could be acquired due to adsorption of water on fractal grains during Earth's accretion. The ocean water (and its D/H ratio) could be a result of mixing of water from several sources.

The results of the calculations of migration of planetesimals from the feeding zone of Jupiter and Saturn showed that the probabilities of collisions of such planetesimal with the Earth and its embryo of mass equal to $0.1m_E$ (where m_E is the Earth mass) were about $2 \cdot 10^{-6}$ and $4 \cdot 10^{-7}$, respectively. We concluded that during the growth of the mass of the Earth's embryo up to $0.5m_E$, the amount of water delivered to the embryo from the feeding zone of Jupiter and Saturn could be about 30% of the total mass of water delivered to the Earth from this feeding zone.

In our calculations of the migration of the objects which initially moved in cometary-type Jupiter-crossing orbits, the probability p_E of a collision of an object during its dynamical lifetime with the Earth exceeded $4 \cdot 10^{-6}$ (Ipatov & Mather 2004, 2006). For planetesimals from the zones with a width of 2.5 AU, the values of p_E for e_o=0.3 and for e_o=0.05 did not differ much. Ipatov & Mather (2004) mentioned that the value of p_E for one body can be greater than the sum of values for thousands other bodies with similar initial orbits if the orbit of the body became Earth-crossing for a long time. The same result was obtained in our recent calculations. For one run with 250 bodies, the value of p_E (which is calculated as the mean value per one body) was $2.6 \cdot 10^{-3}$ for a_{om}=7.5 and e_o=0.3, and it was $9.8 \cdot 10^{-4}$ for a_{om}=10 AU and e_o=0.05. If we exclude the series of calculations with such larger values of p_E, then p_E typically was about $5 \cdot 10^{-6}$ for a_{om}=5 AU, about $3 \cdot 10^{-6}$ for a_{om}=7.5 AU, and about $2 \cdot 10^{-6}$ for a_{om}=10 AU. The 'a_{om}' series of runs show that the values of p_E could exceed $2 \cdot 10^{-6}$ for the feeding zone of Jupiter and Saturn. In principle, p_E could be greater by a factor of several than $2 \cdot 10^{-6}$, if many thousands of planetesimals migrated to the Earth, and among them there were planetesimals that spent long times in Earth-crossing orbits. For a_{om} from 12.5 to 25 AU, the values of p_E were mainly between 10^{-6} and $3 \cdot 10^{-6}$, i.e., differed not much from the value $2 \cdot 10^{-6}$ mentioned above. For a_{om} from 27.5 to 40 AU, the values of p_E on average exceeded $5 \cdot 10^{-7}$. The main growth of p_E in our calculations was during 20 Myr. However, for bodies initially located beyond Saturn's orbit, p_E could slowly grew even after 200 Myr. For times of migration of planetesimals from beyond the orbit of Saturn, we also need to add time during which massive embryos of Uranus and Neptune reached their present orbits. Results of our calculations of migration of the embryos of Uranus and Neptune from 8 and 10 AU under the gravitational influence of planets and planetesimals to the present orbits of Uranus and Neptune were presented e.g. in Ipatov (1991, 1993).

The total mass of water delivered to the Earth from beyond Jupiter's orbit could be comparable with the mass of the Earth's oceans, which is $2.25 \cdot 10^{-4} m_E$. At $p_E = 2 \cdot 10^{-6}$, for the total mass of planetesimals in the feeding zone of Jupiter and Saturn equal to $100 m_E$, and planetesimals consisted half in water, the planetesimals from this feeding zone could deliver about a half of Earth's ocean water. Another half of the water could come from more distant regions. Most of the water that was delivered from such distant regions to the Earth's embryo came when its mass was not small (e.g., was mainly greater than $0.5 m_E$).

In series JS, the ratio of the mass of water delivered to a planet to the mass of the planet for the Earth was smaller by a factor of 2, 1.25, and 1.3 than that for Mars, Venus and Mercury, respectively. For series JN, the above factor equaled to 3.4, 0.7, and 0.8, respectively. This mass fraction would result in relatively large ancient oceans on Mars and Venus. The estimates are indicative of the presence of ancient oceans on Mars and Venus, which could partially survive deep under the surface, as on Mars, (Usui 2017), or were lost in the course of evolution, as on Venus (Marov & Grinspoon 1998, chapter 9).

In the calculation series JS and JN, the probability p_M of a collision of a planetesimal during its dynamical lifetime with the Moon varied from $7 \cdot 10^{-8}$ to $2.7 \cdot 10^{-7}$; and, in three calculation series with 250 planetesimals, it was approximately $1.2 \cdot 10^{-7}$. In the considered calculation series, the ratio p_E/p_M of the probability of a collision of a planetesimal with the Earth to that with the Moon was mainly in the range from 16 to 17. In the JS and JN calculation runs with 250 planetesimals, the ratio p_E/p_M varied from 16.53 to 16.9 and from 16.08 to 16.74, respectively. For comparison, the squared ratio of the radii of the Earth and the Moon is 13.48. The mass of planetesimals migrated from beyond Jupiter's orbit and collided with the Moon could be smaller than that with the Earth by the factor not more than 20. However, the fraction of water and volatiles that left the Moon at collisions was greater than that for the Earth. For migrating objects, which initial orbits were close to those of Jupiter-family comets, the ratio p_E/p_M varied mainly from 15.2 to 17.6. For asteroids from 3/1 resonance with Jupiter and for comets with initial eccentricity equal to 0.975, p_E/p_M reached 18.6 and 15.2, respectively.

The work was supported by the grant of Russian Science Foundation N 17-17-01279 (migration of bodies to the Moon) and by the Program of Fundamental Studies of the Presidium of RAS N 17 (migration of bodies to the terrestrial planets).

References

Drake, M. & Campins, H. 2006, in: D. Lazzaro, S. Ferraz-Mello, & J. A. Fernandez (eds.), *Asteroids, Comets, & Meteors*, Proc. IAU Symp. 229 (Cambridge: Cambridge Univ. Press), p. 381

Hallis, L. J., Huss, G. R., Nagashima, K., Taylor, G. J., Halldrsson, S. A., Hilton, D. R., Mottl, M. J., & Meech, K. J. 2015, *Science*, 350, 795

Ipatov, S. I. 1991, *Soviet Astron. Letters*, 17, 113

Ipatov, S. I. 1993, *Solar System Research*, 27, 65

Ipatov, S. I. 2018, *The Ninth Moscow Solar System Symposium 9M-S3* (Space Research Institute, Moscow, Russia, October 8-12, 2018), 9MS3-SB-11, p. 104, https://ms2018.cosmos.ru/

Ipatov, S. I. & Mather, J. C. 2004, *Annals of the New York Academy of Sciences*, 1017, 46, http://arXiv.org/format/astro-ph/0308448

Ipatov, S. I. & Mather, J. C. 2006, *Adv. Sp. Res.*, 37, 126, http://arXiv.org/format/astro-ph/0411004

Ipatov, S. I. & Mather, J. C. 2007, in: G. B. Valsecchi, D. Vokrouhlicky, & A. Milani (eds.), *Near-Earth Objects, Our Celestial Neighbors: Opportunity and Risk*, Proc. IAU Symp. No. 236 (Cambridge: Cambridge Univ. Press), p. 55, http://arXiv.org/format/astro-ph/0609721

Marov M. 2018, in: P. Read, *et al.* (eds), *The Formation and Evolution of the Solar System*, Oxford Research Encyclopedia of Planetary Science (Oxford University Press), id.2

Marov, M. Ya. & Grinspoon, D. H. 1998, *The Planet Venus* (New Haven: Yale University Press), 442 p

Marov, M. Ya. & Ipatov, S. I. 2018, *Solar System Research*, 52, 392

Petit, J.-M., Morbidelli, A., & Chambers, J. 2001, *Icarus*, 153, 338

Usui, T. 2017, *Nature*, 552, 339

Discussion

KHAIBRAKHMANOV: How long does it take for Earth to gain its ocean with contemporary mass?

MAROV: Frankly, it is difficult to answer. There is evidence that the Earth took its oceans as early as 3.8-4.0 Gya, that is in a good coincidence with the LHB (late heavy bombardment) period. My point is that it is the time when volatile-rich bodies migrating from the outskirts of the Solar System fell down on the formed crust and were responsible for the late veneer formation on the Earth's surface, filling up the ancient oceans.

Origins: From the Protosun to the First Steps of Life
Proceedings IAU Symposium No. 345, 2018
B. G. Elmegreen, L. Viktor Tóth & M. Güdel, eds.

doi:10.1017/S174392131800858X

Neutral atmospheric escape in the Solar and extrasolar planetary systems

Dmitry V. Bisikalo[1] and Valery I. Shematovich[2]

[1]Institute of astronomy RAS, Moscow, Russia
email: bisikalo@inasan.ru

[2]Institute of astronomy RAS, Moscow, Russia
email: shematov@inasan.ru

Abstract. New data obtained by space missions to various objects in the Solar system and observations of the outer Solar system and exoplanets by space and ground-based telescopes allowed us to conclude that the atmospheric escape plays an important role in the evolution of the terrestrial planets in the Solar system. We present the recent results of application of the kinetic approach to the problem of neutral escape from planetary atmospheres. As an example, the recent measurements by Mars Express and MAVEN spacecraft are compared with the calculations of neutral escape with the aim to understand the atmospheric loss at Mars. Also the recent calculations of the mass-loss rates of the hot Neptune and Jupiter atmospheres are presented.

Keywords. stars: flare, planets and satellites: formation, ultraviolet: solar system, etc.

1. Introduction

Studies of the evolution of atmospheres of Solar System planets and planetary systems of other stars have significantly advanced during recent decades. Progress has been achieved owing to unique measurement data obtained by spacecraft during their flights to different objects in the Solar System, observations of the outer bodies of the Solar system and exoplanets using space and ground-based telescopes, and the development of mathematical simulation methods. These results have shaped a new basis for the physics of the upper atmosphere, i.e., the aeronomy that is directly linked to planetary cosmogony. In particular, they have allowed to infer a conclusion that dissipation of a planetary atmosphere (or atmospheric escape) plays an important role in the evolution of planets primarily of terrestrial planets of the Solar System, although many details of those events are not fully clear and remain the topic of active discussions (Shematovich & Marov 2018). Observations and theoretical models of the exoplanet atmospheres exposed to the extreme fluxes of X-ray and UV radiation of the parent star provide a remarkable opportunity to test theoretical understanding of the key processes - thermal and non-thermal escape. Such improved models of the neutral escape will lead to a better understanding of the paleoclimate and the evolution of the of primary and secondary atmospheres of terrestrial planets (Massol *et al.* 2016). It is known, that a prediction of escape rates due to thermal and non-thermal processes defines the long-term evolution of planetary atmospheres, therefore atmospheric escape has long been a subject of interest in the comparative planetology, and in recent years especially in understanding of the evolution of exoplanet atmospheres.

2. Neutral Oxygen Loss at Mars

The Martian thermosphere, exosphere, and ionosphere form the tenuous upper boundary - hot planetary corona the region through which all energy and matter from the Sun that encounters Mars must either pass or be deflected. The Martian neutral corona, ionosphere, and crustal magnetic fields interact directly with the incident solar wind plasma, forming a highly dynamic induced magnetosphere that influences the structure and variability of the upper atmosphere, and results in a rich array of plasma processes. Further, the energy input to the atmosphere from the Sun and solar wind energizes atmospheric particles, giving some of them sufficient energy to escape the planet. The close proximity of the solar wind to the Martian ionosphere, as well as the considerable extent of the exosphere relative to the size of the planet, leads to a number of atmospheric escape processes for both neutral and ionized particles. Neutral particles may have sufficient energy to escape on their own by virtue of their thermal distribution (Jeans escape), may receive the energy necessary to escape from solar radiation and subsequent chemical reactions (photochemical escape), or may receive the energy to escape from collisions with particles entering the atmosphere from above (e.g. by sputtering and/or precipitation). Escape rates of neutral particles have been indirectly estimated using a combination of measurements of the atmospheric reservoirs for escape and models (Shematovich & Marov 2018, Jakosky *et al.* 2018). The total atmospheric escape from both ions and neutrals varies with the same drivers observed to modify the solar wind interaction region. This variation can be extrapolated to estimate the total atmospheric loss over Martian history using scaling arguments, or by tuning simulations to match the present-day variation (Jakosky *et al.* 2018).

Neutral O escaping flux due to the photochemistry is about $(0.5-1.0)\times10^7$ cm^{-2}s^{-1} depending on the solar activity level (Shematovich & Marov 2018, Jakosky *et al.* 2018). Recently the neutral O escaping flux due to the H^+ and H precipitation was estimated by the values about $(0.7-30.0)\times10^5$ cm^{-2}s^{-1} (Shematovich 2017) when the energy spectra of the precipitating protons measured by MEX ASPERA-3 instrument at low solar activity level were used as the inputs at upper boundary. We apply the updated kinetic Monte Carlo model of the proton and hydrogen atom precipitation into the Martian upper atmosphere (Bisikalo *et al.* 2018) to estimate the atmospheric escape of hot oxygen atoms induced by precipitation. As the input at model upper boundary placed at 500 km the energy spectra of penetrating into the atmosphere high-energy particles measured by MAVEN SWIA instrument (Halekas *et al.* 2015) we used. It was found that O escaping flux is changing in the range $(1.7-0.8)\times10^7$ cm^{-2}s^{-1} for the MAVEN SWIA spectra of the precipitating protons from magnetosheath (Case 1 in Fig. 1) and hydrogen atoms from the undisturbed solar wind (Case 2 in Fig. 1) respectively. It is seen, that the neutral oxygen escape due to the H^+ and H precipitation can be competitive with the photochemical escape for the extreme solar events.

3. Atmospheric Mass Loss at Hot Neptunes and Jupiters

Observations of exoplanets using ground-based and space telescopes operating in the IR and UV wave ranges enable obtaining first estimates of such very important atmospheric characteristics as the composition and thermal state. Currently, the formation, stability, and evolution of the atmospheres of exoplanets are under active exploration using both observations with ground-based and space telescopes and mathematical simulation. A number of important results regarding the nature of the exoplanets have been obtained, and unique specific features were discovered in the planetary systems of single and binary stars that contain bodies substantially different from Solar System planets such as hot Neptunes and Jupiters, i.e., giant planets in close-in orbits around their host stars. To investigate the atmospheric mass-loss rate for such exoplanets we had developed the aeronomic model (Ionov, Pavlyuchenkov & Shematovich 2018). This one-dimensional aeronomic model of the upper atmosphere of the giant planet was applied to study the reaction of the atmosphere of a hot Jupiter HD 209458b to an additional heating caused

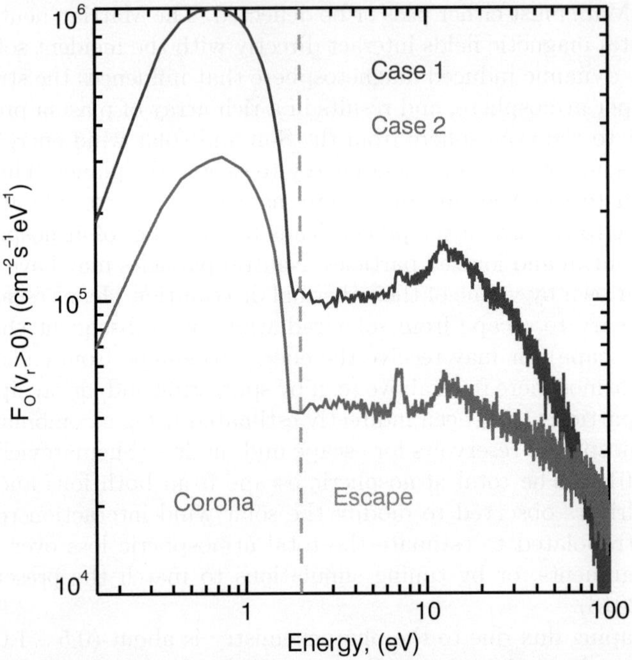

Figure 1. Energy distribution functions of the upward moving hot O atoms at height 300 km were calculated using the kinetic Monte Carlo model of the interaction of energetic H^+ and H particles with the upper atmosphere of Mars. The MAVEN SWIA measurements of proton spectra in the magnetosheath and in the undisturbed solar wind were used as the inputs to the model.

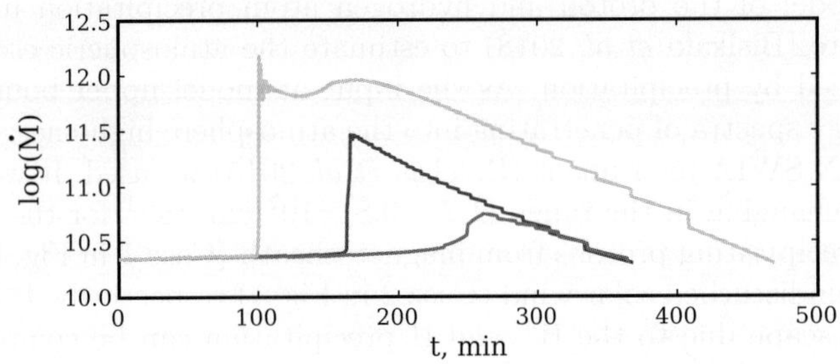

Figure 2. Time evolution of the mass-loss rates (in g s^{-1}) obtained at the exobase for the atmosphere of a hot Jupiter like HD 209458b during the three considered superflares. The red, blue, and green lines correspond to the mass-loss rates obtained considering flares emitting an XUV flux of 10, 100, and 1000 times that of the quiet star, respectively.

by the forcing of a stellar flare. It was found that the absorption of additional energy of the stellar flare in the extreme ultraviolet leads to the local heating of the atmosphere, accompanied by formation of two shock waves, propagating in the upper atmosphere (Bisikalo *et al.* 2018). The development of shock waves significantly affects atmospheric loss. We present this in Fig. 2, which shows the mass-loss rates at the exobase as a function of time for three considered superflares. For the case of the quiet star, we find a mass-loss rate of $2{\times}10^{10}$ g s^{-1} in the spherical symmetry approach, which increases by 1.9, 4, and 17.5 times during a flare emitting an XUV flux that is 10, 100, and 1000 times that of the quiet star, respectively. Since the excess mass loss occurs for 130, 210, and 380 minutes, the additional atmospheric loss corresponds to $3{\times}10^{14}$, $1{\times}10^{15}$, and $8{\times}10^{15}$ g for the three considered flares, respectively.

Depending on the frequencies of the flares the total mass loss from the atmosphere might therefore significantly increase. For old solar-type stars, taking into account that flares emitting an XUV flux 10, 100, and 1000 times that of the quiet star are believed to happen approximately 1, 0.1, and 0.001 times per year, the effect of flares on the evolution of planetary atmospheres of hot Jupiters can be considered to be negligible. However, the impact of flares may be more pronounced for younger systems. The local effect of the flare can be significant and change the gas dynamic structure of the whole envelope.

4. Conclusion

Studies of extrasolar planets make an immense contribution to explaining dissipation of planet atmospheres and, primarily, terrestrial planets in the early Solar System and thus offering an approach to the reconstruction of their formation history. We can expect that the expansion of the research area far beyond the Solar System and further improvement in mathematical models developed on the basis of exoplanet aeronomy will facilitate better understanding of evolution processes and key problems of planetary cosmogony. We acknowledge the support by Russian Science Foundation (Project 18-12-00447).

References

Bisikalo, D. V., Shematovich, V. I., Gerard, J.-C., & Hubert B. 2018, *J. Geophys. Res.*, 123, 5850

Bisikalo, D. V., Cherenkov, A. A., Shematovich, V. I., *et al.*, 2018, *Astron. Rep.*, 62, 648

Halekas, J. S., Lillis, R. J., Mitchell, D. L., *et al.*, 2015, *Geophys. Res. Lett.*, 42, 8901

Ionov, D. E., Pavlyuchenkov, Y. N., & Shematovich, V. I. 2018, *MNRAS*, 476, 5639

Jakosky, B. M., Brain, D., Chaffin, M., *et al.*, 2018, *Icarus*, 315, 146

Massol, H., Hamano, K., Tian, F., *et al.*, 2016, *Space Sci. Reviews*, 205, 153

Shematovich, V. I., 2017, *Sol. Syst. Res.*, 51, 249

Shematovich, V. I., & Marov, M. Ya. 2018, *Physics - Uspekhi*, 61, 217

Discussion

KEDZIORA-CHUDCZER: Does your model take into account the evolution of the Sun?

BISIKALO: Not exactly, but we can simulate the aeronomic profiles as a function of the stellar flux which corresponds to the evolution.

KEDZIORA-CHUDCZER: How is the atmospheric escape for terrestrial planets modified by the early Sun's lower flux?

BISIKALO: It is a complicated question, because the atmosphere is also changed with age, so I can't give a simple answer.

KHAIBRAKHMANOV: Could you comment on the role of planetary magnetospheres on the mass loss from exoplanet atmospheres?

BISIKALO: Magnetic fields play a significant role in mass loss. Both the internal magnetic field and induced magnetic field prevent the planetary atmosphere from the penetration of solar wind plasma and reduce the mass loss.

Origins: From the Protosun to the First Steps of Life
Proceedings IAU Symposium No. 345, 2018
B. G. Elmegreen, L. Viktor Tóth & M. Güdel, eds.

doi:10.1017/S1743921319002874

Modelling the atmosphere of potential habitable planets

Nicolas Iro

University of Vienna, Dept. of Astrophysics, Türkenschanzstr. 17, 1180 Vienna, Austria
email: nicolas.iro@univie.ac.at

Abstract. The list of planets discovered in the habitable zone of its star is continuously growing. We present a simple one-dimension radiative transfer model in order to better infer on the habitability of such systems. Particular focus is on the TRAPPIST-1 planets (Gillon *et al.* 2017), particularly on planets b, c, d, e and f.

Keywords. planets and satellites: general

1. Introduction

We are modelling several terrestrial planets discovered around the habitable zone of TRAPPIST-1. This system is composed of at least seven Earth-sized planets (from 0.77 to 1.15 $R_{\oplus}$, Grimm *et al.* 2018) orbiting a dwarf M8 star. Their orbital periods range from 1.5 to 9.2 days, receiving from 35% to 4 times the average flux received by Earth from the Sun. Morley *et al.* (2017) and Lincowski *et al.* (2018) presented one-dimension models of the TRAPPIST-1 planets, including thermal structure and planetary spectra with assumptions on their composition. Turbet *et al.* (2018) presented climate modelling of the TRAPPIST-1 planets.

The main stellar, orbital and planetary parameters are summarised in Table 1.

2. Model

<u>*Radiative-convective model*</u>

We use the one-dimension radiative-convective model from (Marcq *et al.* 2017). It has been developed for applications to magma ocean planets, with atmospheric mixures of H_2O, CO_2 and N_2. This model solves the longwave radiative transfer using k-correlated calculations. The k-correlated coefficient is computed by KSPECTRUM (Eymet *et al.* 2016) and the radiative transfer is solved using DISORT (Stamnes *et al.* 1988). See Marcq *et al.* (2017) for further details.

The shortwave radiation is not taken into account. Instead, the incoming stellar radiation is parameterised by $(1 - A)S_{\star}$, where A takes into account A_{clear} and A_{cloud} (both free parameters). The top of the atmosphere temperature can be either calculated or set as a free parameter. The calculation of this temperature, in the absence of consistent heating calculation gives unrealistic results. As a consequence, we treated it as a free parameter.

In order to compare our results to Morley *et al.* (2017), we also compute for each planet an Earth-like composition as well as a Venus-like composition (see Table 2 for the model parameters we used).

Table 1. Summary of the planetary system parameters.

	Parameters	TRAPPIST-1 b	c	d	e	f	References
Star	Type	M8					*a*
	T_{eff} [K]	2560					*b*
	R [$R_\odot$]	0.0117					*c*
Orbit	P [Days]	1.51	2.42	4.05	6.10	9,21	*b*
	a [AU]	0.0115	0.0158	0.0223	0.0293	0.0385	*d*
	S [$S_\odot$][1]	4	2.1	1.1	0.6	0.35	Calculated
Planet	M [$M_\oplus$]	1.02	1.16	0.30	0.77	0.93	*d*
	R [$R_\oplus$]	1.12	1.10	0.77	0.91	1.04	*d*
	ρ [$\rho_\oplus$]	0.73	0.88	0.62	1.02	0.82	*d*
	g [$g_\oplus$]	0.81	0.97	0.48	0.93	0.85	*d*
	T_{eq} [K][2]	400	342	288	251	219	*b*

Notes:
[1] S: Stellar constant in units of the Solar constant at 1 AU.
[2] Assuming a zero albedo.
[a] Liebert & Gizis (2006)
[b] Gillon *et al.* (2017)
[c] Gillon *et al.* (2016)
[d] Grimm *et al.* (2018)

Table 2. Summary of the simulation parameters.

Parameters	Earth case	Venus case
P_{surf} [*bar*]	1	100
N_2	78%	3.4%
CO_2	0.035%	96%
H_2O	1%	0.3%

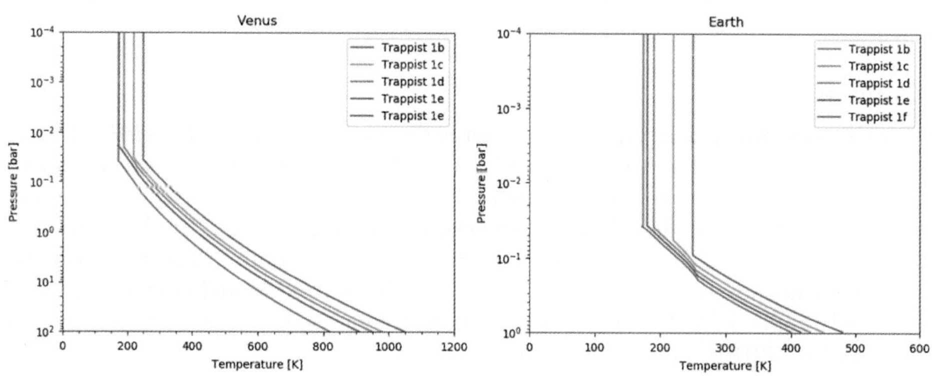

Figure 1. Thermal profiles of the Trappist 1 modelled planets in the Venus (*Left*) and Earth (*Right*) atmosphere setups.

3. Results

Thermal profiles

Figure 1 can be compared with Figure 4 of Morley *et al.* (2017) for the Venus case. They are overall similar. One difference however is that Morley *et al.* (2017) fixed the radiative zone upper boundary to $P = 0.1$ bar and iterated T_{surf} until $T_{\text{eff}} = T_{\text{eq}}$. Here we do not fix the boundary between troposphere and mesosphere (see Section 2) but have one more free parameter (T_{top} and T_{surf}). So we fixed the top of atmosphere temperature

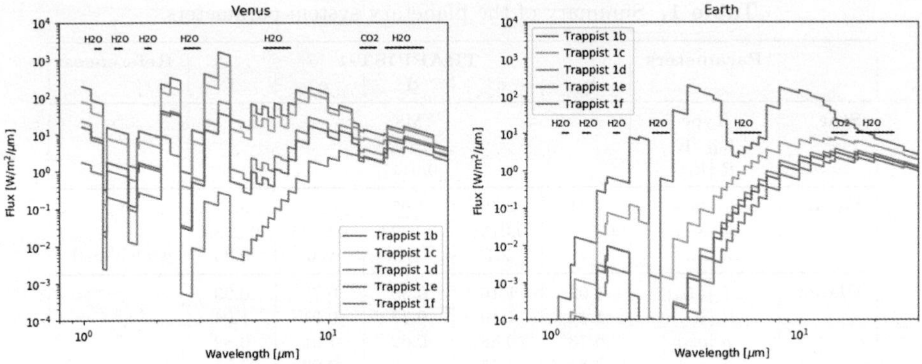

Figure 2. Planetary spectra.

to match Morley *et al.* (2017) and iterated $T_{\rm surf}$ until $T_{\rm eff} = T_{\rm eq}$. The fact that the radiative/convective boundary layer is here allowed to vary makes our surface temperatures less spread.

Considering the difference of irradiation between TRAPPIST-1e and TRAPPIST-1f, the very similar temperature profile makes us wonder if the radiative transfer is valid for low temperatures.

Spectra

Figure 2 can be compared with Figure 11 of Morley *et al.* (2017) for the Earth case. With a lower resolution in our case, they are qualitatively similar. One can recognize the surface temperature blackbody with some H_2O and CO_2 absorption bands. As expected, in the Venus case, the atmosphere is thick enough that the emission spectrum is totally different than a blackbody at the surface temperature, especially at wavelengths shorter than 4 μm.

4. Conclusion

We used a one-dimension radiative transfer code in order to model TRAPPIST planets. This allowed us to calculate thermal profiles as well as thermal emission spectra for these planets. The code is relatively light which will allow us in the future to integrate in e.g. 3D general circulation model. However, the shortwave radiation is yet to be implemented in a consistent way, which is under way (E. Marcq, private comm). The next step would be to couple this radiation code with our General Circulation Model (Plasim), which has been recently applied (with its original radiation scheme) to the planet Proxima Centauri b (Galuzo *et al.* 2019).

References

Eymet, V., Coustet, C., & Piaud, B. 2016, *J. of Physics Conference Series*, 676, 012005

Galuzzo, D., Cagnazzo, C., Berrilli, F., Fierli, F., Giovanelli, L., & Iro, N. 2019 *In Prep.*

Gillon, M., Jehin, E., Lederer, S. M., *et al.* 2016, *Nature*, 533, 221

Gillon, M., Triaud, A. H. M. J., Demory, B.-O., *et al.* 2017, *Nature*, 542, 456

Grimm, S. L., Demory, B.-O., Gillon, M., *et al.* 2018, *A&A*, 613, A68

Liebert, J., & Gizis, J. E. 2006, *PASP*, 118, 659

Lincowski, A. P., Meadows, V. S., Crisp, D., *et al.* 2018, *ApJ*, 867, 76

Marcq, E., Salvador, A., Massol, H., & Davaille, A. 2017, *J. of Geophysical Research (Planets)*, 122, 1539

Morley, C. V., Kreidberg, L., Rustamkulov, Z., *et al.* 2017, *ApJ*, 850, 121

Stamnes, K., Tsay, S.-C., Jayaweera, K., & Wiscombe, W. 1988, *Appl. Optics*, 27, 2502
Turbet, M., Bolmont, E., Leconte, J., *et al.* 2018, *A&A*, 612, A86

Discussion

TRIMBLE: Do any of your 3D models yield 3 Hadley cells per hemisphere?

IRO: One equator to pole in each hemisphere is typical in the case of slow rotating planets. If we increase the rotation rate the Hadley cell would break but this is not the case here.

TRIMBLE: And that is because of phase-locking?

IRO: Yes, the phase locked configuration leads to slow rotation of the planet.

GUENTHER: Temperature of TRAPPIST-1: on the transparency it is 5560K. I guess 2560K.

IRO: Yes, it was obviously a typo!

ANONYMOUS: Have you done any improvements to the traditional model?

IRO: The models we compared to were different ones. We are still in the process of understanding the differences in the 3D models, radiation, composition and cloud parameterization could explain the discrepancies.

Origins: From the Protosun to the First Steps of Life
Proceedings IAU Symposium No. 345, 2018
B. G. Elmegreen, L. Viktor Tóth & M. Güdel, eds.

doi:10.1017/S1743921319002035

The biological impact of superflares on planets in the Habitable Zone

Adriana Valio, Raissa Estrela, Luisa Cabral and Abel Grangeiro

Mackenzie Presbyterian Universtiy, CRAAM, São Paulo, Brazil
email: avalio@craam.mackenzie.br

Abstract. Younger and fully convective stars are much more active than our Sun, producing many superflares. Here we estimate the impact of the superflares UV radiation on living organisms on the surface of orbiting planets in the habitable zone of the star. For this we study two active stars, Kepler-96 (solar type) and TRAPPIST-1 (M dwarf). Kepler-96, with an age of 2.4 Gyr, is at the same stage of the Sun when the first multicellular organisms appeared on Earth. The biological impact of super flares are studied on a hypothetical Earth at 1AU of Kepler-96 and on planets TRAPPIST-1*e*, *f*, and *g* for three atmospheres scenarios: an Archean and Present-day atmospheres with and without ozone. We estimated the survival rates of two bacteria and concluded that life would only survive on the surface of these planets if their atmosphere had an ozone layer, or in shallow waters of an ocean.

Keywords. Astrobiology, Stars: flare, Planetary systems

1. Introduction

Living around an active star can be vary dangerous. Even though the Sun is considered mildly active among other stars, our technological society is frequently affected by its activity. An important manifestation of solar activity are solar flares. Among the disruptions caused by a flare is an increase in hazardous radiation such as ultraviolet (UV) and X-rays, beside energetic particles. This high energy radiation can be lethal to astronauts outside the protection of rockets or the Space Station. Fortunately, common humans on the ground are protected from this ionizing radiation by Earth's atmosphere. But it may not have been so in the past.

Young stars are known to produce much stronger flares, due to an increased dynamo action. Also fully convective stars, such as early M-type stars are very active generating flares sometimes 10,000 times stronger than those of the present Sun. Superflares release significant amount of XUV, EUV, FUV and UV radiation. Depending on the size of the flare, they can cause potential effects on the planetary atmosphere such as atmospheric loss and/or affect the chemical composition of the upper atmosphere. Moreover, protons accelerated in the flare produce odd nitrogen and odd hydrogen in the upper stratosphere and mesosphere that destroy ozone (Segura *et al.* 2010). Last but not least, the energetic radiation from superflares could affect the origin and evolution of life on a planet orbiting in the Habitable Zone of the star.

The goal of this work is to determine if superflares can be dangerous to life present in the surface or an ocean of a hypothetical planet in the habitable zone (HZ) of Kepler-96 and the three planets in the HZ of the TRAPPIST-1 system, planets *e*, *f* and *g*. The next section describes the two stars studied and their orbiting planets, the following Section 3 characterizes the observed flares and estimates the UV flux generated. Three types of planetary atmospheres are assumed for each planet (Section 4) to determine the amount of damaging UV flux that reaches the surface, and thus the biological impact on two

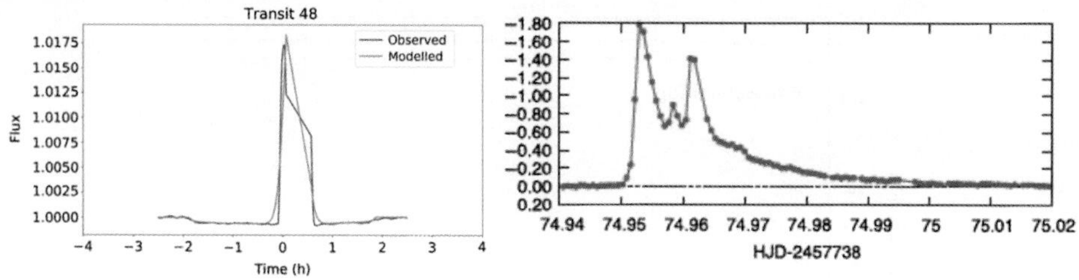

Figure 1. Left: Largest superflare observed in the light curve of Kepler-96 (Estrela & Valio 2018). Right: Largest superflare of TRAPPIST-1 (Segura *et al.* 2010).

bacteria (*D. Radiodurans* and *E. Coli*). A brief discussion of the results and the main conclusions are presented in the last section.

2. Stars and their planets

Kepler-96 is a solar type star, mass of $(1.0 \pm 0.06)M_{\odot}$ and radius of $(1.02 \pm 0.09)R_{\odot}$, with approximately half the age of the Sun, at 2.34 Gyr. It is orbited by a hot Super-Earth (Kepler-96b) of $8.6M_{Earth}$ and $2.67R_{Earth}$ with a period of 16.2385 days. Being a young star, Kepler-96 produced several flares that were observed during the transits of Kepler-96b.

Kepler-96 has an age that corresponds to the end of the Archean Era on Earth, when life here underwent a major change with the emergence of the first multi-cellular organisms. This caused a boom of oxygen on our atmosphere, known as the Great Oxygenation Event, thus creating the ozone layer. Due to its close orbit, the surface temperature of Kepler-96b is estimated to be between 2000 to 3000K, thus it is interesting to evaluate the effect of the UV flux produced by the giant flares of Kepler-96 and the impact this would have on life on the surface of a hypothetical planet located at 1 AU of Kepler-96, within this solar-type star habitable zone.

The other interesting system worth studying under the same aspect is the TRAPPIST-1 planetary system with 7 terrestrial planets. TRAPPIST-1 is an M8 star ($(0.08 \pm 0.009)M_{\odot}$ and $(0.117 \pm 0.004)R_{\odot}$) of 0.5 Gyr age. Of the seven terrestrial planets in orbit, three of them are within the habitable zone of the star: planets e (0.0282 AU), f (0.0371 AU), and g (0.0451 AU).

3. Stellar flares and superflares

Many flares were observed in the Kepler-96 lightcurve, especially during the transits of the Super-Earth. The most intense flare is shown in Figure 1 (left panel). Estrela & Valio (2018) analyzed three of these events and estimated the total energy from each flare by modelling the time profile as a Gaussian. This strongest flare released a total of 1.8×10^{35} ergs. This energy range corresponds to superflares (Maehara *et al.* 2015)

To estimate the amount of UV flux produced by the flare of Kepler-96, a solar-type star, we took the strongest solar flare recorded in modern times, a GOES class X17 flare that occurred on the 28th October 2003. The total energy from this flare was $E = 4 \times 10^{32}$ ergs (Woods *et al.* 2004). During this flare, the solar MUV (200 – 300 nm) flux increased by 12%. Since the total thermal blackbody flux of Kepler-96 and that of the Sun are very similar, we considered that the superflare of Kepler-96 would increase the UV flux by the same amount. Thus applying this same percentage to the superflare of Kepler-96, yields an increase of 5400% in the MUV flux. The final flux was decreased by 75% of the present valued, due to the smaller flux of the young Sun with an age of 2.4 Gyr.

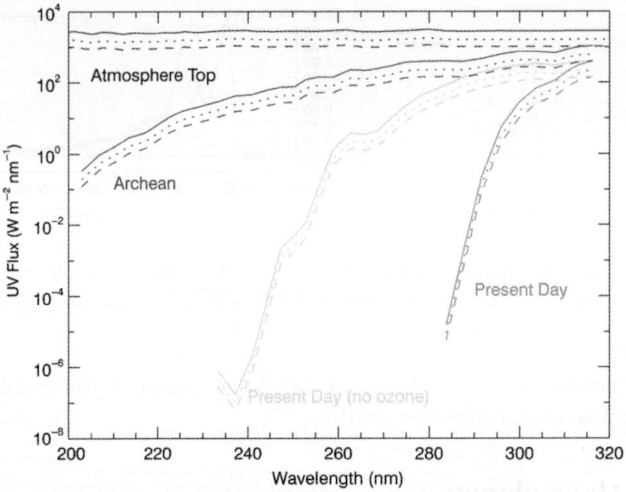

Figure 2. Ultraviolet flux at the top of the atmosphere of planets TRAPPIST-1e (solid), TRAPPIST-1f (dotted), TRAPPIST-1g (dashed) shown in black. The UV flux at the surface transmitted by three atmosphere models: Archean (blue), Present day without Ozone (green), and Present day atmosphere with Ozone (magenta).

The TRAPPIST-1 star is known to be very active, with 47 flares detected with energies between 10^{30} and 10^{33} ergs (Vida *et al.* 2017). The largest flare, depicted in the right panel of Figure 1, released a total energy of 1.24×10^{33} erg over 43 min.

The UV flux contribution from TRAPPIST-1 was taken as the same UV flux fraction measured from a flare of Ad Leo, also an M red dwarf (Segura *et al.* 2010). Hawley & Pettersen (1991) observed a strong flare from the red dwarf Ad Leonis in the visible and ultraviolet. From their Table 6, we calculated the percentage of UV flux with respect to the visible from this flare as 59%. Taking into account the distance of the planets from the star, the total UV flux reaching the top of the atmosphere of the planets are: $3.275 \times 10^{12} erg/m^2$, $1.888 \times 10^{12} erg/m^2$, and $1.278 \times 10^{12} erg/m^2$ for planets e, f, and g, respectively. These fluxes as a function of wavelength are depicted in Figure 2 as black lines for the three planets following O'Malley & Kaltenegger (2017).

4. Planetary atmospheres

Three types of atmosphere were considered: an archean atmosphere and a present day one with and without Ozone. The present day atmosphere is composed of 80% of N_2 and 20% O_2, whereas the Archean atmosphere (3.9 Gyr – 2.5 Gyr) consists of 80% of N_2 and 20% CO_2 (Cnossen *et al.* 2007). Moreover, for Kepler-96, the flux was assumed to be 75% of the present-day solar irradiation.

To analyze the biological impact, we used the UV flux (180 – 300nm) passing through the different atmospheres. Estimates of the UV flux reaching the planetary surface were made using the curves shown in Figure 6 of Cnossen *et al.* (2007) for these three model atmospheres. This figure was adapted to determine the transmitted flux for each model atmosphere. In the case of TRAPPIST-1, a red dwarf star, we did not use the irradiation from the Sun (top curve of the Cnossen *et al.* figure), but rather the spectra proposed by O'Malley & Kaltenegger (2017) (black curves in Figure 2). The net UV flux that reaches the surface of the three TRAPPIST-1 planets in the HZ are shown for the different atmosphere types in Figure 2.

Table 1. Biological effective irradiance, E_{eff} (J/m^2), due to the strongest superflare

Planet	Bacteria	Archean	Present day - no ozone	Present day
Kepler-96 (1 AU)	*E. coli*	1.4×10^4	2.4×10^3	21.5
	D. radiodurans	8.0×10^3	1.3×10^3	7.5
TRAPPIST-1e	*E. coli*	9.0×10^3	962	8.2
	D. radiodurans	5.1×10^3	580	2.5
TRAPPIST-1f	*E. coli*	5.2×10^3	555	4.7
	D. radiodurans	2.9×10^3	334.4	1.5
TRAPPIST-1g	*E. coli*	3.3×10^3	353	3.0
	D. radiodurans	1.9×10^3	212	0.9

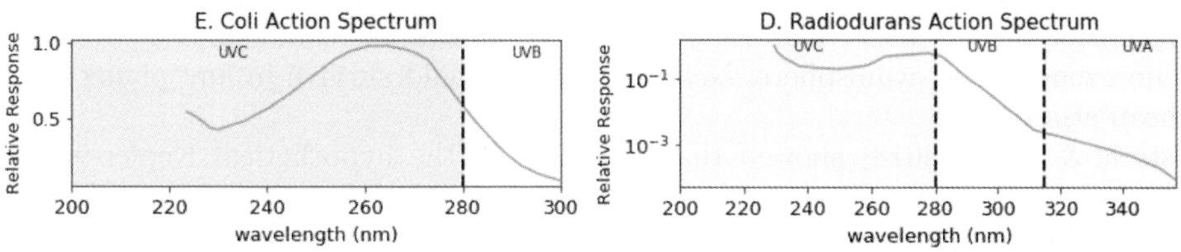

Figure 3. Action spectra, or biological response, for *E. coli* (left) and *D. radiodurans* (right).

5. Biological impact

Superflares will increase the UV flux of the star by a significant fraction. If a planet has an atmosphere with absorbers such as N_2, CO_2, or O_2, the short wavelengths radiation (0.1 – 200 nm) will be attenuated through the atmosphere. UVB and UVC, the most hazardous radiation, will be partlally or totally absorbed by an Ozone layer. The DNA molecule of living beings is mainly damaged in the UVC and UVB range (MUV $\sim 200 - 300$ nm).

For this study, we chose two bacteria, a very resistant one, *Deinococcus radiodurans*, and a more common one, *Escherichia coli*. Since the response of a biological body varies as function of the wavelength, the biological response effectiveness at different wavelengths has to be estimated. For that, the incident UV flux, $F_{UV}(\lambda)$, is weighted by the action spectrum, $S(\lambda)$, which gives the biological response of different organisms.

$$E_{eff} = \int_{180}^{300} F_{UV}(\lambda)\ S(\lambda)\ d\lambda \tag{1}$$

The action spectra of the two bacteria are shown in Figure 3. The UV dosage for 10% survival of the two bacteria are 22.6 (Gascon *et al.* 1995) and 553 J/m^2 (Ghosal *et al.* 2005) for *E. coli* and *D. radiodurans*, respectively.

To determine the survival of the two bacteria on the surface of planets in the habitable zone of their stars, we have calculated the integral in Eq. 1 for the hypothetical planet in orbit at 1 AU of Kepler-96 and the three TRAPPIST-1 planets in the HZ, e, f, and g, considering the strongest observed flare of both stars (see Figure 1). The results are listed in Table 1.

6. Discussion and conclusions

We analyzed the strongest flare detected in planetary transits of Kepler-96b (Estrela & Valio 2018), with total energy of 1.8×10^{35} ergs, and of TRAPPIST-1 (Vida *et al.* 2017), with 1.24×10^{33} ergs. Both these flares had energies in the range of superflares. We used the strongest recorded solar flare (Woods *et al.* 2004) to model the UV radiation of Kepler-96 flare, whereas the UV contribution of the TRAPPIST-1 flare was taken from a well observed flare of Ad Leo (Hawley & Pettersen 1991), a star of same spectral type.

Next, to determine the amount of UV flux that reaches the surface of the planets, three atmospheric models were considered: an Archean atmosphere and present-day atmospheres with and without Ozone (Cnossen *et al.* 2007). Two bacteria, *E. coli* and the hard X-ray resistant *D. radiodurans*, were chosen to determine their survival rate under the heavy UV bombardment due to the flares. For this study, we calculated the effectiveness response of each bacteria due to the UV flaring flux (see Table 1).

The results of the biological impact suggests that the UV flux from the strongest superflare received by a biological body would only allow the presence of life on the surface of a hypothetical planet at 1 AU of Kepler-96 if there was an atmosphere with Ozone. The same is true for TRAPPIST-1e, the planet in the HZ closer to the star. However, for the two other planets, *f* and *g*, *D. radiodurans* could survive even if there was no ozone in the atmosphere. Neither bacteria could survive in any planet with an Archean atmosphere.

Estrela & Valio (2018) showed that an ocean in the hypothetical Kepler-96 planet in the HZ could protect organisms from the increased UV radiation, allowing life in depths within the photic zone (up to 200m). Maybe this scenario is also applicable to the TRAPPIST-1 planets, that is, an ocean is a necessary condition to protect primitive organisms from the increased UV flux due to superflares, until an ozone layer is formed. In a future work, we will consider the cumulative effects of all superflares from the star.

References

Cnossen, I., Sanz-Forcada, J., Favata, F., Witasse, O., Zegers, T., & Arnold, N. F. 2007, *JGR*, 112, E02008

Estrela, R. & Valio, A. 2018, *Astrobiology*, 18, 1414-1424

Gascon, J., Oubina, A., Perez-Lezaun, A., & Urmeneta, J. 1995, *Curr. Microbiol.*, 30, 177182

Ghosal, D., Omelchenko, M. V., Gaidamakova, E. K., Matrosova, V. Y., Vasilenko, A., Venkateswaran, A., Zhai, M., Kostandarithes, H. M., Brim, H., Makarova, K. S., Wackett, L. P., Fredrickson, J. K., & Daly, M. J. 2005, *FEMS Microbiol. Rev.*, 29, 361375

Hawley, S. L., & Pettersen, B. R. 1991, *ApJ*, 378, 725

Maehara, H., Shibayama, T., Notsu, Y., Notsu, S., Honda, S., Nogami, D., & Shibata, K. 2015, *Earth Planets Space*, 67, 59

OMalley-James, J. T. & Kaltenegger, L. 2017, *MNRAS*, 469, L26L30

Segura, A., Walkowicz, L. M., Meadows, V., Kasting, J., & Hawley, S. 2010, *Astrobiology*, 10, 751771

Vida, K, Kovari, Zs., Pal, A., K. Olah, K, & Kriskovics, L. 2017, *ApJ*, 841, 124-129

Woods, T. N., Eparvier, F. G., Fontenla, J., Harder, J., Kopp, G., McClintock, W. E., Rottman, G., Smiley, B., & Snow, M. 2004, *Geophys. Res. Lett*, 31, L10802

Discussion

GUENTHER: You study the impact of super-flares on the organism, but there are many small flares for each large one. What effect would these small but frequent flares have?

VALIO: Definitely the cumulative effects of smaller super-flares are important. Moreover, the recurrence of super-flares during the recovering phase of mico-organisms can also be damaging. We plan to investigate these issues in the near future.

Origins: From the Protosun to the First Steps of Life
Proceedings IAU Symposium No. 345, 2018
B. G. Elmegreen, L. Viktor Tóth & M. Güdel, eds.

doi:10.1017/S174392131900293X

Stellar activity and winds shaping the atmospheres of Earth-like planets

Theresa Lueftinger[1], Manuel Güdel[1], Sudeshna Boro Saikia[1], Colin Johnstone[1], Beatrice Kulterer[1], Oleg Kochukhov[2] and Kristina Kislyakova[1]

[1]Department of Astrophysics, University of Vienna,
Tuerkenschanzstrasse 17, A-1180 Wien, Austria
emails: theresa.rank-lueftinger@univie.ac.at, manuel.guedel@univie.ac.at, sudeshna.boro.saikia@univie.ac.at, colin.johnstone@univie.ac.at, a01301394@unet.univie.ac.at, kristina.kislyakova@univie.ac.at

[2]Dept. of Astronomy & Space Physics, Uppsala University,
Box 516, SE-751 20 Uppsala, Sweden
email: Oleg.Kochukhov@physics.uu.se

Abstract. Planets orbiting young, active stars are embedded in an environment that is far from being as calm as the present solar neighbourhood. They experience the extreme environments of their host stars, which cannot have been without consequences for young stellar systems and the evolution of Earth-like planets to habitable worlds. Stellar magnetism and the related stellar activity are crucial drivers of ionization, photodissociation, and chemistry. Stellar winds can compress planetary magnetospheres and even strip away the outer layers of their atmospheres, thus having an enormous impact on the atmospheres and the magnetospheres of surrounding exoplanets. Modelling of stellar magnetic fields and their winds is extremely challenging, both from the observational and the theoretical points of view, and only ground breaking advances in observational instrumentation and a deeper theoretical understanding of magnetohydrodynamic processes in stars enable us to model stellar magnetic fields and their winds – and the resulting influence on the atmospheres of surrounding exoplanets – in more and more detail. We have initiated a national and international research network (NFN): 'Pathways to Habitability – From Disks to Active Stars, Planets to Life', to address questions on the formation and habitability of environments in young, active stellar/planetary systems. We discuss the work we are carrying out within this project and focus on how stellar evolutionary aspects in relation to activity, magnetic fields and winds influence the erosion of planetary atmospheres in the habitable zone. We present recent results of our theoretical and observational studies based on Zeeman Doppler Imaging (ZDI), field extrapolation methods, wind simulations, and the modeling of planetary upper atmospheres.

Keywords. Stars, magnetic fields, activity, habitability, planetary atmospheres

1. Stellar magnetic field-, activity- and wind evolution

The activity output (radiation, winds, stellar flares and high-energy particles) that is triggered by the magnetic field of the central star is largely responsible for the erosion of planetary upper atmospheres, whereby the star's X-ray and ultraviolet radiation (XUV) play the most significant role here. This radiation declines in time, whereby the precise evolutionary path crucially depends on the stellar rotational evolution (Tu *et al.* 2015, Fig. 1). A broad, general picture is described by simple, unique decay laws (Güdel *et al.* 1997; Ribas *et al.* 2005; Claire *et al.* 2012) but especially at ages less than a few 100 Myr, these may not apply to individual stars. This age range is crucial, though, for setting

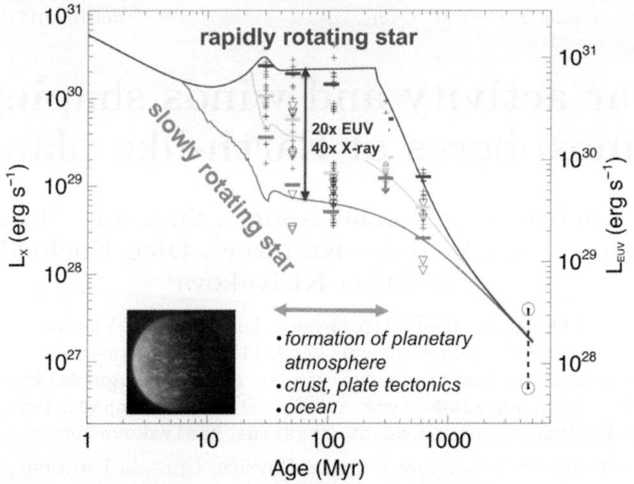

Figure 1. High-energy Sun in time (Tu *et al.* 2015).

the boundary conditions (astrophysically and geophysically) for the co-evolution of star-planetary systems into habitable ones. Stellar wind estimates are confronted with similar issues: although indirect estimations are available (Wood *et al.* 2005), little systematics can be inferred for stars at young ages (Johnstone *et al.* 2015a; Johnstone *et al.* 2015b). It highly depends on the initial conditions of stellar rotation at very young ages, how it changes with time: it can evolve along completely different tracks for many 100 million years. As a result, evolutionary tracks for stellar magnetic activity, short-wavelength radiation and ionized winds can be very different, which implies very different evolutionary scenarios for the erosion of planetary atmospheres.

Within the research network (NFN): 'Pathways to Habitability – From Disks to Active Stars, Planets to Life' we systematically address the evolution that stellar environments undergo during the PMS and young MS phases, addressing the entire radiative and particle output in the context of stellar rotational evolution with a focus on the study of the magnetic-field and wind evolution, as presented in this talk.

2. Magnetic fields and Zeeman Doppler imaging

Complexity and strength of magnetic fields are believed to be a strong function of the internal stellar structure (Morin *et al.* 2010; Gregory *et al.* 2012; Johnstone *et al.* 2014), which has crucial implications for the properties of the stellar wind, the according mass loss rates and the evolving planetary environment. The structure and strength of magnetic fields can be assessed via the Zeeman Doppler Imaging (ZDI) technique (e.g., Donati *et al.* 2008, Lüftinger *et al.* 2010a; Lüftinger *et al.* 2010b).

Built around complex mathematical procedures ZDI has become one of the most powerful astrophysical remote sensing methods. This technique has meanwhile reached a state of maturity in which we can determine strength, distribution, polarity and even polarity reversals of surface magnetic fields. At the same time temperature- or abundance structures on the surface of stars can be reconstructed by inverting time series of high-resolution spectropolarimetric data.

However, we still have only limited knowledge of statistical and evolutionary systematics. Thus, using the NARVAL spectropolarimeter (attached to the Télescope Bernard Lyot, Pic du Midi, France) we have obtained and/or dearchived spectropolarimetric data of a number of young active stars (e.g. π^1 UMa, Xi Boo A, etc.) determining their magnetic field geometry via ZDI and estimating their wind characteristics using field

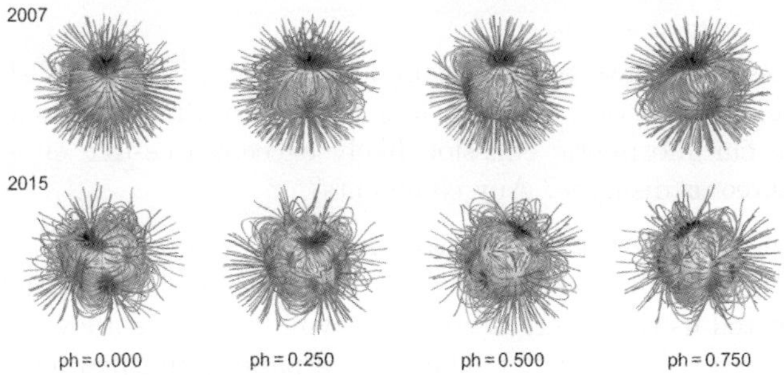

Figure 2. Stellar surface magnetic field distribution of π^1 UMa

extrapolation methods (Lüftinger *et al.*, Kulterer *et al.*, Boro Saikia *et al.* 2019, in prep.). In this context we study systematics in magnetism and winds linked to stellar evolutionary history and relate those to our empirical models predicting wind evolution.

In a collaboration with the Space Sciences Laboratory (Berkeley, USA) and the NASA Goddard Space Flight Center/SEEC (USA) we try to understand, based on our ZDI-maps via data-driven, three-dimensional, magnetohydrodynamic modeling, how eruptive superflares from young solar-like stars (e.g. κ^1 Cet, Lynch *et al.*, submitted) would impact their interplanetary environments.

Acknowledgements

These research projects receive funding from the Austrian Space Application Programme (ASAP) of the Austrian Research Promotion Agency (FFG) within ASAP11, the FWF NFN project S11601-N16, and the sub-project S11604-N16. We also thank the TBL team for their valuable support during the observing runs.

References

Claire, M. W., Sheets, J., Cohen, M., Ribas, I., Meadows, V. S., & Catling, D. C. 2012, *ApJ*, 757, 95, 12

Donati, J.-F., Morin, J., Petit, P., *et al.* 2008, MNRAS, 390, 545-560

Fichtinger, B., Güdel, M., Mutel, R. L., Hallinan, G., Gaidos, E., Skinner, S. L., Lynch, Ch., & Gayley, K. G. 2017, *A&A*, 599A

Gregory, S. G., Donati, J.-F., Morin, J., Hussain, G. A. J., Mayne, N. J., Hillenbrand, L. A., & Jardine, M. 2012, *ApJ*, 755, id. 97, 20

Güdel, M., Guinan, E. F., & Skinner, S. L. 1997, *ApJ*, 483, 947-960

Johnstone, C. P., Jardine, M., Gregory, S. G., Donati, J.-F., & Hussain, G. 2014, *MNRAS*, 437, 3202-3220

Johnstone, C. P., Güdel, M., Lftinger, T., Toth, G., & Brott, I. 2015a, *A&A*, 577, id.27, 22

Johnstone, C. P., Güdel, M., Brott, I., & Lüftinger, T. 2015b, *A&A*, 577, id.28, 25

Lynch, B. J., Airapetian, V. S., Kazachenko, M. D., DeVore, C. R., Lüftinger, T., Kochukhov, O., Rosén, L., & Abbett, W. P. 2019, *Nat. Astr.*, submitted

Lüftinger, T., Kochukhov, O., Ryabchikova, T., Piskunov, N., Weiss, W. W., & Ilyin, I. 2010a, *A&A*, 509, id71, 12

Lüftinger, T., Frhlich, H.-E., Weiss, W. W., *et al.* 2010b, *A&A*, 509, id43, 11

Morin, J., Donati, J.-F., Petit, P., Delfosse, X., Forveille, T., & Jardine, M. M. 2010, *MNRAS*, 407, 2269-2286

Ribas, I., Guinan, E. F., Güdel, M., & Audard, M. 2005, *ApJ*, 622, 680-694

Tu, L., Johnstone, C. P., Güdel, M., & Lammer, H. 2015, *A&A*, 577, idL3, 4

Wood, B. E., Mller, H.-R., Zank, G. P., Linsky, J. L., & Redfield, S. 2005, *ApJ*, 628, L143-L146

Discussion

LINSKY: Mass loss rates have been measured for about 15 G and K stars by the astrosphere technique. Observation of mass loss rates at radio wavelengths is difficult to measure because chromospheric emission likely exceeds free-free emission from stellar winds. Do you agree or disagree? Any comments?

LUEFTINGER: I agree. Getting mass loss rates from radio observations actually only yields upper limits for stellar winds and mass loss rates, which is actually very valuable, as non-detections tell us that there is no stellar wind signature above a certain threshold (obtained by the radio observations). There was a paper published recently by our group in Vienna (Fichtinger *et al.* 2017), where this is described in detail. Via ALMA and VLA radio observations we could determine radio emission and mass loss rate limits of four young solar-type stars.

Origins: From the Protosun to the First Steps of Life
Proceedings IAU Symposium No. 345, 2018
B. G. Elmegreen, L. Viktor Tóth & M. Güdel, eds.

doi:10.1017/S1743921319001704

Outbursts of young Sun-like stars may change how terrestrial planets form

P. Ábrahám[1], Á. Kóspál[1,2], L. Chen[1] and A. Carmona[3]

[1]Konkoly Observatory, Research Centre for Astronomy and Earth Sciences, Hungarian Academy of Sciences, Konkoly-Thege Miklós út 15-17, H-1121, Budapest, Hungary

[2]Max Planck Institute for Astronomy, Königstuhl 17, D-69117 Heidelberg, Germany

[3]Université de Toulouse, UPS-OMP, IRAP, F-31400 Toulouse, France
email: abraham@konkoly.hu

Abstract. While the Sun is nowadays a quiet and well-balanced star, in its first few million years it might have been often out of temper, like those young low-mass stars which episodically undergo unpredictable outbursts. The prototype of one of the two classes of young erupting stars, EX Lupi, had its historically largest outburst in 2008. It brightened by a factor of 30 for six months, due to elevated accretion from the circumstellar disk on to the star. Our group observed the system during the outburst, and discovered the crystallisation of amorphous silicate grains in the inner disk by the heat of the outburst. Our mid-infrared monitoring of the freshly produced crystals revealed that their emission in the inner disk quickly dropped already within a year after the outburst. Here we report on new observations of the 10 μm silicate feature, obtained with the MIDI and VISIR instruments at Paranal Observatory, which demonstrate that within five years practically all forsterite disappeared from the inner disk. We attempt to model this process by an expanding wind that transports the crystals from the terrestrial zone to outer disk regions where comets are supposed to form. Since the eruptions of EX Lup are recurrent, we speculate that the early Sun also experienced similar brightenings, and the forming planetary system might have incorporated some of the mineralogical and chemical yields provided by the outbursts. EX Lup, as a proxy for the proto-Sun, may be a telltale object to understand the origin of molecules and minerals we routinely encounter on Earth.

Keywords. stars: pre-main sequence, circumstellar matter, individual(EX Lup)

1. Eruptive phenomena during early stellar evolution

Over a period of 10^5–10^6 yrs, the material of the circumstellar disks around low-mass pre-main sequence stars is, to a large part, being accreted onto the protostar, building up its final mass. This process, however, is far from being smooth. In addition to the usual daily-weekly fluctuation of the accretion rate observed in T Tauri stars, there are episodic events of enormously increased accretion rate, which lead to temporary brightenings of the system. Young eruptive objects are traditionally categorized into two classes. EX Lupi-type stars (EXors) are defined by their repetitive outbursts lasting several months, while FU Orionis-type objects (FUors) produce decade-long and more energetic outbursts (Hartmann & Kenyon 1996; Audard *et al.* 2014). The optical brightness change is 3–5 mag, corresponding to a 10–100 times increase of the accretion rate from the inner disk onto the star.

2. The 2008 outburst of EX Lup

In 2008 EX Lup, the prototype of the EXor class, underwent its historically largest outburst, brightening by about 5 magnitudes in visual light for a period of six months.

Our group obtained a 5–37 μm spectrum with the InfraRed Spectrograph of the Spitzer Space Telescope on 2008 April 21, shortly after the peak of the outburst. Comparing our spectrum with a pre-outburst observation from 2005, we detected new spectral signatures of crystalline silicate material, mainly forsterite, appeared during the outburst (Fig. 1).

We investigated several possible scenarios for the origin of these crystalline features. One possibility was that the crystalline particles were already present on the disk surface, but due to the higher luminosity during the outburst they became better illuminated and visible. Our calculations, however, demonstrated that the required change in luminosity is significantly higher than observed, thus this explanation was excluded. Another possibility was that crystalline material already existed in the midplane of the disk, and was transported to the disk surface by vertical mixing processes related to the increased turbulence during outburst. The calculated timescales for this process, however, turned out to be much longer than the length of the outburst. Therefore, our conclusion was that the originally amorphous silicate grains were transformed to crystalline grains in the inner disk via annealing, a process highly efficient above 1000 K. The appearance of crystalline forsterite grains in EX Lup was the first direct observation of on-going silicate crystallization in a celestial object (Ábrahám *et al.* 2009).

3. The disappearance of crystalline material

The outburst of EX Lup ended in 2008 September. In order to document any subsequent changes in the silicate features, we obtained post-outburst Spitzer spectra on 2008 October 10 and 2009 April 6 (Fig. 1). While we expected that the crystalline features in the 10 μm emission peak became relatively stronger after the inner disk surface had become fully crystallized, our observations showed that the degree of crystallinity in the 10 μm feature *decreased* with time, while strong peaks related to cold forsterite grains appeared in the 30 μm range. Juhász *et al.* (2012) performed radiative transfer and turbulent mixing calculations to disentangle the effects potentially responsible for decay of crystalline peaks around 10 μm. Their results excluded that vertical mixing could have replaced the freshly formed crystals by amorphous particles from the interior of the disk. Instead, the authors suggested the radial outward transportation of the silicate crystals by a wind driven by the outburst. The appearance of cold forsterite peaks in the 30 μm range supported this scenario.

Aiming to follow the further fate of the crystalline particles after the 2008 outburst and to understand the physical processes in action, we performed new N-band observations of EX Lup on 2016 August 20 using the VLT/VISIR instrument. Moreover, we analyzed archival mid-infrared interferometric observations obtained with the VLTI/MIDI instrument on 2013 May 27/28. Both mid-infrared spectra, plotted in Fig. 1, show similar results: the crystalline spectral features, detected in 2009, have completely disappeared from the 10 μm silicate peak, and the shape of the emission feature is identical with the pre-outburst one representative of amorphous dust grains. The inner disk returned to its original amorphous state just in 5 years.

4. Radiative transfer modelling

In order to test the outward transportation hypothesis of Juhász *et al.* (2012), we performed radiative transfer modeling of EX Lup in its pre-outburst and post-outburst states, using the RADMC3D code. We assumed that all crystalline grains were created during the outburst between 0.3 au and 0.7 au, and later they were driven outward by a disk wind. To model this, we added to the circumstellar disk component an expanding hollow crystal-rich sphere of constant width, and fitted the mass and radius of the sphere to the observed spectra. For details of the modeling process, see Ábrahám *et al.* (2019).

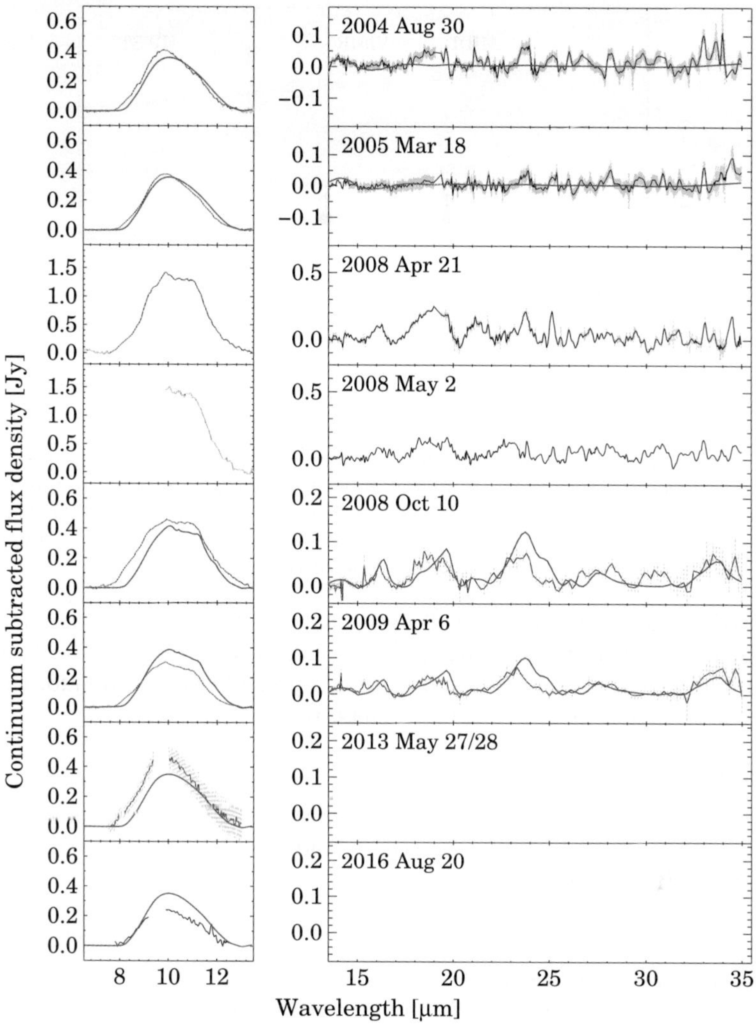

Figure 1. Continuum subtracted spectra of EX Lup at different epochs. Black curves are observations, red curves show our models.

Our best-fit models are overplotted in Fig. 1 with red curves. In these models the shell transports $\sim 2\times10^{23}$ g of forsterite, corresponding to about 10^4 Hale-Bopp-like comets. In order to match the shape of the 10 μm feature in 2008 October and 2009 April, we had to place the shell at r=1.2 au and 1.5 au, respectively. To be consistent with our spectra from 2013/2016, the sphere must be more distant from the star than 2.5 au.

5. Conclusions

With some assumptions, our modeling results can be used to constrain the likely location of the crystalline particles at any date (shaded area in Fig. 2). Comparing the data points for 2008 October and 2009 April, we can derive an average expansion velocity of $\sim$3 km/s, which is a rather low value for stellar winds, but fits into the wind velocity range found by Banzatti *et al.* (2019) in some cases in a sample of T Tauri stars. According to the figure, currently the forsterite grains are situated somewhere between 3.5 au and 7 au. This radial distance range overlaps with the expected location of the water snowline in the EX Lup disk ($\sim$5 au). Thus we may conclude that the crystalline

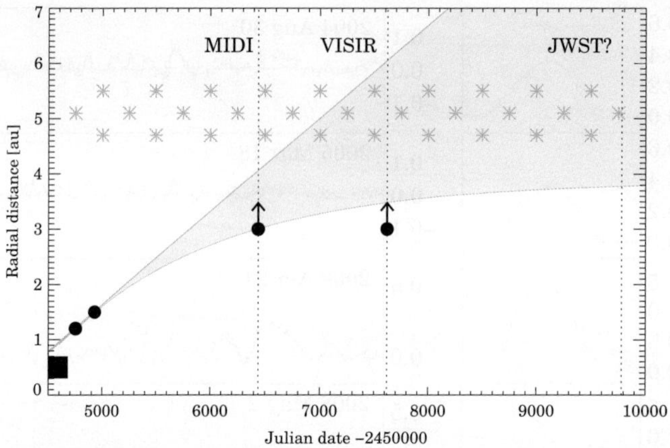

Figure 2. Radial location of the expanding sphere of crystal-rich dust as predicted by our modeling. The black square is the outburst in 2008, the two data points around JD∼4700 and 4900 are the Spitzer Space Telescope observations in 2008 October and 2009 April, while the lower limits are our MIDI and VISIR measurements. Asterisks mark the location of the water snowline. The shaded area is the probable location of the crystalline forsterite grains.

grains created in the 2008 outburst will likely reach and cross the water snowline. Since it is the formation zone of icy planetesimals, we speculate that the forsterite grains could be incorporated into cometary bodies, providing a possible explanation for the enigmatic high crystalline fraction in solar system comets, too.

Acknowledgements

This project has received funding from the European Research Council (ERC) under the European Union's Horizon 2020 research and innovation programme under grant agreement No 716155 (SACCRED).

References

Ábrahám, P., Juhász, A., Dullemond, C. *et al.* 2009, *Nature*, 459, 224
Ábrahám, P., Chen, L., Kóspál, Á. *et al.* 2019, *to be submitted*
Audard, M., Ábrahám, P., Dunham, M. M. *et al.* 2014, in H. Beuther, R.S. Klessen, C. P. Dullemond, & T. Henning (eds.) *Protostars and Planets VI*, Univ. of Arizona Press, 387
Banzatti, A., Pascucci, I., Edwards, S. *et al.* 2019, *The Astrophysical Journal*, 870, 76
Hartmann, L., & Kenyon, S. J. 1996, *Annu. Rev. Astron. Astrophys.*, 34, 207
Juhász, A., Dullemond, C. P., van Boekel, R. *et al.* 2012, *ApJ*, 744, 118

Discussion

KAMP: There is some degeneracy in the 10μm Si-feature concerning size and crystallinity. Did you consider the possibility of grain growth at all in building your scenario to explain the temporal changes seen in your 10μm Si Features?

ÁBRAHÁM: Indeed, both grain growth and crystallization changes the shape of the silicate feature, but in a somewhat different way. In the EX Lup spectrum obtained close to the peak of the outburst on 2008 April 21, we identified a relatively sharp peak at 10 μm, a well-defined shoulder around 11.3 μm, and a smaller peak at 16μm. All these features are characteristic of crystalline forsterite emission, thus in our case grain growth could be excluded as possible explanation for the observed changes in the spectral shape.

Origins: From the Protosun to the First Steps of Life
Proceedings IAU Symposium No. 345, 2018
B. G. Elmegreen, L. Viktor Tóth & M. Güdel, eds.

doi:10.1017/S1743921319001984

Bio-habitability and life on planets of M-G-type stars

Amri Wandel

Racah Institute of Physics, The Hebrew University of Jerusalem
Jerusalem 91904, Israel
email: amri@huji.ac.il

Abstract. The recent detection of Earth-sized planets in the habitable zone of Proxima Centauri, Trappist-1, and many other nearby M-type stars (which consist some 75% of the stars) has led to speculations, whether liquid water and life actually exist on these planets. Defining the bio-habitable zone, where liquid water and complex organic molecules can survive on at least part of the planetary surface, we suggest that planets orbiting M-type stars may have life-supporting conditions for a wide range of atmospheric properties (Wandel 2018). We extend this analysis to synchronously orbiting planets of K- and G-type stars and discuss the implications for the evolution and sustaining of life on planets of M- to G-type stars, in analogy to Earth.

Keywords. Habitable Zone, exoplanets, M-type stars, climate model

1. Introduction

Most of the recently discovered exo-planets orbit Red Dwarf stars (RDs), which constitute about 75% of the stars in the Milky Way galaxy. They are characterized by a luminosity much lower than that of our sun. Consequently, habitable planets (in the Habitable Zone, HZ, defined as enabling surface liquid water in the presence of an adequate atmosphere) orbit closely and hence are likely to be gravitationally locked. The findings of the Kepler telescope have shown that 10–75% of the RDs have habitable Earth to Super-Earth sized planets (the range depends on the precise definition of the HZ, 10% is the percentage for very conservative borders, 75% for wider HZ-borders (see e.g. Batalha *et al.* 2013; Dressing and Charbonneau, 2015), which implies that such planets could be found within less than 10 light years from Earth (Wandel, 2015; 2017). In addition to RDs, K-type stars may also have habitable tidally locked or synchronously orbiting planets (Wandel & Gale, 2019). While some authors point out some disadvantages for the evolution of life on planets orbiting M-type stars, such as high levels of XUV radiation and stellar winds, which may cause atmosphere erosion (Heller, Leconte & Barnes 2011; Leconte *et al.* 2015; Lingam & Loeb, 2017), other works argue that such planets may nevertheless host life as we know it (e.g. Tarter *et al.* 2007; Gale & Wandel, 2017; Wandel & Tal-Or, 2019). They point out that the impact of the high XUV flux received by planets orbiting in the habitable zone of young M-type hosts, may be less important for aquatic life.

Model Predictions of the Climates of RD-planets

A major requirement for life is a surface temperature range which supports liquid water and complex organic molecules. This depends on the irradiation from the host star, but to a large extent also on the planet's atmosphere. Global Circulation Models (GCMs) using radiative transfer, turbulence, convection and volatile phase changes can be used to calculate the conditions on planets, given the properties of their atmospheres. Such 3D climate models of M-dwarf planets suggest the presence of liquid water for a variety of atmospheric conditions (e.g. Pierrehumbert 2011, Wordsworth 2015). Climate modeling studies have shown that an atmosphere only 10% of the mass of Earth's atmosphere can transport heat from the day side to the night side of tidally locked planets, enough to prevent atmospheric collapse by condensation (Joshi *et al.* 1997; Tarter *et al.* 2007; Scalo *et al.* 2007; Heng & Kopparla 2012). On locked planets the water may be trapped on the night side (e.g. Leconte 2013), but on planets with enough water or geothermal heat, part of the water remains liquid (Yang *et al.* 2014). 3D GCM simulations of planets in the habitable zone of M-dwarfs support scenarios with surface water and moderate temperatures (Yang *et al.* 2014; Leconte *et al.* 2015; Owen & Mohanty 2016; Turbet *et al.* 2016; Kopparapu *et al.* 2016; Wolf 2017 to name a few).

While rocky planets with no or little atmosphere, like Mercury, have an extremely high day-night contrast and planets with a thick, Venus-like atmosphere tend to be nearly isothermal, intermediate cases, with up to 10 bar atmospheres, conserve significant surface temperature gradients (e.g. Selsis *et al.* 2011).

We argue here that life clement environments may be supported on many tidally locked planets. Recent calculations (Wandel 2018) suggest that planets orbiting M-type stars may have life supporting temperatures, at least on part of their surface, for a wide range of atmospheric properties. It is possible to extend this treatment to other stellar types (K an G) and show that tidally locked planets may well have conditions allowing photosynthesis (Gale & Wandel 2017). Given the very large number of potential planets, and the possibility of liquid water on at least some of the tidally locked planets, we discuss here characteristics of complex life which may evolve on the tidally locked planets and how they may differ from life on Earth.

Wandel (2018) supplements the usual treatment of the habitable zone (the region around a star where planets can support liquid water, e.g. Kasting, Whitmire & Reynolds, 1993), with the definition of a bio-habitable zone, where surface temperatures can support complex organic molecules, in addition to liquid water. The atmospheric properties that affect the surface temperature distribution are reduced to two factors: circulative heat redistribution and heating (due to atmospheric greenhouse effect and irradiation by the host star). The presence of surface liquid water and life supporting temperatures are analyzed within this two-dimensional parameter space.

Surface Temperature distribution on locked and synchronous planets

We extend the surface temperature model for tidally locked planets to synchronous (like Mercury, which has a spin-to-orbit period ratio of 3:2) and nearly synchronous (e.g. Venus) orbiting planets. In the latter, the spin (day/night) period is relatively long, comparable to the orbital period. We find that the circulative heat transport is enhanced and can be represented by an effective heat redistribution factor (Wandel & Gale 2019)

The planet is defined as being within the bio-Habitable Zone, if its lowest surface temperature is above freezing, that is, $T_{\min} > 273$K, and its highest surface temperature is below the highest temperature which allows complex organics, that is $T_{\max} < 400$K. This is depicted in Fig. 2, below. The lower temperature limit is naturally chosen as the freezing point of water, which is only weakly dependent on pressure, and is widely accepted as a lower limit for organic processes. The surface temperature of a locked planet is fiven by

$$\sigma T^4(\theta) = (1 - A) S F(\theta) \tag{1}$$

where σ is the Stefan Boltzman constant, A is the Bond albedo and F *is a function of the latitude angle, the substellar point defined as* $\theta = 0$. The irradiation or insolation is given by $S = L_*/4\pi a^2$, where a is the distance of the planet from its host star and L_* is host star's luminosity. If the planet is locked and has no horizontal circulation or energy

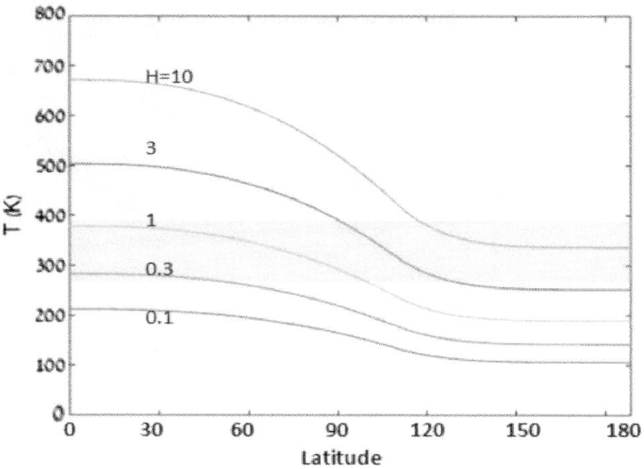

Figure 1. Temperature profiles for several values of the heating factor *H*. A global heat redistribution of $f = 0.2$ is assumed. The liquid water temperature range at 1 bar is indicated by the shaded light blue area.

redistribution, then $F = \cos(\theta)$ for $0 < \theta < 90$ and $F = 0$ for $90 < \theta < 180$. In the opposite case, planets with an efficient horizontal heat spread or those rotating rapidly, are nearly isothermal, $F = 1/4$ and the equilibrium surface temperature is $T_{eq} = [(1 - A)S/4\sigma]^{1/4}$. In the intermediate case we assume that a fixed fraction, f, of the stellar irradiation at each point is redistributed evenly over the whole planetary surface due to circulation in the planet's atmosphere. In this case we have $0 < f < 1$. While in the day hemisphere the heating is modulated by the angle from the substellar point, on the night hemisphere it is homogenous:

$$F(\theta) = \begin{cases} \frac{f}{4} + (1 - f)cos(\theta) & 0 \leq \theta \leq 90^\circ \\ \frac{f}{4} & 90^\circ < \theta \leq 180^\circ \end{cases} \tag{2}$$

It is convenient to combine the albedo, the atmospheric screening a and greenhouse factor H_g with the insolation or irradiation S, into a single dimensionless parameter, which we call the heating factor

$$H = (1 - A)H_g \alpha S/S_\oplus. \tag{3}$$

where $S/S_\oplus$ is the insolation relative to Earth and H_g is the greenhouse factor, i.e. the amount of heat capture of the atmosphere. The lowest and highest temperatures can be written as

$$T_{min} = 278\ (Hf)^{1/4}\ \mathrm{K} \tag{4}$$

$$T_{max} = T_0 = 394\ H^{1/4}(1\text{-}^3/_4f)^{1/4}\ K \tag{5}$$

Surface temperature profiles for various values of the combined heating factor *H* are shown in Fig 1. A moderate local heat transport is assumed (Wandel 2018).

Fig 2 shows the temperature range on a tidally locked planet (eqs 4,5) as a function of the heating factor *H*. From the intersection of the curves with the boundaries of the bio-habitable temperature zone it is possible to derive the habitability range of the heating factor. As shown in the figure by the arrows at the bottom, for $0 < T < 100C$ the bio-habitable range of the greenhouse heating factor turns out to be $0.2 < H_g < 30$. Letters show the four values of the terrestrial planets of the Solar System. Note that Venus, Earth and Mars have average temperatures consistent with the equilibrium solid green curve (isothermal surface, effective f=*1*), while Mercury's temperature range is shown by the vertical arrow.

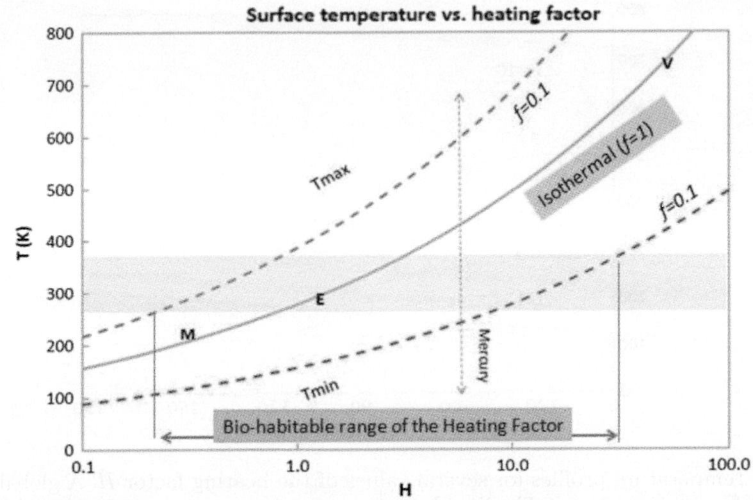

Figure 2. Minimum and maximum surface temperatures of tidally locked planets as a function of the heating factor *H*. The green solid curve shows the equilibrium temperature for an isothermal planet, while the dashed curves show the highest and lowest surface temperatures of a tidally locked planet with 10% global redistribution ($f=0.1$). Also shown are the locations of the four terrestrial planets of our Solar System.

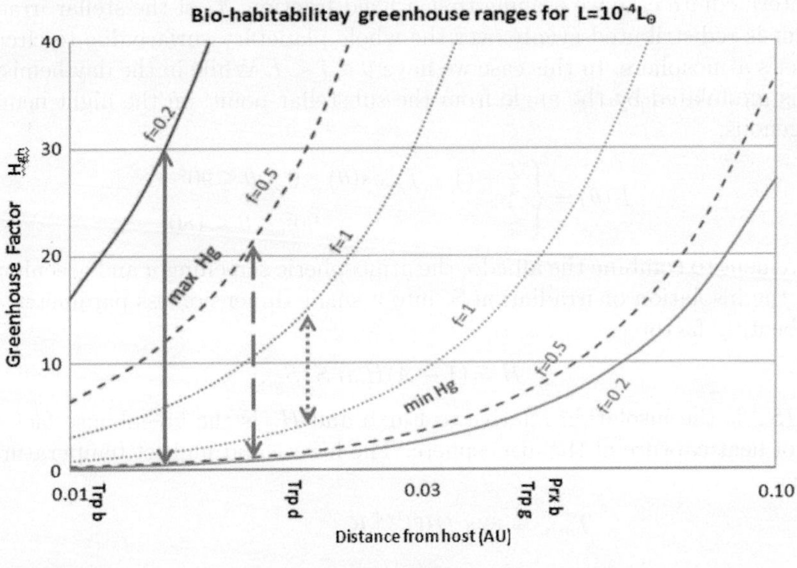

Figure 3. Maximal (red, upper) and minimal (blue, lower) bio-habitable zone boundaries of the greenhouse factor vs. the distance from the host star, for three values of the heat circulative redistribution parameter: $f=0.2$ (solid), 0.5 (dashed) and 1 (dotted). The host luminosity is assumed to be $L=10^{-4}L_{\odot}$. The locations of Proxima b and 3 of the Trappist-1 planets are marked on the x-axis

Fig 3 shows the greenhouse part (H_g) of the heating factor, as a function of the distance from a host star with a luminosity of $10^{-4}L_{\odot}$ (similar to that of Trappist-1), for 3 values of *f*. The locations of several of the Trappist-1 planets are marked on the x-axis. The y-axis shows the optical depth to infrared radiation, roughly indicating the atmosphere's opacity in the "lower wavelength" (near IR), which is related to the greenhouse heating factor by $t_{lw}=H_g-1$.

Conclusion

We have show that tidally locked planets of M-type stars may have temperatures suitable for liquid water and complex organic molecules on at least part of their surface for a wide range of atmospheric properties. Extending this analysis to synchronously orbiting planets, makes these conclusions applicable also to K and perhaps even G-type stars, where the habitable zone extends much further from the host star, and hence habitable planets would less likely be tidally locked, but may still be in nearly synchronous orbits. In analogy to extended daylight latitudes on Earth extreme environments.

References

Batalha, N. M., J. F. Rowe, S. T. Bryson, *et al.* 2013, *ApJS*, 204, 24
Dressing C. D. & Charbonneau D. 2015, *ApJ*, 807, 45
Gale, J. & Wandel, A. 2017, *Int. J. of Astrobiology*, 16 (1), 1
Heller R., Leconte J. & Barnes R. 2011, Tidal obliquity evolution of potentially habitable planets. *Astron & Astrophys.*, 528, A27.
Heng, K., & Kopparla, P. 2012, *ApJ*, 754, 60
Joshi, M. M., Haberle, R. M., & Reynolds, R. T. 1997, *Icarus*, 129, 450
Kasting, J. F., Whitmire, D. P. & Reynolds, R. T. 1993, *Icarus*, 101, 108
Kopparapu, R. K. 2013, *ApJ*, 767, L8
Kopparapu, R. K., Wolf, E. T., Jacob Haqq-Misra J., *et al.* 2016, The inner edge of the habitable zone for synchronously rotating planets around low-mass stars using general circulation models, *Astrophysical J.* 819, 84: 1-14
Leconte, J., Forget, F., Charnay, B. *et al.* 2013, 3-D climate modeling of close-in land planets: Circulation patterns, climate moist bistability, and habitability. *Astronomy & Astrophysics*, 554, A69: 1-17
Leconte, J., Wu, H., Menou, K., & Murray, N. 2015, *Science*, 347, 632
Lingam, M. & Loeb, A. 2017, *ApJ*, 846, L21
Owen, J. E. & Mohanty, S. 2016, *MNRAS*, 459, 4088
Pierrehumbert, R. 2011, *ApJ*, 726, L8
Scalo, J., Kaltenegger, L., Segura, A. *et al.* 2007, *Astrobiology*, 7(1), 85
Selsis, F., Wordsworth, R. D., & Forget, F. 2011, *A&A*, 532, A1
Tarter, J. C., Backus, P. R., Mancinelli, R. L., *et al.* 2007, *Astrobiology*, 7, 30
Turbet, M., Leconte , J., Selsis F., *et al.* 2016, *A&A*, 596, 112
Wandel, A. 2015, *Int. J. of Astrobiology*, 14, 511
Wandel, A. 2017, *Acta Astronautica*, 137, 498
Wandel A 2018, On the bio-habitability of M-dwarf planets, *Astrophysical J*, 858, 165: 1-13
Wandel, A., & Gale, J. 2019, The Bio-habitable Zone and atmospheric properties for Planets of Red Dwarfs, *IJAsB*, in press, arXiv:1907.11098.
Wandel, A. & Tal-Or, L. 2019, On the habitability of Teegardens Star planets, *Astrophysical J Letters*, 800, L21: 1-5, arXiv:1906.07704.
Wolf, E. T. 2017, *ApJ*, 839, L1
Wordsworth, R. 2015, *ApJ*, 806, 180
Yang, Y., Yonggang, L., Yongsun, H. & Abbot, D. S. 2014, *ApJ*, 796, L22

Discussion

MANKO VOLODYMYR: Tidally locked planets almost definitely must undergo the libration (caused by orbital eccentricity or gravitational influence of other planets in the system). Do you know about models exploring these effects?

WANDEL: My model and previous works concentrate on the basic model of a single gravitationally locked planet, with a steady state surface temperature distribution. The effect of eccentricity on other planets may be explored at a future stage of the project.

Origins: From the Protosun to the First Steps of Life
Proceedings IAU Symposium No. 345, 2018
B. G. Elmegreen, L. Viktor Tóth & M. Güdel, eds.

doi:10.1017/S1743921319001637

Characterization and Properties of Earth-like Planets

Dániel Apai[1,2]

[1]Steward Observatory, The University of Arizona
Tucson, AZ, USA
email: apai@arizona.edu

[2]Lunar and Planetary Laboratory, The University of Arizona
Tucson, AZ, USA

Abstract. The search for life in the Universe is intertwined with studies of extrasolar planets aimed at identifying and understanding habitable rocky planets, including those similar in size, bulk composition, planetary environment, and evolution to Earth. The past five years have seen dramatic progress in our understanding of the small (1–4 R_{Earth}) planet population. Here we briefly review key results on the occurrence rates of small planets, the first evidence for compositional diversity of these worlds, early results on the characterization of their atmospheres, and the progress toward finding and interpreting potentially habitable planets orbiting the closest stars. We also briefly highlight next steps in furthering our understanding of the origins and properties of habitable worlds.

Keywords. astrobiology, planetary systems, planetary systems: formation

---

Perhaps no other field of astrophysics has seen a progress as dramatic as the studies of sub-jovian exoplanets over the past decade. Advances are particularly exciting in the exploration of potentially rocky and habitable planets, which lays the foundation for one of the grandest experiments conceivable: the search for signatures of life in planetary systems beyond our own. This contribution aims to review the latest advances and exciting next steps in our quest to understand the diversity of small (1–4 R_{Earth}) exoplanets, with special emphasis on the identification and characterization of potentially habitable rocky worlds – worlds that may be Earth-like.

1. Occurrence Rate of Small Planets

One of the most fundamental and strategically important questions in the search of Earth-like planets is the occurrence rate of Earth-sized, habitable zone planets around sun-like stars. This value – referred to as η_{Earth} – is central not only to our understanding of exoplanetary systems, but also to identify the optimal strategy for discovering and characterizing candidate Earth-like planets. A high η_{Earth} value may make a single combined search–and–characterization approach viable, while a low η_{Earth} would likely necessitate a two-stage approach.

NASA's Kepler mission has transformed the field of exoplanets through its capability of robustly detecting small exoplanets (e.g., Batalha 2014). Kepler's primary science goal was to determine η_{Earth}. To measure this value it searched for transiting planets around ~170,000 stars. Kepler has detected hundreds of small planets, including many Earth-sized and even sub-Earth-sized planets. Yet, the total sample of known Earth-sized planets in habitable zones of sun-like stars remains small due to the limited duration of

Kepler's primary mission. Therefore, most estimates of η_{Earth} are based on extrapolations of samples dominated by larger-than-Earth and closer-in planets, often in the habitable zones of lower-mass stars. As different groups' estimates rely on different assumptions and on different versions of the Kepler data set, different η_{Earth} estimates have emerged reflecting broad range of possible answers.

Although the exact occurrence rates of Earth-sized planets around sun-like stars remains to be determined, we can conclude with reasonably high confidence that η_{Earth} cannot be small (i.e., $\eta_{Earth} > 10^{-2}$), and may even be close to unity.

Importantly, Kepler detection statistics has revealed important differences between the architectures of planetary systems around higher- and lower-mass stars. Dressing & Charbonneau (2015) noted that the occurrence rates of small planets around M-type stars is higher than that in the combined sample of F, G, and K-type host stars. Mulders *et al.* (2015a) showed that, in fact, short-period, small planet occurrence rates are a monotonic function of host star mass – surprisingly, F, G, K, and M-type stars host small planets at progressively increasing rates (i.e., inversely proportionally with stellar mass). Mulders *et al.* (2015b) also found that not only does the average number of short-period small planets increase with decreasing stellar mass, but that planetary systems around later spectral type stars also have more solid mass in their planets than those around earlier spectral type stars.

While the community discussion on η_{Earth} has not yet been settled conclusively, a meta-study has been carried out by the Study Analysis Group 13 of the NASA EXOPAG† to seek a consensus range. This study (Belikov *et al.* 2017) contrasted the methodology, assumptions, and datasets that underpinned the different estimates. The meta-study included results and methodology from many individual groups (including, for example, Petigura *et al.* 2013, Kopparapu 2013, Dressing & Charbonneau 2015, Mulders *et al.* 2015a, Burke *et al.* 2015). The three key findings of the SAG13 study are as follows: (1) The precise definition of η_{Earth} – such as host spectral type range, planet radius range, and habitable zone boundaries – has major impact on the derived η_{Earth} values. (2) With this in mind the SAG13 defined their own η_{Earth} (for 0.5–1.5 R_{Earth}, G spectral type, extended habitable zone boundaries per Kopparapu 2014). (3) With consistent methodology and based on same or similar datasets, multiple individual teams find similar η_{Earth} values, albeit with significant uncertainty. The SAG13 team suggests adopting an $\eta_{Earth} \sim 0.6$, but cautions that the value is sensitive to the factors described above.

These are profound results as they suggest that Earth-sized habitable planets in the Galaxy may number between tens of millions to tens of billions (including stars of all spectral types). They also mean that it is likely that habitable zone earth-sized planets are relatively common in the solar neighborhood.

As this manuscript is being prepared for submission, the revolution in small exoplanet discovery is entering its next stage with NASA's TESS mission. Over time, it is likely that the planet detections and insights gained from the TESS mission will help to further refine our best estimates for η_{Earth}.

2. Compositional Diversity of Small Planets

While the nature of most small planets remains poorly known, limited but nevertheless valuable insights could be gained through the determination of their bulk densities. Bulk density can be determined for planets that are transiting *and* also detected via stellar radial velocity (RV) measurements: while the former enables relative measurements of the planet's radii, the latter provides relative measurements of the planets'

† https://exoplanets.nasa.gov/exep/exopag/overview/

masses. As both measurements are more challenging for smaller planets than for larger ones, the sample of small planets with both measurements is a small, but slowly growing one.

Bulk density measurements allow placing constraints on the compositions of the planets. For example, Earth's bulk density is $\rho = 5.5$ g cm^{-3}, consistent with its interior dominated by an iron-rich core ($\rho = 9.9 - 13.1$ g cm^{-3}, 32.5w%) and silicate mantle and crust ($\rho = 4.5$g cm^{-3}, 67.47 w%). Neptune, on the other hand, has a much lower density ($\rho = 1.64$ g cm^{-3}), consistent with its volatile-rich composition (H_2: 80 wt%, He: 19 wt%, CH_4: 1 wt%). Extending this comparison to exoplanets, thus, can provide powerful insights into their bulk compositions. It is important, however, to note that bulk density is a degenerate measure of bulk composition: for most densities – even when considering only three key components – an infinite number of compositions can lead to the same density (Rogers & Seager 2010a). In fact, bulk density measurements are most constraining for outliers, i.e., very low or very high density worlds, where either high-density (silicates, iron) or low-density (hydrogen/helium, water) components must dominate. The range of compositions consistent with a measured bulk density and its uncertainties can be narrowed down if specific formation pathways can be identified (e.g., Rogers & Seager 2010b)

The California–Kepler Survey has studied 2,205 Kepler exoplanet host stars with high-resolution optical spectroscopy to refine the stellar radii, often the largest uncertainty in these Kepler-detected exoplanets' radii (as transits measure R_p/R_s, planet radius relative to the stellar radius). The spectroscopic characterization enabled an improvement in the precision with which planets' radii are measured, leading to a median uncertainty of 12%. This improved precision revealed a pattern in planet size distribution that was indiscernible in previous datasets (Fulton *et al.* 2017). A striking feature seen in this distribution is a bimodal size distribution for small planets. The depleted planet occurrence rate around radii of $R_p = 1.6\ R_\oplus$ (often referred to as the *Fulton gap*), is a highly significant feature. The presence of the Fulton gap has been confirmed by subsequent studies (e.g., Hsu *et al.* 2018), including a study that further characterized the host stars by combining spectroscopic classification with GAIA-measured distances (Fulton *et al.* 2018). The presence of such a gap has been predicted by atmospheric photoevaporation models (e.g., Lopez & Fortney 2014, Chen & Howard 2016, Owen & Wu 2017) and are also explained by models in which tenuous atmospheres can be lost due to heating by the planetary core (Ginzburg *et al.* 2018).

Particularly exciting examples of small planets in the solar neighborhood are found in the TRAPPIST-1 system (Gillon *et al.* 2016, Gillon *et al.* 2017). This system includes seven, roughly Earth-sized planets. The relative sizes of the planets can be derived from their transit depths as measured in the 3.6 μm Spitzer/IRAC band (less sensitive to stellar heterogeneities than shorter wavelengths measurements, Rackham *et al.* 2018). This multi-planet system is so tightly packed that very significant transit timing variations are observable, which allow for refined planet mass measurements (e.g., Agol & Deck 2016). Due to the complexity of the system and the strengths of the gravitational interactions between the planets, finding masses from transit timing variations is a highly non-linear and challenging task; on the other hand, the gradually increasing temporal baseline and increasing number of transits observed enables increasingly precise models. A detailed study by Grimm *et al.* (2018) modeled the accumulated transit timing variation data and derived relatively high-precision mass estimates for the planets. Combined with the planet radii measurements, these allowed the comparisons of the bulk densities of the TRAPPIST-1 worlds. The relatively high densities of planets *c* and *e* suggest that these worlds are rocky, while the other planets are more likely to have significant gaseous envelopes or large mass fractions of water.

3. Atmospheric Characterization

Exoplanets that transit their host stars also offer an opportunity to study the host star's light as it filters through the planets' atmospheres (e.g., Seager & Deming 2010, Heng & Kitzmann 2018). At wavelengths where planets' atmospheres can absorb the planets will be effectively larger in size (optical depth $\tau = 1$ is reached higher in the atmosphere). Therefore, by measuring the wavelength-dependence of the planets' size (transit depth), key absorbers can be identified or inferred.

However, such wavelength-dependent modulations of the time-dependent intensity modulations introduced by planetary transits (small signals in themselves) are very challenging to measure. While ground-based telescopes have proved to be powerful tools in studying the atmospheres of hot jupiters, sub-jovian planets remained accessible to only a few instruments that combined sensitivity and photometric stability (or ability to measure and correct for instrumental and atmospheric systematics).

The Hubble Space Telescope's WFC3 instrument's near-infrared camera has proved to be a particularly powerful instrument for transit spectroscopy. Reaching essentially photon-noise-limited performance, the atmospheres of. even small planets' can be explored for water vapor absorption. Kreidberg *et al.* (2014) has demonstrated the feasibility of co-adding spectral data from twelve transits of the sub-neptune GJ 1214b, while efficiently correcting for detector and telescope systematics. de Wit *et al.* (2016) and de Wit *et al.* (2018) used the same instrument to explore, for the first time, the transmission spectra of habitable zone, approximately Earth-sized planets, targeting six worlds in the TRAPPIST-1 system. These studies found no evidence for water absorption (in the planets' atmospheres) but, at least in the case of GJ1214, offered strong evidence for high-altitude hazes that weaken near-infrared molecular absorption features.

Given the successes of HST/WFC3 infrared spectroscopy of transiting planets, it may be tempting to think that by carefully correcting for instrument systematics, future instruments will allow us to reach photon-noise-limited precision for small transiting planets. However, this may not be the case: there are important astrophysical systematics that may limit the precision of transit spectroscopy before photon noise limit can be reached. An important source of such systematics are the host stars themselves. It has been long recognized that starspots – even if not occulted by the planet – will impact the transmission spectra (e.g., Pont *et al.* 2008, Sing *et al.* 2016). This effect was often considered likely negligible or correctable for transiting hot jupiters. However, with the increasing precision demanded by small exoplanets, this systematics has been revisited by Rackham *et al.* (2018). This study described the *transit lightsource effect*, the contamination of the transiting exoplanet transmission spectra due to a spectral difference between the integrated spectrum of the transit chord (actual lightsource) and the disk-integrated spectrum of the host star (the observable property). *Any* heterogeneity on the star will introduce a spectral difference between the two, thereby contaminating the transmission spectrum (see Figure 1). Understanding and correcting for the transit light source effect is essential for future characterization of small exoplanets (Apai *et al.* 2018).

Important examples of the transit light source effect were described in visual spectra of the sub-neptune GJ 1214b (Rackham *et al.* 2017) and in the approximately Earth-sized planets orbiting TRAPPIST-1 (Zhang *et al.* 2018). The latter study showed that the combined dataset of ground-based photometry, HST/WFC3 spectra, and Spitzer/IRAC transit depth measurement can be best modeled with stellar contamination due to hotter faculae and large starspots on TRAPPIST-1. In a follow-up study, Wakeford *et al.* (2018) explored a number of possible configurations for the stellar heterogeneity and how these would affect the planet's apparent transmission spectra.

The Transit Light Source Effect

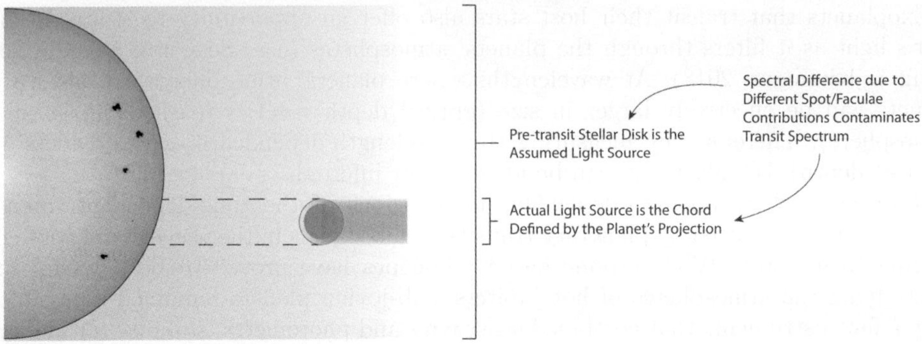

Figure 1. The transit light source (TLS) effect is the contamination of exoplanet transmission spectra. The effective contamination occurs because the light source of the transmission spectroscopy (transit light chord) has a spectra different from the disk-integrated spectrum of the star (directly observable quantity). TLS can introduce apparent slopes and molecular features in the transmission spectra of planets; the effect is particularly problematic for late-type host stars and small planets. Modified from Rackham *et al.* (2018).

4. Understanding Potentially Habitable Worlds in our Neighborhood

The focus in the search for habitable planets is increasingly shifting to stars close to the Solar System: While Kepler has explored systems typically between 0.5 to 1.5 kpc in order to provide statistical insights into the small planet population, studies aimed at discovering exoplanets that can be characterized in detail must look for targets within tens of parsecs from the Sun. NASA's TESS mission is expected to yield over ten thousand new exoplanet candidates (e.g., Sullivan *et al.* 2015, Barclay *et al.* 2018), many within a few hundred parsecs. Many of the TESS candidates will orbit bright enough host stars to allow characterization via transit spectroscopy. Simultaneously, increasingly sensitive ground-based transit surveys with 1m and larger telescopes (e.g., SPECULOOS†, EDEN‡) target nearby late red dwarf stars, where their telescopes can detect earth-sized transiting planets.

As transit searches are only sensitive to the subset of planetary systems with orbital planes close to edge on (as viewed from Earth), we must turn to other planet detection methods to complete our census of habitable planets in the solar neighborhood. Progress in precision radial velocity measurements recently enabled the detection of small-amplitude (m/s-level) velocity modulations in red dwarfs. This capability led to the detections of small planet candidates around the nearby stars Proxima Centauri b (Anglada-Escudé, G. *et al.* 2016), Barnard's Star (Ribas *et al.* 2018), and Ross 128 (Bonfils *et al.* 2018). Perhaps most exciting among these is Proxima. Not only is it the closest star to the Solar System, but its planet also orbits in the habitable zone and, even considering the activity of its host star and a likely tidally-locked state, it may be habitable (e.g., Meadows *et al.* 2018, Carone *et al.* 2018).

Proxima Cen b is a good example for the challenges we face when attempting to characterize habitable exoplanets in the next decades. Detection of these worlds remains challenging and – for the foreseeable future – we will always have only partial information on *very complex* systems, that may even include alien ecosystems. While our characterization of individual planets will remain greatly incomplete, we may be able to *interpret*

† https://www.speculoos.uliege.be/
‡ http://project-eden.space

these worlds by combining specific measurements of the planet with *statistical contextual understanding* of small planets in general. In fact, through the thousands of planet detections emerging from Kepler and (soon) from TESS, and from microlensing (from WFIRST), over the next decades we shall develop a very good statistical understanding of the small planet population.

Bayesian probabilistic assessment provides a path forward to combine specific information on a planet with probabilistic priors that can represent the properties of the parent population. A study by Bixel & Apai (2016) demonstrated this principle on Proxima b and showed how planet occurrence rates measured for M-dwarf host stars by Kepler can help refine posterior probability distributions on Proxima b's mass and composition. Probabilistic Bayesian assessment of planets' nature can also include statistical predictions from planet formation and planet evolution (including interior and atmospheric evolution). The same or similar approach will likely be essential for interpreting future observations of small exoplanets, whether detected through direct imaging, transits, microlensing, or radial velocity surveys.

Important contextual information can be gained from planet formation and evolution models, as many properties of the small planets will not be directly observable even with powerful new telescope facilities. However, models that can explain the observed properties of the planet populations will provide a good basis to predict properties such as bulk composition or orbital evolution pathways, at least in a probabilistic sense. Indeed, multiple groups are developing large-scale planet formation simulations to aid such studies (e.g., PlanetS in Switzerland and the EOS/NExSS team in the US). With large databases of simulated planetary systems, statistical comparisons to the planet surveys is becoming possible. An important new tool enabling such comparisons is *EPOS* (Mulders *et al.* 2018), which simulates the entire Kepler mission (complete with its sensitivity, observing strategy, identification and confirmation biases) in a fraction of a second. Equipped with this tool, simulated hypothetical planet populations can be compared against the observations and models can be tested, optimized, and verified. The interpretation of observations of individual nearby exoplanets will greatly benefit from statistically verified planet formation/evolution models.

5. Next Steps

As rapid progress in astrobiology and exoplanet exploration continues, the quest for habitable planets and the search for life beyond the Solar System emerge as likely science goals for major ground- and space-based astronomical projects. Planets – especially ones with a prominent biosphere – are likely to be very complex systems. Identifying and interpreting them via purely remote sensing will require multi-disciplinary, comprehensive, and coordinated research. The characterization itself will require powerful new telescopes and instruments. Multiple reports with community input have been written over the past three years to identify science questions, opportunities, and critical milestones in this process. For example, the EXOPAG SAG reports reviewed science opportunities afforded by transiting exoplanets (Cowan *et al.* 2015), science questions for direct imaging missions (Apai *et al.* 2017), possible biosignatures and their interpretation (e.g., Meadows *et al.* 2018, Schwieterman *et al.* 2018, Catling *et al.* 2018, Fujii *et al.* 2018, Walker *et al.* 2018). Two new NASA-commissioned reports laid out a science strategy for studies of exoplanets and astrobiology (see NAS website). These and similar reports provide comprehensive overviews of the future of the field. Here we only highlight three key questions central to the search for life on other planets.

First, in considering future exoplanet missions aimed at understanding habitable exoplanets, when setting the required sample size of planets to be studied, it is paramount to consider the complexity of rocky planets. If too small samples of planets are studied, the risk of incomplete understanding – or even misinterpretation – of the observations will be high.

Second, understanding the stellar contamination emerging from the transit lightsource effect will be important to fully exploit the transit-based data – our only probes of the atmospheres of small planets prior to direct imaging missions.

Third, development of planet formation and planetary system evolution models that can reproduce every properties of the planet populations – as observed through transits, radial velocity, direct imaging, and microlensing – will likely be essential to correctly predict and, later, to interpret the diversity of small exoplanets.

Acknowledgements

The results reported herein benefited from collaborations and/or information exchange within NASA's Nexus for Exoplanet System Science (NExSS) research coordination network sponsored by NASA's Science Mission Directorate.

References

Anglada-Escudé, G., Amado, P. J., Barnes, J., *et al.*, 2016, *Nature*, 536, 437
Apai, D., Cowan, N., Kopparapu, R., *et al.*, 2017, *EXOPAG SAG15 Report*, arXiv:170802821A
Apai, D., Rackham, B. V., Giampapa, M. S., *et al.*, 2018, *NAS White Paper*, arXiv:180308708A
Agol, E., & Deck, K., 2016, *ApJ*, 818, 177
Barclay, T., Pepper, J., & Quintana, E. V., 2018, *ApJS*, 239, 2
Batalha, N. M., 2014, *PNAS*, 11112647B
Belikov, R., *et al.*, 2017, *EXOPAG Report*, https://exoplanets.nasa.gov/exep/exopag/sag/#sag13
Bixel, A. D., & Apai, D., 2016, *ApJ*, 836, 31
Bonfils, X., Astudillo-Defru, N., Díaz, R., *et al.*, 2018, *A&A*, 613, 9
Burke, C. J., Christiansen, J. L., Mullally, F., *et al.*, 2015, *ApJ*, 809, 19
Carone, L., Keppens, R., Decin, L., & Henning, Th., 2018, *MNRAS*, 473, 4672
Catling, D. C., Krissansen-Totton, J., Kiang, N. Y., *et al.*, 2018, *Astrobiology*, 18, 709
Chen, H., & Rogers, L. A., 2016, *ApJ*, 831, 180
Cowan, N., Greene, T., & Angerhausen, D., 2015, *PASP*, 127, 311
Dressing, C. D., & Charbonneau, D., 2015, *ApJ*, 807, 45
Fulton, B. J., *et al.* 2017, *AJ*, 853, 122
Fulton, B. J., *et al.*, 2018, *AJ*, 156, 264
Fujii, Y., Angerhausen, D., Deitrick, R., *et al.*, 2018, *Astrobiology*, 18, 739
Gillon, M., Triaud, A. H. M. J., Demory, B-O. *et al.*, 2016, *Nature*, 533, 221
Gillon, M., Triaud, A. H. M. J., Demory, B-O. *et al.*, 2017, *Nature*, 542, 456
Ginzburg, S., Schichtling, H. E., & Sari, R., 2018, *MNRAS*, 476, 759
Grimm, S. L., Demory, B.-O., Gillon, M., *et al.*, 2018, *A&A*, 613, 68
Hsu, D. C., Ford, E. B., Ragozzine, D., & Morehead, R. C., 2018, *AJ*, 155, 205
Kiang, N. Y., Domagal-Goldman, S., Parentau, M. N., *et al.*, 2018, *Astrobiology*, 18, 619
Heng, K., & Kitzmann, D., 2017, *MNRAS*, 470, 2972
Kopparapu, R. K., 2013, *ApJ*, 767, 8
Kopparapu, R. K., Ramirez, R. M., *et al.*, 2014, *ApJ*, 787, 29
Kreidberg, L., Bean, J. L., Désert, J-M. *et al.*, 2014, *Nature*, 505, 69
Lopez, E. D., & Fortney, J. J., 2014, *ApJ*, 792, 1
Meadows, V. S., Arney, G. N., Schwieterman, E. W., *et al.*, 2018, *Astrobiology*, 18, 133
Mulders, G. D., Pascucci, I., Apai, D., & Ciesla, F. J., 2018, *AJ*, 156, 24
Mulders, G. D., Pascucci, I., & Apai, D., 2015a, *ApJ*, 798, 112
Mulders, G. D., Pascucci, I., & Apai, D., 2015b, *ApJ*, 814, 130
Owen, J. E., & Wu, Y., 2017, *ApJ*, 847, 29
Petigura, E. A., Howard, A. W., & Marcy, G. W., 2013, *PNAS*, 11019273P
Pont, F., Knutson, H., Gilliland, R. L., *et al.*, 2008, *MNRAS*, 385, 109

Rackham, B. V., Espinoza, N., Apai, D., *et al.*, 2017, *ApJ*, 834, 151
Rackham, B. V., Apai, D., & Giampapa, M. S., 2018, *ApJ*, 853, 122
Ribas, I., Tuomi, M., Reiners, A., *et al.*, 2018, *Nature* 563, 365
Rogers, L. A., & Seager, S., 2010a, *ApJ*, 712, 974
Rogers, L. A., & Seager, S., 2010b, *ApJ*, 716, 1208
Seager, S., & Deming, D., 2010, *ARAA*, 48, 631
Schwieterman, E. W., Kiang, N. Y., Parentau, M. N., *et al.*, 2018, *Astrobiology*, 18, 663
Sing, D., Fortney, J. J., Nikolov, Nikolay, *et al.*, 2016, *Nature*, 529, 59
Sullivan, P. W., Winn, J. N., Berta-Thompson, Z. K., *et al.*, 2015, *ApJ*, 809, 77
Wakeford, H. R., Lewis, N. K, & Fowler, J., 2018, *Astron. J.*, in press, arXiv:1811.04877
Walker, S. I., Bains, W., & Cronin, L., 2018, *Astrobiology*, 18, 779
de Wit, J., Wakeford, H. R., Gillon, M., *et al.*, 2016, *Nature*, 537, 69
de Wit, J., Wakeford, H. R., Lewis, N. K, *et al.*, 2016, *Nat. Astr.*, 2, 214
Zhang, Z., Zhou, Y., Rackham, B. V., & Apai, D., 2018, *AJ*, 156, 178

Discussion

GÜNTHER: Have all rocky planets the composition like the Earth, or are there planets with a Mercury-like composition as derived from density measurements of exoplanets?

APAI: For most small planets with available bulk density measurements these are consistent with rocky (broadly Earth-like) or more volatile-rich composition. To my knowledge, there has been one Kepler-detected relatively massive planet with initial radial velocity measurements suggesting iron-like density.

KEDZIORA-CHUDCZER: Can you explain the featureless spectra for TRAPPIST- 1 by the existence of hazes or clouds in the atmosphere rather than stellar variability due to spots?

APAI: A featureless spectrum alone could be explained by a lack of atmosphere or by high-altitude hazes. However, the data show that the TRAPPIST-1 spectra are not featureless, but show features *opposite* in sign from what the planetary absorption could produce – i.e., it shows an inverse water absorption feature. Only stellar contamination is consistent with such a feature.

Origins: From the Protosun to the First Steps of Life
Proceedings IAU Symposium No. 345, 2018
B. G. Elmegreen, L. Viktor Tóth & M. Güdel, eds.

doi:10.1017/S1743921319001558

Exoplanet host-star properties: the active environment of exoplanets

John P. Pye[1], David Barrado[2], Rafael A. García[3], Manuel Güdel[4], Jonathan Nichols[1], Simon Joyce[1], Nuria Huélamo[2], María Morales-Calderón[2], Mauro López[2], Enrique Solano[2], Pierre-Olivier Lagage[3], Colin P. Johnstone[4], Allan Sacha Brun[3], Antoine Strugarek[3], Jérémy Ahuir[3] and On behalf of the ExoplANETS-A Consortium

[1]Department of Physics & Astronomy, University of Leicester, Leicester, LE1 7RH, U.K.
email: pye@le.ac.uk

[2]Depto. Astrofísica, Centro de Astrobiología (INTA-CSIC), ESAC Campus, Camino Bajo del Castillo s/n, 28692, Villanueva de la Cañada, Spain

[3]AIM, CEA, CNRS, Université Paris-Saclay, Université Paris Diderot, Sorbonne Paris Cité, F-91191 Gif-sur-Yvette, France

[4]University of Vienna, Dept. of Astrophysics, Türkenschanzstr. 17, 1180 Vienna, Austria

Abstract. The primary objectives of the ExoplANETS-A project are to: establish new knowledge on exoplanet atmospheres; establish new insight on influence of the host star on the planet atmosphere; disseminate knowledge, using online, web-based platforms. The project, funded under the EU's Horizon-2020 programme, started in January 2018 and has a duration ~ 3 years. We present an overview of the project, the activities concerning the host stars and some early results on the host stars.

Keywords. methods: data analysis, catalogs, stars: activity, stars: atmospheres, stars: coronae, (stars:) planetary systems, stars: winds, outflows, ultraviolet: stars, X-rays: stars

---

1. Introduction

Seven institutes† in Europe have combined their expertise in the field of exoplanetary research to develop the European Horizon-2020 ExoplANETS-A‡ project under the coordination of CEA Saclay. In the framework of the project, novel data calibration and spectral extraction tools, as well as novel retrieval tools, based on 3D models of exoplanet atmospheres, will be developed to exploit archival data from space- and ground-based observatories, and produce a homogeneous and reliable characterization of the atmospheres of transiting exoplanets. Additionally, to model successfully the exoplanet atmosphere, it is necessary to have a sound knowledge of the host star. To this end, we will collect a coherent and uniform database of the relevant properties of host stars from online archives (e.g. XMM-Newton, Gaia) and publications. These exoplanet and host-star catalogues will be accompanied by computer models to assess the importance

† CEA Saclay, Paris, France; CAB-INTA, Madrid, Spain; MPIA, Heidelberg, Germany; University College London, U.K.; University of Leicester, U.K.; SRON, Utrecht, NL; Universitat Wien, Austria

‡ *Exoplanet Atmosphere New Emission Transmission Spectra Analysis*; https://cordis.europa.eu/project/rcn/212911_en.html; The ExoplANETS-A project has received funding from the EU's Horizon-2020 programme; Grant Agreement no. 776403.

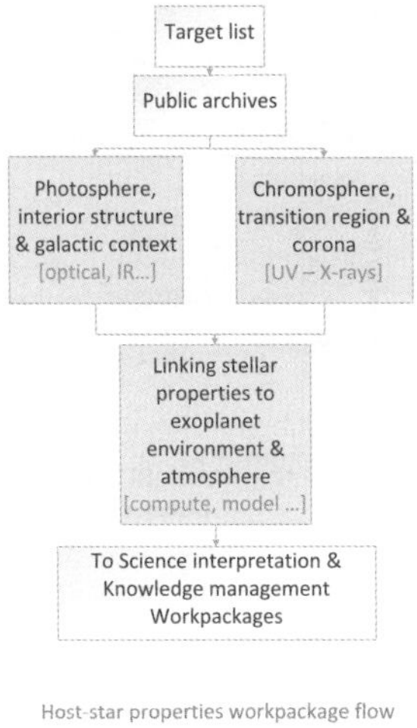

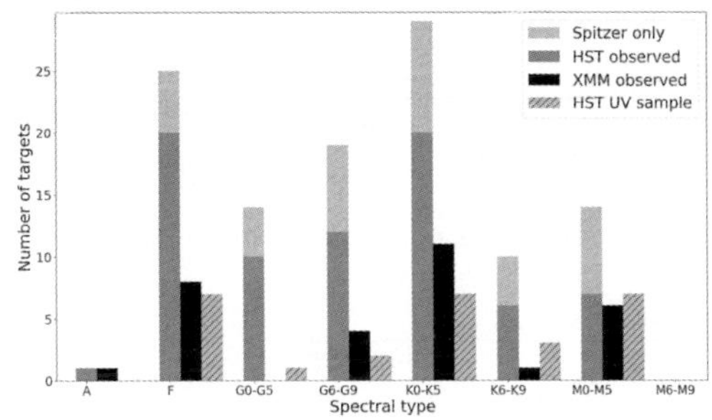

Figure 1. *(a, left).* Schematic of the flow of data through the host-stars workpackage, and interfaces to the overall project. *(b, right).* The target sample, in terms of the distribution of host stars with spectral type, showing the breakdown by archival-observation types.

of star–planet interactions, for example the 'space weather' effects of the star on its planetary system. The knowledge gained from this project will be published through peer-reviewed scientific journals and modelling tools will be publicly released.

The project has six work packages (WPs); the focus in this paper is on the WP 'Host-star properties: the active environment of exoplanets'. Fig.1(a) illustrates the flow of data through the host-stars WP, and interfaces to the overall project.

We outline the activities concerning the host stars, and present early results from the host-star investigations, including the basic stellar observational and physical properties, and indications of future observations needed to maximize the coverage of the target list of ~ 100 stars. We also discuss some of the modelling aspects.

2. The host-stars activities

In addition to compiling the observational data, we will where necessary, e.g. in the EUV, interpolate, extrapolate and scale spectral and other information, to cover gaps in the observational data, in order to provide the full XUV spectral range for modelling the host-star influence on the exoplanet atmosphere (e.g. Nemec *et al.* 2019). This work will be accompanied by computer models to assess the importance of star–planet interactions, for example the 'space weather' effects of the star on its planetary system; We will also model the possible evolutionary scenarios for the stellar activity over the star's lifetime, in order to gain insight into the past environment of the exoplanet (e.g. Johnstone *et al.* 2015).

3. The sample

The sample of exoplanets and host stars considered by the project comprises all transiting-exoplanet systems observed by HST or Spitzer. Currently, this corresponds to 135 exoplanets, of which 85 have HST data; the associated number of stars is 113, with 76 having HST data (Fig.1(b)). For the host stars, the primary online archival databases to be used are HST (for UV spectra), XMM-Newton (for X-ray), Gaia (for astrometric properties) and SIMBAD (for spectral types etc). The searches also include: GALEX

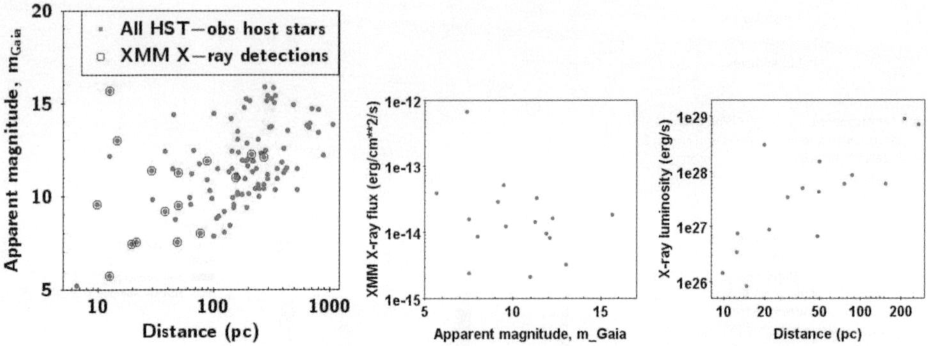

Figure 2. Examples of host-star properties, plotted from the catalogue content.

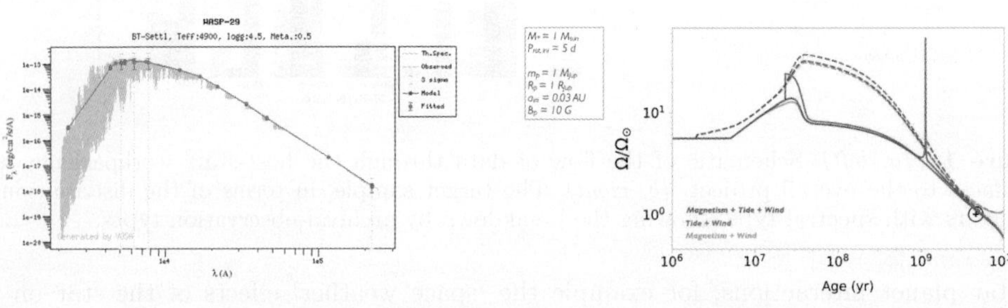

Figure 3. *(a, left).* SED fitting using VOSA. Blue spectrum represents the theoretical model (Allard *et al.* 2012) that best fits while red dots represent the observed photometry. *(b, right).* Stellar rotation rate (days) as a function of log(age, years) for 3 cases: wind + Alfén wings torques (light blue curve), wind + tides (red) and wind + tides + Alfvén wings torques (dark blue) (Ahuir 2018; see Benbakoura *et al.* 2019 for details of the ESPEM code used for the study).

(UV photometry); Chandra, ROSAT and other X-ray catalogues, together with results retrieved from the published literature. At X-ray wavelengths, most of the detections are from the 3XMM (DR8) serendipitous source catalogue (Rosen *et al.* 2016).

In summary, to date (November 2018), we have the following statistics:

• X-ray (3XMM-DR8 cat.): 31 stars observed, with 17 detections in the public archive, and the outcomes awaited for the others;

• UV photometry (GALEX, GR6 cat.): 70 stars detected (51 in the HST sample)

• UV spectra (HST COS and/or STIS): 26 stars observed

• X-ray observation & UV photometry: 23 stars

• X-ray observation & UV spectra: 20 stars

All the stars with XMM observations also have HST visible/IR spectra.

Fig. 2 illustrates a few of the stellar parameters from our initial assessments to date.

In order to fill in gaps in the stellar parameter space (e.g. to have adequate UV and X-ray measurements across spectral types) we are planning observing proposals, principally to HST and XMM-Newton.

4. Determination of the physical parameters of the host stars

A coherent and uniform determination of the stellar properties is essential to avoid any bias in the final results; use of Virtual Observatory (VO) tools facilitates this goal. For example, we have used VOSA (http://svo2.cab.inta-csic.es/theory/vosa/) to build observational Spectral Energy Distributions (SED) and compare them with theoretical SEDs (Fig. 3(a)) to obtain physical parameters such as effective temperature, stellar radius and luminosity. Effective temperatures range from 3000 to 6500K. We have also used VOSA to search for infrared excess in our sources, with negative outcome.

Medium/high resolution spectra for 48 of our sources observed with FEROS, HARPS, FORS 1-2, X-SHOOTER, FLAMES and UVES were gathered from the ESO archive. Visible-light photometry from Kepler/K2 and TESS will also be used to help characterise the stars in terms of rotational and aperiodic variability. One third of our sources have Kepler light-curves while TESS will probably provide data for the remainder.

The ExoplANETS-A database of stellar properties and spectral models will be made publicly available, via a web-based interface, by the end of 2020.

5. Scientific interpretation of the data: star–planet interactions

Most exoplanets live around active stars, whose magnetism and intense activity can have a direct impact on habitability conditions and more generally on the exosphere of the planet. Intense wind and storms can lead to atmospheric loss. Further, a large fraction of exoplanets live close to their host star (within 10–20 solar radii), to a point where in many cases they orbit within the star's Alfvén surface. This has direct consequence on star–planet interaction and planet migration, as magnetic torques through Alfvén wings directly connecting the planet to its host star can occur. We have developed both ab-initio 3D MHD simulation of such close-in systems (Strugarek *et al.* 2015, Strugarek *et al.* 2017) as well as a simpler 1D secular evolution model of star–planet systems (Benbakoura *et al.* 2019). In Fig.3(b) we show the evolution of a star–planet system subject to intense magnetic torques through direct Alfvén wings connection (on top of tides and wind effect) and compare it to the case with magnetic torques and stellar wind only (light blue curve) or tides + wind (red curve). We note that adding all the effects leads to a quick demise of the planet, that impacts the star's rotation rate. Such Alfvén wings can also lead to hot spots on the star's surface, impacting directly their light-curve and hence the transit curves. We intend to develop 3D and secular models for the most important systems identified in the ExoplANETS-A project, as well as obtaining spectropolarimetric magnetic maps of the host star in order to model the stellar wind and assess as accurately as possible the space environment around these exoplanetary systems.

References

Ahuir, J. 2018, "Evolution orbitale des exoplanètes par couples magnétiques et de marée", Master Thesis, University of Paris XI

Allard, F., Homeier, D., & Freytag, B. 2012, *Phil. Trans. RS A*, 370, 2765

Benbakoura, M., Réville, V., Brun, A. S., Le Poncin-Lafitte, C., & Mathis, S. 2019, *A&A*, 621, A124

Johnstone, C. P., Güdel, M., Stökl, A., Lammer, H., *et al.* 2015, *ApJ*, 815, L12

Nemec, N.-E., Güdel, M., Lüftinger, T., & Johnstone, C. P. 2019, in: B. G. Elmegreen, L. V. Tóth & M. Güdel (eds.), *Origins: from the Protosun to the First Steps of Life; Proc. IAU Symp. 345*, this volume

Rosen, S. R., *et al.* 2016, *A&A*, 590, A1

Strugarek, A., Bolmont, E., Mathis, S., Brun, A. S., Réville, V., Gallet, F., & Charbonnel, C. 2017, *ApJ*, 847, L16

Strugarek, A., Brun, A. S., Matt, S. P., & Réville, V. 2015, *ApJ*, 815, 111

Origins: From the Protosun to the First Steps of Life
Proceedings IAU Symposium No. 345, 2018
B. G. Elmegreen, L. Viktor Tóth & M. Güdel, eds.

doi:10.1017/S1743921319002047

Early life on Earth: Tracing the chemical path from non-living to living

Addy Pross

Department of Chemistry, Ben Gurion University of the Negev, Be'er Sheva 84105, Israel
email: pross@bgu.ac.il

Abstract. Amongst the most tantalizing questions in science are those relating to the life issue. What is it, how did it emerge, does it exist beyond our planet? In this review some central themes that have governed this debate over past decades will be described. Through the newly proposed Persistence Principle, it is argued that material stability can be achieved through either kinetic or thermodynamic means, opening up the possibility for life to be understood as a kinetic, rather than a thermodynamic, phenomenon. That insight allows the evolutionary process from inanimate to animate to be understood as one that was initiated with the emergence of a kinetically stable prebiotic replicative chemical system. Such a chemical system, once established, was able to evolve and complexify toward increasingly stable replicative forms, toward life. With a clearer understanding of what life is, the search for life in the universe can become more clearly directed.

Keywords. Dynamic kinetic stability, origin of life, persistence principle, prebiotic chemistry.

1. Introduction

Though physicists, chemists and biologists have been addressing the origin of life problem in a scientific manner for a century now, it is remarkable how limited the progress has been. Physics and chemistry have been unable to adequately characterize the life phenomenon, with the result that biology remains a science having to address its central issues in biological terms, rather than in more fundamental physical-chemical terms. Indeed, that disconnect between physics/chemistry and biology has had a serious detrimental effect on our understanding of the origin of life (OOL) process. As a result, the question of how a prebiotic physical-chemical system could have become transformed into a biological one continues to be one of modern science's major challenges, an on-going challenge to both the physical and biological sciences.

Without a doubt the experiment that set the OOL field alight came about in 1952, when Stanley Miller, a chemistry graduate student at the University of Chicago, carried out what is arguably the most famous experiment in prebiotic chemistry (Miller 1953). Miller took a mixture of the gases thought to have made up the atmosphere of the prebiotic earth – methane, hydrogen, ammonia, and water vapor, and subjected it to an electric discharge, thereby simulating the effect of primordial lightening. The result was unexpected and dramatic; the murky mixture that resulted was found to contain a range of amino acids. The fact that amino acids, one of the key molecular building blocks of life, could be so readily formed under (presumed) prebiotic conditions seemed to signify that resolution of the OOL mystery was at hand.

Triggered by Miller's landmark experiment, several decades were spent exploring the prebiotic chemical landscape. That effort proved highly fruitful and could be thought of as part of the ahistoric approach to the problem (Fry 2000; Sutherland 2015). Plausible reactions able to produce the other main biotic building blocks - lipids, sugars, and nucleotides (from which nucleic acids are constructed), were uncovered (Sutherland 2015). Though some of these reactions, as in the case of nucleotide synthesis, required quite creative chemical thinking, it was striking that within a few decades, feasible synthetic pathways for many of life's molecular building blocks, under presumed prebiotic conditions, were uncovered. But over time it became increasingly apparent that the OOL problem is much deeper than discovering novel synthetic routes to prebiotically significant materials. In this review I will present what appears to be the essence of the OOL problem: how living things were at all able to emerge from an inanimate world, and the current status of the problem.

2. Ahistorical Nature of the OOL Riddle

Consider, you gather a group of the world's top synthetic chemists and biochemists and ask them to synthesize a simple living thing. For this synthetic exercise all required chemical materials, reagents and catalysts, whether simple or exotic, whether easy or hard to synthesize, would be supplied off the shelf, so knowledge of how those building blocks happened to appear some 3.5-4 billion years earlier would not be required. And creating the required reaction conditions? In principle, no problem there either. Any equipment able to create any desired reaction conditions would be supplied.

So how to begin? The materials chosen for the synthetic attempt would presumably be those biotic materials found within all living things - amino acids, nucleotides, sugars, lipids, and so on. But what reaction conditions should be applied? A reasonable choice would be to choose the reaction conditions that led to life's emergence some 3.5-4 billion years ago? But what were they? Over the years a variety of possible locations for life's emergence have been proposed – warm ponds, hydrothermal vents under the seas, hot springs, clay surfaces, to name key ones (Fry 2000). Worryingly, all proposals make interesting points and are thought-provoking, but none are really falsifiable, a necessary condition for a scientific hypothesis to be useful. If fact, somewhat awkwardly, we seem to have come full circle. The location debate was initiated with Darwin's 'warm little pond' (as proposed in his 1871 letter to a friend, Joseph Hooker) and has ended up in Mulkijanian's 'hot spring' (Mulkidjanian *et al.* 2012). The irony is hard to ignore. More significantly, however, none of the proposals help inform us what life is and why it emerged. The problem appears to lie elsewhere.

3. Physical Basis for Change

All physico-chemical processes are directed, and the general principle that governs that directedness is the Second Law of Thermodynamics. That's why making an omelette from eggs is easy but the doing the reverse is a lot harder. Nature's direction is from thermodynamically less stable to thermodynamically more stable. However, for the OOL process – the transition from life's building blocks to simplest life – the reaction appears to have gone the wrong way, toward the creation of a thermodynamically *unstable* system, one which depends on a continual supply of energy for that counter-thermodynamic process to be thermodynamically feasible. How could a non-equilibrium chemical system, maintained in that non-equilibrium chemical state through a continual supply of energy, have come about? Carrying out traditional chemical reactions is unlikely to lead us into this new dimension of chemical behavior. Just as exploring a 2-dimensional surface won't get you airborne and into the third dimension, undertaking traditional chemical processes is unlikely to fortuitously get you into such an energy-fueled, homeostatic non-equilibrium state. Chemists have been carrying out a wide range of chemical reactions for several hundred years now, but the kind of chemistry that leads to life-like entities does not come about. That is the reason the great physicists of the 20th century were deeply puzzled by the life phenomenon. Clearly some out-of-the-box chemical thinking would be required to get you into that very special physical-chemical state. Given these difficulties with the synthetic approach to resolving the OOL problem, let us now consider the physical/conceptual approach to the problem.

Beyond the apparent incompatibility between the Second Law of Thermodynamics and the OOL process, there is an added puzzling aspect to the OOL question. How can the inordinate complexity of all living things, even the simplest ones, be explained? Even though mathematicians and physicists are struggling to define what complexity is, as the famous quote goes, we know it when we see it. Life is incomprehensively complex (Adami 2002). Dawkins stated it succinctly in the very opening line of one of his texts: *we animals are the most complicated things in the known universe* (Dawkins 1996). Indeed, seventy years of molecular biology have brought that message home in the clearest of terms. But the reason for that complexity remains obscure. Why did simple become complex? And to muddy the waters further, not all biologists agree with the statement that complexification increased over evolutionary time. However, the realization that life's complexity emerged is itself significant, as it can open an additional means in our attempts to understanding the life process. Maybe the answer to the 'why life' question can be found by answering the 'why complexity' question. Can we somehow conceptually strip away the levels of complexity to reach some inner core, to uncover life's essence? Though life is certainly complex, could it be that the life principle is simple?

4. The Replication Reaction

Thinking along those lines, in the 1970s, Tibor Gánti, a Hungarian theoretical biologist, composed an abstract model, the Chemoton model, in which he attempted to characterize minimal life (Gánti 2003). In the model he identified living things as being composed of three sub-systems, integrated into a functional whole. His model did not mention specific molecular forms, but that is part of its appeal. It attempts to describe how any living thing would be constructed, just like an architect's drawings of a building would not need to specify the precise materials from which the building would be composed. The three components that Gánti identified, though slightly modified over the years were: (a) a self-replicating informational system, (b) a metabolic system able to supply energy and chemical building blocks, and (c) a physical compartment which encloses the cell's components. That conceptual division opened up new mechanistic horizons for experimental chemists. Reaching out for simplicity, chemists could now seek out chemical systems able to generate these capabilities separately. Indeed, through the discovery in the 1960s that certain molecules were able to self-replicate, a new direction in the origin of life studies was initiated, one whose focus was on *molecular replication*.

Replication is central to the life process, and DNA replication, as part of that general process, is key, as it is the means by which genetic information is transmitted from one generation to the next. However the process of DNA replication is a highly complex one, requiring the cellular environment for it to take place. How could such a complex multi-step process have gotten started? The first hint for how it could have taken place came in a 1967 landmark experiment, when Sol Spiegelman, a creative molecular biologist at the University of Illinois, carried out a molecular replication reaction in a test-tube (Mills *et al.* 1967). Spiegelman's experiment was conducted on an RNA molecule, a nucleic acid cousin to DNA, meaning that it is also a chain-like molecule constructed from a sequence of nucleotides. However, in contrast to DNA whose sugar component within the DNA nucleotide building blocks is deoxyribose, that in RNA nucleotides is ribose. The difference is important in that compared to DNA, which prefers to exist in a double stranded structure, the famous double helix, RNA prefers to adopt a single stranded form. That single-stranded structure means that RNA is more reactive than DNA, leading to quite different biochemical function and behavior. And that's where Spiegelman's dramatic test-tube experiment comes in.

Spiegelman discovered that mixing an RNA molecule with individual A, G, C and U nucleotides (together with an enzyme to catalyze the reaction), resulted in the RNA molecule being able to make copies of itself, much as DNA does in the cell, but in this case the process is carried out in a simple test-tube reaction. In other words RNA is a self-replicating molecule, much like DNA, but significantly, the replication reaction is able to take place outside of the highly complex environment of the living cell. Thus, to illustrate, placing an RNA molecule of a particular sequence, say AAGUCCUGAUCCUG, in a test tube with activated building blocks, A, G, U, C, and a catalyst, would result in the formation of many RNA molecules with precisely the same AAGUCCUGAUCCUG sequence.

But an even more dramatic result was to come. Spiegelman discovered that occasionally exact RNA replication did not take place. Sometimes copying errors occurred. Instead of a particular nucleotide, say G, fitting in at a certain position on the RNA template, another nucleotide, say A, would latch into place instead. As a result the RNA copy ended up being slightly different to the original. In biological parlance, a mutation had occurred. And indeed, when that replication process was carried out multiple times, by continually transferring the product of replication from one test-tube to another, an evolutionary process was revealed. Indeed, after 74 batch transfers the original RNA molecule, originally some 4000 nucleotides long, actually shortened to a fraction of its original length, just some 550 nucleotides long.

The conclusion was unambiguous: polymeric molecules with a sequence structure, as exemplified by RNA, were able, under the right conditions, to both replicate *and* evolve. That groundbreaking work was then followed up by Manfred Eigen, Peter Schuster, Leslie Orgel and others, both experimentally and theoretically (Eigen 1992). Evolution appeared to be not just a biological phenomenon, but a chemical one as well. Replicating molecules, at least certain ones, could evolve. That result appeared to resolve a long-standing dilemma in the origin of life thinking. Chemically speaking all living things are based on a dual nucleic acid-protein system. Both structurally and functionally, life's exceptional operational capabilities derive from that integrated capability. But that dual nucleic acid-protein system creates a classic 'chicken and egg' dilemma. In order for a cell to replicate, and the DNA genome within the cell to replicate as part of that general process, both nucleic acids and proteins are required. Without proteins the cell's DNA cannot be replicated, and without the coded information written into the nucleic acid sequence, the required proteins cannot be formed. Neither can come into being without the other. So how could this dual system have come into existence?

5. The RNA World

Through consideration of the Spiegelman experiment, those three grand figures, Leslie Orgel, Francis Crick and Carl Woese, offered a solution to the dilemma (Robertson & Joyce 2012). They postulated that the more complex dual DNA-protein world was preceded by a simpler world, in which RNA alone formed the basis of life at the time. RNA, being more reactive than its DNA cousin, and being present in nature as a single strand polymer, might possess some catalytic activity. Thus in that earlier world, it would have been able to carry out both replicative *and* catalytic functions, though less effectively than the dual DNA-protein system. Indeed, in the early 1980s Sidney Altman and Robert Cech confirmed experimentally that nucleic acids were able to undertake catalytic function, though, as anticipated, less effectively than proteins (Stark *et al.* 1978; Kruger *et al.* 1982). Such RNA molecules were termed *ribozymes* to express both their RNA basis and their enzymatic capability.

Reassuringly, following Max Delbruck's advice of looking carefully into cell structure to reveal the cell's history, support for the RNA-world view can be found. One key organelle, the ribosome, the exquisite molecular machine which manufactures proteins, is itself composed of several nucleic acid and protein molecules. But more careful inspection reveals an interesting feature of ribosome structure. The active core of the ribosome, where the function of generating proteins is carried out, is composed solely of RNA molecules. Those protein molecules which are present in the ribosome are only found in the ribosome periphery, not in its active core. That observation led to the conclusion that early in evolution, the primal ribosome was an RNA construction, and the incorporation of protein molecules into the structure was due to evolutionary 'fine-tuning' which came about at a later time. The existence of RNA's catalytic capabilities, together with such historic information as manifest in the ribosome structure, lent further support for the RNA-world proposal.

But that's where the good news ends. The RNA world proposal runs into a deep problem that was raised earlier. If the RNA-world thesis proposes that early life was based on RNA chemistry and was based on RNA's special chemical ability to undergo self-replication, then the study of the RNA replication reaction should throw light on the emergence of life process. At this point, however, the story becomes more problematic. The reality is that a half century of RNA study since those heady days of the 1960s have failed to reveal any indication of an evolutionary process from a simpler replicating system to a more complex and more life-like system. As discussed above, Spiegelman's RNA molecules, when subjected to continual rounds of replication, did show an evolutionary process, but toward *shorter* RNAs, i.e., away from life, not toward it. Though the extensive RNA studies of past decades have offered valuable technological benefits, no evolutionary process toward greater complexity, toward life, has been observed (Joyce 2015). Molecular evolution was directed by that ubiquitous Second Law - toward the equilibrium state, not toward life. Clearly, some crucial element was missing. In the OOL context, here was yet another dead-end.

Which leads us to the issue of metabolism, the second key component in Gánti's chemoton model (Gánti 2003). Metabolic cycles are central to the functioning of all living things. Two well-known examples are the citric acid cycle, which leads to the oxidation of acetate to carbon dioxide, and the urea cycle, which converts toxic ammonia to less harmful urea. Such cycles are not chemical cycles in the usual sense in that they are highly complex and are enzymatically controlled. The question then arises: could simple non-enzymatic cycles have appeared spontaneously on the prebiotic earth as precursors to the complex cycles of extant life? But in contrast to our discussion on replication, in which simple replication reactions are possible having been observed experimentally, the emergence of cyclic chemical organization is more problematic.

For life to have emerged from such a cycle, it would have needed to be autocatalytic. Though the term 'autocatalytic' and 'self-replicating' sound very different, they overlap to a considerable degree. The former term expresses the idea that a molecule or molecular system catalyzes its own formation, but that signifies that the system has induced a copy of itself to appear. The 'replication' term is generally used when a particular entity catalyzes its own formation directly, such as in a molecular replication reaction via a template mechanism. In contrast, the term autocatalysis is generally used for those cases where the replication comes about indirectly, through cycle formation, as is discussed below. So though the terms are quite different, in practical terms they often lead to the same result – the copying process of some entity.

In the context of the origin of life, the idea that life might have emerged from an autocatalytic cycle was first raised by Stuart Kauffman in the 1980s (Kauffman 1986). If there exists a set of molecules, or molecular aggregates, A, B, C, D, and E, and if A is able to catalyze the formation of B, B is able to catalyze the formation of C, and so on through to E, and if E is able to catalyze the formation of A, thereby closing the cycle, the cycle then becomes *autocatalytic*. Catalytic closure means the system as a whole is able to catalyze its own formation. Accordingly, if the building blocks for the formation of A, B, C, D, and E, are readily available, then the quantities of all the components within the cycle will grow exponentially. The process of autocatalysis effectively leads to a process of replication.

6. Dynamic Kinetic Stability (DKS)

As is well known, exponential growth of any kind is unsustainable. Such growth cannot be maintained as resources are quickly depleted. Thus for a replicating system to be stable over time it needs to be in a dynamic state, in a non-equilibrium steady-state, where the system's rate of formation and its rate of degradation are maintained roughly in balance. But for such a system to be maintained over time, the degraded system needs to be continually reactivated. A simple equation, which describes such a cyclic kinetic state, is shown in eq 1:

$$dX/dt = kMX - gX \tag{1}$$

where X is the replicator concentration, M is the concentration of building blocks from which X is composed (for simplicity assumed to be of one kind), and k and g are rate constants for replicator formation and decay, respectively. The kMX term represents the rate of replicator formation from its building blocks, M, while gX represents the replicator's first order rate of decay. A steady state population, a state which is effectively 'stable', can be achieved and maintained provided dX/dt remains close to zero. A simple physical metaphor to illustrate the cyclic nature of the replicative process is a water fountain. A fountain is generated when an energy source pumps water out of the fountain nozzle, after which it falls back into the reservoir, before being recycled once again (Pross 2016).

Two points should be noted. First, if the replicating system can be maintained over time, its stability is not thermodynamic, but rather kinetic – the stability kind is *dynamic kinetic stability* (DKS) (Pross 2016). The stability of the system is manifest in the stability of the population of replicators, not in the individual replicators, which are transient, as they are continually being turned over, like water in the fountain. Second, a further condition for stability is that a continual supply of energy is provided, its role being to reactivate the building blocks so that the replication reaction can be continually maintained (Pross 2016).

But in order to explain an evolutionary process for such a DKS system, an added feature is crucial. The possibility of imperfect replication – mutation – must exist. Once such mutations take place, those that are favorable in that they lead to more persistent replicators (greater DKS), will end up replacing replicators of lower stability (lower DKS). That result is implicit in the mathematics of eq 1. The direction of change – the evolutionary direction – will not be toward greater thermodynamic stability, but toward greater DKS. Of course for such a process to occur the replicating system must be structured so that it is evolvable. For nucleotide-based replicators, sequence variation indeed allows for an evolutionary process (Pross 2011).

In recent years considerable experimental support for the DKS concept has been reported. A variety of energy-fueled chemical systems, able to maintain themselves over time, have been discovered and, what's more, these are found to have very different properties to thermodynamically-controlled systems (Pascal & Pross 2015). Effectively a new dimension in chemical reactivity, a kinetic dimension, has been discovered. The discovery of that new dimension is quite dramatic in chemical terms as it offers a physico-chemical

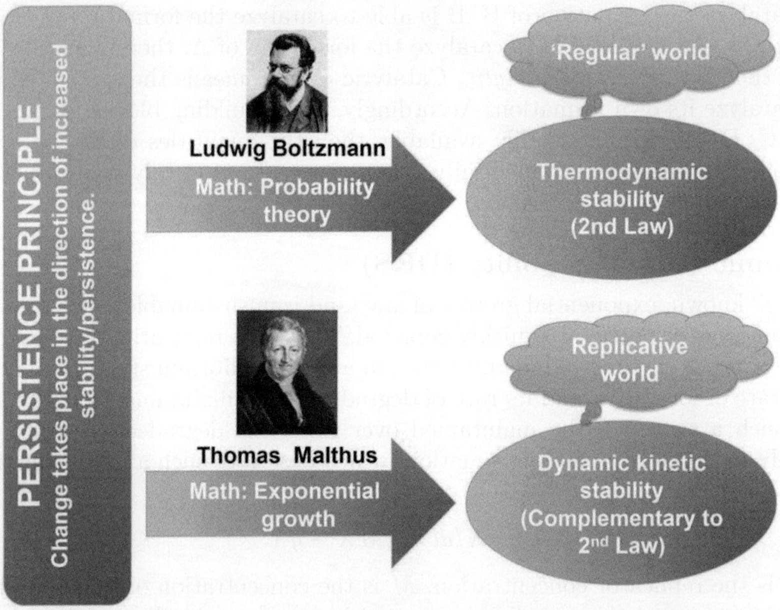

Figure 1. Schematic diagram expressing the persistence principle and the two mathematical formulations through which stability/persistence may be expressed: (a) Boltzmann's probabilistic formulation leading to thermodynamic stability, and, (b) Malthusian exponential growth leading to dynamic kinetic stability. Diagram taken from Pascal & Pross 2015.

framework for all living processes, one that has been absent till now (Pascal & Pross 2019).

7. The Persistence Principle

The realization that physico-chemical systems may become persistent for kinetic, not just thermodynamic reasons, leads to the formulation of a Persistence Principle (Pascal & Pross 2015). That principle may be stated as follows: *systems will tend from less stable/persistent to more stable/persistent forms*, or, more concisely, all matter is *driven toward more persistent forms.* The statement is logically grounded in that *less* persistent systems are, well, less persistent, meaning they will, by definition, tend to change, while *more* stable/persistent systems, again by definition, are *less* likely to undergo change. In simplest terms, changing things will change until they change into things that don't.

Of course the Second Law of Thermodynamics expresses that same directive – toward persistent forms, but through thermodynamic considerations. The Second Law seeks persistent forms through achieving the systems' equilibrium state. Significantly then, there exist two mathematical descriptions able to lead toward persistence – the Malthusian mathematics of exponential growth and Boltzmann's mathematics of statistical mechanics. Strikingly, however, each of these two mathematical kinds leads to different material consequences, with very different material expression – animate and inanimate. These two distinct paths toward greater persistence are illustrated in Fig. 1.

8. Concluding Remarks

By considering the stability concept from a kinetic perspective, rather than the more traditional thermodynamic perspective, we learn that the life phenomenon is not the inexplicable form of matter that it appears to be, one oddly incompatible with the

Second Law. A kinetic approach, based on logical/mathematical considerations, indicates that material stability can rest on an extra-thermodynamic (kinetic) base, not just on a thermodynamic one. That kinetic perspective reveals a new dimension of chemical potentiality that has only recently been discovered (Pascal & Pross 2019). That in turn leads to the new insight that life processes are but a particular expression of that kinetic chemical dimension. With a physical-chemical understanding of what life is, it is now feasible to offer a life definition: life is a self-sustained kinetic replicative network of chemical reactions whose evolutionary roots lie in an energy-fueled prebiotic replicative chemical system. The identity of that prebiotic system will likely remain unknown as the historic record of that event is long gone. But once such a replicating system emerged from the diverse environment of the early earth, the logic underpinning the persistence principle switches to one based on the mathematics of exponential growth, rather than on the equilibrium mathematics of statistical mechanics, and that switch expresses itself through the emergence of a distinctly different kind of material form - life.

As a final note, life's inordinate complexity can now be seen to derive directly from this kinetic perspective. In the functional world of replication, greater stability has been found to be enhanced by greater complexity: increasing complexity for increasing stability (Pross 2013). Accordingly, the evolutionary process leads to an increase in both stability *and* complexity. Life's mystery lies in the contingent conditions that would have enabled a primal dynamic steady-state replicative system to emerge, rather than on the nature of the evolutionary process itself. Systems chemistry laboratories are currently working on attempts to synthesize such chemical systems.

With regard the question as to whether life exists beyond our planet, the preceding discussion opens up a new approach to addressing that question. Once experiments on earth are able to outline how readily energy-fueled chemical systems able to establish dynamic replicative networks can be formed, those earth-bound experiments will have provided preliminary indications as to the likelihood of life's existence in the cosmos at large. Almost a century after Alexander Oparin, a Russian biochemist, and J.B.S. Haldane, a British evolutionary biologist, (Oparin 1952; Haldane 1929) raised the origin of life problem as one that science could potentially solve, that most imponderable of scientific questions may finally be closer to resolution.

Acknowledgements

The author is indebted to Prof. Robert Pascal for continuing discussions regarding the general role of kinetics in evolution. This publication reports the results of work undertaken within the COST actions CM1304 Emergence and Evolution of Complex Chemical Systems, and TD 1308 Origins.

References

Adami, C. 2002, *BioEssays*, 24, 1085–1094

Dawkins, R. 1996, *The blind watchmaker, Norton, NY*

Eigen, M. 1992, *Steps toward life,* Oxford UP, NY

Fry, I. 2000, *The emergence of life on earth,* Rutgers UP, NJ

Gánti, T. 2003, *The principles of life,* Oxford UP, Oxford

Haldane, J. B. S. 1929, *The Rationalist Annual* 148, 3–10

Joyce, G. F. 2015, *J. Mol. Evol.,* 81, 146–149

Kauffman, S. 1986, *J. Theor. Biol.*, 119, 1-24

Kruger, K., Grabowski, P. J., Zaug A. J., Sands, J., Gottschling D. E., & Cech, T.R. 1982, *Cell* **31,** 147-57

Miller, S. L. 1953, *Science,* 117, 528

Mills, D. R., Peterson, R. L., & Spiegelman, S. 1967, *Proc. Nat. Acad. Sci. USA* 58, 217

Mulkidjanian, A. Y., Bychkov, A. Y., Dibrova, D. V., Galperin, M. Y. & Koonin, E. V. 2012, *Proc. Nat. Acad. Sci. USA* 109, E821-E830
Oparin, A. I. 1952, *The origin of life*, Dover, NY
Pascal, R., & Pross, A. 2015, *Chem. Commun.* 5, 16160
Pascal, R., & Pross, A. 2019, *J. Syst. Chem.* 7, 1-8
Pross, A. 2011, *J. Syst. Chem.* 2, 1-14
Pross, A. 2013, *J. Mol. Evol.* 76, 185-91
Pross, A. 2016, *What is life?* 2nd Ed., Oxford UP, Oxford
Robertson, M. P. & Joyce, G. F. 2012, *The Origins of the RNA World*, in Cold Spring Harb Perspect Biol. 4, a003608
Stark B. C., Kole, R., Bowman, E. J. & Altman, S. 1978, *Proc. Nat. Acad. Sci. USA* 75, 3717-21
Sutherland, J. D. 2015, *Angew. Chem. Int. Ed.* 54, 2–20

Discussion

CUNTZ: Based on your work - what is your recommendation toward astrobiologists to optimize their efforts in the search for life in the Universe?

PROSS: To be honest - a very difficult problem. In some sense we need to progress more with what life is before we can seriously deal with seeking it elsewhere.

Origins: From the Protosun to the First Steps of Life
Proceedings IAU Symposium No. 345, 2018
B. G. Elmegreen, L. Viktor Tóth & M. Güdel, eds.

doi:10.1017/S1743921319002138

Unveiling the whole from its parts or to see the forest for the trees

Eduardo Janot-Pacheco[1], Marina Rachid[1], Philippe Bendjoya[2], Armando Domiciano[2], Caroline Antunes-Rosa[3], Marcelo Emilio[4] and Claudia Lage[3]

[1]Institute of Astronomy, Geophysics and Atmospheric Sciences,
So Paulo University, Brazil
email: eduardo.janot@iag.usp.br

[2]Université Côte d'Azur, Observatoire de la Côte d'Azur,
CNRS, Laboratoire Lagrange, France
email: Philippe.BENDJOYA@unice.fr

[3]Carlos Chagas Filho Institute of Biophysics,
Rio de Janeiro Federal University, Brazil
email: lage@biof.ufrj.br

[4]Ponta Grossa State University, Brazil
email: marcelo_emilio@yahoo.com

Abstract. If biologically complex molecules as DNA were present in the extraterrestrial targets, their spectral signatures would be rather difficult to be unambiguously identified. As a matter of fact, the molecular array of a single nucleobasis will generate a tangled spectral signature. On the other hand, a part of it, e.g. bands due to the group of HNCO of guanine may have been detected but associated to smaller molecules, e.g. isocianic acid (HNCO). However, if comprised in a nucleobasis, its detection would be misinterpreted. Five key transitions were preliminarly selected for either purines and pyrimidines that should be observed together in the same target. If this happens, it may be that we are detecting the whole from its parts.

Keywords. ISM: molecules, ISM: lines and bands, astrochemistry

1. Introduction

We live in a molecular universe. In a spiral galaxy as the Milky Way (MW), interstellar molecular clouds are the bigger and heavier baryonic components of the spiral system. The largest molecular clouds are visible to the naked eye, appearing as dark patches against the brighter background of the MW. They can be seen all over the magnificent projection of the galaxy disk on the sky, mainly from the southern hemisphere.

Biology tell us that more than 95 % of all life forms on Earth is composed by H, O, C N, P and S, sources of the known CHONPS anagram. Moreover, the genetic material of all life forms is encoded in huge molecules, formed by the long sequences of deoxyribonucleic or ribonucleic acids (DNA and RNA).

2. Molecular factories

It is well known that interstellar molecular clouds and circumstellar envelopes harbour ice grains that can act as factories of molecular synthesis. Indeed, atoms and molecules can meet and react at the surface of the ices present in the clouds, the cosmic rays and energetic photons acting as the necessary energy source for ionization and chemical reactions (e.g. Marshall & Sadeghpour 2016).

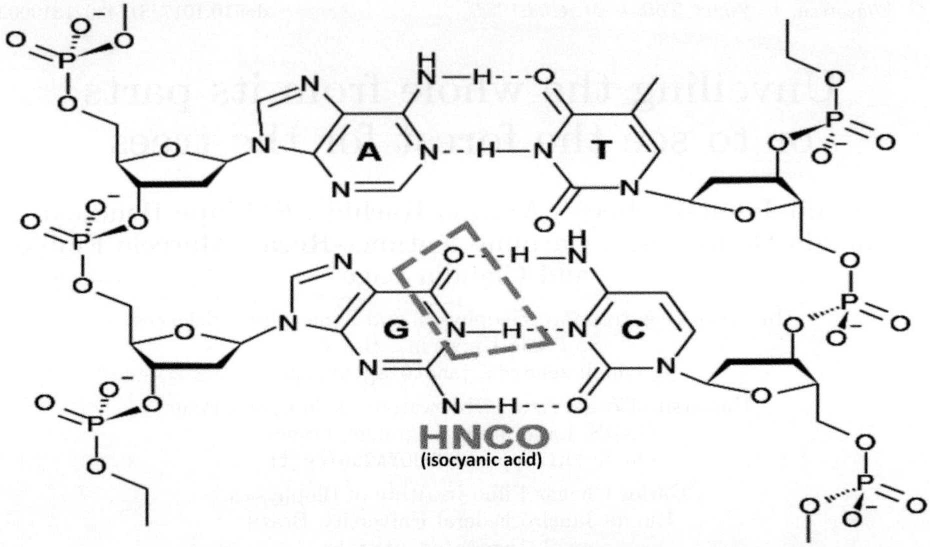

Figure 1. A small segment of a DNA molecule is depicted. The spectrum of the whole molecule is quite complex and entangled. The independent vibrational signal of a local isocyanic acid is perhaps detected, but not interpreted as coming from a part of the DNA molecule.

Surprisingly, complex, pre-biotic molecules used in biochemistry on Earth are found in the interstellar medium, in circumstellar envelopes, planetary atmospheres and cometary surfaces, asteroids and meteorites, and interplanetary dust particles (Ehrenfreund & Cami 2010).

3. An organic friendly universe

We live in an organic-friendly universe: there are pre-biotic molecules everywhere in the MW, e.g., in interstellar clouds and circumstellar envelopes of forming stars, where tens of organic molecules up to 13 atoms have been detected in radiobands (https://www.astro.uni-koeln.de/cdms) (excluding PAHs; these compounds will not be considered in this paper). Relatively complex organic molecules (Dimethyl ether and methyl formate) have also been observed in nearby galaxies as the Magellanic Clouds (Sewilo *et al.* 2018).

Powner, Gerland & Sutherland (2009), Jeilani *et al.* (2016) and Hudson *et al.* (2012) have shown that primitive abiogenic reactions in the primordial soup of simple molecules as hydrogen cyanide (HCN), glycoaldehide and formamide may generate life building blocks as pyrimidines and purines. Thus, relatively straightforward prebiotic chemical reactions may form the robust code of the biosphere on Earth, which is based on nucleic acids. Why then, it is so difficult to address their detection in astrophysical targets? At least a part of the answer could be based on the fact that a) their characteristic IR bands are complex and often overlap with the features due to other species, making it difficult to safely identify them and b) the best wavelength range to collect spectra of complex molecules, the mid-IR, is only accessible by means of expensive, cooled spectroscopes.

4. Detecting the whole from its parts?

Our work hypothesis is that maybe only partial (i.e, local) transitions coming from big molecules have been identified, but are not interpreted as such. Could it be that we are seeing the trees and missing the forest as a whole? In Figure 1 a small part of a DNA is shown. The sugar-phosphate backbones can be seen at the extreme left and right

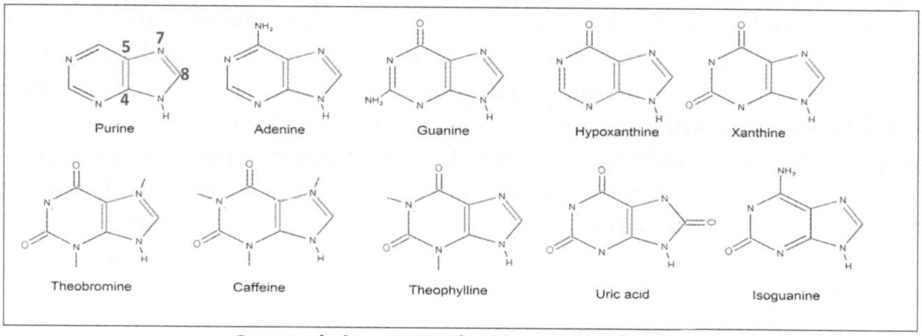

Spectral signatures for the purine family
in-plane ring vibration ("breathing" movement): 9 lines between 14.4 and 6.7 μm
out-of-plane ring vibration ("butterfly" movement): lines at 43.47 and 34.96 μm
$N^7 - C^5$ in-ring bond: line at 8.10 μm
$N^7 - C^8$ in-ring double bond: 5 lines between 6.75 and 8.40 μm
$C^4 - C^5$ in-ring double bond: 3 common lines at 6.37, 6.25 and 5.93 μm

Figure 2. Five proposed preliminary spectral characteristics of the purine family (sketched on the top) (approximate wavelengths). If they are detected together in the same astronomical target we may be seeing the forest (the nucleobases) for the trees (the local transitions within them).

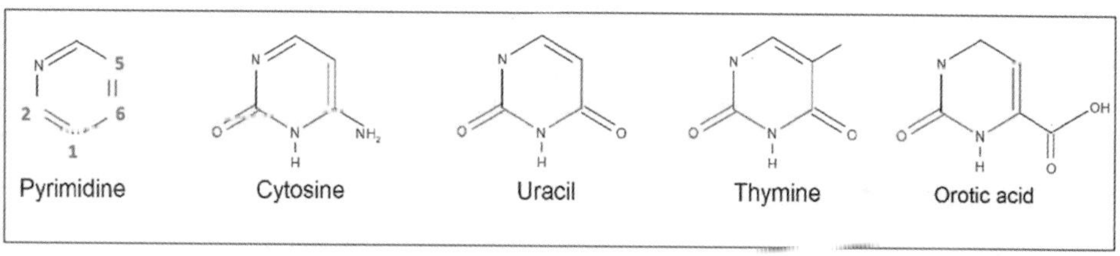

Spectral signatures for the pyrimidine family
Out-of-plane ring vibration ("chair" movement): 3 lines between 43.10 – 51.54 μm
$C^5 - C^6$ in-ring double bond: lines at 5.94 and 6.19 μm
$N^1 - C^6$ in-ring single bond: line at 6.79 μm
$C^2 - N^1$ in-ring bond: lines at 6.79 and 13.18 μm
$C^2 - O$ double bond: lines between 5.62 and 5.88 μm

Figure 3. The same for the pyrimidine family (on the top).

and the alternate nitrogen containing bases (pyrimidines and purines) are seen towards the center. A HNCO fragment is highlighted in the base pairing between a guanine (G) and a cytosine (C). The independent vibrational feature of an isocyanic acid is perhaps detected, but not interpreted as coming from a part of the bigger structure, as a guanine molecule. Nucleic acids may be present in the interstellar medium, but their spectra are quite complex and entangled. Within the spectral forest of IR or radio spectra, parts of such a complex molecule could be detected as simple, local transitions. Although nucleic acids may be there, we only see parts of the whole.

We propose to define spectral signatures of the two nucleobases families (purines and pyrimidines) by selecting a number of local transitions within the molecules. In order to enhance the signal, the chosen transitions should be present in all types of nucleobases (there are 5 types of pyrimidines and 10 types of purines). Five transitions were preliminarily proposed for each family and shown in Figures 2 and 3. If they are detected together at the same astronomical target, we are perhaps seeing the forest (the nucleobases) through detection of the trees (local transitions within these nucleobases).

References

Ehrenfreund, P., & Cami, J. 2010, *Cold Spring Harb Perspective Biol*, doi: 101101/cshperspect.a002097

Hudson, J. S., Eberle, J. F., Vachhani, R. H., Rogers, L. C., Wade, J. H., Krishnamurthy, R., & Springsteen, G. 2012, *Angewandte Chemie International Edition*, 51, 5134

Jeilani, Y. A., Williams, P. N., Walton, S., & Nguyen, M. T. 2016, *Physical Chemistry Chemical Physics*, 18, 20177

Marshall D. W. & Sadeghpour, H. R. 2016, *MNRAS* , 455, 2889

Powner, M. W., Gerland, B. & Sutherland, J. D. 2009, *Nature*, 459, 239

Sewilo, M., Indebetouw, R., Charnley, S. B., Zahorecz, S., Oliveira, J. M., van Loon, J. Th., Ward, J. L., Rosie Chen, .-H., Wiseman, J, Fukui, Y., Kawamura, A., Meixner, M., Onishi, T., & Schilke, P. 2018, *ApJ Letters*, 853, 2

Discussion

KEDZIORA-CHUDCZER: Can you comment on the chirality of these discussed compounds?

JANOT-PACHECO: Indeed, chirality may play a certain role in the transitions of complex organic molecules. We intend to include this aspect of the problem in the future.

Origins: From the Protosun to the First Steps of Life
Proceedings IAU Symposium No. 345, 2018
B. G. Elmegreen, L. Viktor Tóth & M. Güdel, eds.

doi:10.1017/S1743921319001789

Simulating putative Enceladus-like conditions: The possibility of biological methane production on Saturn's icy moon

Ruth-Sophie Taubner[1,2], Patricia Pappenreiter[3], Jennifer Zwicker[4], Daniel Smrzka[4], Christian Pruckner[1], Philipp Kolar[1], Sébastien Bernacchi[5], Arne H. Seifert[5], Alexander Krajete[5], Wolfgang Bach[6], Jörn Peckmann[4,7], Christian Paulik[3], Maria G. Firneis[2], Christa Schleper[1] and Simon K.-M. R. Rittmann[1]

[1]Archaea Biology and Ecogenomics Division, Department of Ecogenomics and Systems Biology, University of Vienna, 1090 Vienna, Austria
email: ruth-sophie.taubner@univie.ac.at

[2]Institute of Astrophysics, University of Vienna, 1180 Vienna, Austria

[3]Institute for Chemical Technology of Organic Materials, Johannes Kepler University Linz, 4040 Linz, Austria

[4]Department of Geodynamics and Sedimentology, Center for Earth Sciences, University of Vienna, 1090 Vienna, Austria

[5]Krajete GmbH, 4020 Linz, Austria

[6]Geoscience Department, University of Bremen, 28359 Bremen, Germany

[7]Institute for Geology, Center for Earth System Research and Sustainability, University of Hamburg, 20146 Hamburg, Germany

Abstract. In this study (Taubner *et al.* 2018), three different methanogenic archaea (*Methanothermococcus okinawensis*, *Methanothermobacter marburgensis*, and *Methanococcus villosus*) were tested for metabolic activities and growth under putative Enceladus-like conditions, including high pressure experiments and tests on the tolerance towards potential gaseous and liquid inhibitors detected in Enceladus' plume. In particular, *M. okinawensis*, an isolate from a deep marine trench (Takai *et al.* 2002), showed tolerance towards all of the added inhibitors and maintained methanogenesis even in the range of 10 to 50 bar. Further, we were able to show that H_2 production based on serpentinization may be sufficient to fuel such methanogenic life on Enceladus. The experiments revealed that methanogenesis could, in principle, be feasible under Enceladus-like conditions.

Keywords. astrobiology, planets and satellites: Enceladus, methods: laboratory

---

1. Introduction

Enceladus is one of the most remarkable objects in the Solar System. Besides its spectacular plume located at the South Polar Terrain, it is the origin of the plume particles which makes Enceladus one of the hot spots in Astrobiology these days: a global subsurface liquid water ocean (Thomas *et al.* 2016). The molecules detected in the plume (Waite *et al.* 2009, 2017), e.g. water (H_2O), carbon dioxide (CO_2), or methane (CH_4), in combination with the possibility of hydrothermal vents at the ocean floor (Hsu *et al.* 2015) led to the hypothesis that this icy moon might be habitable for life as we know it. The detection of molecular hydrogen (H_2) in the plume (Waite *et al.* 2017) corroborates this hypothesis, and serpentinization of the chondritic core of Enceladus is a potential source of H_2. The most promising area on Enceladus where life may exist is at the seafloor of the ice-covered ocean, where the pressure was estimated to be 35 to 80 bar (Hsu *et al.* 2015, Taubner *et al.* 2018).

Possible life forms on Enceladus must be chemoautotrophic, i.e. independent of the presence of organic matter, and would need to be anaerobic, i.e. independent of oxygen. Hydrogenotrophic methanogens, i.e. archaea that metabolize H_2 and CO_2 to produce CH_4 and H_2O, are among the organisms that meet these characteristics. In this study we raise the question if some of the CH_4 detected in the plume might originate from biological activity.

2. Experimental Design

The first step of the presented study (Taubner *et al.* 2018), was to screen the literature for suitable strains. As presented elsewhere (Hsu *et al.* 2015), the temperature close to the hydrothermal vents on Enceladus' ocean floor could be relatively high. Therefore and because of the higher growth rate, we decided to focus on (hyper-) thermophilic methanogens. After some preliminary tests, we decided to focus on the three methanogenic strains *Methanothermococcus okinawensis*, *Methanococcus villosus*, and *Methanothermobacter marburgensis*. The latter strain originates from a waste water treatment plant (Schönheit *et al.* 1980), whereas the other two strains were isolated from a shallow submarine hydrothermal system (*M. villosus*, Bellack *et al.* 2011) and a deep marine trench (*M. okinawensis*, Takai *et al.* 2002), respectively. To prioritise one of the above mentioned strains, we tested these organisms for tolerance towards potential gaseous and liquid inhibitors detected in Enceladus' plume (formaldehyde, ethene, methanol, and high ammonia levels). Only one strain, namely *M. okinawensis*, showed continuous and stable growth. We therefore selected this strain for the final high-pressure experiments. These experiments were performed in a high pressure system consisting of either a 0.7 or 2 L stirred Büchi reactor, a heating system, a vacuum pump, a transfer vessel (for inoculum and special media compounds), and a connection to a computer. The reactor can be controlled via remote-control and the appended gas mixture can be easily varied. We tested the methane production of *M. okinawensis* at different pressure values (ca. 10, ca. 25, and ca. 50 bar) with a Enceladus-like gas phase (ca. 30-40 Vol.-% N_2, ca. 40-55 Vol.-% H_2, ca. 5-7 Vol.-% CO_2, ca. 3-5 Vol.-% CO, ca. 3-6 Vol.-% C_2H_4) at 65°C. We were able to show that *M. okinawensis* was able to perform H_2/CO_2 conversion and CH_4 production under the above described Enceladus-like conditions. Even at the highest applied pressure of 50 bar, a CO_2 to CH_4 conversion of more than 70% was observed.

3. Serpentinization

In parallel, serpentinization-based H_2 production rates were modelled using the computer code PHREEQC (Parkhust *et al.* 1999) as part of this study to determine if serpentinization reactions on Enceladus can support a rate of H_2 production that is high enough to sustain autotrophic, hydrogenotrophic methanogenic life. Depending on the applied mineral content, a H_2 production rate of approximately 5 to 50 nmol g^{-1} L^{-1} d^{-1} was estimated (Taubner *et al.* 2018), which may be sufficient to serve as a substrate for biological CH_4 production on Enceladus.

4. Conclusion

This work (Taubner *et al.* 2018) presents the first systematic astrobiological study on methanogenic physiology under high pressure and may enable a new assessment of potential habitats in the Solar System, especially icy moons. This study does neither prove that there is life nor that there is an origin of life setting on Enceladus. However, it showed that methanogenesis could, in principle, be feasible under Enceladus-like conditions and that some of the CH_4 detected in Enceladus' plume could, in principle, be of biological origin. It also points out the importance of planetary protection regulations for future space missions to icy moons like Enceladus and leads to a rethinking of how we should define a habitable zone in a planetary system.

References

Bellack, A., Huber, H., Rachel, R., Wanner, G. & Wirth, R. 2011, *Int. J. Syst. Evol. Microbiol*, 61, 1239

Hsu, H.-W., Postberg, F., Sekine, Y., *et al.* 2015, *Nature*, 519, 207

Parkhurst, D. L. & Appelo, C. A. J. 1999, *Water-Resources Investigations Report*, 99, 4259

Schönheit, P, Moll, J. & Thauer, R.K. 1980, *Archives of Microbiology*, 127(1), 59

Takai, K., Inoue, A. & Horikoshi, K. 2002, *Int. J. Syst. Evol. Microbiol*, 52(4), 1089

Taubner, R.-S., Pappenreiter, P., Zwicker, J., *et al.* 2018, *Nat. Commun.*, 9, 748

Thomas, P. C., Tajeddine, R., Tiscareno, M. S., Burns, J. A., Joseph, J., Loredo, T. J., Helfenstein, P. & Porco, C. 2016, *Icarus*, 264, 37

Waite, Jr., J. H., Lewis, W. S., Magee, B. A., *et al.* 2009, *Nature*, 460, 487

Waite, J. H., Glein, C. R., Perryman, R. S., *et al.* 2017, *Science*, 356, 155

Discussion

WANDEL: Could you comment on the difference between the two best solar system candidates, Europa and Enceladus?

TAUBNER: The most obvious difference is the larger volume of Europa and its subsurface ocean, which leads to a way-higher pressure at the seafloor of Europa compared to Enceladus.

WANDEL: In particular about the prospects for contact of the water body with the silicate bottom?

TAUBNER: According to current models, both Enceladus' and Europa's oceans are in direct contact with the silicate core.

FAN ZHANG: Have you tried to slowly introduce toxicity to the Earth organisms and see if they can evolve to tolerate the environment that didn't pass your selection criteria and get more options?

TAUBNER: This work is planned, not yet done.

Origins: From the Protosun to the First Steps of Life
Proceedings IAU Symposium No. 345, 2018
B. G. Elmegreen, L. Viktor Tóth & M. Güdel, eds.

doi:10.1017/S1743921319002205

Towards astrobiological experimental approaches to study planetary UV surface environments

Ximena C. Abrevaya[1,2], Martin Leitzinger[3], Oscar J. Oppezzo[4], Petra Odert[3], G. Juan M. Luna[1,2], Manish Patel[5], Ana F. Forte-Giacobone[4,6] and Arnold Hanslmeier[3]

[1]Instituto de Astronomía y Física del Espacio (UBA-CONICET)
Ciudad Autónoma de Buenos Aires, Argentina.
email: abrevaya@iafe.uba.ar

[2]Facultad de Ciencias Exactas y Naturales, Universidad de Buenos Aires
Ciudad Autónoma de Buenos Aires, Argentina.

[3]Institute of Physics, IGAM, University of Graz, Austria.

[4]Departamento de Radiobiología, Comisión Nacional de Energía Atómica. Buenos Aires, Argentina.

[5]School of Physical Sciences, The Open University, Milton Keynes, MK7 6AA, U.K.

[6]Departamento de Ingeniería, Universidad Nacional de Tres de Febrero. Buenos Aires, Argentina.

Abstract. The stellar ultraviolet radiation (UVR) has been studied in the last decade and has been found to be an important factor to determine the habitability of planetary surfaces. It is known that UVR can be a constraint for life. However, most of the studies of UVR and habitability have missed some fundamental aspects: i) Accurate estimation of the planetary atmospheric attenuation, ii) The biological inferences used to represent the impact of the stellar UVR on life are theoretical and based on the action spectrum (for DNA or microorganisms) or considering parameters as the "lethal dose" obtained from non-astrobiological experiments. Therefore, the conclusions reached by previous studies about the UVR habitability of planetary bodies may be inaccurate. In this work, we propose how to address these studies in a more accurate way through an interdisciplinary approach that combines astrophysics, microbiology, and photobiology and by the use of specially designed laboratory experiments.

Keywords. stars: activity, ultraviolet: stars, methods: laboratory, astrobiology

1. Introduction

Astrobiology is a relatively new and multidisciplinary field of science that focused on the research about the origin, evolution and distribution of life in the universe. The main international roadmaps for research in astrobiology, as NASA astrobiology Roadmap and The European AstRoMap project (supported by the European Commission Seventh Framework Programme), have among their objectives the characterization of habitable environments in the universe considering the evolution of life and its physical environment that includes radiation (DesMarais *et al.* 2008, Horneck *et al.* 2016). Even though classically the habitability zone (HZ) is defined by the presence of liquid water on the surface of the planet, there are several factors that can constrain the existence of life on planetary bodies. Among these factors, radiation appears to be crucial because habitability is strongly dependent on stellar activity (Hanslmeier 2018). In general: (i) The radiation received by the planetary body and the plasma environment provided by the parent star, play a role on the evolution of the planet and its atmosphere (e.g: Khodachenko *et al.* 2007), (ii) Radiation may have a role in the origin of life, as could be part of photochemical reactions necessary for the polymerization of prebiotic organic molecules (e.g: Ranjan *et al.* 2017), (iii) Radiation can damage cells and therefore can be detrimental for life (e.g: Kielbassa *et al.* 1997). These effects on life can be beneficial or detrimental depending mainly on the radiation flux and wavelength. In particular, radiation at UV wavelengths (200-400 nm) have multiple targets at cellular level and can reach the surface of a planet. Therefore, UVR can be a constraint for life (Abrevaya & Thomas 2017) and the effects of radiation on life will depend on activity, evolution and spectral type of the parent star. The variability in stellar activity is also another important factor for habitability and can be particularly observed as sudden and energetic explosive events known as "flares", originated in magnetic processes affecting all the layers of the stellar atmosphere (photosphere, chromosphere and corona), which heat the stellar plasma and accelerate its protons, electrons and heavy ions to velocities near the speed of light. Studies related to stellar activity comprise energetic events as stellar cosmic rays, flares, coronal mass ejections (CMEs), enhanced coronal X-rays, and enhanced chromospheric UV emission, among others (e.g, Ayres *et al.* 1997; Gershberg 2005; Scalo *et al.* 2007). These analyses also included young solar analogs, in the context of the early Earth (Ribas *et al.* 2010).

The determination of UV habitable environments has been an active field of research in the last 10 years (see e.g.: Cnossen *et al.* 2007; Rugheimer *et al.* 2015; O'Malley-James & Kaltenegger 2017; Howard *et al.* 2018). However, although several authors have analyzed UV radiation fluxes in order to determine the UV habitability of planetary bodies, most of these studies have missed some fundamental aspects. In general: (i) They did not assume the presence of a planetary atmosphere or assumed atmospheric compositions that may not be accurate enough considering the spectral type and stellar activity which may lead to an inaccurate estimation of UV fluxes (fluence rates) reaching the planetary surface. (ii) The estimation of the UV habitable zone through biologically relevant UV fluxes has been done using theoretical calculations based on extrapolations of the action spectrum (DNA or microorganisms) or considering parameters as the "lethal dose". The use of the DNA action spectrum is not accurate from the biological point of view because the irradiation effects on isolated DNA molecules does not represent the complex and multiple effects of radiation on cells. Additionally, when considering extrapolations from microorganisms action spectrum or lethal doses from the literature, some important information is missing as the stage of the growth and physiological conditions of the microbial population. Moreover this information is taken in general from experimental works that were done at low fluence rates and this may not represent some astrophysical scenarios. Therefore, the conclusions reached by previous studies about the UVR habitability of planetary bodies may be inaccurate. In this work we propose how to address these problems by using an interdisciplinary astrobiological approach, that includes astrophysics in combination with microbiology and photobiology and by using specially designed laboratory experiments.

2. Effects of UVR on life

It is possible to distinguish two main effects of UVR on life : (i) Direct effects, when radiation is affecting prebiotic chemistry or cells (e.g.: microorganisms). (ii) Indirect effects, when radiation is affecting the planetary atmosphere, and therefore life on the planet (Abrevaya 2013; Abrevaya & Thomas 2017). If we consider direct effects, UVR can be very harmful and even lethal to living beings, as it is capable to damage DNA but also damaging other cellular components through direct or indirect mechanisms

(Kielbassa *et al.* 1997). However, if the doses are not lethal, they could induce mutations and increase the genetic variability, thus providing new raw material for all sorts of selective pressure. At the same time, higher doses of UVR can also act as a selective pressure itself, leading to the appearance of organisms adapted to live under UVR stress, such as those with pigments for instance (Scalo *et al.* 2007; Wynn-Williams *et al.* 2002). Additionally, UVR could have played an important role during the polymerization of the first organic molecules in the chemical evolution of the prebiotic era on the Earth (Sagan & Khare 1971; Chyba & Sagan 1992; Pestunova *et al.* 2005). Additionally, among the indirect effects we find that UVR can affect the photochemistry at atmospheric level. For example, very recently, new photochemical models have determined that Earth-like atmospheres (ozone) are less feasible for exoplanets in the HZ of some main sequence stars as M dwarfs (García-Sage *et al.* 2017). To determine the influence of UVR on life on planetary bodies, previous studies have been using extrapolations from the DNA action spectrum (Cockell 1998, 1999, 2000; Buccino *et al.* 2006, 2007; Cuntz *et al.* 2010; Rugheimer *et al.* 2015), and microorganisms action spectra (O'Malley-James & Kaltenegger 2017; Estrela & Valio 2018). Nevertheless, the action spectrum represents the relative effectiveness of radiation of different wavelengths to produce a given biological effect and does not represent the chances of survival of a microbial population, but at which wavelengths the radiation is more effective producing an effect. Additionally, DNA is not the only target that can be affected during radiation damage, as proteins and other cellular structures can also be affected. Other studies used extrapolations from parameters as the lethal dose (LD) (Howard *et al.* 2018), that represent the amount of radiation that is necessary to kill a percentage of microorganisms of the population (e.g.: LD90, the percentage necessary to kill the 90 percent of the population). This value is obtained as a parameter from the exponential region of the survival curve for a particular microorganism. However the shape of the curve will depend on the fluence rate of radiation, stage of the growth and physiological conditions of the microbial population under consideration. In general, the probability to suffer damage or to survive to radiation is dependent on several physiological mechanisms as DNA repair and shielding strategies or photoprotection mechanisms to cope with radiation (e.g.: pigments). These mechanisms are particularly important in radioresistant microorganisms. The overall biological effect will depend, therefore, on the balance between photoprotection, damage and DNA repair. Therefore, experimental approaches are needed to determine more accurately the chances of life to arise or develop in a planetary context.

3. New approaches and future work

Very recently Ranjan *et al.* (2017), clearly postulate the lack of experimental approaches to study UV habitability in exoplanets, emphasizing also the need to further develop experimental studies under laboratory conditions. In our studies we propose a two step interdisciplinary approach (Abrevaya 2013): a first step with astrophysical analysis to provide the data of UVR fluence rates on the planetary surfaces that will be used in a second step to reproduce UVR planetary scenarios on laboratory conditions to perform biological experiments.

Step 1: estimation of planetary atmospheric attenuation and planetary UVR surface fluxes.

The amount of UVR reaching the surface of planetary bodies (200-400 nm) will depend on the stellar emission and the atmospheric features of the planetary bodies being considered. Therefore, we will use the stellar radiation fluxes from the observations already available in the IUE and FUSE databases to compute the 1D-radiative transfer modeling of atmospheres (adapted from a previous model in Patel *et al.* 2002), which will be applied considering different atmospheric compositions. These compositions will be chosen according to the current knowledge about photochemical modelling obtained from the literature.

Step 2: determination of radiation constraints for life by using experimental approaches The UVR fluence rates obtained in step 1 will be used to simulate radiation environments in laboratory experiments. In these experiments several kinds of microorganisms will be exposed to different UVR fluences (in wavelenghts between 200 - 400 nm) to evaluate biological effects and potential lethality of the microbial population under different fluence rates, that will depend on the planetary scenario to be considered. For these experiments we will take into consideration: (i)Physiological condition and stage of the growth of the microbial population: we will irradiate liquid cultures of microorganisms in stationary phase, as an approximation to the physiological condition of microorganisms in the environment, (ii)Potential attenuation of radiation by environmental factors or UV absorbing compounds: we will irradiate the microorganisms in a "worst case scenario", avoiding any external factor that can attenuate radiation and with continuous stirring to prevent shielding, (iii)Evaluation of DNA repair capabilities and survival: we will recover the irradiated cultures under "dark" and under "light" conditions, to evaluate the photorreactivation of the DNA repair system in the microorganisms under evaluation. The survival will be evaluated by the plate counting technique.

This will allow us to determine a potential range of UVR fluence rates and fluences according to stellar spectral type, stellar age and planetary atmospheric conditions where it would be possible to find life as we know it.

The proposed approaches are also currently being developed as part of two different interdisciplinary projects: the BioSun and the EXO-UV projects (Abrevaya *et al.* 2014a,b), aiming to expand and improve previous studies, by the incorporation of experimental laboratory approaches.

References

Abrevaya, X. C. & Thomas, B. C. 2017, *In: R.Gordon and A. Sharov (eds.) Habitability of the Universe before Earth. Elsevier*, 531 pp

Abrevaya, X. C., Hanslmeier, A., Leitzinger, M., *et al.* 2014a, *RevMexAA*, 44, 144

Abrevaya X. C., Leitzinger, M., Odert P., *et al.* 2014b, *AbSciCon Abstracts*, 7376

Abrevaya, X. C. 2013, *BAAA*, 56, 113

Ayres, T. R. 1997, *J. Geophys. Res.*, 102, 1641

Buccino A. P., Lemarchand G. A., & Mauas P. J. D. 2007, *Icarus*, 192, 582

Buccino A. P., Lemarchand G. A., & Mauas P. J. D. 2006, *Icarus*, 183, 491

Chyba, C. F. & C. Sagan 1992, *Nature*, 355, 125

Cnossen I, Sanz-Forcada J, Favata F, *et al.* 2007, *J. Geophys. Res.*, 112, E02008

Cockell, C. S. 1998, *J. theor. Biol.*, 193, 717

Cockell, C. S. 1999, *Icarus*, 141, 399

Cockell, C. S. 2000, *Icarus*, 30, 467

Cuntz, M., Guinan, E. F., Kurucz, R. L. 2010, *Proceedings IAU Symposium*, 264, 1

DesMarais D. J., Nuth J. A., Allamandola L. J., *et al.* 2008, *Astrobiology* 8, 715

Estrela, R. & Valio, A. 2018, arXiv:1708.05400

García-Sage, K., Glocer, A., Drake, J. J., *et al.* 2017, *ApJL* 844, L13

Gershberg, R. E. 2005, *In: Solar-Type Activity in Main-Sequence Stars, Elsevier*, 494 pp

Hanslmeier, A. 2018, *In: Planetary Habitability and Stellar Activity. World Scientific Press, Singapore, New York, London*, 264 pp

Horneck, G., Walter, N., Westall, F., *et al.* 2016, *Astrobiology*, 16, 201

Howard, W. S., Tilley, M. A., Corbett, H., *et al.* 2018, *ApJL* 860, L30

Khodachenko, M. L., Ribas, I., Lammer, H., *et al.* 2007, *Astrobiology* 7, 167

Kielbassa, C., Roza, L., & Epe, B. 1997, *Carcinogenesis*, 18, 811

O'Malley-James, J. T. & Kaltenegger, L. 2017, *MNRAS Lett.*, 469, L26

Patel, M. R., Zarnecki, J. C., & Catling, D. C. 2002, Planet. *Space Sci.*, 50, 915
Pestunova, O., Simonov, A., Snytnikov, V. *et al.* 2005, *Adv. Space Res.* 36, 214
Ranjan, S., Wordsworth, R., & Sasselov, D. D. 2017, *ApJ* 843, 110
Ribas, I., Porto de Mello, G. F., Ferreira, L. D., *et al.* 2010, *ApJ*, 714, 384
Rugheimer, S. A., Segura, & L. Kaltenegger 2015, *ApJ* 806, 137
Sagan, C. & Khare, B. 1971, *Science*, 173, 417
Scalo, J., Kaltenegger, L., Segura, A. *et al.* 2007, *Astrobiology* 7, 85
Wynn-Williams, D. D., Edwards, H. G. M., Newton, E. M., *et al.* 2002, *Int. J. Astrobiol.* 1, 39

Discussion

ARTUR DE LA VILLARMOIS: Do you think that the presence of an atmosphere will change your results?

ABREVAYA: Yes, in the sense of the fluency rate and wavelengths of UVR that are reaching the surface, and therefore, the amount of UVR that the microorganisms will receive. This depends on the atmospheric opacity, that's why we consider the atmospheric features in the radiative transfer modelling, in the first step of our approach.

JOHNSTONE: Can you extend the experiment to see if the survivors repopulate and are then more resistant to the radiation?

ABREVAYA: Yes, this will be part of our future experiments.

BENZ: Superflares are very rare. Ten times smaller flares are about 100 times more frequent and a hundred times smaller flares are 10^4 times more frequent. How will this change your results?

ABREVAYA: We will consider that in our experiments, this is another very important factor to be analyzed. We need to do more experiments to answer this question.

Origins: From the Protosun to the First Steps of Life
Proceedings IAU Symposium No. 345, 2018
B. G. Elmegreen, L. Viktor Tóth & M. Güdel, eds.

doi:10.1017/S1743921318008219

Dynamical environments of MU69: a state of chaotic clearing

José Lages[1] **and Ivan I. Shevchenko**[2]

[1]Institut UTINAM, Observatoire des Sciences de l'Univers THETA, CNRS, Université de Bourgogne Franche-Comté, Besançon 25030, France
email: jose.lages@utinam.cnrs.fr

[2]Institute of Applied Astronomy, Russian Academy of Sciences, Saint Petersburg 191187, Russia
email: ivan.i.shevchenko@gmail.com

Abstract. The second (after Pluto) plausible target object for the New Horizons mission is 2014 MU69. It is a classical TNO, a primordial contact binary. Identifying any material in the vicinities of a target object is of an especial concern for planning cosmic fly-byes, as it is hazardous for a space probe. Luckily, no such material has been reported for MU69 up to now. The point of our report is that this lucky absence is just a dynamical consequence of the physical nature of MU69. Spinning gravitating dumbbells create zones of dynamical chaos around them, and this has a *clearing effect*: any material put in orbits around a rotating dumbbell (e.g., any material ejected from its surface) cannot be long-lived in such zones; it either escapes into space, or returns to the parent body's surface. As the orbiting matter is removed in this way, a spinning gravitating dumbbell clears its vicinities. We show that MU69 is able to create such a clearing, making itself a safe and hospitable target for a space mission. Therefore, the guest probe is expected to be safe on arrival.

Keywords. Celestial mechanics, minor planets, asteroids, Kuiper Belt.

1. Introduction

After a number of adventures, including accomplishment of a HST survey, the second (after Pluto) plausible target object for the New Horizons mission was successfully discovered in 2014, named 2014 MU69. Later on, it has turned out to be a classical TNO, a primordial contact binary with lobe sizes of 20 and 18 km (Stern 2017; Parker *et al.* 2017). Such a dumbbell shape is typical for TNOs. Lightcurve observations are under way, aimed to retrieve the rotation period (Benecchi 2016). Identifying any material in the vicinities of a target object is of an especial concern for planning cosmic fly-byes, as it is hazardous for a space probe. Luckily, no such material has been reported for MU69 in a meticulous survey (Kammer *et al.* 2017).

2. Chaotic clearing

The point of our report is that this lucky absence is just a dynamical consequence of the physical nature of MU69. Spinning gravitating dumbbells create zones of dynamical chaos around them (Lages, Shepelyansky & Shevchenko 2017), and this has a *clearing effect*: any material put in orbits around a rotating dumbbell (e.g., any material ejected from the surface) cannot be long-lived in this zone; it either escapes into space, or returns to the parent body's surface (Lages, Shevchenko & Rollin 2018). As the orbiting matter is removed in this way, a spinning gravitating dumbbell clears its vicinities. A well-known example of analogous "cleansing" is the formation of the Wisdom gap in the co-orbital

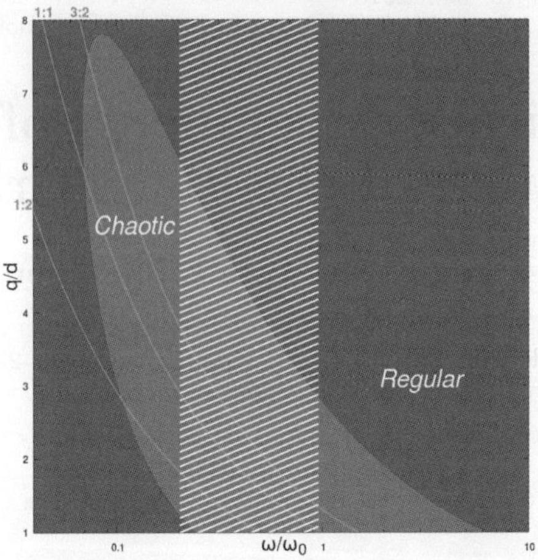

Figure 1. Extents of the chaotic zone around a contact binary as a function of the binary's rotation rate ω. Red (blue) colour indicates chaos (stability). White shaded area delimits the range of typical rotation rates of the Kuiper belt objects, according to Thirouin *et al.* (2014).

neighbourhood of a planet. One of the conditions in the definition of a planet implies the co-orbital cleansing (IAU General Assembly 2006).

The Wisdom chaotic gap is formed by the overlap of the accumulating first-order mean-motion resonances in the co-orbital neighbourhood of a planet; whereas the circum-dumbbell chaotic zone is formed by the overlap of the accumulating integer and half-integer spin-orbit resonances with the rotating dumbbell (Lages, Shevchenko & Rollin 2018). In the both cases, any material injected in the chaotic zones is subject to an unlimited chaotic diffusion in the eccentricity, to be removed, finally, out of these areas. In Figure 1, the extents of the chaotic zone, where the clearing takes place, are represented graphically in the "dumbbell rotation rate – particle's pericentric distance" diagram. The rotation rate is in the units of its critical value, equal to the contact binary's Keplerian rate of rotation, corresponding to disintegration of the binary. The pericentric distance is in units of the binary's size, defined as the distance between the mass centers of its components. The form and size of the chaotic domain depends on the binary's mass ratio only weakly, as shown in Lages, Shevchenko & Rollin (2018); therefore, using the parameters of MU69 obtained in any future observations, the extents of the chaotic clearing can be easily evaluated.

3. Conclusions

Tantalizingly, the chaotic-clearing phenomenon affects both targets of the New Horizons mission, but in different ways: Pluto is not able to cleanse any radial neighbourhood of its orbit, and on this reason it was deprived of the planetary status (IAU General Assembly 2006); conversely, MU69 is able to create a clearing, but of another (circum-body) kind, making itself a safe and hospitable target for a space mission. The rendezvous will take place on 31 December 2018 or 1 January 2019. The object of destination, as we point out, chaotically cleanses its environments; therefore, the guest probe is expected to be safe on arrival.

Acknowledgements

I.I.S. was supported by the Russian Science Foundation (project No. 19-12-11010).

References

Benecchi, S. 2016, "The lightcurve of New Horizons encounter TNO 2014 MU69." HST Proposal id. 14627. Available at: `http://archive.stsci.edu/`

Kammer, J., *et al.* 2017, "Probing the Hill sphere of 2014 MU69 with HST FGS." 49th Meeting of the AAS Division for Planetary Sciences. Abstract 216.04

Lages, J., Shepelyansky, D. L., & Shevchenko, I. I. 2017, *AJ*, 153, 272

Lages, J., Shevchenko, I. I., & Rollin, G. 2018, *Icarus*, 307, 391

Parker, A. H., *et al.* 2017, "Multiplicity of the New Horizons extended mission target (486958) 2014 MU69." 49th Meeting of the AAS Division for Planetary Sciences. Abstract 504.04

Stern, A. 2017, "The PI's Perspective: The Heroes of the DSN and the 'Summer of MU69'." Available at: `http://pluto.jhuapl.edu/News-Center/`

Thirouin, A., Noll, K. S., Ortiz, J. L., & Morales, N. 2014, *A&A*, 569, A3

IAU 2006 General Assembly: Result of the IAU Resolution votes 2006, IAU0603 – Press Release. Available at: `http://www.iau.org/news/pressreleases/detail/iau0603/`

Origins: From the Protosun to the First Steps of Life
Proceedings IAU Symposium No. 345, 2018
B. G. Elmegreen, L. Viktor Tóth & M. Güdel, eds.

doi:10.1017/S1743921318008220

Resonant multi-lane patterns in circumbinary young debris disks

Tatiana V. Demidova and Ivan I. Shevchenko

Pulkovo Observatory, Russian Academy of Sciences,
196140 Saint Petersburg, Russia
email: proxima1@list.ru

Abstract. Formation of resonant multi-lane patterns in circumbinary young debris disks with planets is considered in a set of representative massively simulated models. We find that the long term-stable resonant patterns are generically formed, shepherded by embedded planets. The patterns are multi-lane, i.e., they consist of several concentric rings. Statistical dependences of their parameters on the planetary parameters are recovered. Relevant additional massive simulations of planetesimal disks in systems with parameters of Kepler-16, 34, and 35 are accomplished and described. We find that co-orbital patterns generically form in systems with moderate orbital eccentricities of the binary's and planetary orbits (like in Kepler-16 and 35 cases).

Keywords. Methods: n-body simulations, planetary systems: protoplanetary disks.

A massive planet, if present in a planetesimal disk of a single or a binary star, induces the formation of various resonant large-scale patterns in the disk (see, e.g., Demidova & Shevchenko 2016; Tabeshian & Wiegert 2016; and references therein). The most prominent co-orbital pattern is more long-term stable for the systems with central binary stars than for the systems with central single stars (Demidova & Shevchenko 2016, 2018; Demidova 2018). To analyze the properties of the co-orbital patterns, we simulate a grid of models of circumbinary disks and planets in binary star systems. In the models, a planet in a circular orbit is embedded in a debris disk of passively gravitating planetesimals, initially also in circular orbits (co-planar with the planetary orbit) around the host star. The number of the planetesimals is 20000. The time length of each simulation is 10^5 orbital periods of the binary.

Formation of a ring-shaped pattern, co-orbital with the planet, in the debris disk, is a well-known phenomenon. The pattern consists of planetesimals moving in the "tadpole" and "horse-shoe" orbits. For circumbinary systems, its properties were studied in Demidova & Shevchenko (2018) and Demidova (2018). To characterize the pattern, we compute the average surface density along the radius from the barycenter, using a numerical procedure proposed in Demidova & Shevchenko (2018). The local surface density as a function of the orbital period always has a pronounced maximum at the 1:1 mean-motion resonance and pronounced minima at the 2:1 and 1:2 mean-motion resonances with the planet (Fig. 1). The co-orbital ring pattern corresponds to the 1:1 resonance.

Let a_p denote the planet's orbital radius, and μ_p denote the ratio of the planet's mass and the sum of masses of the central binary and the planet. Using a numerical procedure of Morrison & Malhotra (2015), we estimate the interior and exterior borders of the cleared zone as $\Delta a_\mathrm{cl,in} \approx 2.08\mu_\mathrm{p}^{0.35} a_\mathrm{p}$ and $\Delta a_\mathrm{cl,ext} \approx 1.71\mu_\mathrm{p}^{0.33} a_\mathrm{p}$. The clearing time turns out to be given by $T_\mathrm{cl} \approx 0.0088\mu_\mathrm{p}^{-1.56}$, in units of the orbital period of the central binary. The derived formulas for $\Delta a_\mathrm{cl,ext}$ and T_cl are quite similar to those given in Morrison & Malhotra (2015), and the formula $\Delta a_\mathrm{cl,in}$ has somewhat greater values of the power-law index and the pre-function coefficient.

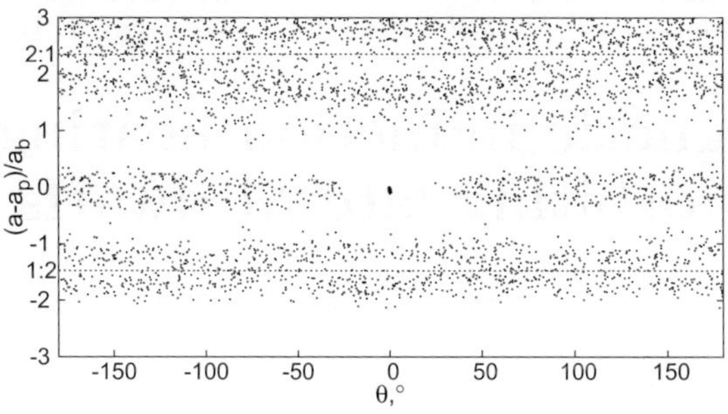

Figure 1. An evolved orbital distribution of planetesimals near the planetary orbit. Here stellar masses are $M_1 = M_2 = M_\odot$, and the planet mass is $m_\mathrm{p} = 3M_\mathrm{J}$, where $M_\odot$ and M_J are the Solar and Jovian masses. The locations of the 2:1 and 1:2 mean-motion resonances with the planet are dashed. a_b is the size of the binary, a_p is the radius of the planetary orbit, and a are the current radial positions of the planetesimals; θ are the current relative angular positions of the planetesimals, as viewed from the barycenter.

Our numerical simulations in models with the parameters of the Kepler-16, 34, and 35 systems (as given in Doyle *et al.* 2011 and Welsh *et al.* 2012) show that the formation of a stable ring co-orbital with the planet is plausible for the Kepler-16 and 35 systems. Concerning Kepler-34, the large orbital eccentricities of the stellar binary and the planet are able to prevent the formation of the pattern. In Morrison & Malhotra (2015), the clearing time of the co-orbital chaotic zone is estimated as $T_\mathrm{cl} \approx 0.024\mu^{-3/2}$, in units of the planetary period. For Kepler-34, in our simulations we find $T_\mathrm{cl} \approx 10^4$, in accord with this formula.

Observations of stellar debris disks often reveal rather sharp contrast edges of the resonant ring structures; see, e.g., Carrasco-González *et al.* (2016). We note that in planetary rings, quite analogously, the edges of annular patterns are often revealed to be extremely contrast (Borderies *et al.* 1982; Fridman & Gorkavyi 1999; Shepelyansky *et al.* 2009), and this can be explained by synchronization processes in the presence of resonant shepherding bodies (Shepelyansky *et al.* 2009). Other contrast-boosting mechanisms can also be active (see Shepelyansky *et al.* 2009 and references therein). The processes that may boost the contrast of pattern borders should be taken into account in simulations of disk structures; we leave it for a future work.

Acknowledgements

This work was supported in part by the Russian Foundation for Basic Research (Project no. 17-02-00028).

References

Borderies, N., Goldreich, P., & Tremaine, S. 1982, *Nature*, 299, 209

Carrasco-González, C., Henning, Th., Chandler, C. J., *et al.* 2016, *ApJ* (Letters), 821, L16

Demidova, T. V. 2018, *Solar System Research*, 52, 180

Demidova, T. V., & Shevchenko, I. I. 2016, *MNRAS* (Letters), 463, L22

Demidova, T. V., & Shevchenko, I. I. 2018, *Astron. Lett*, 44, 119

Doyle, L. R., Carter, J. A., Fabrycky, D. C., *et al.* 2011, *Science*, 333, 1602

Fridman, A. M., & Gorkavyi, N. N., 1999 *Physics of Planetary Rings. Springer, Berlin*

Morrison, S., & Malhotra, R. 2015, *ApJ*, 799, 41

Shepelyansky, D. L., Pikovsky, A. S., Schmidt, J., & Spahn, F. 2009, *MNRAS*, 395, 1934

Tabeshian, M., & Wiegert, P. A. 2016, *ApJ*, 818, 159

Welsh, W. F., Orosz, J. A., & Carter, J. A. 2012, *Nature*, 481, 475

Origins: From the Protosun to the First Steps of Life
Proceedings IAU Symposium No. 345, 2018
B. G. Elmegreen, L. Viktor Tóth & M. Güdel, eds.

doi:10.1017/S1743921318008232

Electromagnetic induction heating of planets orbiting late M dwarfs

Kristina G. Kislyakova[1,2]

[1]Department of Astrophysics, University of Vienna,
Türkenschanzstrasse 17, A-1180, Vienna, Austria
email: kristina.kislyakova@univie.ac.at

[2]Space Research Institute, Austrian Academy of Sciences,
Schmiedlstrasse 6, A-8042 Graz, Austria

Abstract. We propose induction heating of planetary interiors as an energy source in the planetary mantles. Induction heating arises when a changing magnetic field induces currents in a conducting planetary mantle which then dissipate to heat the planet, mostly within an upper layer called the skin depth. This physical process can play a role in planetary interiors around strongly magnetized stars such as low mass M dwarfs with kG magnetic fields, which are common among these stars.

Keywords. planets and satellites: general, stars: magnetic fields, stars: late-type

1. Introduction

Internal energy sources play a big role in the evolution of exoplanets, their interiors and habitability. Important energy sources include: radioactive decay, mantle differentiation, core and inner core formation, and tidal heating. We show that planets orbiting magnetically active M stars are also subject to induction heating, which can similarly to tidal heating change the temperature profile of the planetary mantle and sometimes even melt it, and lead to increased volcanic activity. For induction heating to be substantial, the planet has to experience constant variations of the stellar magnetic field along its orbit. Fig. 1 illustrates how large scale variable magnetic field can be generated around the planet if the host star has a dipole dominated magnetic field.

2. Model and results

We consider a planet to be a sphere made up of concentric layers with uniform conductivity (Kislyakova *et al.* 2017). We solve the induction equation in every layer and calculate the energy dissipation. Late M dwarfs have measured magnetic fields which often exceed some kilogauss (Shulyak *et al.* 2017), which make electromagnetic induction effects very important for planets orbiting them. We use the interior model CHIC (Noack *et al.* 2016) to model magmatic effects and estimate the increase of volcanic outgassing due to an additional heating source.

Fig. 2 shows the energy release calculated for an Earth-mass, Earth-sized planet orbiting WX UMa, a very active late M dwarf with a mass of 0.1 $M_\odot$ and a dipole-dominated magnetic field of 7.3 kG (Shulyak *et al.* 2017). For inclinations higher than ≈30° induction heating is very efficient for a variety of orbital distances including the habitable zone (HZ) of this star. Strong heating of planetary interiors can lead to formation of magma oceans beneath the planetary surface (Kislyakova *et al.* 2017). The middle of the HZ of WX UMa is located at 0.046 au.

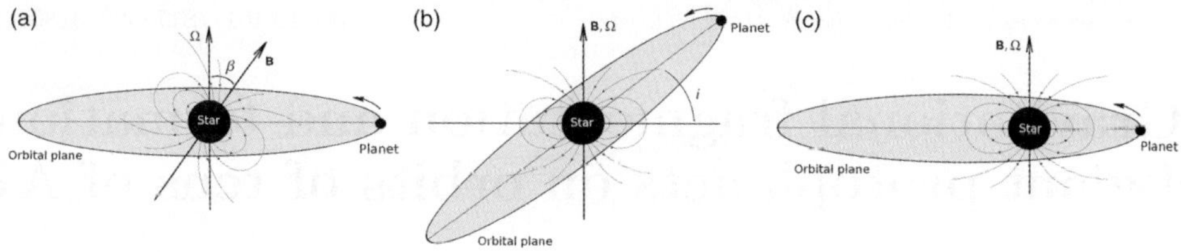

Figure 1. Three possible mechanisms to generate a varying magnetic field along the planetary orbit due to an inclination of the stellar dipole field (a), planetary orbit (b), and orbital eccentricity (c).

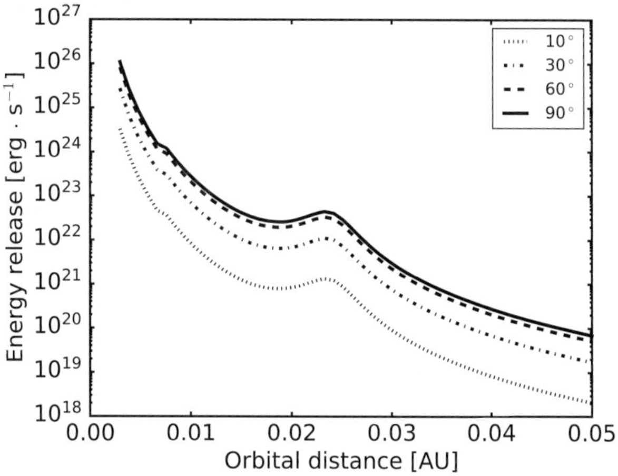

Figure 2. Energy release inside a putative Earth-sized, Earth-mass planet orbiting WX UMa on an inclined orbit (Kislyakova *et al.* 2018). We assumed an Earth-like electrical conductivity profile for the planetary mantle.

3. Implications for planetary evolution and habitability

Volcanoes can eject gases composed mainly of SO_2, CO_2, H_2O, and S_2. The exact composition depends on the redox state and volatile abundances in planetary interiors, and also on the surface pressure (Gaillard *et al.* 2014). On planets orbiting late M dwarfs, volcanic activity would resemble the one on Io, the most volcanically active body in the solar system. Severe volcanic activity can be detrimental for habitable conditions as it can lead to acid rains, ozone layer depletion, and abrupt warming from greenhouse gas emissions. However, excess outgassing can also replenish the atmosphere lost during the early active phase of the evolution of the host star. Our results are presented in more details in (Kislyakova *et al.* 2017, 2018).

References

Gaillard, F., & Scaillet, B. 2014, *EPSL*, 403, 307

Kislyakova, K. G., Noack, L., Johnstone, C. P., *et al.* 2017, *Nature Astronomy*, 1, 878–885

Kislyakova, K. G., Fossati, L., Johnstone, C. P., *et al.* 2018, *ApJ*, 858, 105

Noack, L., Rivoldini, A., & Van Hoolst, T. 2016, *IJASM*, 9(1/2), 66

Shulyak, D., Reiners, A., Engeln, A., *et al.* 2017, *Nature Astronomy*, 1, 0184

Origins: From the Protosun to the First Steps of Life
Proceedings IAU Symposium No. 345, 2018
B. G. Elmegreen, L. Viktor Tóth & M. Güdel, eds.

doi:10.1017/S1743921318008244

Gravitational fragmentation and formation of giant protoplanets on orbits of tens of AU

Vardan G. Elbakyan[1] **and Eduard I. Vorobyov**[1,2]

[1]Research Institute of Physics, Southern Federal University, Roston-on-Don, 344090 Russia
email: vgelbakyan@sfedu.ru

[2]University of Vienna, Department of Astrophysics, Vienna, 1180, Austria
email: eduard.vorobiev@univie.ac.at

Abstract. Migration of dense gaseous clumps that form in young protostellar disks via gravitational fragmentation is investigated to determine the likelihood of giant-planet formation. We show that gaseous clumps that form in the outer regions of the disk (>100 au) through disk fragmentation often migrate toward the central star on timescales from a few thousand to few tens of thousands of years. The tidal mass loss helps the clumps to significantly slow down or even halt their inward migration at a distance of a few tens of AU from the protostar.

Keywords. stars: formation, planetary systems: protoplanetary disks

1. Introduction

The importance of disk fragmentation as a likely mechanism for giant planet and brown dwarf formation has been reinforced by the detection of wide-orbit (from several tens to hundreds of astronomical units) planetary and brown dwarf companions to low-mass stars (e.g., Marois *et al.* 2008; Kalas *et al.* 2008; Lafreniére *et al.* 2010).

Here we investigate the process of clump migration in a gravitationally unstable disk using high-resolution numerical hydrodynamics simulations in the thin disk limit (see Vorobyov & Basu 2015 for details). We focus on the properties of one of the gaseous clumps (IF1) formed in the disk via gravitational fragmentation and investigate the details of its inward migration until it is destroyed by the action of stellar tidal torques.

2. Dynamics of fragment IF1

Figure 1 shows the gas surface density distribution in and around IF1 during its inward migration towards the protostar. Three different time instances are shown and every panel has progressively smaller spacial scales to better resolve the fragment.

As IF1 approaches the star, its size shrinks because of the shrinking Hill radius and part of its material starts streaming away along the tidal arms. As a result, the fragment loses a large fraction of its initial mass. Finally, the rotational motion inside IF1 almost disappears ($t = 3.39$ kyr), turning into a sheer outflow. At this stage, IF1 ceases to be gravitationally supported and tidal torques tear apart the fragment.†

Figure 2 presents the mass, radial distance, and central temperature of IF1 as a function of time. The vertical dashed line indicates the time instance ($t = 1.8$ kyr) when the second collapse to planetary densities is supposed to take place (not modelled in our simulation). We found that if IF1 had collapsed at $t = 1.8$ kyr, it would have formed a protoplanet

† The animation of clump inward migration can be found at http://www.astro.sfedu.ru/animations/accretion.mp4

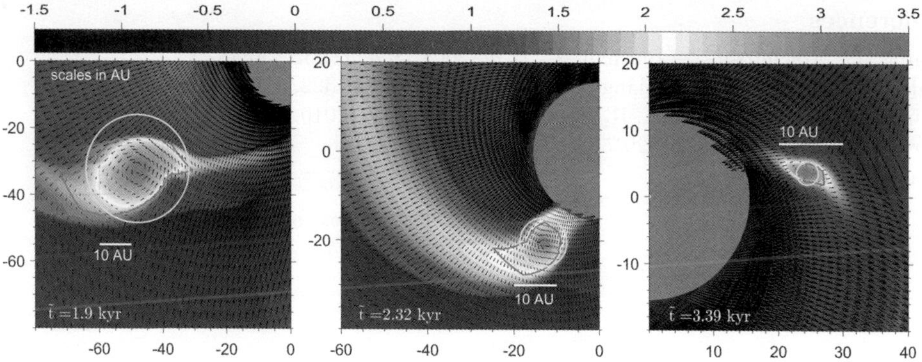

Figure 1. Zoomed-in view on IF1 during its inward migration towards the protostar. The black arrows show the gas velocity field superimposed on the gas surface density distribution. The shrinking spatial scales are used to better resolve the fragment. The yellow circles mark the Hill radius of the fragment and the red curves outline the fragment shape. The color bar shows the gas surface density in log g cm^{-2}.

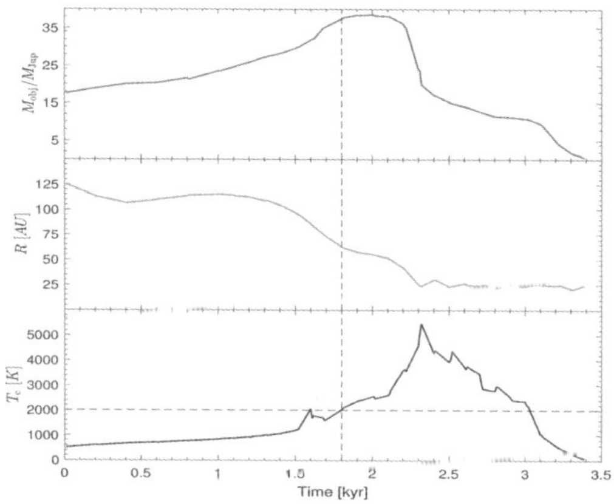

Figure 2. Mass of IF1 (top panel), radial distance of IF1 (middle panel), and central temperature of IF1 (bottom panel), all vs. time. The horizontal dashed line shows a threshold temperature of 2000 K, above which molecular hydrogen dissociates and the second collapse is supposed to ensure. The vertical dashed line shows the time instance when the central temperature of IF1 reaches the threshold value.

with a mass of 0.81 $M_{\rm Jup}$. We expect that during the subsequent inward migration, the protoplanet may survive and settle at an orbit of approximately a few tens of AU.

3. Conclusions

Our findings can be summarized as follows:

• Gaseous clumps that form in the outer disk regions often migrate toward the central star and lose most of their diffuse envelopes through tidal torques.

• The tidal mass loss helps the clumps to significantly slow down or even halt their inward migration at a distance of a few tens of AU from the protostar.

• Central part of the clumps may further experience the second collapse down to planetary densities through the dissociation of molecular hydrogen at T >2000 K.†

† More details can be found in Vorobyov & Elbakyan, 2018, A&A, 618, A7.
This work was supported by the Russian Science Foundation grant 17-12-01168.

References

Marois, C., Macintosh, B., Barman, T., *et al.* 2008, *Science*, 322, 1348
Kalas, P., Graham, J. R., Chiang, E., *et al.* 2008, *Science*, 322, 1345
Lafreniére, D., Jayawardhana, R. & van Kerkwijk, M. H. 2010, *ApJ*, 719, 497
Vorobyov, E. I., & Basu, S. 2015, *ApJ*, 805, 115

Origins: From the Protosun to the First Steps of Life
Proceedings IAU Symposium No. 345, 2018
B. G. Elmegreen, L. Viktor Tóth & M. Güdel, eds.

doi:10.1017/S1743921318008256

Pebble accretion onto planets in turbulent discs

Wilhelm Kley[1], Giovanni Picogna[2] and Moritz H. R. Stoll[1]

[1]Institut für Astronomie & Astrophysik, Universität Tübingen, Morgenstelle 10, D-72076 Tübingen, Germany
email: wilhelm.kley@uni-tuebingen.de

[2]Universitätssternwarte, Universität München, Scheinerstr.1, D-81679 München, Germany
email: picogna@usm.lmu.de

Abstract. Planets form in protoplanetary accretion discs around young protostars. These discs are driven by internal turbulence and the gas flow is not laminar but has stochastic components. For weakly ionised discs the turbulence can be generated purely hydrodynamically through the vertical shear instability (VSI). Embedded particles (dust/pebbles) experience a hydrodynamic drag and drift inward radially and are stirred up vertically by the turbulent motion of the disc. We study the accretion of particles onto a forming planet embedded in a VSI turbulent protoplanetary disc through a series of 3D hydrodynamical simulations for locally isothermal discs with embedded planets in the mass range from 5 to 100 Earth masses ($M_\oplus$).

Keywords. hydrodynamics; planetary systems: formation, protoplanetary disc

1. Particles in the disc

To calculate the dynamics of the disc we perform hydrodynamical simulations of a 3D section of a disc in spherical coordinates, using the PLUTO code. The disc is initially in an equilibrium state which is unstable to the VSI, generating turbulent motions for which we measure an effective $\alpha_\mathrm{d} = 5 \cdot 10^{-4}$. Particles of 10 different sizes (or stopping times) are embedded into the flow (10^5/bin) under the action of aerodynamic drag forces (SK16). These particles are stirred up by the VSI motions as displayed in Fig. 1, where we compare our results (crosses) to other models (circles).

2. Accretion of particles by embedded planets

In a second step we add planets of different masses to the simulations and monitor the accretion of the particles onto them. First we determined the *accretion efficiency* which is the number of accreted particles onto the planet divided by the number of particles that would otherwise drift across the location of the planet in an unperturbed disc, $P_{eff} = \dot{M}_\mathrm{acc}/\dot{M}_{drift}$, and second the *isolation mass*, which is the critical planet mass above which particle accretion is greatly reduced. Fig. 2 shows that the number of accreted particles is highest around unit stopping time, even though most of the particles drift past the planet in this size range. The isolation mass lies between 10 and 20 $M_\oplus$, above this planet mass the number of accreted particles around $\tau_\mathrm{s} \approx 1$ is essentially shut off. Due to the concentration of particles near the midplane (see Fig. 1) the results for the VSI and laminar disc models are similar in our case.

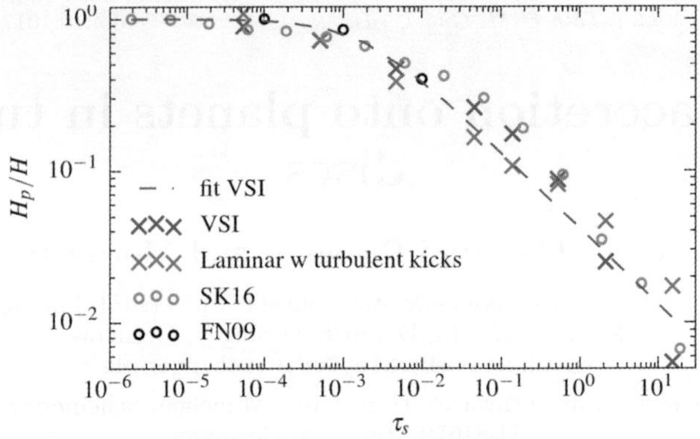

Figure 1. Vertical distribution (particle over gas scale height) of particles with different Stokes numbers τ_s. Compared are VSI and laminar (using α_d and stochastic particle kicks) models. Results of MHD (FN09) and previous VSI (SK16) runs are superimposed (adapted from PSK18).

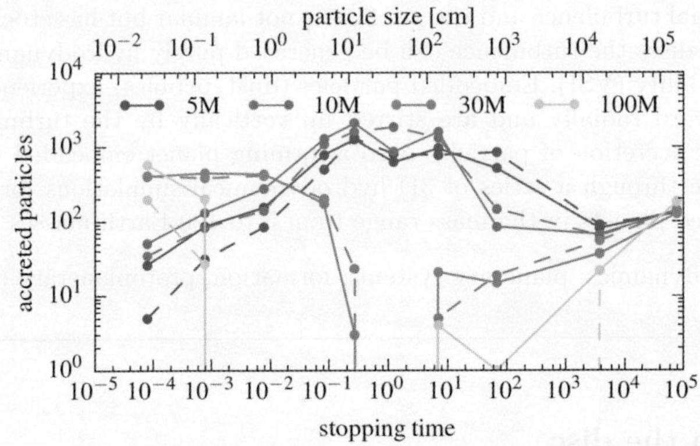

Figure 2. Accreted particles (integrated over 50 planetary orbits) as a function of the particle stopping time with the approximate size on the top x-axis (assuming a planet location at 5au). The VSI (solid line) and α-disc model (dashed line) are compared for different planet masses (in $M_\odot$). Adapted from PSK18.

3. Summary

The angular momentum transport generated by the VSI turbulence corresponds to $\alpha_d = 5 \cdot 10^{-4}$. The particle dynamics and accretion properties are similar for particles in the VSI turbulent disc and those in laminar case with stochastic kicks. For the small mass planets (i.e. $5 - 10 M_\oplus$), well-coupled particles with $\tau_s = 1$, which have a size of about one meter at this planet location, we find an accretion efficiency $P_{\rm eff} \approx 1.6 - 3\%$, a result which compares well to LJ12. However, the fast inward drift for $\tau_s = 1$ particles makes them the most effective for rapid growth, leading to mass doubling times of about 20,000 yr provided there is a steady influx of pebbles. For masses between 10 and 30 $M_\oplus$ the core reaches the pebble isolation mass and the particles are trapped at the pressure maximum just outside of the planet, shutting off further particle accretion.

References

Fromang, S. & Nelson, R. P., 2009, *A&A*, 496, 597 (FN09)
Lambrechts, M. & Johansen, A., 2012, *A&A*, 572, A107 (LJ12)
Picogna, G., Stoll, M. H. R. & Kley, W., 2018, *A&A*, 616, A116 (PSK18)
Stoll, M.H.R. & Kley, W., 2016, *A&A*, 594, A57 (SK16)

Origins: From the Protosun to the First Steps of Life
Proceedings IAU Symposium No. 345, 2018
B. G. Elmegreen, L. Viktor Tóth & M. Güdel, eds.

doi:10.1017/S174392131800827X

Circumstellar disk fragmentation and the origin of massive planetary companions, brown dwarfs, and very low-mass stars

M. B. N. Kouwenhoven[1], Yun Li[2], D. Stamatellos[3] and S. P. Goodwin[4]

[1]Department of Mathematical Sciences, Xi'an Jiaotong-Liverpool University, 111 Ren'ai Rd., Suzhou Dushu Lake Science and Education Innovation District, Suzhou Industrial Park, Suzhou 215123, P.R. China
email: t.kouwenhoven@xjtlu.edu.cn

[2]Department of Astronomy, School of Physics and Astronomy, Shanghai Jiao Tong University, Shanghai 200240, P.R. China

[3]Jeremiah Horrocks Institute for Mathematics, Physics & Astronomy, University of Central Lancashire, Preston PR1 2HE, UK

[4]Department of Physics and Astronomy, University of Sheffield, Sheffield S3 7RH, UK

Abstract. The low-mass end of the initial mass function remains poorly understood. In this mass range, very low-mass stars, brown dwarfs, and massive planets are able to form through a variety of physical processes. Here, we study the long-term evolution of disk-fragmented systems around low-mass stars, for the epoch up to 10 Myr (the typical lifetime of an embedded cluster) and up to 10 Gyr (the age of the Milky Way). We carry out N-body simulations to study the decay of disk-fragmented systems and the resulting end products. Our simulations indicate rapid decay and frequent physical collisions during the first 10 Myr. We find that disk fragmentation provides a viable mechanism for explaining hierarchical triple systems, the brown dwarf desert, single and binary brown dwarfs, and very low-mass binary systems in the solar neighbourhood.

Keywords. Stars: formation, stars: low mass, stars: brown dwarfs, galaxy: solar neighbourhood, planets and satellites: general

1. Introduction

Brown dwarfs, very low-mass stars, and very massive planets (hereafter, low-mass objects; LMOs) are among the most abundant objects in the Milky Way. Their faint luminosities make it challenging to study physical and kinematic properties of even the nearest neighbours of the Sun. These LMOs cover an interesting transition region between (or including) the highest-mass planets that can be formed through core-accretion and the lowest-mass stars that can form through core collapse in dense molecular clouds. Various mechanisms have been proposed to explain the origin of LMOs in the Galactic field, including extrapolations of the above-mentioned mechanisms into the LMO regime, and the mechanism of circumstellar disk fragmentation. The LMOs discovered in our Galaxy likely represent a combination of all these processes. In this work we explore the hypothesis that all LMOs are a result of the disk fragmentation process. The study presented here is described in detail in Li *et al.* (2015) and Li *et al.* (2016).

2. Modelling the outcome of circumstellar disk fragmentation

Stamatellos & Whitworth (2009) carried out an ensemble of smoothed-particle hydrodynamics simulations to study disk fragmentation among $0.7\,M_\odot$ stars with massive circumstellar disks. Their simulations are carried out until most gas is dissipated. We carry out 9000 N-body simulations, for which the initial conditions are statistically identical to the outcome of Stamatellos & Whitworth (2009)'s simulations. We carry out simulations of these few-body systems up to 10 Gyr. The results are analysed in detail for the period up to $t = 10$ Myr, which corresponds to the time in the embedded cluster, and subsequently up to 10 Gyr, roughly the age of the field star population in the Milky Way.

3. Implications for the Galactic field star population

Most of the disk-fragmented systems are highly unstable and decay within a million years, after which the number of merging and escape events reduces exponentially with time, until a roughly stable situation is reached after a Gyr. Despite the large initial number ($3-11$) of companions per system, $\sim 30\%$ of the primary stars end up as single, $\sim 60\%$ with one companion, $\sim 10\%$ with two companions, and a small fraction ($\lesssim 0.1\%$ with more than two companions. Most companions escape, while $20-25\%$ of the host stars experience physical collisions with companions. Planetary-mass and brown-dwarf companions rarely obtain orbits smaller than 50 au, while low-mass stars do appear in this regime. Systems with more than one companions remaining show both planetary-type and hierarchical configurations, with period ratios in the range $10-10^4$ and $10-10^3$, respectively. The hierarchical systems (with companions orbiting a common center of mass in orbit around the host star) are often highly inclined or retrograde. Ejected companions typically have velocities below $5\,\mathrm{km\,s^{-1}}$. Among these, $1.5-5\%$ escapes as low-mass binary systems with semi-major axes of $10-50$ au that often contain one or two brown dwarfs.

Our results contribute to explaining the properties of the stellar population in the solar neighbourhood. Among the 77 known objects within 5 pc of our Sun, 88% are of stellar type late-K or M or brown dwarfs. Among these 77 objects, 40 are single, 13 are in binaries, and 4 are in triple systems. The vast majority of these binaries and multiples consists of, or contains, very low-mass stars and brown dwarfs. Notable examples include GJ 1245 and EZ Aquari, which are both very low-mass triples, and Epsilon Indi, which consists of a main-sequence star with a double low-mass companion. Other examples of such hierarchical double-companion systems include HIP68532 and HIP69113 (Kouwenhoven *et al.* 2005, 2007). Such types of systems are regularly observed in our simulations, but are difficult to form through conventional (core-collapse) star formation processes. The challenge of explaining the origin of the stellar and binary population still exists, but disk fragmentation provides a viable explanation for the abundance, binary/triple properties, and kinematics of very low-mass stars and brown dwarfs in the solar neighbourhood.

Acknowledgements

M.B.N.K. acknowledges support from the National Natural Science Foundation of China (grant 11573004). This research was supported by the Research Development Fund (grant RDF-16-01-16) and the Research Conference Fund of Xi'an Jiaotong-Liverpool University (XJTLU).

References

Kouwenhoven, M. B. N., Brown, A. G. A., Zinnecker, H., Kaper, L., & Portegies Zwart, S. F. 2005, *A&A*, 430, 137

Kouwenhoven, M. B. N., Brown, A. G. A., Portegies Zwart, S. F., & Kaper, L. 2007, *A&A*, 474, 77

Li, Y., Kouwenhoven, M. B. N., Stamatellos, D., & Goodwin, S. P. 2015, *ApJ*, 805, 116

Li, Y., Kouwenhoven, M. B. N., Stamatellos, D., & Goodwin, S. P. 2016, *ApJ*, 831, 166

Stamatellos, D., & Whitworth, A. P. 2009, *MNRAS*, 392, 413

Origins: From the Protosun to the First Steps of Life
Proceedings IAU Symposium No. 345, 2018
B. G. Elmegreen, L. Viktor Tóth & M. Güdel, eds.

doi:10.1017/S1743921318008281

Planet-disk interactions in HD 169142? Tracing ellipticity, structures, and offsets

Gesa H.-M. Bertrang and Henning Avenhaus

Max Planck Institute for Astronomy, Königstuhl 17,
69117 Heidelberg, Germany
email: bertrang@mpia.de; avenhaus@mpia.de

Abstract. In Bertrang *et al.* (2018), we present new data of the protoplanetary disk surrounding the Herbig Ae/Be star HD 169142 obtained in the very broad-band (VBB) with the Zurich imaging polarimeter (ZIMPOL), a subsystem of the Spectro-Polarimetric High-contrast Exoplanet REsearch instrument (SPHERE) at the Very Large Telescope (VLT). Our Polarimetric Differential Imaging (PDI) observations probe the disk as close as 0.″03 (3.5 au) to the star and are able to trace the disk out to ~1.″08 (~ 126 au). We find an inner hole, a bright ring bearing substructures around 0.″18 (21 au), and an elliptically shaped gap stretching from 0.″25 to 0.″47 (29 – 55 au). Outside of 0.″47, the surface brightness drops off, discontinued only by a narrow annular brightness minimum at ~ 0.″63 – 0.″74 (74 – 87 au). These observations confirm features found in less-well resolved data as well as reveal yet undetected indications for planet-disk interactions, such as small-scale structures, star-disk offsets, and potentially moving shadows.

Keywords. (stars:) planetary systems: protoplanetary disks, hydrodynamics, polarization, radiative transfer, scattering, methods: numerical, techniques: high angular resolution, techniques: polarimetric, magnetic fields

---

1. Introduction

Circumstellar disks around young stars are the key to understand star and planet formation (e.g., Williams & Cieza 2011). However, due to new high-contrast and high-sensitivity instruments, the study of star and planet formation is undergoing a rapid change with new spectacular observations of structures in protoplanetary disks such as gaps (e.g., Quanz *et al.* 2013), spirals, and sub-structures that appear in the scattered light (e.g., Avenhaus *et al.* 2017, 2018), where observations at long wavelengths reveal signs of dust traps (e.g., Long *et al.* 2018). These recent observations provide a glimpse at a sophisticated picture of disk evolution and dispersal, which in turn regulates the origin of planetary systems. In this work, we present new observations at unprecedented spatial resolution observed as well as a new model of the protoplanetary disks HD 169142.

2. Overview

We present PDI observations on unprecedented spatial resolution of the circumstellar disks around the Herbig Ae/Be star HD 169142 obtained with SPHERE/ZIMPOL (Fig. 1). Our analysis of the data reveal an intrinsically elliptical geometry (Fig. 2), small-scale substructures (Fig. 3), and potentially moving shadows in this disk (Fig. 3). We deepen our analysis by applying radiative transfer simulations and find a new model description for HD 169142: a flat disk. The features we detect in this object are indicative for planet-disk interactions and consistent with a set of 3 planets as shown by our hydrodynamical simulations (Fig. 4).

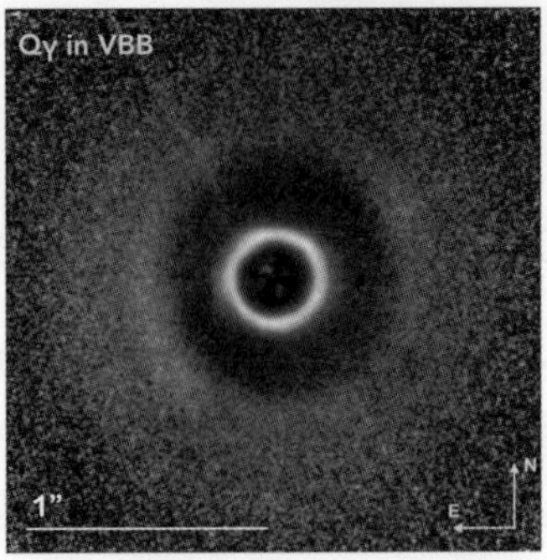

Figure 1. The gap which separates the bright inner ring from the outer disk is of elliptical shape [and not annular, as it was described before in Quanz *et al.* (2013)]. It can be explained by a circular outer disk which is observed under an inclination of 13°. We find that both the inner ring and the outer disk are offset relative to the central star. Our radiative transfer model shows that the residual scattered light detected inside the gap is indeed a consequence of the convolution with the PSF. Our model supports a gap devoid of any dust (Bertrang *et al.* 2018).

3. Implications

Intrinsically elliptical disk geometry.

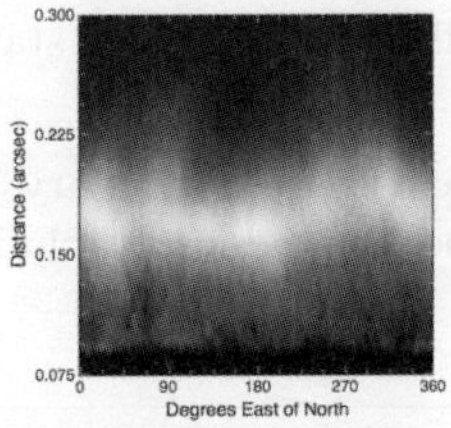

Figure 2. Surface brightness distribution of the inner ring. **Puzzling about the inner ring is the inclination of the disk of 13° - why is the ring appearing as a ring?** In an in-depth analysis, we find that the double-dip structure of the surface brightness distribution strongly indicates an intrinsically elliptical shape of the inner "ring", to our knowledge, the first of its kind - an actual ring would dip only once (Bertrang *et al.* 2018). A later finding in MWC758 (Dong *et al.* 2013) presents another case of this dust geometry which generally points strongly towards planet-disk interactions.

Moving shadows and substructures in the inner ring.

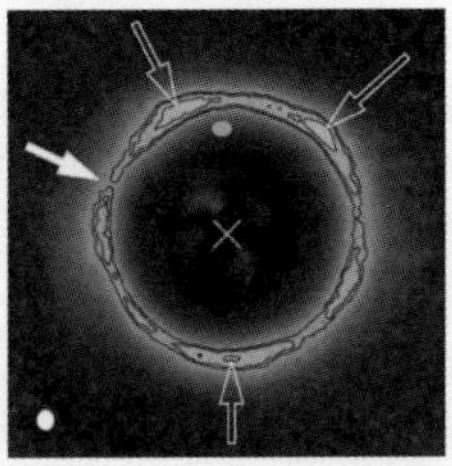

Figure 3. The bright inner ring reveals clear substructures, indicative of interactions with the (proto)planetary candidate (green ellipse). Additionally, observations with SPHERE/ZIMPOL in P95 (Bertrang *et al.* 2018) and with NACO in P89 (Quanz *et al.* 2013) reveal a dip in the surface brightness in the inner ring of HD 169142, at 50° (filled white arrow; P95) and 80° (P89), respectively. If this is the same dip then its angular velocity is ∼ 16 km/s. A likely explanation is a giant planet or a brown dwarf which is surrounded by a small accretion disk, located at 6 au from the central star.

The new model: Flat disk with intrinsically elliptical shape and three planets.

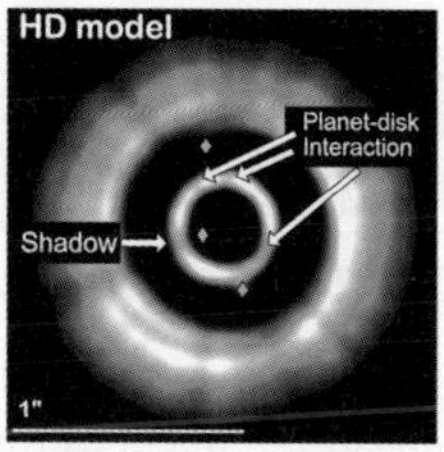

Figure 4. The new model, a flat disk, fits all, the SED, as well as the spatially resolved SPHERE/ZIMPOL and ALMA data very well (Bertrang *et al.* 2018). Our model contains one planet (10 M_J) at 15 au, and two planets (each 1 M_J) at 35 au, resp. 45 au (green diamonds). We find that the innermost planet casts a shadow on the inner ring, while the two outer planets create density bumps in the inner ring due to gravitational interaction. Also the offsets and the intrinsically elliptical shape we find in the data are explained by the planet-disk interactions in our model.

References

Avenhaus, H. *et al.* 2017, *AJ*, 154, 33
Avenhaus, H. *et al.* 2018, *ApJ*, 863, 44
Bertrang, G. H.-M. *et al.* 2018, *MNRAS*, 474, 5105
Dong, R. *et al.* 2018, *ApJ*, 860, 124
Long, F. *et al.* 2018, *ApJ*, in press
Quanz, S. *et al.* 2013, *ApJ*, 766, L2
Williams, J. P. & Cieza, L. 2011, *ARA&A*, 49, 67

Origins: From the Protosun to the First Steps of Life
Proceedings IAU Symposium No. 345, 2018
B. G. Elmegreen, L. Viktor Tóth & M. Güdel, eds.

doi:10.1017/S1743921318008293

Analysis of KOI 2700b, the second exoplanet with a comet-like dusty tail – selected results

Zoltán Garai[1] and Peter Dolinský[2]

[1]Astronomical Institute, Slovak Academy of Sciences,
05960 Tatranská Lomnica, Slovak Republic
email: `zgarai@ta3.sk`
[2]Slovak Central Observatory,
Komárňanská 134, 94701 Hurbanovo, Slovak Republic
email: `peter.dolinsky@suh.sk`

Abstract. The *Kepler* object KOI 2700b was discovered recently as the second exoplanet with a comet-like dusty tail. We aimed at verifying the disintegrating-planet scenario of KOI 2700b by modeling its light curve and to put constraints on various tail and planet properties, as was done in the case of KIC 12557548b. Here, we describe some selected results of our analysis.

Keywords. planets and satellites: general, techniques: photometric, scattering

---

1. Introduction and Data analysis

The *Kepler* object KOI 2700b (KIC 8639908b) was discovered by Rappaport *et al.* (2014) as the second exoplanet with a comet-like dusty tail. It exhibits a distinctly asymmetric transit profile, likely indicative of the emission of dusty effluents and reminiscent of KIC 12557548b (Kepler 1520b), the first exoplanet with a comet-like dusty tail, discovered by Rappaport *et al.* (2012). We aimed at verifying the disintegrating-planet scenario of KOI 2700b by modeling its light curve and to put constraints on various tail and planet properties, as was done in the case of KIC 12557548b by Budaj (2013).

We used the publicly available *Kepler* data from the quarters 1 – 17 in the form of SAP-fluxes. Only the long cadence data were used to construct the light curve. These are 64 842 observations with an exposure time of about 30 min. *Kepler* observations were reduced in a similar manner to in Budaj (2013). We obtained the phase-folded and binned transit light curve of KOI 2700b, which we iteratively modeled using the radiative-transfer code SHELLSPEC, described in Budaj & Richards (2004). During the modeling we applied selected species (γ-alumina, enstatite, forsterite, olivine with 50% magnesium and 50% iron and pyroxene with 40% magnesium and 60% iron) and dust particle sizes (0.01, 0.1 and 1 micron in radii). Mie absorption and scattering on spherical dust grains with realistic dust opacities, phase functions, and finite radius of the source of the scattered light were taken into account.

2. Selected results and Conclusions

We modeled the comet-like tail as part of a ring around the parent star. The geometrical cross-section of the ring is monotonically enlarging from the planet to the end of the ring. In our calculations we assumed a dust tail with a cross-section of 0.05×0.05 $R_\odot$ at the beginning and 0.09×0.09 $R_\odot$ at its end. Since there is still ample room for a contribution to the light curve from the solid body of the planet, we also included it in the model, defined as a dark, non-transparent sphere and located at the beginning of the ring. Our

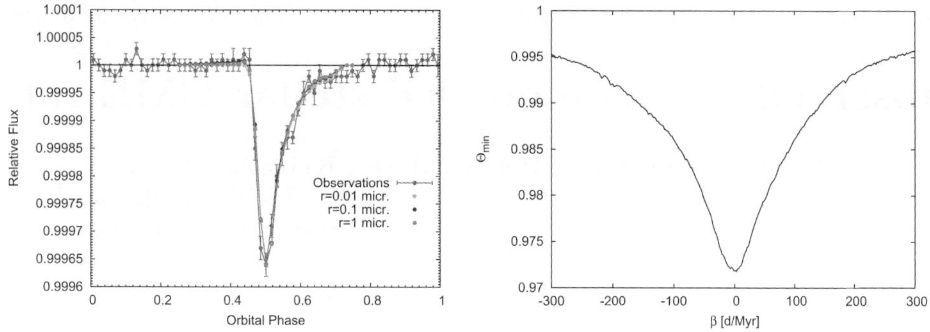

Figure 1. Model light curves calculated for 0.01-micron, 0.1-micron, and 1-micron grains of γ-alumina (left), and search for a long-term orbital period change of KOI 2700b (right).

average fitted value for the dimension of the planet is $R_{\rm p}/R_{\rm s} = 0.008$ ($R_{\rm p} = 0.497$ $\rm R_{\oplus}$, or 3172 km). Since the modified ring model of the tail satisfies the observations well, we confirm the disintegrating-planet scenario of KOI 2700b (Fig. 1).

Furthermore, we derived some interesting features of KOI 2700b and its comet-like tail. As an example, we describe search for a long-term orbital period change of KOI 2700b. Perez-Becker & Chiang (2013) suggest that disintegrating planets undergo negligible orbital period variations due to the evaporation process. During this step we examined this prediction via searching for possible long-term changes of the orbital period of KOI 2700b. For this purpose we used the method of phase dispersion minimization, as per Stellingwerf (1978), application PDM2, version 4.13†, which uses bin structure 50/2. The detrended data were used for this purpose, see Budaj (2013). We assumed that the period changes linearly:

$$P = P_0 + \beta t, \tag{2.1}$$

where β is a dimensionless value, but is often expressed in days/million years (d/Myr). The output from the analysis is a curve (Fig. 1), which shows the dependence of $\Theta_{\rm min}$ as a function of β. The term $\Theta_{\rm min}$ is a dimensionless statistical parameter, see Stellingwerf (1978). The minimum value of $\Theta_{\rm min}$ indicates the speed of the period change – β. We fitted a parabola to this curve and obtained $\beta = 1.3 \pm 3.2$ d/Myr, which means that there is no significant evidence for the long-term orbital period change during the time span of the *Kepler* observations. This result is in agreement with the prediction, presented by Perez-Becker & Chiang (2013). The error was estimated by means of Monte Carlo simulations. We generated and analyzed 100 artificial datasets with the same standard deviation as the original data. For more details and results see Garai (2018).

References

Budaj, J. 2013, *A&A*, 557, 72
Budaj, J., & Richards, M. T. 2004, *Contributions Astron. Obs. Skalnate Pleso*, 34, 167
Garai, Z. 2018, *A&A*, 611, 63
Perez-Becker, D., & Chiang, E. 2013, *MNRAS*, 433, 2294
Rappaport, S., Levine, A., Chiang E., *et al.* 2012, *ApJ*, 752, 1
Rappaport, S., Barclay, T., DeVore, J., *et al.* 2014, *ApJ*, 784, 40
Stellingwerf, R. F. 1978, *ApJ*, 224, 953

† Actual version of the software package is available at `http://www.stellingwerf.com`.

Origins: From the Protosun to the First Steps of Life
Proceedings IAU Symposium No. 345, 2018
B. G. Elmegreen, L. Viktor Tóth & M. Güdel, eds.

doi:10.1017/S174392131800830X

Observed versus modelled stellar CME rates

Martin Leitzinger[1], Petra Odert[1], Krisztian Vida[2], Levente Kriskovics[2], Eike W. Guenther[3], Heidi Korhonen[4], Robert Greimel[1], Arnold Hanslmeier[1] and Helmut Lammer[5]

[1]University of Graz, Institute of Physics/IGAM, NAWI Graz
Universitätsplatz 5, 8010 Graz, Austria
email: martin.leitzinger@uni-graz.at

[2]Konkoly Observatory, MTA CSFK, Konkoly Thege Miklós út 15-17, 1121 Budapest, Hungary

[3]Thüringer Landessternwarte Tautenburg, Karl-Schwarzschild-Observatorium,
Sternwarte 5, 07778 Tautenburg, Germany

[4]Dark Cosmology Centre, Niels Bohr Institute, University of Copenhagen,
Juliane Maries Vej 30, 2100 Copenhagen

[5]Space Research Institute, Austrian Academy of Science, Schmiedlstraße 6, 8042 Graz, Austria

Abstract. Stellar coronal mass ejections (CMEs) may play an important role in stellar and planetary evolution, therefore the knowledge on parameter distributions of this energetic activity phenomenon is highly relevant. During the last years several attempts have been made to detect stellar CMEs of late-type main-sequence and pre main-sequence stars from dedicated optical spectroscopic observations. Up to now only a handful of distinct stellar CME detections are known which contradicts the results from stellar CME modelling, which predict higher CME rates. We report on dedicated ongoing and future observational attempts to detect stellar CMEs and discuss the observational results with respect to the results from stellar CME modelling.

Keywords. stars: activity, stars: late-type, stars: chromospheres

---

To gain knowledge on stellar CME parameters a logical approach is to deduce them from observations. Up to now only a handful of stellar CMEs has been detected. This poses the question if CMEs are a rare phenomenon on stars? The majority of the stellar CME detections from literature have used the signature of stellar material being ejected into the stellar astrosphere, which is Doppler-shifted flux seen either in emission or absorption in the blue wings of Balmer lines, as in many cases the CME core is a filament with a strong signal in the Balmer lines. Moreover, nearly all of the so far known stellar CMEs were detected by chance, except for the study by Guenther & Emerson (1997). Therefore our team initiated search programs for stellar CMEs (Leitzinger *et al.* (2014), Korhonen *et al.* (2017)). The data we analyzed to date did not show signatures of CMEs, but we are continuing with those programs and developed in addition the idea of a flare alert system, which is based on long-term photometric monitoring at Observatory Lustbühel Graz (OLG) of a high number of late-type main-sequence stars. If a flare is detected, an alert is sent to involved observatories doing then the spectroscopic follow up. In this way up to several thousands of stars can be investigated simultaneously.

To estimate the expected occurrence rates of CMEs on other stars, one may extrapolate relations between flares and CMEs from the Sun and combine them with stellar flare rates. This has been done recently by Aarnio *et al.* (2012), Drake *et al.* (2013), and Odert *et al.* (2017). Thereby it is possible to compare the existing stellar CME detections

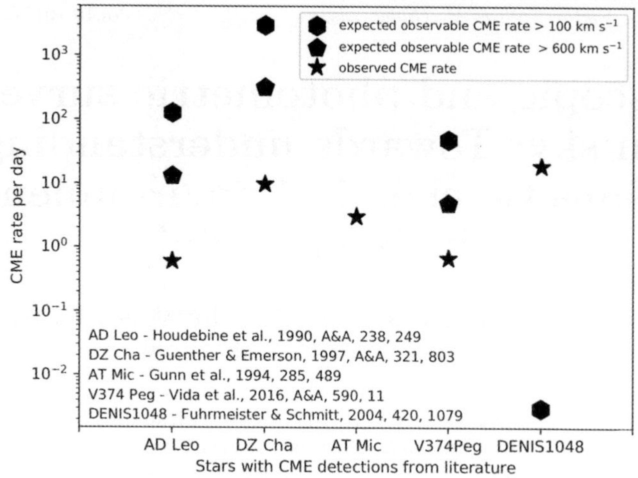

Figure 1. Expected observable (pentagon and hexagon symbols) versus observed CME rates from literature (black star symbols). The pentagon and hexagon symbols correspond to expected observable CME rates for CMEs with >100 km s^{-1} and >600 km s^{-1} ($v_{los} > v_{esc}$), respectively. The expected observable CME rates are calculated with the CME model by Odert *et al.* (2017) and its updated version (Odert *et al.*, in preparation). All expected observable CME rates exceed the observed ones except for the DENIS1048 event. This is mainly caused by the fact that this probable slow stellar CME event was detected in only 1.3 h of total observing time and extrapolation of the model to an M9 star is in this case not representative.

with the expected CME rates. As for now, we know three events from literature which have a high probability to represent indeed CMEs, because the projected velocities of the events were larger than the escape velocity of the corresponding stars. Two further events showed lower velocities, but have the typical signature of stellar CMEs. In Fig. 1 we show the results of this comparison. As one can clearly see the observed CME rates are far below the predicted rates. This might be related to the following reasons: **(a)** *Model ingredients:* Extrapolating the solar flare-CME relations (association rate, CME mass-flare energy relation, CME velocity-flare energy relation) to stars with higher activity levels and different spectral types may lead to overestimations. Observations are needed to confirm if these relations hold or break down for other parameter ranges. **(b)** *Limitations of the observational methodology:* Only the most luminous (i.e. massive) CMEs may be observable in optical spectra as distinct extra-emissions/absorptions. Solar erupting filaments can be seen up to several solar radii - those would be detectable with the Doppler method if the filament is not dissolved too quickly into the wind. The duration of CMEs may thus be too short because of signal dilution during propagation and expansion.**(c)** *Physics:* Not all CMEs may escape on other stars because of a strong overlaying magnetic field (Alvarado-Gomez *et al.* (2018)).

We want to conclude here that besides the reasons given above probably alternative methods need to be developed to study CMEs on stars.

References

Aarnio, A. N., Matt, S. P., & Stassun, K. G 2012, *ApJ*, 760, 9
Alvarado-Gomez, J. D., Drake, J. J., Cohen, O. *et al.* 2018, *ApJ*, 862, 93
Drake, J. J., Cohen, O., Yashiro, S., & Gopalswamy, N. 2013, *ApJ*, 764, 170
Guenther, E. W. & Emerson, J. P. 1997, *A&A*, 321, 803
Korhonen, H., Vida, K., Leitzinger, M., Odert, P., & Kovacs 2017, *IAUS*, 328, 198
Leitzinger, M., Odert, P., Greimel, R., *et al.* 2014, *MNRAS*, 443, 898
Odert, P., Leitzinger, M., Hanslmeier, A., & Lammer, H. 2017, *MNRAS*,472, 876

Origins: From the Protosun to the First Steps of Life
Proceedings IAU Symposium No. 345, 2018
B. G. Elmegreen, L. Viktor Tóth & M. Güdel, eds.

doi:10.1017/S1743921318008347

Spectroscopic and photometric survey of the northern sky: Towards understanding of the Galactic chemical environment in the Solar vicinity

Š. Mikolaitis[1], **G. Tautvaišienė**[1], **A. Drazdauskas**[1], **R. Minkevičiūtė**[1], **L. Klebonas**[1,2], **V. Bagdonas**[1], **E. Pakštienė**[1] **and R. Janulis**[1]

[1]Institute of Theoretical Physics and Astronomy, Vilnius University, Sauletekio av. 3, 10257, Vilnius, Lithuania
email: `sarunas.mikolaitis@tfai.vu.lt`

[2]Mathematisch-Naturwissenschaftliche Fakulttat, Universittat Bonn, Wegelerstrase 10, 53115 Bonn, Germany

Abstract. High-resolution spectra for all bright ($V < 8$ mag) and cooler than F5 spectral class dwarf stars were observed in two fields with radii of 20 degrees (centered at $\alpha(2000) = 161.03^{\circ}$ and $\delta(2000) = 86.60^{\circ}$ and at $\alpha(2000) = 265.08^{\circ}$ and $\delta(2000) = 39.58^{\circ}$) towards the northern ecliptic pole. They coincide with two of the preliminary ESA PLATO fields which also will be targeted by the NASA TESS mission. We use high-resolution spectra obtained with the VUES spectrograph mounted on the 1.65 m telescope at the Moletai Astronomical Observatory of the Institute of Theoretical Physics and Astronomy, Vilnius University. In total we observed 405 stars. Spectroscopic atmospheric parameters and abundances of 23 neutral and ionised atomic species were determined for 261 slowly rotating stars (up to 15 $\mathrm{km\,s^{-1}}$). 73 % of stars were analysed spectroscopically for the first time. We also derived stellar ages and orbital parameters to draw a chemical picture of the Solar vicinity.

Keywords. stars: abundances

1. Introduction

A spectroscopic characterisation is available for less than 30 % of bright stars in the Solar neighbourhood leaving us with missing information about the Galactic environment in this region of space. This is very unfortunate for astero-seismic and planetary studies. For example, new space missions (e.g. NASA TESS or ESA PLATO) will perform an in-depth analysis of large fields of the sky-sphere searching for extraterrestrial planets around bright stars that are similar to the Sun. In order to achieve their goals, the space missions need a full characterisation of observational objects. However, large spectroscopic surveys that have the greatest input in studying properties of Solar environment and far beyond, usually exclude brightest nearby stars. Thus here we present results of our observations and data analysis. Majority (83 %) of investigated stars are in the TESS object lists and all of them are in the preliminary PLATO fields.

2. Observations and method of analysis

Spectra of programme stars were observed with the Vilnius University Echelle Spectrograph (VUES) designed and constructed at the Exoplanet Laboratory of the Yale University (Jurgenson *et al.* 2016) and mounted on the f/12 1.65 meter Ritchey-Chretien

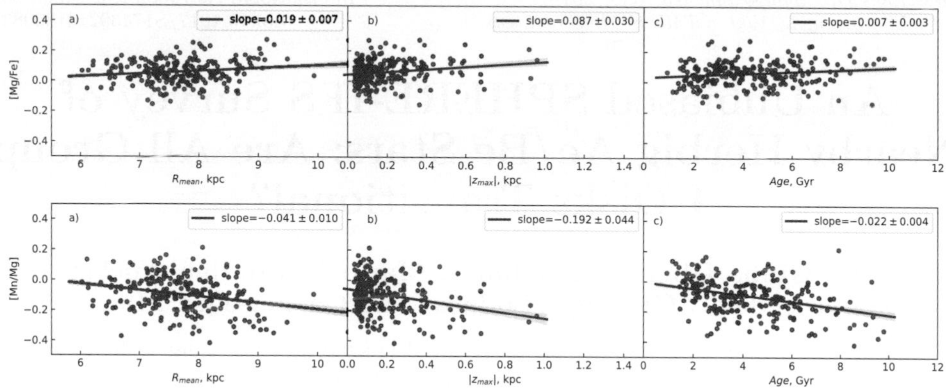

Figure 1. [Mg/Fe] and [Mn/Mg] ratios as a function of [Fe/H], $R_{\rm mean}$, and $|Z_{max}|$.

telescope at the Molėtai Astronomical Observatory of the Institute of Theoretical Physics and Astronomy, Vilnius University. For observations we used a resolving power $R \approx 60\,000$ and the wavelength range from 4000 to 8800 Å. During our observations in 2016–2017, we obtained spectra of 405 FGK dwarf stars. After the primary spectral revision and identification of spectroscopic double-line binaries and fast-rotating stars, the further analysis was performed for 261 stars. We analysed the spectra using a differential model atmosphere technique. For more information about observations and determination of stellar atmospheric, kinematic, and orbital parameters see Mikolaitis *et al.* (2018). Abundances of NaI, MgI, AlI, SiI, SiII, SI, KI, CaI, CaII, ScI, ScII, TiI, TiII, VI, CrI, CrII, MnI, FeI, FeII, CoI, NiI, CuI, ZnI were determined as described by Mikolaitis *et al.* (2017).

3. Results

The elemental abundance trends (e.g. Figure 1) in our sample of stars resemble the typical α- or iron-peak-element behaviour (e.g. Adibekyan *et al.* 2012 or Bensby, Feltzing, & Oey 2014). From the kinematical point of view, almost all our stars belong to the thin disk substructure of the Milky Way. The derived galactocentric metallicity gradient is -0.062 ± 0.017 dex kpc^{-1} (2.5 σ significance), the vertical metallicity gradient is -0.179 ± 0.064 dex kpc^{-1} (3 σ significance), the age-metallicity gradient is -0.025 ± 0.007 dex kpc^{-1}. Gradients comply with the latest inside-out thin disk formation models, including those with stellar migration taken into account (e.g. Toyouchi & Chiba 2014; Minchev, Chiappini, & Martig 2013, 2014).

Acknowledgements

We thank the Research Council of Lithuania (LAT-08/2016).

References

Adibekyan, V. Z., Sousa, S. G., Santos, N. C. *et al.* 2012, *A&A*, 545, A32

Bensby, T., Feltzing, S., Oey, M. S. 2014, *A&A*, 562, A71

Jurgenson, C., Fischer, D., McCracken, T. *et al.* 2016, *Journal of Astronomical Instrumentation*, 05, 02, 1650003

Minchev, I., Chiappini, C., Martig, M. 2013, *A&A*, 558, A9

Minchev, I., Chiappini, C., Martig, M. 2014, *A&A*, 572, A92

Mikolaitis, Š., de Laverny, P., Recio-Blanco, A. *et al.* 2017, *A&A*, 600, A22

Mikolaitis, Š., Tautvaišienė, G., Drazdauskas, A., *et al.* 2018, *PASP*, 130, 074202

Toyouchi, D., Chiba, M. 2014, *ApJ*, 788, 89

Origins: From the Protosun to the First Steps of Life
Proceedings IAU Symposium No. 345, 2018
B. G. Elmegreen, L. Viktor Tóth & M. Güdel, eds.

doi:10.1017/S1743921318008359

An Unbiased SPHERE-IFS Survey of Nearby Herbig Ae/Be Stars: Are All Group I Disks Transitional?

Maria Giulia Ubeira Gabellini[1,2], Mario E. van den Ancker[2], Davide Fedele[3], Giuseppe Lodato[1] and Carlo Felice Manara[2]

[1]Dip. di Fisica, Universitá Degli Studi di Milano, Via Celoria, 16, Milano, I-20133, Italy
email: maria.ubeira@unimi.it, giuseppe.lodato@unimi.it

[2]ESO, Karl-Schwarzschild-Strasse 2, D-85748 Garching bei München, Germany
email: mvandena@eso.org, cmanara@eso.org

[3]INAF - Osservatorio Astrofisico di Arcetri, Largo E. Fermi 5, I-50125 Firenze, Italy
email: fedele@arcetri.astro.it

Abstract. Using a novel method for speckle noise suppression from SPHERE-IFS data, we performed a systematic survey for disks in 22 Herbig Ae/Be stars, spatially resolving five disks and detecting seven new companion candidates. The fraction of sources with spatially resolved disks is systematically higher in the Meeus *et al.* (2001) group I sources, showing that disks are indeed more easily seen in scattered light in this sub-class of Herbig stars, consistent with the interpretation of group I sources having large gaps in their disks.

Keywords. Techniques: high angular resolution, protoplanetary disks, stars: pre-main-sequence

---

1. Introduction

One of the main goals of current planet-formation studies is to find planets still embedded in a proto-planetary disk, in order to catch the planet formation process as it happens (e.g. Testi *et al.* 2014). These young planets are expected to imprint signatures of their formation, such as cavities, gaps or asymmetries, which can be observable in the scattered light.

Maaskant *et al.* (2013) have suggested that the large near-infrared excesses observed in the Spectral Energy Distributions (SEDs) of Meeus *et al.* (2001) group I Herbig Ae/Be stars can be explained by inferring the presence of a large gap in their disk. In contrast to this, the disks surrounding Meeus *et al.* (2001) group II sources, which exhibit a more continuous SED, would not have such large gaps according to this hypothesis.

A strong prediction of this scenario is that the outer edge of the dust gap should be visible in scattered light for the group I sources, but not for the group IIs. We here analyze coronagraphic near-infrared (0.95–1.65 μm) Integral Field Spectrograph (IFS) data for an unbiased sample of 22 Herbig stars obtained with SPHERE, the extreme adaptive optics system at the VLT (Beuzit *et al.* 2008), in order to test this prediction.

2. A novel method for speckle noise suppression in IFS data

We have obtained SPHERE coronagraphic near-infrared integral-field spectrograph (IFS) data for 22 nearby Herbig Ae/Be stars. One of the main issues in detecting faint disks next to bright stars in adaptive optics images is residual speckle noise due to static wavefront aberrations within the adaptive optics system. Traditional methods to correct

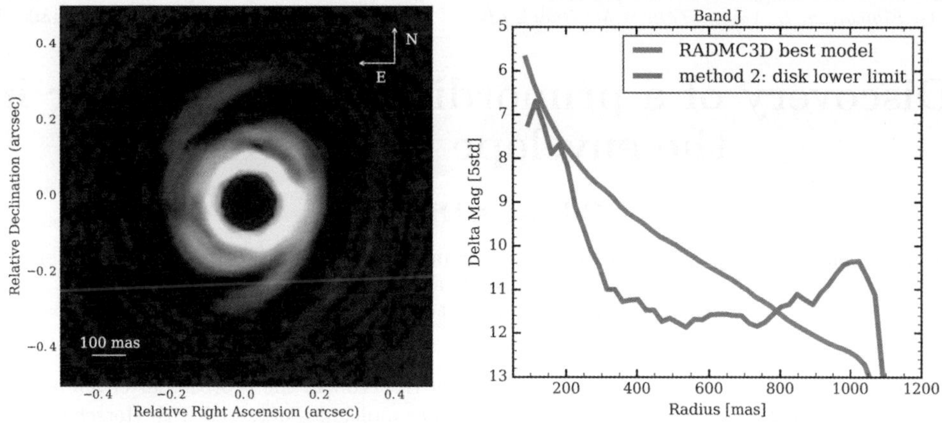

Figure 1. Speckle-corrected SPHERE-IFS image of HD 100453 in the J-band (left) and J-band contrast curve compared to a RADMC3D model for HD 100453 (right). The central dark spot in the image is due to the obscuration of the central star by the coronagraph.

for speckle noise, such as angular differential imaging (ADI), filter out radially symmetric extended structures, such as disks.

Therefore we employ a novel method for correcting for speckle noise in our IFS data, based on the following ingredients: (1) the position of a speckle is moving in a known way with wavelength, (2) the position of a real source or a disk component is stable at different wavelengths, and (3) as speckles are caused by uncorrected light from the central star, the spectral slope of speckle noise is known. Using the above ingredients, a map of speckle noise can be created for each of the 39 wavelengths, and subtracted from the data to create a corrected image in which the signal of the circumstellar disk is revealed (Fig. 1).

3. Conclusions

Using the method outline above, we have detected five spatially extended disks, and seven new companion candidates within our sample of 22 Herbig stars. The fraction of sources with detected disks in 0.6 ± 0.3 in the group I sources and 0.1 ± 0.1 in the Group II sample, showing that disks are indeed more easily spatially resolved in scattered light in Meeus *et al.* Group I sources, as predicted by the Maaskant *et al.* scenario. We will further test the models of Maaskant *et al.* (2013) using RADMC3D (Dullemond 2012) to perform a more quantitative test between the observed and predicted surface brightness of the disk in the Y, J, and H-bands as a function of radius (Fig. 1). In addition, we are currently performing follow-up observations on the detected companion candidates to confirm that they are physically connected to the Herbig star.

References

Beuzit, J.-L., Feldt, M., Dohlen, K., *et al.* 2008, in *Proc. SPIE*, Vol. 7014, 701418
Dullemond, C.P., *et al.* 2012, *Astrophysics Source Code Library*, record ascl:1202.015
Maaskant, K.M., Honda, M., Waters, L.B.F.M., *et al.* 2013, *A&A*, 555, 64
Meeus, G., Waters, L.B.F.M., Bouwman, J., *et al.* 2001, *A&A*, 365, 476
Testi, L, *et al.* 2014, in *Protostars and Planets VI*, pp. 339–361

Origins: From the Protosun to the First Steps of Life
Proceedings IAU Symposium No. 345, 2018
B. G. Elmegreen, L. Viktor Tóth & M. Güdel, eds.

doi:10.1017/S1743921318008360

Discovery of a primordial water reservoir in the envelope of HH 211

Odysseas Dionatos

Department of Astrophysics, University of Vienna, Türkenschanzstrasse. 17, 1180 Vienna, Austria
email: odysseas.dionatos@univie.ac.at

Abstract. We report on the detection of a rich water reservoir in the protostellar envelope of the Class 0 source HH211. In striking contrast to all other molecules detected with Herschel/PACS, water emission peaks around the central source where both ortho and para forms are detected. The measured ortho-to-para ratio of just 0.65 indicates formation of water-ice at very low temperatures and a non-destructive photo-desorption process around the protostar. While part of the H_2O emission is likely related to collisional excitation, the centralized morphology around the protostar suggests that radiative excitation is also significant, despite the fact that radiation appears to have a very different impact on the H_2O molecules when compared to the terminal outflow shocks. The very low ortho-to-para ratio suggests that H_2O around the protostar originates from primordial envelope material that has never been thermally processed before.

Keywords. stars: formation, ISM: molecules, ISM: abundances, ISM: individual (HH 211).

1. Introduction

Observations of the abundance ratio of nuclear spin isomers of H_2O can possibly provide clues about the formation conditions and the thermal processing of molecules. For temperatures above 50 K the H_2O ortho-to-para ratio (OPR) is ~ 3 while for lower temperatures the ratio tends to zero. Therefore low OPR can be associated with the formation of H_2O at very low temperature environments. Processes leading to the desorption of water into the gas phase can alter the low-temperature OPR and modify it toward the thermal equilibrium value of 3. Non-destructive photo-desorption of H_2O is believed to preserve the original formation value of OPR (van Dishoeck *et al.* 2013), therefore providing clues about the thermal history of the molecule. We here report on the detection of an unusually low water - OPR in the envelope of HH 211, indicating that the water originates from primordial, unprocessed envelope material.

2. An extremely low ortho-to-para ratio for water

In Fig. 1, we present examples of Herschel/PACS maps for the ortho- and para- forms of water (left and right panels, respectively). Water appears to be, in most cases, centralized around the protostar, as discussed extensively in Dionatos *et al.* (2018). Ortho transitions often display some weaker emission toward the SE and NW bowshocks, with the SE emission being in most cases stronger in comparison to the NW. An excitation diagram for the water emission around the protostellar source is presented in Fig. 2. While there are just three para-H_2O lines detected (marked as filled circles in Fig. 2), their distribution in the excitation diagram follows closely the distribution of the ortho-H_2O lines. Performing a separate excitation analysis, column densities are found to be $\sim$3.9 and 6.1×10^{12} cm^{-2} for the ortho and para forms, respectively. Given the similar excitation conditions, any non-thermal excitation and opacity issues would affect the ortho and para forms in the

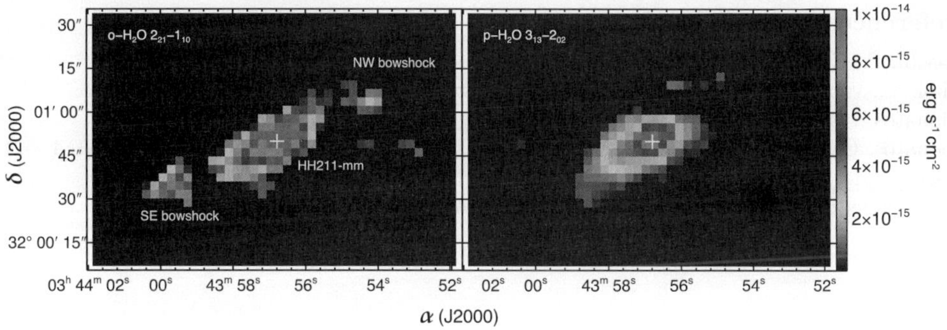

Figure 1. Herschel/PACS maps showing examples of ortho- and para- H_2O transitions.

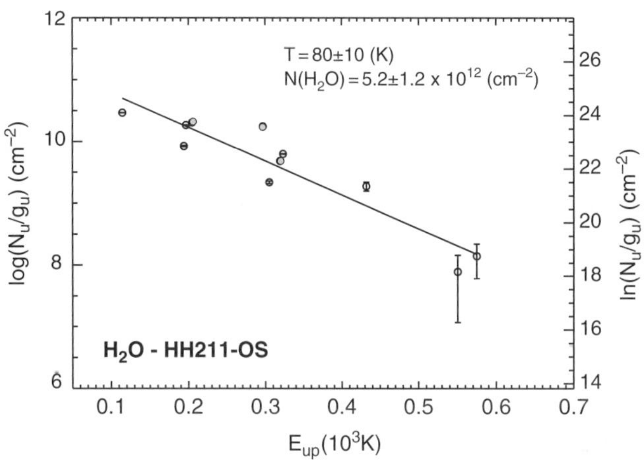

Figure 2. Excitation diagram for the water ortho- and para- transitions (open and filled circles, respectivelly) detected around HH 211-mm.

same way, so that their column density ratio can provide a measure of their abundance ratio which we find $\sim$ 0.65. The non-detection of any para-H_2O lines in either the SE or NW positions suggests that the OPR likely reaches the equilibrium value of 3 at the terminal shocks, as it has also been measured for H_2 (Dionatos *et al.* 2010).

The OPR estimated around HH211 is one of the lowest values measured in the surroundings of an embedded protostar, and is within errors comparable to the OPR$\sim$0.77 observed in the outer disk of TW Hya (Hogerheijde *et al.* 2011). It is also much lower than expected based on the temperatures of $\sim$90 K derived from the excitation analysis. This inconsistency between the gas temperature and very low OPR values has been noticed also in the Orion PDR (Choi *et al.* 2014). Water around embedded protostars, or even pre-stellar cores, has been proposed to be the product of photodesorption from dust grains in the parental envelope (Caselli *et al.* 2012). Given that OPR can indeed provide any information about the thermal history of the gas, then the low OPR in the surroundings of the protostar can only be interpreted through processing of primordial material that has never been thermally altered before. At the shock positions in contrast, dust temperature must have been increased multiple times through successive shocks, so that the post-shock formation and subsequent adsorption of water onto/from the dust grains reflects processing at higher temperatures.

References

Caselli, P., Keto, E., Bergin, E. A., *et al.* 2012, *ApJ*, 759, L37
Choi, Y., van der Tak, F. F. S., Bergin, E. A., & Plume, R. 2014, *A&A*, 572, L10
Dionatos, O., Ray, T., & Güdel, M. 2018, *A&A*, 616, A84
Dionatos, O., Nisini, B., Cabrit, S., Kristensen, L. & Pineau Des Forêts, G. 2010, *A&A*, 521, 7
Hogerheijde, M. R., Bergin, E. A., Brinch, C., *et al.* 2011, *Science*, 334, 338
van Dishoeck, E. F., Herbst, E., & Neufeld, D. A. 2013, *Chem. Rev.*, 113, 9043

Origins: From the Protosun to the First Steps of Life
Proceedings IAU Symposium No. 345, 2018
B. G. Elmegreen, L. Viktor Tóth & M. Güdel, eds.

doi:10.1017/S1743921318008384

Heating mechanisms in accretion disks around young stellar objects

Natália F.S. Andrade, Rafael Rechiche de Campos and Vera Jatenco-Pereira

Instituto de Astronomia, Geofísica e Ciências Atmosféricas (IAG), Universidade de São Paulo, Rua do Matão 1226, São Paulo, SP, Brazil
email: natalia.fernanda.andrade@usp.br

Abstract. Accretion disks are observed around young stellar objects such as T Tauri stars. In order to complete the star formation, particles in the disk need to loose angular momentum in order to be accreted into the central object. The magneto-rotational instability (MRI) is probably the mechanism responsible for a magneto-hydrodynamic (MHD) turbulence that leads to disk accretion, which implies the disk particles to be coupled with the magnetic filed lines. As the temperature in the disk is low, we considered, besides the viscous heating mechanism often included in the models by means of the α - prescription, the damping of Alfvén waves as an additional heating source. In particular, we show that the mechanism derived that couples the turbulent and non-linear damping mechanisms of Alfvén waves proved to be very efficient, generating temperatures almost one order of magnitude higher than those mechanisms considered independently.

Keywords. stars: pre–main-sequence, stars:formation, MHD, plasma

1. Introduction

Accretion disks are commonly associated with young stellar objects (YSOs) and are strongly related with the evolution process of such protostars. Those structures can be defined as rotationally supported and are known to be formed of both gas and dust (Armitage 2011). Those circumstellar structures are responsible for accreting mass onto the central object, until the star is massive enough to propitiate its entry in the Main Sequence (MS). This mass transference towards the central star is accompanied by a net transference of angular momentum outwards. The most promising mechanism of angular momentum transference in accretion disks is the Magneto-Rotational Instability (MRI), proposed by Balbus & Hawley (1991). This instability, responsible for promoving the turbulence and angular momentum transport in disks, requires high levels of ionization in order to be effective, since the particles need to be coupled with the magnetic field lines. It is known that the T Tauri accretion disks exhibit very low temperatures and ionization degrees. Therefore, an extra heating mechanism, besides the viscous dissipation proposed by Shakura & Sunyaev (1973), must be present in those structures. The effects of the damping of Alfvén waves as heating sources in T Tauri accretion disks were previously studied by different authors (e.g. Vasconcelos *et al.* 2000). In the present work, we consider the coupling of two of the mechanisms proposed by Vasconcelos *et al.* (2000), the turbulent and non-linear damping, whereas the first transfer energy from large to small scales due to the interaction of an Alfvén wave with a turbulent cell (Hollweg 1987), the latter consists of the interaction between two Alfvén waves of opposite directions, originating a sound wave, which will be rapidly dissipated, releasing energy (Lagage & Cesarsky 1983).

Table 1. We show the temperatures, for three radial distances of the disk, generated by the damping mechanisms of Alfvén waves: non-linear, turbulent and both coupled, for a disk of $M_* = 0.5\, M_\odot$, $\dot{M} = 10^{-8}\, M_\odot/yr$, $\alpha = 0.01$, $\mu = 2.33$ and $f = 0.002$.

Damping mechanism	T(K) r = 0.1 (UA)	T (K) r = 10 (UA)	T (K) r = 99.95 (UA)
non-linear	274.34	15.47	3.79
turbulent	1016.14	56.79	10.89
coupled	2827.85	194.87	30.39

2. The model

In the present work, we have adopted the α-prescription, as proposed by Shakura & Sunyaev (1973), in the definition of viscosity. We have also assumed that the disk is optically thick, geometrically thin, and that it can be divided in layers, in agreement with the model suggested by Gammie (1996).

Additionally, since the disk is optically thick, it is reasonable to assume that the temperature generated is given by the black-body law, $T = (D/\sigma)^{1/4}$, with D representing the energy dissipated, $D = \int_0^H \Phi\,\Gamma/v_A$, where σ is the Stefan-Boltzmann constant, H is the disk scale height, Φ is the wave energy flux, Γ is the damping rate and v_A is the Alfvén velocity.

By the maximization of both turbulent and non-linear damping mechanisms, we found the following prescription for the damping rate of the coupled mechanism:

$$\Gamma = \frac{(c_s/v_A)\rho < \delta v^2 > + B^{3/2} < \delta v^2 >^{1/2}}{B}, \tag{2.1}$$

where c_s is the sound velocity, v_A is the Alfvén velocity, ρ is the volumetric density, $< \delta v^2 >$ is the quadratic velocity dispersion and B is the magnetic field strenght.

3. Results

From Table 1, we show that even for small wave fluxes, denoted by the magnitude of the f parameter, the coupled mechanism, significantly, heats the disk to temperatures necessary for the IMR to be effective.

4. Conclusions

In this work, we show that the coupled mechanism proposed is quite efficient in heating the disk, when compared to the turbulent and non-linear damping mechanisms, yeilding to high temperatures even for small wave flux (see Table 1).

In the sequence, we intend to consider the damping of magneto-acoustic waves as an extra source for the disk heating.

References

Armitage, P. J. 2011, *ARAA*, 49, 195
Balbus, S. A. & Hawley, J. F. 1991, *ApJ*, 376, 214
Gammie, C. F. 1996, *ApJ*, 457, 355
Hollweg, J. V. 1987, *ApJ*, 312, 880
Lagage, P. & Cesarsky, C. 1983, *A&A*, 125, 249
Shakura, N. I. & Sunyaev, R. A. 1973, *A&A*, 24, 337
Vasconcelos, M. J., Jatenco-Pereira, V. & Opher, R. 2000, *ApJ* 534, 967

Origins: From the Protosun to the First Steps of Life
Proceedings IAU Symposium No. 345, 2018
B. G. Elmegreen, L. Viktor Tóth & M. Güdel, eds.

doi:10.1017/S1743921318008396

Resolving the origin of hydrogen-line emission in YSOs with near-infrared interferometry

Alexander Kreplin[1], Edward Hone[1], Larisa Tambovtseva[2], Karl-Heinz Hofmann[3] and Stefan Kraus[1]

[1]Department of Physics and Astronomy, University of Exeter,
Stocker Road, Exeter EX4 4QL, UK
email: a.kreplin@exeter.ac.uk,

[2]Pulkovo Observatory of RAS,
Pulkovskoe Shosse 65, St Petersburg 196140, Russia

[3]Max Planck Institut für Radioastronomie,
Auf dem Hügel 69, 53121 Bonn, Germany

Abstract. The origin of the Brγ-line emission in Herbig Ae/Be stars is still an open question and might be related e.g., to a disc wind or the stellar magnetosphere. The study of the continuum and Brγ-emitting region of Herbig Ae/Be stars with high-spectral and high-spatial resolution gives great insights into the sub-au scale hydrogen gas distribution.
We observed the Herbig Be star MWC 120 with the VLTI/AMBER instrument in different spectral channels across the Brγ line with a spectral resolution of $R \sim 1500$. Using radiative transfer modeling we found a radius of the line emitting region of ~0.4 au that is only two times smaller than the K-band continuum region. This is consistent with a disc wind scenario rather than an origin of magnetospheric emission.
We present near-infrared AMBER (R~12000) observations of the Herbig B[e] star MWC297 in the Brγ-line. We found that the near-infrared continuum emission is ~3.6 times more compact than the expected dust-sublimation radius, possibly indicating the presence of highly refractory dust grains or optically thick gas emission in the inner disk. Our velocity-resolved channel maps marking the first time that kinematic effects in the sub-AU inner regions of a protoplanetary disk could be directly imaged.

Keywords. stars: individual (MWC 120, MWC 297); pre-main-sequence; stars: imaging; stars: winds, outflows; accretion, accretion disks; radiative transfer; techniques: high angular resolution; techniques: interferometric; circumstellar matter; stars: emission-line, Be

1. High-angular resolution observations of Herbig Ae/Be stars

Jets and outflows play a fundamental role in star formation and the physics of the accretion-ejection process. While the hot circumstellar dust (T~1500 K) can be studied in the near-infrared continuum, gas and its kinematics can be studied in lines, in particular, the Brγ line (λ=2.1661μm).

Comparison of the visibilities, differential and closure phases in the continuum and the line emission region of Herbig Ae/Be star MWC 120 (Fig. 1, top left) with geometric and radiative transfer disc wind models (Fig. 1, top left, channel maps) reveal that the Brγ emission arises from a 2 times more compact region (0.15 au) than the K-band continuum (0.35 au), consistent with emission dominantly emerging from an extended disk wind. A comparison with other Herbig Ae/Be star shows that for stars of late spectral type (A1–B9), the derived inner radii of the disc-wind base can be approximately described by the corresponding Alfvén radii for each target (Fig. 1, top right; Kreplin *et al.* 2018).

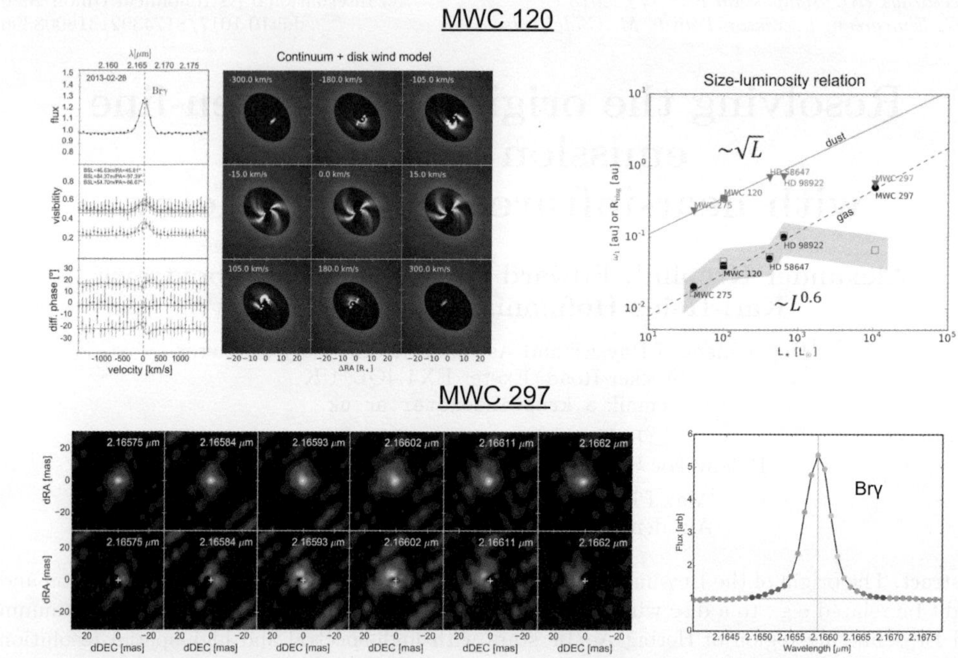

Figure 1. Top: Comparison of the MWC 120 measurements (data points) with the best-fitting disc-wind model (from top to bottom: spectrum, visibilities, differential phases). Right: Size-luminosity relation of a Herbig Ae/Be target sample showing the continuum dust emission size (green) and the inner disk wind radii (black). The inner disk wind base radius (black squares) might be proportional to the Alfvén radius (red squares) for late-spectral type stars (A1-B9). **Bottom:** First velocity-resolved reconstructed images (left) across the Brγ line (right) for a young star (MWC 297) at infrared wavelengths (IRBis algorithm; Hofmann *et al.* 2014). Top panels: continuum+line emission; Bottom panels: line emission only (i.e. continuum subtracted)

High spectral resolution VLTI/AMBER observations (R=12000) of the Herbig Be star MWC 297 show a very compact K-band continuum region ($\sim$0.8 au) compared to the expected dust sublimation radius of $\sim$3 au. The modelled Brγ emitting region is $\sim$2.3 times larger than the continuum region. We applied a disk-wind model and were able to fit the Brγ-line profile (Fig. 1, bottom right), as well as the high-resolution spectro-interferometry and spectro-astrometry data (see Hone *et al.* 2017). Our model-independent channel maps have an angular resolution of 2 mas and represent the first velocity-resolved image obtained for a young star in the infrared (Fig. 1, bottom left).

Acknowledgements

We acknowledge funding from a Science and Technology Facilities Council Rutherford Fellowship (ST/J004030/1), Rutherford Grant (ST/K003445/1) and European Research Council Starting Grant (Grant Agreement No. 639889).

References

Kreplin, A., Tambovtseva, L., Grinin, V., Kraus, S., Weigelt, G., & Wang, Y. 2018, *MNRAS*, 476, 4520

Hone, E., Kraus, S., Kreplin, A., Hofmann, K.-H., Weigelt, G., Harries, T., & Kluska, J. 2018, *A&A*, 607, 17

Hofmann, K.-H., Weigelt, G., & Schertl, D. 2014, *A&A*, 565, 48

Origins: From the Protosun to the First Steps of Life
Proceedings IAU Symposium No. 345, 2018
B. G. Elmegreen, L. Viktor Tóth & M. Güdel, eds.

doi:10.1017/S1743921318008402

A physically motivated core definition applied to dust emission observations of the Pipe nebula

Birgit Hasenberger[1] **and João Alves**[1,2]

[1]Department for Astrophysics, University of Vienna,
Türkenschanzstraße 17, 1180 Vienna, Austria
email: birgit.hasenberger@univie.ac.at, joao.alves@univie.ac.at

[2]Radcliffe Institute for Advanced Study, Harvard University
10 Garden Street, Cambridge, MA 02138, USA

Abstract. Dense cores represent a critical stage in the star-formation process, but are not physically well-defined entities. We present a new technique to define core boundaries in observations of molecular clouds based on the physical properties of the cloud medium. Applying this technique to regions in the Pipe nebula, we find that our core boundaries differ from previous analyses, with potentially crucial implications for the statistical properties of the core sample.

Keywords. methods: data analysis, ISM: clouds, (ISM:) dust, extinction, ISM: structure, ISM: individual (Pipe nebula), submillimeter

1. Introduction

The investigation of dense cores allows us to gain insights into the physical mechanisms that are relevant at the early stages of the star-formation process. Observationally, cores are commonly defined by their 2D morphology in maps of molecular clouds – without explicitly imposing constraints on the physical properties of the region defined as a core (Pineda *et al.* 2009, Smith *et al.* 2008). We develop a core extraction scheme that is physically motivated and thus requires us to go beyond projections in the plane of the sky. Dust emission observations of the Pipe nebula are a suitable basis for our study due to the quiescent nature of the region and the quality of the available data (Hasenberger *et al.* 2018, Forbrich *et al.* 2009, Alves *et al.* 2008).

2. Estimating 3D density and temperature

We utilise *Herschel-Planck* flux maps of the Pipe nebula region (Hasenberger *et al.* 2018) for our analysis. From these maps, we estimate 3D flux distributions based on two assumptions: For each structure at each flux level, a) the extent along the line of sight is similar to the extent in the 2D projection, and b) the 3D radial distribution is approximated well by applying a variant of the inverse Abel transform (Abel 1826) to the 2D projection. The overall 3D flux distribution is then the sum of the contributions from each structure in each flux level. We derive estimates for the 3D density and temperature distributions by fitting the 3D flux distributions with a modified blackbody spectrum.

We applied this method to a number of subregions in the Pipe nebula. As a first test of our technique, we compare temperatures in dense cores derived from NH_3 line emission observations (Rathborne *et al.* 2008), the 2D dust map, and the 3D dust map (see Fig. 1). We generally find good agreement between the line-emission and 3D dust values, corroborating our assumptions and technique.

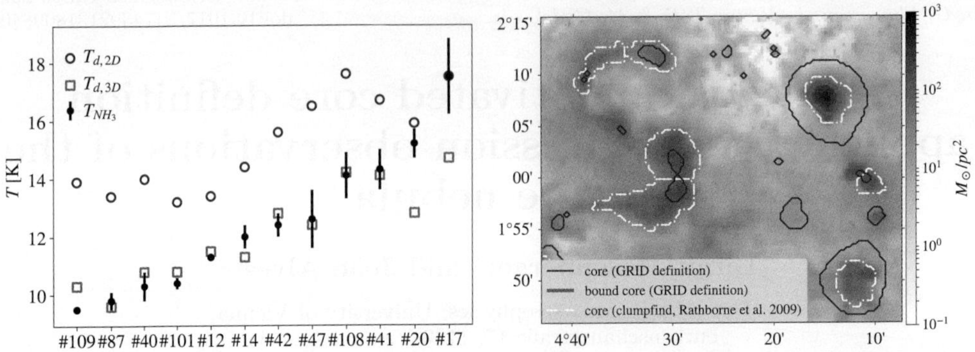

Figure 1. *Left:* Comparison of core temperatures derived using different methods, with the x-axis indicating the core number as given by Alves *et al.* (2007). *Right:* Map of the column-density distribution derived from the 3D density estimate for the Smoke subregion of the Pipe nebula, with different core boundary definitions shown as contours.

3. Physically motivated core boundaries

With 3D density and temperature estimates at hand, we are able to construct 3D maps of gravitational (E_g) and thermal energy (E_{th}). Following the gravitational identification (GRID) procedure (Gong *et al.* 2011), GRID cores are defined as individual wells in the gravitational potential, and bound GRID cores as areas where $E_g > E_{th}$. A preliminary analysis of the subregions indicates that only few areas in the Pipe nebula qualify as bound GRID cores (see Fig. 1), which might have critical implications for the statistical properties of the core sample, for example the core mass function.

4. Conclusions and outlook

We use dust emission observations of the Pipe nebula to develop a new, physically motivated core extraction scheme. By estimating the 3D distribution of density and temperature, we are able to define core boundaries based on the balance between gravitational and thermal energy. Our preliminary analysis suggests that this method yields a core sample significantly different from the results of traditional core-extraction techniques. In order to deepen our understanding of dense cores, we aim to apply this procedure to the entire Pipe nebula and other molecular clouds in the future.

References

Abel, N. H. 1826, *Journal für die reine und angewandte Mathematik*, 1, 153-157

Alves, J., Lombardi, M., & Lada, C. J. 2007, *A&A*, 462, L17-L21

Alves, J., Lombardi, M., & Lada, C. J. 2008, *Handbook of Star Forming Regions, Volume II*, 415

Forbrich, J., Lada, C. J., Muench, A. A., Alves, J., & Lombardi, M. 2009, *ApJ*, 704, 292-305

Gong, H., & Ostriker, E. C. 2011, *ApJ*, 729, 120

Hasenberger, B., Lombardi, M., Alves, J., Forbrich, J., Hacar, A., & Lada, C. 2018, *A&A*, 620, A24

Pineda, J. E., Rosolowsky, E. W., & Goodman, A. A. 2009, *ApJL*, 699, L134L138

Rathborne, J. M., Lada, C. J., Muench, A. A., Alves, J. F., & Lombardi, M. 2008, *ApJSS*, 174, 396-425

Rathborne, J. M., Lada, C. J., Muench, A. A., Alves, J. F., Kainulainen, J., & Lombardi, M. 2009, *ApJ*, 699, 742-753

Smith, R. J., Clark, P. C., & Bonnell, I. A. 2008, *MNRAS*, 391, 10911099

Origins: From the Protosun to the First Steps of Life
Proceedings IAU Symposium No. 345, 2018
B. G. Elmegreen, L. Viktor Tóth & M. Güdel, eds.

doi:10.1017/S1743921318008487

Simple Hydrides (OH and CH) Trace the Dark Molecular Gas

Di Li[1,2], Ningyu Tang[1,2], Hiep Nguyen[3,4], J. R. Dawson[3,4], Carl Heiles[5], Pei Wang[1] and the PRIMO collaboration[6]

[1]National Astronomical Observatories, Chinese Academy of Sciences, Beijing 100012
email: dili@nao.cas.cn

[2]CAS Key Laboratory of FAST, Beijing, Chinese Academy of Sciences

[3]Department of Physics and Astronomy and MQ Research Centre in Astronomy, Astrophysics and Astrophotonics, Macquarie University, NSW 2109, Australia

[4]Australia Telescope National Facility, CSIRO Astronomy and Space Science, PO Box 76, Epping, NSW 1710, Australia

[5]Department of Astronomy, University of California, Berkeley, 601 Campbell Hall 3411, Berkeley, CA 94720-3411

[6]Pacific Rim Interstellar Matter Observers†

Abstract. Emission lines from CO and HI are the standard tracers of molecular and atomic interstellar medium, respectively. In the past two decades, a consensus has formed that a substantial fraction of Galactic molecular gas evades detection by these two tracers, thus giving rise to the empirical concept of dark molecular gas (DMG). Largely based on the experience and evidence garnered from the Arecibo Millennium survey, we have formed an international consortium, the Pacific Rim Interstellar Matter Observers (PRIMO), to pursue alternative tracers of DMG, particularly absorption against background radio sources (quasars). PRIMO have carried out observing programs at Arecibo, JVLA, Delingha 13.7m, ATCA and ALMA, among others. Our observations reveal abundant hydrides, namely OH and CH, in DMG clouds. The historical difficulty of mapping OH can be explained by the measured OH excitation temperature

$$f(T_{\rm ex}^{\rm OH}) \propto \frac{1}{\sqrt{2\pi}\sigma}\exp[-(\ln(\mathrm{T}_{\rm ex}^{\rm OH}) - \ln(3.4\ \mathrm{K}))^2/(2\sigma^2)],$$

which is a modified log-normal function peaking close to the numerical value of the L-band Galactic continuum background (synchrotron + CMB). Both OH and CH are shown to be better tracers of molecular hydrogen than CO, particularly in the intermediate extinction regions ($A_v \sim$ 0.05-2 magnitude), where DMG dominates.

Keywords. astrochemistry, ISM: atoms, ISM: molecules, Galaxy: abundances, radio lines: ISM

1. Overview

Hydrogen is the most abundant element in the universe. The atomic (HI) and molecular (H_2) form of hydrogen are key ingredients of the interstellar medium (ISM). Discovered in 1951, the λ21cm hyperfine structure transition of HI traces the total quantity of atomic ISM reasonably well, owing to its low opacity and commonly achievable excitation conditions. H_2, however, is illusive due to its low mass and lack of permanent dipole, which means that its lowest rotational levels are both forbidden and lie too high above ground state to be readily excited in molecular clouds. The discovery of CO, combining the third- (oxygen) and fourth- (carbon) most abundant elements, was considered a major

† http://ism.bao.ac.cn/primo

astronomical discovery of the 20th century. A key caveat of using CO to trace H_2 is its large opacity; the widely used empirical CO intensity to H_2 column density conversion factor, the X-factor, faces the intrinsic challenge of relating surface brightness to total quantities summed over the full depth of a cloud. The viral argument (Solomon *et al.* 1987) has to be invoked. It is thus not surprising to see variation and uncertainty in the X-factor, and voluminous literature exists on this topic (e.g. Genzel *et al.* 2015).

A growing consensus is the underestimation of the total neutral gas mass as estimated from $N_{\rm HI}+XW_{\rm CO}$ (where $N_{\rm HI}$ is the atomic column density under the optically thin assumption, $W_{\rm CO}$ is the velocity-integrated CO intensity and X is the X-factor). Grenier *et al.* (2005) analyzed diffuse γ-ray emission from EGRET attributable to cosmic-ray interactions with hydrogen nuclei. Compared to the total hydrogen estimated from $N_{\rm HI}+XW_{\rm CO}$, wide-spread gamma-ray-derived hydrogen excess was found and dubbed 'dark gas'. Planck Collaboration *et al.* (2011) later showed an apparent excess in dust opacity in multiple Planck bands for gas with intermediate extinction. The self-shielding threshold of H_2 is about a decade smaller than that of CO (van Dishoeck & Black 1988). A natural expectation from PDR models is thus a layer of H_2 gas without stable CO, the canonical abundance of which is $[\rm CO]/[\rm H_2]\sim 10^{-4}$, meaning all carbon is tied in this molecule. We believe that the 'missing' molecular gas in this intermediate extinction layer ($A_v \sim 0.05$-2 magnitude) is composed largely of dark molecular gas (DMG).

2. Hydrides Trace DMG

CH was the first interstellar molecule ever discovered; OH was the first discovered in the radio band. These simple hydrides are expected to be abundant in DMG. Despite their importance, however, large scale mapping of the extended ISM in either is rare. While the Λ-doubling transitions of OH and CH in L- and C-bands are accessible, the lines tend to be extremely weak compared to (say) CO. The complexity of their excitation is also an issue: their level populations are strongly affected by the rotational levels in the IR bands (Fig. 1). In cold gas or star forming regions, the molecules can be readily 'pumped' to IR levels, then decay readily into preferred sub-levels of the ground state, making excitation states of Λ-doubling transitions non-thermal. This was recognized immediately after the discovery of OH (Cook 1966), and to-date most studies of OH have been focussed on bright, compact OH maser sources. While masing in the DMG is much weaker, without the validity of LTE assumptions, converting flux to column density is challenging.

We tackle the excitation of hydrides from two angles. First, we utilize the quasar absorption analysis procedures developed by Heiles & Troland (2003). In their Arecibo Millennium survey, multiple off-source (OFF) positions were observed in addition to each on-souce (ON) sightline with a bright quasar in the beam. An international consortium, the Pacific Rim Interstellar Matter Observers (PRIMO), were formed in 2013 to expand such absorption studies to multiple bands. PRIMO has managed to obtain time from Arecibo, JVLA, ATCA, Delingha 13.7m, ALMA, and others, and has confirmed the abundant existence of OH in DMG. Based on OH absorption measurements, Planck dust opacities, and optical dust reddening, Nguyen *et al.* (2018) published systematic measurements of OH abundance across randomly sampled sightlines in the Arecibo sky, finding $[\rm OH]/[\rm H_2]\sim 10^{-7}$, which was stable (to within the uncertainties) in the range $\rm N(H_2)\sim 0.1$–$10\times10^{21}\ cm^{-2}$. Li *et al.* (2018) measured main-line OH excitation temperatures, finding an empirical trend,

$$\rm f(T_{ex}^{OH}) \propto \frac{1}{\sqrt{2\pi}\sigma}\exp\left[-(\ln(T_{ex}^{OH}) - \ln(3.4\ K))^2/(2\sigma^2)\right]. \tag{2.1}$$

The OH excitation temperature $\rm T_{ex}^{OH}$ peaks close to the Galactic emission background, providing a natural explanation for the difficulty of mapping OH in emission.

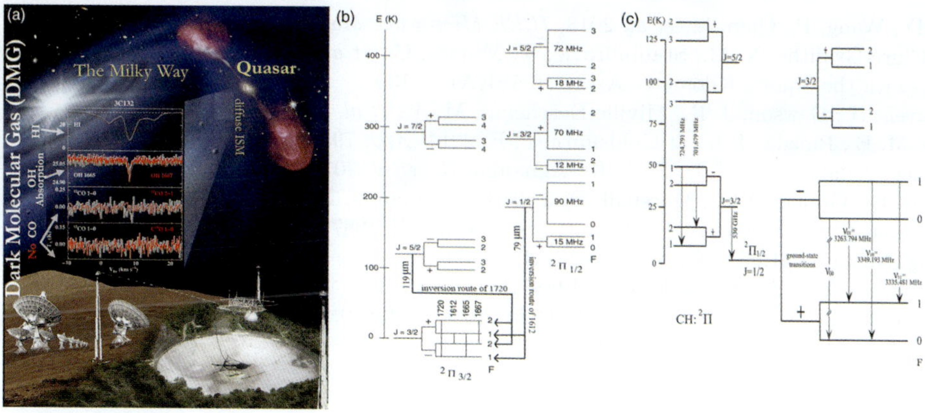

Figure 1. (a): Illustration of the absorption concept toward continuum sources. The spectra of DMG toward 3C132 are shown. The energy structures of OH and CH are shown in (b) and (c), respectively.

Second, we model the emission/absorption of hydrides in regions with other, independent constraints on their physical and chemical conditions, where the excitation problem can be solved by generating profiles under specific assumptions, made consistent with all available data sets, including other spectral species, 2MASS extinction, etc. Through a combined analysis of HI narrow self-absorption (HINSA), C^+, and OH, Tang *et al.* (2017) revealed a correlation between C^+ and OH in DMG clouds. A target with better constraints is the apparent transition edge in the Taurus molecular cloud (Goldsmith *et al.* 2008). Orr *et al.* (2014) published a self-consistent geometry and chemistry model of the regions, relying largely on CI and C^+ observations. We carried out Arecibo L- and C-band observations of the same region, and utilizing Orr's model, fitted to the excitation and abundance of OH and CH throughout the transition zone ($A_v \sim 0.2$–2). Xu *et al.* (2016) found an evolution in OH abundance from 8×10^{-7} to 1×10^{-7} with increasing extinction in the DMG (transition) region, while the fraction of DMG (DGF) was found to decrease with increasing extinction as

$$\mathrm{DGF} = 0.9 \times \exp\left[-\left(\frac{A_v - 0.79}{0.71}\right)^2\right]. \tag{2.2}$$

These trends are borne out by recent analysis of diffuse γ-ray data from Fermi (Remy *et al.* 2018). Xu & Li (2016) also compared the X-factors of hydrides with that of CO. The measured X-factors of OH and CH have much less scatter than that of CO in the transition zone, consistent with the fact that CO is evolving more dramatically in this intermediate extinction (below its self-shielding threshold) region.

In summary, simple hydrides such as OH and CH are better tracers of H_2 than CO in regions with dark molecular gas. Quasar absorption measurements have been demonstrated to provide an accurate handle of the excitation conditions of hydride transitions in radio bands. The upcoming, significantly more sensitive radio telescopes, FAST and SKA in particular (Li *et al.* 2018; McClure-Griffiths *et al.* 2015), will enable systematic measurement of hydrides throughout the Galactic ISM, thus quantify the so-called dark molecular gas.

References

Cook, A. H. 1966, *Nature*, 211, 503

Genzel, R., Tacconi, L. J., Lutz, D., *et al.* 2015, *ApJ*, 800, 20

Li, D., Wang, P., Qian, L., *et al.* 2018, *IEEE Microwave Magazine*, 19, 112
McClure-Griffiths, N. M., Stanimirovic, S., Murray, C., *et al.* 2015, in Advancing Astrophysics with the Square Kilometre Array (AASKA14), 130
Nguyen, H., Dawson, J. R., Miville-Deschênes, M.-A., *et al.* 2018, *ApJ*, 862, 49
Orr, M. E., Pineda, J. L., & Goldsmith, P. F. 2014, *ApJ*, 795, 26
Planck Collaboration, Ade, P. A. R., Aghanim, N., *et al.* 2011, *A&A*, 536, A19
Remy, Q., Grenier, I. A., Marshall, D. J., & Casandjian, J. M. 2018, *A&A*, 611, A51
Solomon, P. M., Rivolo, A. R., Barrett, J., & Yahil, A. 1987, *ApJ*, 319, 730
Tang, N., Li, D., Heiles, C., *et al.* 2017, *ApJ*, 839, 8
van Dishoeck, E. F., & Black, J. H. 1988, *ApJ*, 334, 771
Xu, D., Li, D., Yue, N., & Goldsmith, P. F. 2016, *ApJ*, 819, 22
Xu, D., & Li, D. 2016, *ApJ*, 833, 90

Origins: From the Protosun to the First Steps of Life
Proceedings IAU Symposium No. 345, 2018
B. G. Elmegreen, L. Viktor Tóth & M. Güdel, eds.

doi:10.1017/S1743921318008499

Carbon and oxygen abundances in dwarf stars of the Solar neighbourhood

G. Tautvaišienė, R. Minkevičiūtė, E. Stonkutė, A. Drazdauskas and Š. Mikolaitis

Institute of Theoretical Physics and Astronomy, Vilnius University,
Sauletekio av. 3, 10257, Vilnius, Lithuania
email: grazina.tautvaisiene@tfai.vu.lt

Abstract. Stars and planets form from the same material, thus some of their properties are expected to be inter-connected. In order to characterise exoplanets, we need to investigate the planet-hosting stars. Carbon and oxygen are quite abundant and play an important role in stellar interiors by generating energy in thermonuclear reactions. Abundances of C and O may influence water availability on exoplanets. The C/O ratio also controls an amount of carbides and silicates that can be formed. Thus, we are performing a uniform study of C/O ratios in bright stars ($V < 8$ mag) located towards the northern ecliptic pole which will be targeted by the TESS and JWST space missions. The first results for a sample of 140 stars analysed are presented.

Keywords. stars: abundances

1. Introduction

Numerous studies have been performed during last years in order to determine a chemical composition of extrasolar planets (see review by Madhusudhan 2018). Since planets and their host stars are formed within the same environment their composition is expected to be interconnected (Suárez-Andrés *et al.* 2017 and references therein). Elemental ratios in stellar atmospheres are important because they govern a distribution and formation of chemical species in the protoplanetary disc. Mg/Si governs a distribution of silicates while an amount of carbides and silicates formed in planets is controlled by the C/O ratio (Thiabaud *et al.* 2015). Information of carbon and oxygen abundances serve for testing and improving planetary formation models, as C/O ratio is one of the key parameters to define structures of planets.

2. Observations and method of analysis

Spectra of programme stars were taken with the Vilnius University Echelle Spectrograph (VUES) designed and constructed at the Exoplanet Laboratory of the Yale University (Jurgenson *et al.* 2016) and mounted on the f/12 1.65 meter Ritchey-Chretien telescope at the Molėtai Astronomical Observatory of the Institute of Theoretical Physics and Astronomy, Vilnius University. A resolving power of this spectrograph is $R \approx 60\,000$ and the wavelength range is from 4000 to 8800 Å. For more information about observations see Mikolaitis *et al.* (2018).

During our observations in 2016–2017, we obtained 365 spectra of 213 FGK spectral type dwarf stars. After the primary spectral revision and identification of spectroscopic double-line binaries and fast-rotating stars, the further analysis was performed for 140 stars, for 47 of which we determined C/O ratios. Unfortunately, a large fraction of spectra were suspended from further investigation due to weakness or telluric contamination of oxygen lines.

We analysed the spectra using a differential model atmosphere technique described in Mikolaitis *et al.* (2018). Stellar atmospheric parameters were determined using traditional equivalent width (EW) based methods. EWs were measured using the DAOSPEC (Stetson & Pancino 2008) software. Stellar atmospheric parameters were computed using the 10th version of the MOOG code (Sneden 1973) using a grid of MARCS stellar atmosphere models (Gustafsson *et al.* 2008).

A spectral synthesis method was used to derive carbon and oxygen abundances. We used the forbidden line at 6300.3 Å for the O abundance determination. Two C_2 molecular bands at 5135 Å and 5635 Å were used to determine C abundances. All calculations were differential with respect to the Sun. Solar elemental abundances were taken from Grevesse *et al.* (2007). The Vienna Atomic Line Data Base (VALD, Piskunov *et al.* 1995) was used for preparation of input data. Abundances of C and O, we investigated in unison, since they are bound together by the molecular equilibrium.

3. Results

The median C/O ratio of our sample is 0.44 ± 0.04(r.m.s.). The Solar value, which we derived from spectra obtained with the same VUES spectrograph, in our study is 0.54. Thus, the distribution of C/O ratios are slightly shifted to lower values compared to Solar. Only four stars (HD 133002, HD 96511, HD 110010, and HD 135143) have C/O ratios higher than 0.65 which are necessary to form carbon rich rocky planets (Moriarty *et al.* 2014).

We investigated C/O ratios as a function of age, effective temperature and metallicity. There are no indications of C/O trends as a function of age and T_{eff}, however C/O values slightly increase with metallicity, which is in agreement with other studies (e.g. Suárez-Andrés *et al.* 2018; Brewer & Fischer 2016).

Acknowledgements

This research is funded by the European Social Fund according to the activity 'Improvement of researchers' qualification by implementing world-class R&D projects of Measure No. 09.3.3-LMT-K-712 (the Lithuanian Science Council grant No. 09.3.3-LMT-K-712-01-0103).

References

Brewer, J. M., & Fischer, D. A. 2016, *ApJ*, 831, 20

Grevesse, N., Asplund, M., & Sauval, A. J. 2007, *SSRv*, 130, 105

Gustafsson, B., Edvardsson, B., Eriksson, K., *et al.* 2008, *A&A*, 486, 951

Jurgenson, C., Fischer, D., McCracken, T., *et al.* 2016, *Journal of Astronomical Instrumentation*, 5, 1650003-239

Madhusudhan, N. 2018, arXiv:1808.04824

Mikolaitis, Š., Tautvaišienė, G., Drazdauskas, A., *et al.* 2018, *PASP*, 130, 074202

Moriarty, J., Madhusudhan, N., & Fischer, D. 2014, *ApJ*, 787, 81

Piskunov, N. E., Kupka, F., Ryabchikova, T. A., Weiss, W. W., & Jeffery, C. S. 1995, *A&AS*, 112, 525

Sneden, C. A. 1973, *Ph.D. Thesis*

Stetson, P. B., & Pancino, E. 2008, *PASP*, 120, 1332

Suárez-Andrés, L., Israelian, G., González Hernández, J. I., *et al.* 2018, *A&A*, 614, A84

Suárez-Andrés, L., Israelian, G., González Hernández, J. I., *et al.* 2017, *Highlights on Spanish Astrophysics*, IX, 567

Thiabaud, A., Marboeuf, U., Alibert, Y., Leya, I., & Mezger, K. 2015, *A&A*, 580, A30

Origins: From the Protosun to the First Steps of Life
Proceedings IAU Symposium No. 345, 2018
B. G. Elmegreen, L. Viktor Tóth & M. Güdel, eds.

doi:10.1017/S1743921318008517

Phenomenon of the wave ordering and the diagram "mean density – global period" in the Solar planetary system

Mykhaylo Skulskyy

Department of Physics, Lviv Polytechnic National University,
Bandera str. 12, 79013, Lviv, Ukraine
email: mysky@polynet.lviv.ua

Abstract. It is shown that spatial ordering of the Solar planetary system can be described by simple wave algorithms. It is detected that the dependence "mean density – global period" reflects the ordering of this system and has signs of an evolutionary diagram.

Keywords. solar system: general.

One can show that for the Solar system exist a clear dependence of $a = f(n)$ (a is the semi-major axis of the planet orbit and n is the sequence number of the planet), which is approximated by two straight lines with a bifurcation point near the asteroid belt. This predicts some differences in the spatial organization of the inner and outer groups of planets. This fact is correlated with the principles of ordering of these planet groups in the scale $L_0 = cP_0 = 19.24$ AU (here c is the speed of light and the period $P_0 = 160$ min) that were published in (Kotov *et al.* 1985). These principles are defined as follows: $2\pi a = L_0/n$ for orbits of inner planets and $2a = nL_0$ for orbits of outer planets (here n is a natural number). Since L_0 – scale has the dimension of a wave, both these principles were transformed into a wave form (Skulskyy 2015). Then, the spatial ordering for outer planets can be represented in a form: $a = n\lambda/2$. Calculated distances from the Sun to outer planets and well-known dwarf planets in the wavelengths were as follows: Jupiter – $\lambda/4$, Saturn – $\lambda/2$, Uranus – $2\lambda/2$, Neptune – $3\lambda/2$, Pluto – $4\lambda/2$, Eris – $7\lambda/2$. This corresponds to the definition of standing waves that are formed as a result of the interference of counter-directed coherent waves in the medium whose linear dimensions are multiple to $\lambda/4$ or $\lambda/2$. The principle of the orbit ordering for inner planets was transformed into a form $2\pi a = m\lambda'$ with the step $\lambda' = (1/12)\lambda/2$ and $m = 3, 6, 8, 12$. In standing waves the orbital architecture of inner planets from Mercury to Mars can be presented as $\lambda/8 : \lambda/4 : \lambda/3 : \lambda/2$. The length of the Mars orbit is directly equal to the length of the standing wave $\lambda_{SW} = \lambda/2$ as to the fundamental harmonic. The logically expected a planet with an orbit length equal to λ was not formed on the position of the asteroid belt (at the bifurcation point of the above dependence $a = f(n)$). The spatial ordering of two groups of planets, which is described by two interrelated wave algorithms, does not look random (Skulskyy 2015).

In an attempt to find out some aspects of the nature of this wave phenomenon, an attention was drawn to the proximity of the observed 160-minute oscillations of the Sun with its eigen global 167-minute oscillations of the Sun. Considering this fact the own global oscillations of all planets were calculated in (Skulskyy 2015), and a resonance relation for the three groups of planets and the Sun was discovered. Here, the results of such calculations for major bodies of the Solar system (including objects which may

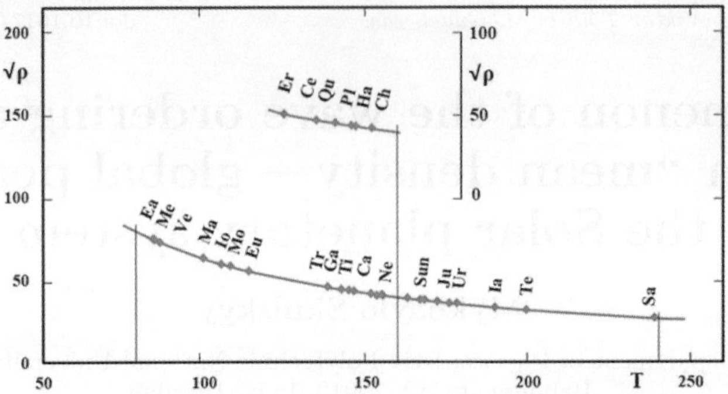

Figure 1. Dependence between periods of global oscillations and mean densities for the main objects of the Solar system (indicated by the first two letters). Planets and their satellites (Moon, Io, Europa, Ganimede, Callisto, Tethys, Dione, Rhea, Titan, Iapetus, Ariel, Umbriel, Titania, Oberon, Triton, Charon – some of them due to the density in the center of the curve are not marked) are given on the lower curve. Transneptunian objects Quaoar, Haumea, Pluto, Charon, Eris and also Ceres are given in the upper fragment. Vertical lines correspond to positions with $T = 80$, 160, 240 min.

still be in a hydrostatic equilibrium with a diameter of up to 500 km and a mass of their bodies up to 10^{21} kg) are presented. Oscillations of celestial objects with symmetric mass distribution and homogeneous gravitational field can be considered as eigen modes of central vibrators. The formula which is used to estimate such periods of celestial objects with mass M and radius R is well known: $T = 2\pi(R^3/GM)^{1/2}$ (here G is the gravitational constant). It was transformed to the form of $T = (3\pi/G\rho)^{1/2}$, where ρ is the mean density of the celestial body. As the result, the functional dependence of $T = f(\rho^{1/2})$ in a wide range of the parameter (which here is in kg/m^3) was formed for all main objects of the Solar system. As is shown in the Figure 1, this dependence has been formed of a smooth curve which is limited by the values of T-periods in the range from 80 to 240 min. Celestial objects and their groups on the curve $T = f(\rho^{1/2})$ take certain places. There is quite compact group of transneptunian dwarf planets, with Pluto and Charon among them. A somewhat isolated group of satellites of the planets with high density includes Io, Moon and Europe. These objects with a rocky surface are very close satellites to their planets and have a warmed up and dense core. Significant tidal forces provide the heating of their subsoils with subsequent sealing. In addition, the highest average density of the Earth can be explained with the tidal forces caused by the relatively close distance to the Moon. These forces can be responsible for the location and movement of celestial objects along the curve "mean density – global period" as a certain evolutionary diagram. The resulting diagram does not depend on the observed period of $P_0 = 160$ min. One can now talk about the coherence of the eigen oscillations of planets with existing eigen 167-minute oscillation of the Sun. Without the interpretation of obtained results one can conclude that wave and gravity factors are represented in obvious interconnections and create an indivisible phenomenon.

References

Kotov V. A., Kuchmi C. 1985, *Izvestija of Crimean Astrophys. Obs.*, 72, 1, 199

Skulskyy, M. Yu. 2015, *Science and Education a New Dimension. Natural and Technical Sciences*, III(5), 41, 63

Origins: From the Protosun to the First Steps of Life
Proceedings IAU Symposium No. 345, 2018
B. G. Elmegreen, L. Viktor Tóth & M. Güdel, eds.

doi:10.1017/S1743921318008530

Formation of S-type planets in close binaries: tidal capture of circumbinary planets

Jianghui Ji[1] **and Yan-Xiang Gong**[1,2]

[1]CAS Key Laboratory of Planetary Sciences, Purple Mountain Observatory, Chinese Academy of Sciences, Nanjing 210008, China
email: jijh@pmo.ac.cn

[2]College of Physics and Electronic Engineering, Taishan University, Taian 271000, China
email: yxgong@pmo.ac.cn

Abstract. In this work, we proposed a possible mechanism for the formation S-type planet in close binaries (0.5 au $< a_B <$ 3 au). Numerical simulations showed that the maximum capture probability is $\sim 10\%$, which can be comparable to the tidal capture probability of hot Jupiters in single star systems. The capture probability is related to binary configurations. Furthermore, we find that S-type planets with retrograde orbits can be naturally produced via capture process. These planets on retrograde orbits can help us distinguish in situ formation and post-capture origin for S-type planet in close binaries. The forthcoming missions (PLATO or TESS) will provide the opportunity and feasibility to detect such planets.

Keywords. binaries (including multiple): close, planetary systems: formation, methods: n-body simulations.

1. Introduction

As of today, about 130 exoplanets were found in the binary star system. Over 100 of them are S-type planets, and others are P-type planets. S-type planets have not yet been found in the binaries with an orbital period P < 1000 days. The space missions, e.g., PLATO or TESS, will offer the opportunity to discover them. However, according to the classical theories of planetary formation, in situ formation of S-type planets in close binaries is very challenging. Compared with the difficulty in formation of S-type planets in close binaries, it is comparatively easy to yield P-type planets around them. Observations provide evidence that P-type planets around close binaries would be common in the universe (Armstrong *et al.* 2014). Here we aim to explore the formation of S-type planets in close binaries by considering the scenario of scattering of P-type planets and tidal interactions from stars, thereby turning P-type orbit into S-type orbit. Convergent migration of two adjacent circumbinary planets in a multi-planet system will cause planetary scattering (Pierens & Nelson (2008), Kley & Haghighipour (2015)).

2. Tidal capture scenario

From the point of view of CRTBP, there exists a Jacobi constant satisfying the condition $C_J > C_J(L_2)$ for orbits to remain stable. It is impossible for the planet to penetrate the forbidden area around the binary and approach either of the stars. However, in a multiple CBP system, C_J of a planet is not perfectly conserved when one considers the gravitational perturbations of the other planets. Especially the change of C_J is significant during close encounters with other planets. Planet-planet scattering can bring about a sudden decrease in C_J, to make $C_J < C_J(L_2)$. When this condition is satisfied, the planet

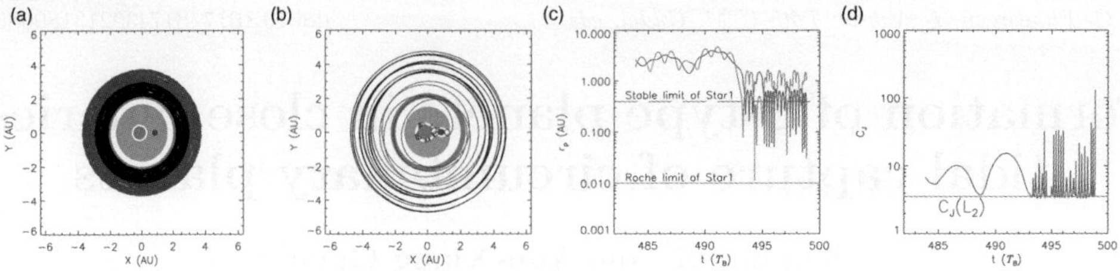

Figure 1. An example of scattering induced tidal capture of circumbinary planet. (a) Two circumbinary planets are modeled as test particles. The thick red line is the unstable boundary outside the binary. The brown regime is the zero velocity surface of $C_J = 4$ for reference. (b) The two planets have identical initial conditions as in Panel (a), but the scattering between two planets has been considered. (c) Red line is time varying distance between the captured planet and the secondary. The blue line shows the distance of the same planet from the primary. (d) The Panel shows the Jacobi constant of the captured planet varies with time.

can approach either of the binary. If the planet is scattered far enough away from the other planets, then its C_J might maintain a value on appreciable time-scales. Again, the motion of the scattered planet can be dictated by the CRTBP as long as the perturbations from the other planets are negligible. If we do not consider other dissipations such as tides, the scattering lead to a *temporary capture*. In most cases planets will be scattered out of the system or collide with two stars. However, tidal interaction from the star becomes more important when the periastron of the planet is very close to either of binary. Therefore, tidal effects between the planet and the star will further diminish the orbital energy of the planet, thereby producing $C_J > C_J(L_2)$ again. Under such circumstance, a *permanent capture* of the planet can form. Figure 1 shows an example of this scenario using an artificial two-planet system.

3. Conclusions

We study the scattering scenario of 3 circumbinary planets in the systems (see Gong & Ji (2018)). The major conclusions are summarized as follows: The scattering induced tidal capture mechanism can turn a P-type orbit into an S-type orbit. S-type planets in close binaries with SMA of $0.5 - 3$ au can be yielded through this capture scenario. The formation probability varies with the mass ratio and eccentricity of the binary. The maximum capture probability is more than 10%. Retrograde S-type planets can be produced via tidal capture mechanism. They can be adopted as an important indicator that distinguishes in situ formation and post-capture formation scenario.

Acknowledgments

This work is supported by National Natural Science Foundation of China (Grants No. 11773081, 11573018), the Foundation of Minor Planets of Purple Mountain Observatory.

References

Armstrong D. J., Osborn H. P., Brown D. J. A., Faedi F., Gómez Maqueo Chew Y., Martin D. V., Pollacco D., & Udry S. 2014, *MNRAS*, 444, 1873

Gong Y.-X., Ji J. H. 2018, *MNRAS*, 478, 4

Kley W., Haghighipour N. 2015, *A&A*, 581, A20

Pierens A., Nelson R. P. 2008, *A&A*, 478, 939

Origins: From the Protosun to the First Steps of Life
Proceedings IAU Symposium No. 345, 2018
B. G. Elmegreen, L. Viktor Tóth & M. Güdel, eds.

doi:10.1017/S1743921318008542

The life cycle of large carbonous molecules in the interstellar medium

Tao Chen[1,2]

[1]Leiden University, Leiden Observatory, Niels Bohrweg 2, NL-2333 CA Leiden, Netherlands

[2]School of Engineering Sciences in Chemistry, Biotechnology and Health, Department of Theoretical Chemistry & Biology, Royal Institute of Technology, 10691, Stockholm, Sweden
email: chen@strw.leidenuniv.nl

The interstellar medium (ISM) contains a rich molecular inventory, where various chemical reactions can take place. This work review the recent quantum chemical computations and laboratory experiments on the study of fragmentation and molecular growth of large carbonous molecules in the ISM.

One of the most abundant carbonous molecules in the ISM are polycyclic aromatic hydrocarbons (PAHs), which lock up 10% – 20% of the carbon in the ISM and are commonly observed in a variety of astronomical objects, ranging from the Galaxy to extragalactic regions (Tielens 2008). Beside PAHs, fullerenes have been identified in the ISM (Cami *et al.*, 2010). And ~ 5 ppm of C/H, i.e., $\sim 1.9\%$ of the total interstellar C are locked on the interstellar graphene (Chen, Li, & Zhang 2017). Interestingly, both experimental and theoretical studies prove that fullerenes and graphene flakes can be formed from large PAHs through dehydrogenations and isomerizations (Berné & Tielens 2012; Zhen *et al.* 2014).

Figure 1 shows some possible carbonous reactants and their reaction pathways in the ISM. In brief, the gas-phase molecules can be ionized, isomerized, dissociated following the impacts of ions, atoms, or photons (Chen *et al.* 2014; Stockett *et al.* 2014, 2015; Zhen *et al.* 2016). Regarding ionizations, the quantum chemical calculations show that the ionization energies follow linear dependencies as functions of charge state for different size of PAHs. And the total ionization and fragmentation cross sections in ion-PAH collisions display markedly different size dependencies for pericondensed and catacondensed PAH molecules, reflecting differences in their first ionization energies. PAHs in charge states above certain limits may survive in experimental time scales due to the presence of reaction barriers (Holm *et al.*, 2011).

About dissociations, we find two types of fragmentation processes: statistical and non-statistical fragmentations. In statistical fragmentation processes, the losses of $C_{2n}H_x$ dominate the mass spectra (Chen *et al.* 2015). Non-statistical fragmentations refer to direct knock-out of atoms from a molecule, and single C-losses are commonly observed in such process (Stockett *et al.* 2014). The fragmentation of molecules lead to the formation of small molecules or isomerize to more stable structure, such as C_{60} or C_{70} (Zhen *et al.* 2014).

Beside fragmentation, small molecules may conglomerate a weakly bonded cluster, and form large molecules in the cluster following statistical or non-statistical fragmentations (Zettergren *et al.* 2013; Delaunay *et al.* 2015; Zhen, Chen, & Tielens 2018). In such reaction cycle, molecular hydrogens (Chen *et al.* 2015) and new carbonous molecules are produced, e.g., the bowl-shape molecules are found in the photodissociation experiement of bisanthenequinone cations (Chen *et al.* 2017), dumbbell C_{118} and C_{119} are formed

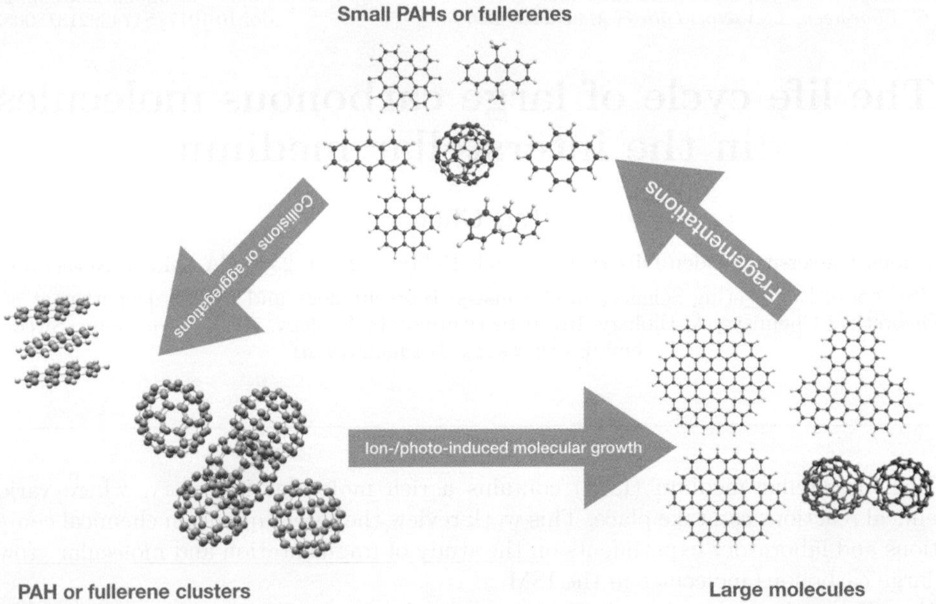

Figure 1. The life cycle of polycyclic aromatic hydrocarbons (PAHs) in the ISM. Small PAH molecules may form weakly bonded clusters through aggregations or soft collisions among molecules. Large molecules can be formed in the clusters rapidly following the impacts of ions, atoms, or photons. On the other hand, large PAHs may also dissociate to small molecules through statistical fragmentation or non-statistical fragmentation processes.

inside clusters of C_{60} by collision with α particles (Zettergren *et al.* 2013), etc. These reactions enrich the molecular inventory in the ISM.

References

Berné, O., & Tielens, A. G. 2012, *Proceedings of the National Academy of Sciences*, 109, 401

Chen, T., Zhen, J., Wang, Y., Linnartz, H., & Tielens, A. G. 2017, *Chemical Physics Letters*, 692, 298

Chen, X. H., Li, A., & Zhang, K. 2017, *The Astrophysical Journal*, 850, 104

Chen, T., Gatchell, M., Stockett, M. H., *et al.* 2014, *The Journal of chemical physics*, 140, 224306

Chen, T., Gatchell, M., Stockett, M. H., *et al.* 2015, *The Journal of chemical physics*, 142, 144305

Delaunay, R., Gatchell, M., Rousseau, P., *et al.* 2015, *The journal of physical chemistry letters*, 6, 1536

Stockett, M. H., Zettergren, H., Adoui, L., *et al.* 2014, *Physical Review A*, 89, 032701

Stockett, M. H., Gatchell, M., de Ruette, N., *et al.* 2015, *International Journal of Mass Spectrometry*, 392, 58

Tielens, A. G. G. M. 2008, Annu. Rev. Astron. *Astrophys.*, 46, 289

Zettergren, H., Rousseau, P., Wang, Y., *et al.* 2013, *Physical review letters*, 110, 185501

Zhen, J., Castellanos, P., Paardekooper, D. M., Linnartz, H., & Tielens, A. G. 2014, *The Astrophysical Journal Letters*, 797, L30

Zhen, J., Chen, T., & Tielens, A. G. 2018, *The Astrophysical Journal*, 863, 128

Zhen, J., Castillo, S. R., Joblin, C., *et al.* 2016, *The Astrophysical Journal*, 822, 113

Origins: From the Protosun to the First Steps of Life
Proceedings IAU Symposium No. 345, 2018
B. G. Elmegreen, L. Viktor Tóth & M. Güdel, eds.

doi:10.1017/S1743921318008554

Volatile organic compounds stability at exoplanetary atmospheres

Thiago Monfredini[1], Manoel Gustavo P. Homem[2] and Douglas Galante[1]

[1]Brazilian Synchrotron Light Laboratory (LNLS/CNPEM)
Rua Giuseppe Maximo Scolfaro, 10000, CEP 13083-100, Campinas, SP, Brazil
email: thiago.monfredini@lnls.br

[2]Department of Chemistry (DQ/UFSCar)
Rod. Washington Luís, km 235 - SP-310, CEP 13565-905, São Carlos, SP, Brazil

Keywords. methods: laboratory, molecular processes, molecular data, astrochemistry, techniques: spectroscopy, ultraviolet: stars

1. Introduction

Species such as isoprene and other terpenoids are reactive volatile organic compounds (VOCs). They can be quite reactive under terrestrial conditions, as they react very rapidly with -OH radicals in the atmosphere, resulting in short chemical lifetimes (minutes to hours). However, they are emitted in significant amounts by trees and other vegetations. So, the emission of biogenic VOCs plays an important role also in the photophysics and photochemistry of the atmosphere, affecting its oxidative capacity (de Souza *et al.* 2007; Sharkey *et al.* 2008; Gu *et al.* 2017). As they are biogenic products, in different conditions from Earth they could constitute relevant biosignature in the atmospheres of exoplanets (Seager *et al.* 2012).

In the context of the relation star-planet, it is important to understand the interaction of XUV photons from the star inciding upon the planetary atmosphere: how its XUV photons interact with these biosignatures, in terms of photoabsorption and ionization efficiency in the upper atmosphere of the planet.

2. Methodology

The absolute photoabsorption of isoprene was measured in the range of 13 to 20 eV of photon energy. The source of UV monochromatic photons was the TGM (Toroidal Grating Monochromator) beamline of the Brazilian Synchrotron Light Laboratory (LNLS/CNPEM), at Campinas, São Paulo (Brazil). It provides photons with flux up to 10^{10} photons s^{-1} and resolution of 0.05 eV. The beamline was operated with a toroidal grating of Au, with 1800 l/mm, working in the range of 13 to 100 eV of photon energy. A neon-gas filter was used to avoid harmonic contamination (Cavasso-Filho *et al.* 2007).

In order to obtain the photoabsorption measurements, we developed a cylindrical ionization chamber with four stages (four copper plates, same size) to signal detection. It works with a base pressure of 10^{-3} mbar and it is filled with the sample in gas-phase until it reaches a pressure of ∼0.1 mbar. A voltage of 60 V is applied in the cylindrical plate. A capacitance manometer (MKS Baratron®) was used for pressure monitoring inside the chamber. Digital pico-ammeters (Keithtley Instruments) were connected to

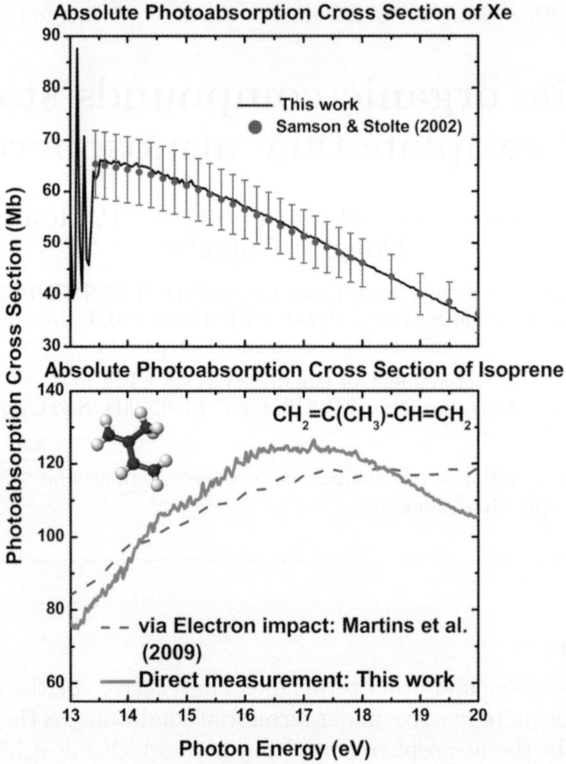

Figure 1. The upper panel shows the photoabsorption cross section of Xe-gas (calibration) in good agreement with Samson & Stolte (2002). The bottom panel shows our results in comparision with Martins *et al.* (2009), estimated via electron impact technique.

each stage to measure the signal response to the incident photon beam upon the gas. The calibration was achieved through measurements of Xe, based in the absolute photoabsorption cross-section data from Samson & Stolte (2002). The sample of isoprene was purchased commercially from Sigma-Aldrich, with a purity of 99.5%. We estimate a typical uncertainty of 10% in the cross section evaluation.

3. Discussion

With this technique, we could determine the absolute photoabsorption cross-section of the gases analyzed. Fig. 1 shows two cases: with isoprene and with Xe. The latter has a very known profile and can be used as a reference. The bottom panel of the Fig. 1 shows the isoprene measurements with the ionization chamber (this work) and with electron impact technique (Martins *et al.* 2009). The differences between then lies down the uncertainty limit of 10% of our data, suggesting a good agreement. With this methodology, also associated with mass-spectrometry techniques, it is possible to determine other important parameters of organic molecules in harsh conditions, such as the ionization efficiency and its absolute photoionization cross-section.

Acknowledgments

This research was supported by the Brazilian agencies CNPq, CAPES (Finance Code 001), FAPESP (Grant No. 2015/08258-2), and by LNLS/CNPEM.

References

Cavasso-Filho, R. L., Homem, M. G. P., Fonseca, P. T. & de Brito, N. A. 2007, *Review of Sci. Instrum.*, 78, 115104

de Souza, G. G. B., Coutinho, L. H., Nunez, C., Bernini, R., Castilho, R. B. & Lago, A. F. 2007, *J. of Phys.: Conf. Series*, 88, 012205

Gu, D., Guenter, A. B., Shilling, J. E., Yu, H., Huang, M., Zhao, C., Yang, Q., Martin, S. T., Artaxo, P., [...], Santos, F.C., Leng, G. & Hu, Z. 2017, *Nature communications*, 8, 15541

Martins, G., Ferreira-Rodrigues, A. M., Rodrigues, F. N., de Souza, G. G. B., Mason, N. J., Eden, S., [...] & Limão-Vieira, P. 2009, *Phys. Chem. Chem. Phys.*, 11, 11219

Samson, J. A. R. & Stolte, W. C. 2002, *J. Elec. Spectrosc. Rel. Phenomena*, 123, 265

Seager, S., Schrenk, M. & Bains, W. 2009, *Astrobiology*, 12, 61

Sharkey, T. D., Wiberley, A. E. & Donohue, A. R. 2008, *Annals of Botany*, 101, 5

Origins: From the Protosun to the First Steps of Life
Proceedings IAU Symposium No. 345, 2018
B. G. Elmegreen, L. Viktor Tóth & M. Güdel, eds.

doi:10.1017/S1743921318008566

Soil Characterization: Martian Analogues

Thiago Monfredini[1], Verônica Teixeira[1], Fabio Rodrigues[2] and Douglas Galante[1]

[1]Brazilian Synchrotron Light Laboratory (LNLS/CNPEM)
Rua Giuseppe Maximo Scolfaro, 10000, CEP: 13083-100, Campinas, SP, Brazil
email: `thiago.monfredini@lnls.br`, `douglas.galante@lnls.br`

[2]Institute of Chemistry (IQ/USP)
Av. Prof. Lineu Prestes, 748, CEP: 05508-000, São Paulo, SP, Brazil

Keywords. astrobiology, methods: laboratory, Earth, planets and satellites: (Mars)

1. Introduction

The investigation of microbial life and/or inorganic biosignatures on Mars has been done mainly by the study of analogue environments on Earth. The nature of the ferriferous regions in Brazil seems to be suitable, as a substantial sources of extremophiles (Duarte *et al.* 2012; Camêlo *et al.* 2018) and potential source of inorganic biosignatures. This work is aimed to improve the knowledge needed to test the possibility of present or evidence of past life on Mars by characterizing samples of the iron-rich soil of the region of Diamantina, located at the center of Brazil. This characterization consists of examining the atomic and molecular compositions and the structural, mechanical and optical properties of the soil. This is done by applying different experimental techniques, such as thermo-gravimetry and X-ray spectroscopy. The molecular and atomic structure had been measured through, both, X-ray diffraction and fluorescence at LNLS (Brazilian Synchrotron Light Laboratory - Campinas). Also, the volatile content was investigated through thermo-gravimetry in a high-temperature furnace.

The sample of soil was extracted from region nearby the Red river, at the city of Diamantina - MG, Brazil. In the present work, we show the results of three different techniques applied to examine the sample: X-ray powder diffraction, x-ray fluorescence, both measured in two beamlines at the LNLS, and thermo-gravimetry, at the chemical lab of the LNLS.

2. Experimental Procedures

X-ray powder diffraction: The XRD1 beamline at the LNLS is a dedicated X-ray diffraction beamline. Monochromatization is performed using a double-crystal Si (111) monochromator. The system is calibrated to use a well determined energy of 12 keV, with of photon flux of 7.8×10^9 photons s^{-1}. It was used NIST standard reference materials (SRM): Si (SRM 640d) and LaB_6 (SRM 660b). The SRM as well as the sample were placed, respectively, inside 0.3 and 0.7 mm-diameter borosilicate capillaries, which are fixed in ferromagnetic stainless steel holders. They are connected with a magnetic tip attached to the three-circle N3050-P1 diffractometer, which is able to rotate, spinning the sample at $\sim$300 r.p.m. during the measurements. A MYTHEN 24K detector was used. All measurements were done at room temperature (Carvalho *et al.* 2016).

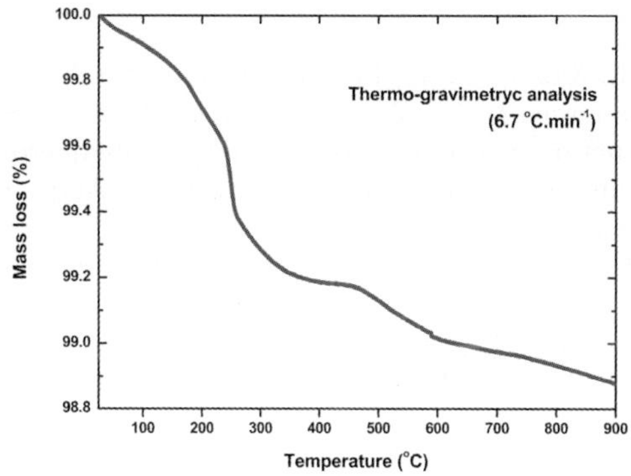

Figure 1. Thermo-gravimetryc analysis of a soil sample from Diamantina-MG (Brazil), suggesting its natural highly dry state.

X-ray fluorescence: The XRF beamline at the LNLS is a dedicated X-ray fluorescence spectroscopy beamline. It is equipped with a silicon (111) channel-cut crystal monochromator. The measurements were performed with the micro-XRF setup, which include an optical microscope with motorized zoom, and X, Y, Z, Θ_z sample stages that are fully remote controlled. The beamline operates between 5 to 20 keV of photon energy. It was used the white-beam mode, with $\sim 1\times 10^{10}$ photons s^{-1}. The fluorescence detection is performed by energy-dispersive solid-state detectors, equipped with HPGe and Si(LI) crystals. For soil analysis, NIST SRM 2711a, containing 25 elements in 50 g of dried, powered soil was used. With this technique, it is possible to detect fluorescence of the electronic K-shell of elements with $Z > 13$ (Perez *et al.* 2002).

Thermo-gravimetry: Thermo-gravimetric analysis (TGA) is a method of thermal analysis in which the mass changes of a sample (in the scanning mode) are measured as a function of increasing temperature, with constant heating rate. The mass loss indicate the mass of the volatile content of the sample. It was applied a ramp of temperature to examine the sample of 6.7^oC min^{-1}, up to a temperature of 900^oC.

3. Results and Discussion

The complementarity of the XRD and XRF techniques suggest the presence of quartz (SiO_2) and hematite (Fe_2O_3). These crystalline patterns on the Mars surface are widely known (Morris *et al.* 2000; Ming *et al.* 2008). Also, there is no evidence of the presence of phyllosilicate minerals. Thermo-gravimetry analysis (Fig. 1) showed weight losses of $\sim$1 wt% at 600^oC. For comparison, from Allen *et al.* (1998), JSC Mars-1, a well-known Martian analog soil, has a large volatile content, with losses of 21.1 wt% in the same conditions. Viking in situ experiments of the Martian soil released 1.0 wt% water from samples heated at 500^oC (Allen *et al.* 1998), which suggest that the Martian regolith is extremely dry (Biemann *et al.* 1977). So, our results show good evidence that Diamantina's soil can be a good candidate for Martian simulant.

Acknowledgments

This research was supported by the Brazilian agencies CNPq, CAPES (Finance Code 001) and by LNLS/CNPEM.

References

Allen, C. C., Jager, K. M., Morris, R. V., Lindstrom, D. J., Lindstrom, M. M. & Lockwood, J. P. 1998 in: R. G. Galloway, & S. Lokaj (eds.), *Space 98*, Proc. Sixth Int. Conf. Expos. on Engineering, Construction, and Operations in Space (Albuquerque), p.469

Biemann, K., Oro, J., Toulmin, P., Orgel, L. E., Nier, A. O., Anderson, D. M., [...] & Lafleur, A. L. 1977, *J. Geophysical Res.* 82, 4641

Camêlo, D. L., Gilkes, R. J., Leopold, M., Silva, A. C. & Vidal-Torradoa, P. 2018, *Catena*, 170, 397

Carvalho, A. M. G., Araújo, D. H. C., Canova, H. F., Rodella, C. B., Barrett, D. H., Cuffini, S. L., Costa, R. N. & Nunes, R. S. 2016, *J. Synchrotron Rad.* 23, 1501

Duarte, R. T. D., Nóbregal, F., Nakayama, C. R. & Pellizari, V. H. 2012, *Int. J. Astrobiology*, 11, 325

Ming, D. W., Gellert, R., Morris, R. V., Arvidson, R. E., Brückner, J., Clark, B. C, Cohen, B. A., dUston, C., Economou, T. [...] & Zipfel, J. 2008, *J. Geophysical Res.* 113, E12S39

Morris, R. V., Golden, D. C., Bell III, J. F., Shelfer, T. D., Scheinost, A. C., [...], Ming, D. W., Allen, C. C. & Britt, D. T. 2000, *J. Geophysical Res.* 105, 1757

Pérez, C. A., Bueno, M. I. S., Neuenshwander, R. T., Sánchez, H. J., & Tolentino, H. 2002, *Review of Sci. Instrum.* 73, 1563

Origins: From the Protosun to the First Steps of Life
Proceedings IAU Symposium No. 345, 2018
B. G. Elmegreen, L. Viktor Tóth & M. Güdel, eds.

doi:10.1017/S1743921318008578

Star Clusters Triggered by GS242-03+37

Jan Palouš and Soňa Ehlerová

Astronomical Institute, Academy of Sciences of the Czech Republic,
Boční II-1401, 141 00 Praha 4, Czech Republic
email: palous@ig.cas.cz

Abstract. We explore the Milky Way supershell GS242-03+37. We argue that the observed HI distribution can be explained as an expanding structure about 100 Myr old powered with a modest energy released by an OB association. The formation of star clusters has been triggered less than 30 Myr ago when the ISM density in the supershell increased due to the galactic differential rotation. The observed age sequence of young star clusters is related to the evolution of the column density during the supershell expansion.

Keywords. ISM: structure, ISM: bubbles, Galaxy: disk, open clusters and associations: general

1. Introduction

The interstellar medium (ISM) is composed of several components: very hot X-ray gas, warm ionized HII, warm and atomic HI, and cold and dusty molecular medium. It is highly structured from a relatively smooth hot component, through a diffuse atomic medium, and to more compact molecular clouds. Stars are formed in very dense and cold cores inside of molecular clouds. Surveys of HI in the Milky Way and in other nearby galaxies discovered shells and bubbles: Heiles (1979), McClure-Griffiths *et al.* (2002), Ehlerová & Palouš (2005, 2013); Bagetakos *et al.* (2011). Some of these shells are expanding and collecting ISM into narrow sheets of higher density and lower temperature. We analyze the connection between HI shells, molecular clouds and young star clusters: if the star clusters could have been created out of molecular clouds whose formation have been triggered by expanding shells. The statistical correlation between shells and CO distribution in the Milky Way show increased frequency of molecular clouds near the high density shell walls (Ehlerová & Palouš 2016) proving that at least some of the clouds form out of gas collected by shells.

In this contribution, we explore the giant supershell GS242-03+37 discovered by Heiles (1979). The detailed description of it is given by McClure-Griffiths *et al.* (2006) with the estimates of its age and energy necessary for its formation. Here we propose a different scenario of its formation and evolution including the effect of the galactic differential rotation. We propose for GS242-03+37 a much larger age of about 100 Myr and a much smaller energy necessary for its formation corresponding to a few tens of supernovae. The details of our analysis are described in Ehlerová & Palouš (2018).

2. GS242-03+37 and Galactic Differential Rotation

GS242-03+37 is a giant supershell of a kpc size. In any model of formation and evolution of such large structure we need to include the effect of the galactic differential rotation. This was explored by the hydrodynamical simulations of expanding shells using infinitesimaly thin shell apoproximation adopted in the code RING described by Ehlerová & Palouš (2018). The shape of GS242-03+37 and its observed expansion velocity may be explained as a combination of original expansion velocity decelerated

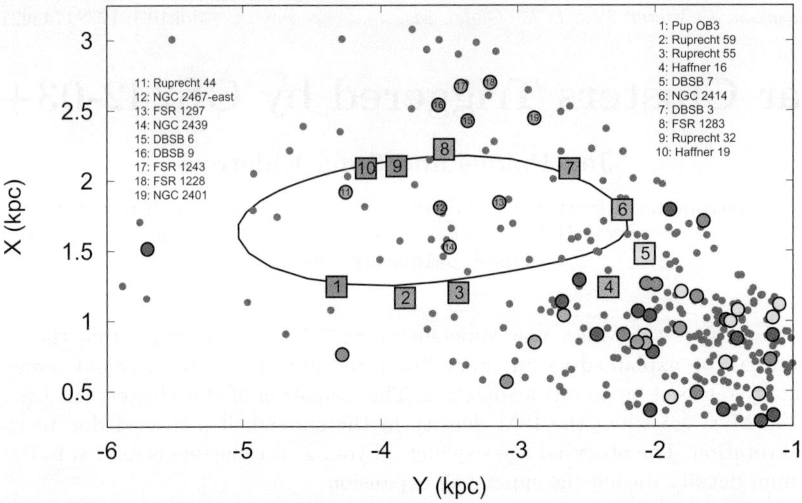

Figure 1. Supershell (solid line) and star clusters (age < 120 Myr - colors, age > 120 Myr - black, inside of the wall - squares). Projection to the galactic (X, Y) plane.

due to accretion of an ambient ISM and the galactic differential rotation. An implication is that it needs about 100 Myr to reach its actual shape in the differentially rotating galactic disk. The energy necessary for its formation is only about 1.3 10^{52} erg since its observed velocity extent contains also the galactic differential rotation. The large age opens the question on its coherence. We argue that its position in the outer part of the Milky Way in the region near to the corotation with spiral arms causes low frequency of perturbations, which makes its long survival possible.

3. GS242-03+37 versus Star Clusters

The positions of young (< 120 Myr) open star clusters correlate with the shape of GS242-03+37. This is shown in Fig. 1, where we see that star clusters reside preferentially in supershell walls. On the top of it, the older clusters are seen near its tips and younger clusters on its sides. We explain it as an effect of the mass accummulation, which is more effective at the supershell tips compared to its walls.

References

Bagetakos, I., Brinks, E., Walter, F., *et al.* 2011, *AJ*, 141, 23
Ehlerová, S., & Palouš, J. 2005, *A&A*, 437, 101
Ehlerová, S., & Palouš, J. 2013, *A&A*, 550, 23
Ehlerová, S., & Palouš, J. 2016, *A&A*, 587, 5
Ehlerová, S., & Palouš, J. 2018, *A&A*, 619, 101
Heiles, C. 1979, *ApJ*, 229, 533
McClure-Griffiths, N.M., Dickey, J.M., Gaensler, B.M., & Green, A.J. 2002, *ApJ*, 578, 176
McClure-Griffiths, N.M., Ford, A., Pisano, D.J., *et al.* 2006, *ApJ*, 638, 196

Origins: From the Protosun to the First Steps of Life
Proceedings IAU Symposium No. 345, 2018
B. G. Elmegreen, L. Viktor Tóth & M. Güdel, eds.

doi:10.1017/S1743921318008591

The common origin of family and non-family asteroids: Implications for meteorites and NEAs

Stanley F. Dermott[1], Dan Li[2] and Apostolos A. Christou[3]

[1]University of Florida, Department of Astronomy, Gainesville, FL 32611, USA
email: sdermott@ufl.edu

[2]National Optical Astronomical Observatory Tucson, AZ, USA
email: dli@noao.edu

[3]Armagh Observatory and Planetarium, College Hill, Armagh BT61 9DG, UK
email: Apostolos.Christou@armagh.ac.uk

Abstract. The observed size-frequency distributions (SFDs) of the five major asteroid families in the Inner Main Belt (IMB), defined by Nesvorný (2015) using the Hierarchical Clustering Method (Zappala *et al.* 1990), are distinctly different and deviate significantly from the linear log-log relation described by Dohnanyi (1969). The existence of these differences in the SFDs, and the fact that the precursor bodies of the major families have distinctly different eccentricities and inclinations, provides an explanation for the observations that the mean sizes of both the family and the non-family asteroids are correlated with their mean proper eccentricities and anti-correlated with their mean proper inclinations. We deduce from this, and from the fact that the SFDs of the family and the non-family asteroids are almost identical, that the family and most of the non-family asteroids in the IMB have a common origin (Dermott *et al.* 2018).

Keywords. Asteroids, minor planets

1. Observations

Further insight into the origin and evolution of the IMB is gained by separating the asteroids into three groups: (1) family asteroids as defined by Nesvorný (2015); (2) halo asteroids that appear to be closely associated with the family asteroids; and (3) all other non-family asteroids that we designate as non-halo asteroids. The five major families in the IMB (Flora, Vesta, Nysa, Polana and Eulalia) are clearly surrounded by halos of asteroids. While this suggests that these halo asteroids originate from the major families, because of the overlap of the families and their halos in proper orbital element space, use of the Hierarchical Clustering Method Zappala *et al.* (1990) to assign halo asteroids to a particular family is difficult or impossible. However, if we do not attempt to assign a halo asteroid to a particular family, then it is possible to argue that halo asteroids are family asteroids and to separate family asteroids and their surrounding halos from all other asteroids. Note, our chief aim here is not to separate the family and halo asteroids, or to assign a halo asteroid to a particular family, but to define a set of non-halo asteroids that are well separated from the family asteroids and their halos. Some characteristics of the family, halo and non-halo asteroid groups are shown in Figure 1 (Dermott *et al.* 2018).

2. Implications

These new results imply that we must seek explanations for the differing characteristics of the various meteorites and near-Earth asteroids originating from the IMB in

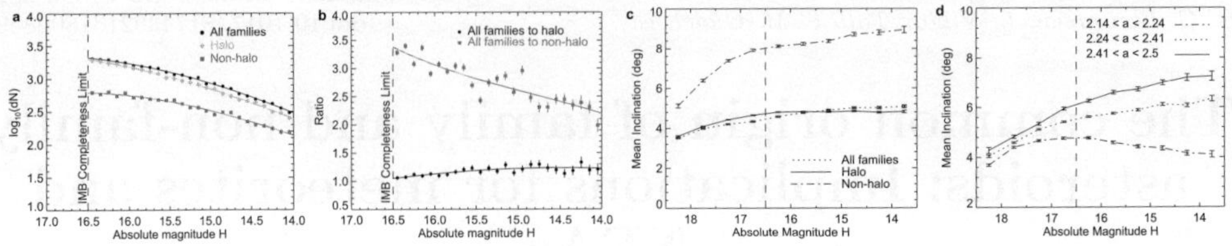

Figure 1. a, Size frequency distributions (SFDs) of the three asteroid groups in the IMB: family, halo and non-halo. b, Ratios of the number of asteroids in a box of width dH=0.1 for all-family to halo, and for all-family to non-halo asteroids. c, Dependence of the mean proper inclination on absolute magnitude for the family, halo and non-halo asteroids. d, Dependence of the mean proper inclination on absolute magnitude for the non-family asteroids in the IMB in three ranges of semi-major axis.

the evolutionary histories of a few, large, precursor bodies. Our findings support the model that asteroids formed big through the gravitational collapse of material in a protoplanetary disk (Johansen *et al.* 2003), but we cannot rule out the formation of very small asteroids as any small primordial bodies would have now been lost from the belt. Our conclusions are based on observations of the present asteroid belt. However, if the initial asteroid belt consisted of a comparatively small number of large bodies formed in a near-coplanar, protoplanetary disk, then even after the excitation of the asteroid orbital eccentricities and inclinations, the initial asteroid collision rate, and the rate of supply of small bodies to the inner solar system, would have been correspondingly low and would only have increased to larger values at a later date. This slow increase in the early asteroid collision rate could have implications for the Late Heavy Bombardment.

References

Dermott, S. F., Li, D., Christou, A. A., Kehoe, T. J. J., & Robinson, J. M. 2018, *Nat. Astron.*, 2, 549

Dohnanyi, J. S. 1969, *J. Geophys. Res.*, 74, 2531

Johansen, A., Oishi, J. S., Mac Low, M-M., Klahr, H., Fleming, T., & Youdin, A. 2007, *Nature*, 448, 1022

Nesvorný, D. 2015, EAR-A-VARGBDET-5-NESVORNYFAM-V3.0. NASA Planetary Data System

Zappala, V., Cellino, A., Farinella, P., & Knežević, Z. 1990, *Astron. J.*, 100, 2030

Origins: From the Protosun to the First Steps of Life
Proceedings IAU Symposium No. 345, 2018
B. G. Elmegreen, L. Viktor Tóth & M. Güdel, eds.

doi:10.1017/S1743921318008608

Retention of Small Charged Dust in Planet Forming Disks

Vitaly Akimkin

Institute of Astronomy, Russian Academy of Sciences
119017, Pyatnitskaya str. 48, Moscow, Russia
email: akimkin@inasan.ru

Abstract. Dust evolution in disks around young stars is a key ingredient for the global disk evolution and accompanying planet formation. The mutual sticking of initially small grains is not straightforward and can be hampered by several processes. This includes dust grain bouncing, fragmentation, electrostatic repulsion and fast drift to the central star. In this study we aim at theoretical modeling of the dust coagulation coupled with the dust charging and disk ionization calculations. We show that the electrostatic barrier is a strong restraining factor to the coagulation of micron-size dust. While the sustained turbulence helps to overcome the electrostatic barrier, dust fluffiness limits this opportunity. Coulomb repulsion may keep a significant fraction of $0.1 - 10\mu$m dust in large regions of protoplanetary disks.

Keywords. accretion, accretion disks, plasmas, turbulence, dust, extinction, cosmic rays

1. Introduction

Dust grains in astrophysical environments are not expected to be electrically neutral (Jung 1937, Spitzer 1941). Among a variety of dust charging mechanisms (Weingartner 2004, Fortov *et al.* 2005), plasma charging, i.e. grain collisions with free electrons and ions, is the most important mechanism in dense dark regions of protoplanetary disks. As electrons are more mobile than ions, the equilibrium grain charge is negative in the case of plasma charging. Photoelectric charging can dominate in illuminated disk upper layers and provide positive grain charge. Okuzumi (2009) showed that Coulomb repulsion of like-charged grains can be strong enough to prevent coagulation of micron-size particles in protoplanetary disk environments.

2. Charged dust coagulation

Two key sources of grain relative velocities are the Brownian motion and turbulence. As the Brownian velocities drop with the grain mass and the grain charge linearly scales with grain radius, then the purely Brownian coagulation will be electrostatically blocked for large enough grains ($0.1 - 10\mu$m; Akimkin 2015). Thus, only non-thermal velocities can be responsible for the overcoming of the electrostatic barrier. If the turbulence arises due to the magneto-rotational instability (MRI), its strength should depend on the gas ionization. The MRI is quenched for low ionization degrees $x_{\rm e} < x_{\rm crit} \simeq 10^{-13}$, that diminishes non-thermal grain velocities in a dead zone. The value of $x_{\rm crit}$ is comparable with the abundance of micron-size grains, which can soak up free electrons, so grains play an important role in MRI development. Thus, dust evolution, grain charging and MRI are deeply interrelated processes and must be considered simultaneously.

We model the charged dust evolution by solving Smoluchowski equation locally for a large set of positions in the vertical 2D cut of a protoplanetary disk with

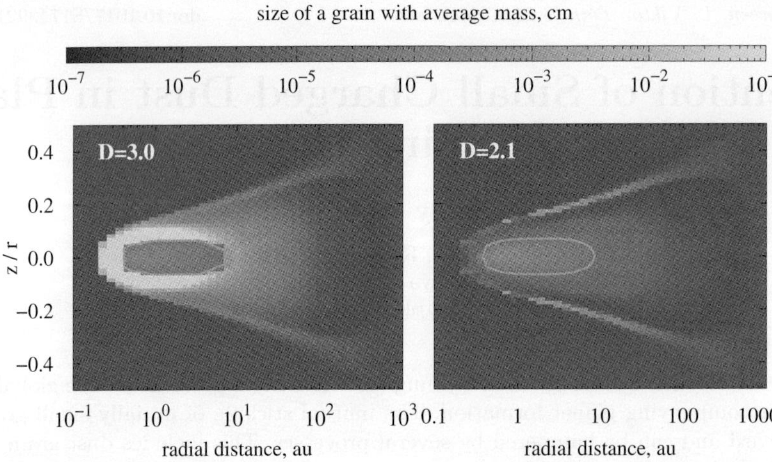

Figure 1. Average grain size a after 10^5 yr of *in situ* dust coagulation for two choices of aggregate fractal dimension $D = 3.0$ and 2.1. Orange solid line shows the location of the dead zone ($\alpha_{\rm dead} = 10^{-6}, \alpha_{\rm active} = 10^{-3}$) at 10^5 yr, which we define as a region with low ionization degree. Dust fragmentation, radial drift and sedimentation are not considered. Small average grain sizes ($< 10\mu$m) trace either regions with strong electrostatic barrier (dead zone for $D = 3.0$, almost the whole disk for $D = 2.1$) or just low-density upper and outer disk parts. Macroscopic grain size ($a > 100\mu$m; yellow colour) corresponds to the charge-free coagulation.

typical parameters. The disk charge structure is calculated from the gas recombination-ionization balance with electron depletion and grain charging self-consistently taken into account (Ivlev *et al.* 2016). The considered ionization sources are cosmic rays, X-rays, and radioactive nuclides. The assumed α-parameter is 10^{-3} for the MRI-active disk and 10^{-6} for the dead zone with $x < 10^{-13}$. We do not consider dust fragmentation, radial and vertical drift, which are important for large mm-size dust. We focus on finding the disk regions where strong electrostatic barrier emerges. Coulomb repulsion is important for micron-size dust that does not typically experience strong drift or destructive collisions. On Fig. 1 we show the average grain size after 10^5 yr of dust coagulation for two cases of dust fractal dimension $D = 3.0$ and 2.1. This is actually the map of the severity of electrostatic barrier, it is the strongest in the dead zone for $D = 3.0$ and almost everywhere in the disk for the fractal dust with $D = 2.1$. The zero charge surface is seen as a bright rim in the disk atmosphere.

The research was supported by the Russian Science Foundation (project No. 17-12-01441).

References

Akimkin, V. V. 2015, *ARep*, 59, 747

Fortov, V. E., Ivlev, A. V., Khrapak, S. A., Khrapak, A. G., & Morfill, G. E. 2005, *Phys. Rep.*, 421, 1

Ivlev, A. V., Akimkin, V. V., & Caselli, P. 2016, *ApJ*, 833, 92

Jung, B. 1937, *AN*, 263, 425

Okuzumi, S. 2009, *ApJ*, 698, 1122

Ormel, C. W., & Cuzzi, J. N. 2007, *A&A*, 466, 413

Spitzer, L., Jr. 1941, *ApJ*, 93, 369

Weingartner, J. C. 2004, *Astroph. of Dust*, 309, 453

Origins: From the Protosun to the First Steps of Life
Proceedings IAU Symposium No. 345, 2018
B. G. Elmegreen, L. Viktor Tóth & M. Güdel, eds.

doi:10.1017/S174392131800861X

Tracing the early planet formation with molecular lines: chemistry of vortex in the protoplanetary disks

Natalia Dzyurkevich[1,2], Wladimir Lyra[2] and Liton Majumdar[3]

[1]Institut für Theoretische Astrophysik, University of Heidelberg,
Albert-Überle Str. 2, Heidelberg, D-69120, Germany
email: dzyurkevich@ita.uni-heidelberg.de

[2]Dept. of Physics & Astronomy & California State University at Northridge
18111 Nordhoff St, Northridge, CA 91330
email: wlyra@csun.edu

[3]Jet Propulsion Laboratory, 4800 Oak Grove Drive, Pasadena, CA 91109
email: neal.j.turner@jpl.nasa.gov

Abstract. The millimeter observations of dust in protoplanetary disks show us spectacular structures like numerous gaps, vortices and spirals. In particular, IRS 48 disk demonstrates a large vortex-like structure. Molecular lines provide information about disks that is complementary to dust continuum observations: formaldehyde was found on the inner edge of the IRS 48 vortex, along with detections of SO_2 and CS isotopes.

We use a reduced chemical network containing main carbon- and sulfur-bearing species to find the molecular species which can be sensitive to the gaps in dust, as well as to accumulation of the dust grains in the vortex. We find that SO molecule is the main reservoir for sulfur in IRS 48, for adopted disk model as in Bruderer *et al.* 2014. While SO is very sensitive to the gap edge, it cannot trace the vortex as it is weakly responding to the local increase in dust. Instead, SO_2 molecule abundance can be expected to drop quickly within the vortex, making it an interesting tracer of dust-trapping structure.

Keywords. Astrochemistry, protoplanetary systems: protoplanetary disks

---

1. Introduction

The transitional disk IRS 48 is very low in mass, only $5.5{\cdot}10^{-4}M_\odot$ in gas and $1.5{\cdot}10^{-5}M_\odot$ in dust mass, and displaying a grand vortex-like asymmetry visible both in dust emission and in molecular lines (Van der Marel *et al.* 2013, van der Marel *et al.* 2014). The presence of the gas in the inner cavity inside of 60 AU is debated (Bruderer *et al.* 2014, van der Marel *et al.* 2016b). Because of the prominence of the vortex, IRS 48 is an ideal laboratory to study how the chemistry is altered locally in the vortex. We propose here that the sulfur-bearing molecules might be useful to detect and to characterize the vortices, which could be potentially applyed for more the disks where vortices are undetectable in millimeter dust emission.

2. Overview

Vortex causes an increase up to 100% in local gas density, and accumulates dust of of sizes for the Stokes numbers between 0.01 and 100. We use compact chemical network as in Dzyurkevich *et al.* (2017), extended to S-bearing species. The abundance of sulfur available for gas-reactions is $n_{S_{\rm init}}/n_{H2} = 7.99 \cdot 10^{-8}$. Fig. 1 shows the most abundant

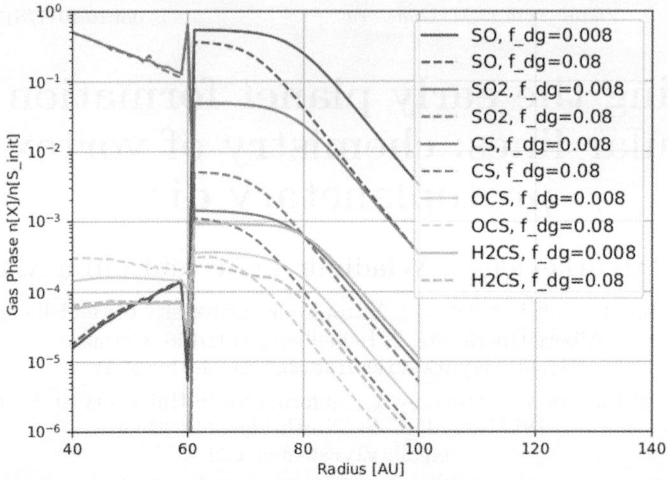

Figure 1. Sulfur-bearing species along radius in IRS 48, relative to the total sulfur abundance.

S-bearing species along the midplane of IRS 48 disk, age 5 Myrs, where solid lines correspond to the plain or opposite-to-vortex part of the disk. Inside 60 AU, all species are in the gas phase because of high temperatures and absence of the dust. Effect of increased dust-to-gas mass ratio on the abundance of S-bearing molecules is shown with dashed lines, as vortex is expected to accumulate dust. Whereas SO is the most abundant molecule, only SO_2 is strongly sensitive to the amount of dust. Species OCS, H_2CS, CS are not tracing the vortex for the gas densities as in IRS 48.

3. Implications

We have performed 1-D chemical modelling of the IRS 48 disk, with the focus on sulfur bearing molecules. We can show that *SO molecule is an efficient tracer of the gas cavity* and is quite abundant, (n[SO]/n[H] from 1.e-9 to 1.e-10). SO molecule will not be able to trace vortex, as it's abundance increases proportionally to gas, and only weakly drops with increased dust-to-gas ratio. *SO_2 is more promising to detect the vortex*, inspite that it is less abundant. SO_2 abundance drops very quickly with increased dust-to-gas mass ratio. In future, we plan to include the formation of H_2S on dust surface. The moleculaes like formaldehide and H_2S can be sensitive the the rotation of the vortex.

References

Bruderer, S., van der Marel, N., van Dishoeck, E. F. & van Kempen, T. A., 2014 *A&A*, 562, 26

Van der Marel, N., van Dishoeck, E. F., Bruderer, S., & van Kempen, T. A., 2014 *A&A*, 563, 113

Van der Marel, N., van Dishoeck, E. F., Bruderer, S., Birnstiel, T., Pinilla, P., Dullemond, C. P., van Kempen, T. A., Schmalzl, M., Brown, J. M., Herczeg, G. J., Mathews, G. S. & Geers, V. 2013, *Science*, 340, 1199

Van der Marel, N., van Dishoeck, E. F., Bruderer, S., Andrews, S. M., Pontoppidan, K. M., Herczeg, G. J., van Kempen, T. & Miotello, A. 2017, *A&A*, 585, 58

Dzyurkevich, N., Commerçon, B., Lesaffre, P. & Semenov, D. 2017, *A&A*, 603, 105

Origins: From the Protosun to the First Steps of Life
Proceedings IAU Symposium No. 345, 2018
B. G. Elmegreen, L. Viktor Tóth & M. Güdel, eds.

doi:10.1017/S1743921318008621

Water delivery to dry protoplanets by hit-and-run collisions

Christoph Burger[1,2], Thomas I. Maindl[1] and Christoph Schäfer[2]

[1]Department of Astrophysics, University of Vienna,
Türkenschanzstraße 17, 1180 Vienna, Austria
email: c.burger@univie.ac.at

[2]Institut für Astronomie und Astrophysik, Eberhard Karls Universität Tübingen,
Auf der Morgenstelle 10, 72076 Tübingen, Germany

Abstract. Final water inventories of newly formed terrestrial planets are shaped by their collision history. A setting where volatiles are transported from beyond the snowline to habitable-zone planets suggests collisions of very dry with water-rich bodies. By means of smooth particle hydrodynamics (SPH) simulations we study water delivery in scenarios where a dry target is hit by a water-rich projectile, focusing on hit-and-run encounters with two large surviving bodies, which probably comprise about half of all similar-sized collisions (Genda *et al.* 2017).

Keywords. Hydrodynamics, methods: numerical, planets and satellites: formation

1. Methods

We base our analysis in this work on existing collision simulations from Burger *et al.* (2018) and re-examine these data under the aspect of a dry target being hit by a water-rich projectile. The scenarios comprise differentiated, self-gravitating, embryo-sized bodies, consisting of an iron core, a silicate mantle, and a water(ice) shell. For details about the SPH code and the material model we refer the reader to Schäfer *et al.* (2016) and Burger *et al.* (2018). Projectile and target have equal initial compositions, 25% iron, 65% silicates and a 10% water mass fraction (wmf). This allows us to track individual contributions from the projectile/target, by tracking the respective SPH particles' origin. For the results presented here we assume a dry target body by tracing only the projectile's water. The differences to a simulation with a really dry target are small (Burger *et al.* 2018), and should be negligible for the global quantities we are interested in.

2. Results and discussion

Post-collision wmf of the initially dry target are plotted in Fig. 1 for various parameter combinations (impact velocity $v/v_{\rm esc}$, impact angle α, $\gamma = M_{\rm proj}/M_{\rm targ}$). Typical hit-and-run collisions (top panels) are compared to head-on collisions (bottom panels). Along with that the water accretion efficiency is illustrated, here defined simply as $\xi_{\rm w} = M_{\rm w,tf}/M_{\rm w,p}$ with water mass on the target fragment and the projectile (see also Burger & Schäfer 2017). We abandon the usual notion of the projectile being the smaller body and switch to a one-body perspective, with an always dry target being hit by a (possibly even more massive) water-rich projectile. In our low-velocity hit-and-run scenarios ($v/v_{\rm esc} = 1.5$, $\alpha = 45°$ in Fig. 1) the target's post-collision wmf is strongly increasing with γ, and this trend continues even for projectiles considerably larger than the target. This behavior is qualitatively different for higher impact velocities ($v/v_{\rm esc} = \{2.5, 3.5\}$, $\alpha = 45°$), where the wmf accreted by the target increases with γ and peaks at mass ratios of roughly 1:1, before

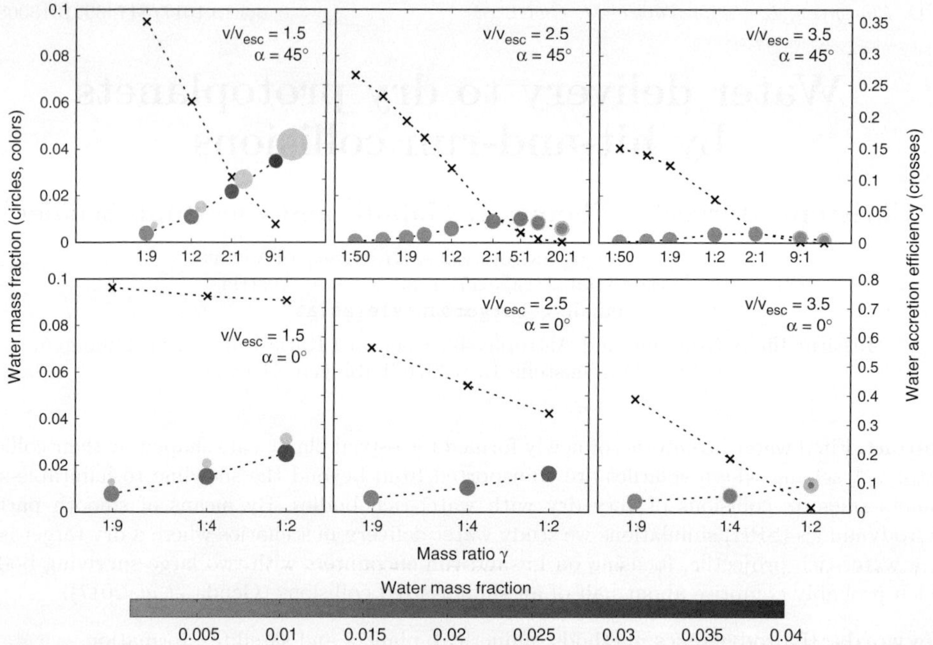

Figure 1. Dry targets being hit by water-rich projectiles (wmf = 0.1), for hit-and-run (top row) and head-on collisions (bottom row). Post-collision wmf are plotted on the left y-axes and color-coded. Water accretion efficiencies ξ_w are plotted on the right y-axes. Grey circles indicate either pre-collision sizes ($\propto$ mass$^{1/3}$), or the projectile body (impacting from the top).

decreasing again, despite the potentially huge water supply in larger projectiles. The water accretion efficiency ξ_w – the fraction of projectile water accreted by the target – is a decreasing function of γ for all scenarios in Fig. 1, and is generally high for small projectile bodies (small γ), but tends to be very low for larger impact velocities and/or mass ratios, indicating that only very little water is transferred to the target.

From the point of view of individual hit-and-run encounters the most water can be delivered to dry target bodies with either low impact velocities, or mass ratios approaching 1:1 for higher v/v_{esc}. However, planet formation is a chaotic process and planets grow from a stochastic history of successive collisions. Therefore it will be rather the (average) water accretion efficiency that determines how much water a growing planet can accrete from the limited amount of water-rich material scattered into its feeding zone. Our results show that this figure is considerably higher for smaller hit-and-run projectiles compared to larger ones, and yet significantly higher for head-on collisions, with water retention up to 80%, compared to at most 35% in our hit-and-run scenarios. It is important to emphasize however, that this does not include the further fate of the (potentially still very water-rich) projectile, after a hit-and-run encounter, therefore it is crucial to consider and track *both* hit-and-run fragments and their volatile inventories.

References

Burger, C. & Schäfer, C. 2017, *Proceedings of the First Greek-Austrian Workshop on Extrasolar Planetary Systems*, 63

Burger, C., Maindl, T.I., & Schäfer, C. 2018, *CeMDA*, 130, 2

Genda, H., Iizuka, T., Sasaki, T., Ueno, Y., & Ikoma, M. 2017, *E&PSL*, 470, 87

Schäfer, C., Riecker, S., Maindl, T.I., Speith, R., Scherrer, S., & Kley, W. 2016, *A&A*, 590, A19

Origins: From the Protosun to the First Steps of Life
Proceedings IAU Symposium No. 345, 2018
B. G. Elmegreen, L. Viktor Tóth & M. Güdel, eds.

doi:10.1017/S1743921318008633

Chemistry in carbon-rich protoplanetary disks: Effect of carbon grain destruction

Chen-En Wei[1], Hideko Nomura[1], Jeong-Eun Lee[2], Wing-Huen Ip[3], Catherine Walsh[4] and T. J. Millar[5,6]

[1]Department of Earth and Planetary Sciences, Tokyo Institute of Technology, 2-12-1 Ookayama, Meguro, Tokyo, 152-8551, Japan

[2]School of Space Research, Kyung Hee University, Seocheon-Dong, Giheung-Gu, Yongin-Si, Gyeonggi-Do, 446-701, Republic of Korea

[3]Graduate Institute of Astronomy, National Central University, No. 300, Zhongda Road, Zhongli Dist., Taoyuan City 32001, Taiwan

[4]School of Physics and Astronomy, University of Leeds, Leeds, LS2 9JT, UK

[5]Astrophysics Research Centre, School of Mathematics and Physics, Queen's University Belfast, University Road, Belfast, BT7 1NN, UK

[6]Leiden Observatory, Leiden University, PO Box 9513, 2300 RA Leiden, The Netherlands

Abstract. The Earth is dramatically carbon poor comparing to the interstellar medium and the proto-sun. The carbon to silicon ratios in inner solar system objects show a correlation with heliocentric distance, which suggests that the destruction of carbon grains has occurred before planet formation. To examine this hypothesis, we perform model calculations using a chemical reaction network under the physical conditions typical of protoplanetary disks. Our results show that, when carbonaceous grains are destroyed and converted into the gas phase and the gas becomes carbon-rich, the abundances of carbon-bearing species such as HCN and carbon-chain molecules, increase dramatically near the midplane, while oxygen-bearing species such as H_2O and CO_2 are depleted. The carbon to silicon ratios obtained by our model calculations qualitatively reproduce the observed gradient with disk radius, but there are some quantitative discrepancies from the observed values of the solar system objects. We adopted the model of a disk around a Herbig Ae star and performed line radiative transfer calculations to examine the effect of carbon grain destruction through observations with ALMA. The results indicate that HCN, $H^{13}CN$ and c-C_3H_2 may be good tracers of this process.

Keywords. astrochemistry, ISM: molecules, solar system: formation

---

The carbon deficit in our solar system is a longstanding problem. There are several possible ways to release the carbon from the solid phase to the gas phase (Finocchi *et al.* 1997; Gail 2002; Siebenmorgen & Krügel 2010; Lee *et al.* 2010). Here, we perform model calculations in two extreme cases: all carbon in carbonaceous grains is either returned to the gas or remains locked in grains. The gas phase elemental abundance ratio satisfies $C/O > 1$ in the former case, while $C/O < 1$ is found in the latter case in our model. Our results show that the differences between two models exist particularly near the midplane of the inner region of the disk where CO is not photodissociated. In the carbon-rich case, the carbon-bearing species are abundant while oxygen-bearing species are depleted.

We calculate the solid carbon to silicon ratio at different disk radii (Table 1). Here we adopt the Solar abundances (C: 2.95×10^{-4} and Si: 3.55×10^{-5}) and assume that the initial gas phase carbon abundance is 7.30×10^{-5}. In the case without carbon grain destruction, the fraction does not change much with radius since the majority of carbon is locked in refractory form. In contrast, when all the carbon is released from the solid

Table 1. The ratio of carbon to silicon in the solid phase at different disk radius and the observation of comet, meteorites, and Earth (Bergin *et al.* 2015).

	without destruction	with destruction		observation value
10 au	6.25	3.62	Comet Halley	5.7
5 au	6.25	3.39	CI	0.71
4 au	6.28	3.39	CM	0.4
3 au	6.28	3.38	CV	0.08
2 au	6.25	3.31	CO	0.07
1 au	6.25	0.04	Earth (BSE)	0.001

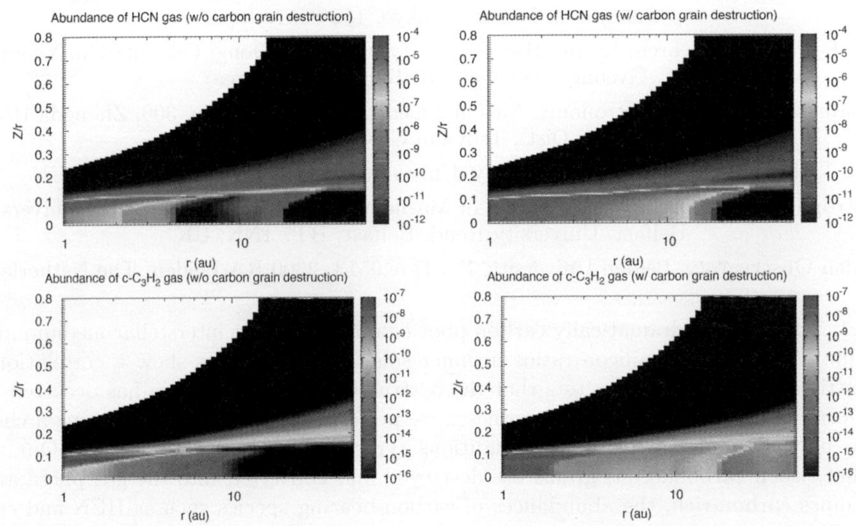

Figure 1. The 2-dimensional abundance distribution of gas-phase molecules for the models without (left) and with (right) carbon grain destruction as a function of radius and height of the disk around a Herbig Ae star.

phase to the gas phase, the fraction varies slightly at 2 au $< r <$10 au and suddenly drops at r=1 au, depending on the fraction of carbon in ice. Though the model produces a radial gradient in the silicon to carbon ratio in the inner solar system, there are some quantitative discrepancies. The model predicts a relatively high carbon fraction in the asteroid belt and too much depletion at 1 au.

To examine whether or not the effect of carbon grain destruction is observable, we predict ALMA observations as well. Our results show that lines emission from T Tauri disks is too weak to probe carbon grain destruction but this effect can be probed through the line ratio of HCN/$H^{13}CN$ and c-C_3H_2 in Herbig Ae disks due to the stronger radiation and larger hot region (Fig. 1). The $H^{13}CN$ emission traces a relatively deeper layer compared to the HCN line and their ratio shows that differences due to this effect may be detectable with ALMA. See Wei *et al.* (2019) for details.

References

Bergin, E. A., Blake, G. A., Ciesla, F., Hirschmann, M. M. & Li, J. 2015, *PNAS*, 112, 8965
Finocchi, F., Gail, H.-P. & Duschl, W. J. 1997, *A&A*, 325, 1264
Gail, H.-P. 2002, *A&A*, 390, 253
Lee, J.-E., Bergin, E. A. & Nomura, H. 2010, *ApJ*, 710, L21
Siebenmorgen, R. & Krügel, E. 2010, *A&A*, 511, A6
Wei, C.-E., Nomura, H. & Lee, J.-E. 2019, *ApJ*, 870, 129.

Origins: From the Protosun to the First Steps of Life
Proceedings IAU Symposium No. 345, 2018
B. G. Elmegreen, L. Viktor Tóth & M. Güdel, eds.

doi:10.1017/S1743921319000061

Obliquity Variability of Terrestrial Planets in the Habitable Zone

Yutong Shan[1] and Gongjie Li[1,2]

[1]Harvard-Smithsonian Center for Astrophysics, The Institute for Theory and Computation, 60 Garden Street, Cambridge, MA 02138, USA
email: yshan@cfa.harvard.edu

[2]Center for Relativistic Astrophysics, School of Physics, Georgia Institute of Technology, Atlanta, GA 30332, USA
email: gongjie.li@physics.gatech.edu

Abstract. Obliquity (axial tilt) and its variability could play an important role in the climate and habitability of a planet. We explore the spin-axis dynamics of two specific habitable zone exoplanets, Kepler-62f and Kepler-186f, using numerical and analytical techniques. Based on our current understanding of their orbital architecture, we find that, in contrast with the typical conditions in the Solar System, Kepler-62f and 186f should have low obliquity variations except in fine-tuned conditions. Extra undetected planetary companions and/or the existence of a satellite could either stabilize or destabilize obliquities at a variety of values.

Keywords. methods: analytical, methods: numerical, planetary systems, celestial mechanics

1. Introduction

The NASA *Kepler* mission has detected thousands of planetary candidates and identified terrestrial planets in the habitable zone (HZ) of their host stars (e.g., Barclay *et al.* 2013, Torres *et al.* 2015). Many of these planets reside in multiplanet systems, where dynamical interactions could affect their orbital and spin evolution, which are important considerations for assessing their habitability.

A planet's obliquity (or axial tilt) is the angle between its spin and orbital axes, which determines the latitudinal distribution of solar radiation on a planet's surface (e.g., Williams & Kasting 1997), therefore affects climate. Evolution in a planet's obliquity is governed by orbital perturbations from its companion planets, as well as torques from the host star and any moons acting on the planetary spin axis (e.g., Ward 1974; Laskar *et al.* 1993). The exact combination of system architecture and intrinsic planetary parameters could lead to a variety of outcomes. In particular, spin-orbit resonances can cause large obliquity variations. For example, without its Moon the Earth's axial tilt would fall into overlaps of resonances and be quite unstable, varying chaotically between 0° and 45° (Laskar *et al.* 1993; Lissauer *et al.* 2012; Li & Batygin 2014).

Exoplanets found in multiplanet systems must contend with possible dynamical instabilities, including obliquity variability. In this contribution, we study the short-timescale obliquity variation of Kepler-62f and Kepler-186f, which are likely rocky transiting planets in the HZ of stars less massive than the Sun (Borucki *et al.* 2013, Quintana *et al.* 2014). Each HZ planet co-inhabits its respective system with at least 4 other small transiting planets in a relatively compact configuration.

2. Obliquity Evolution: Numerical and Analytical Results

Using planetary system parameters from the literature (Borucki *et al.* 2013; Quintana *et al.* 2014; Torres *et al.* 2015) and radius-to-mass relations of Chen & Kipping (2017),

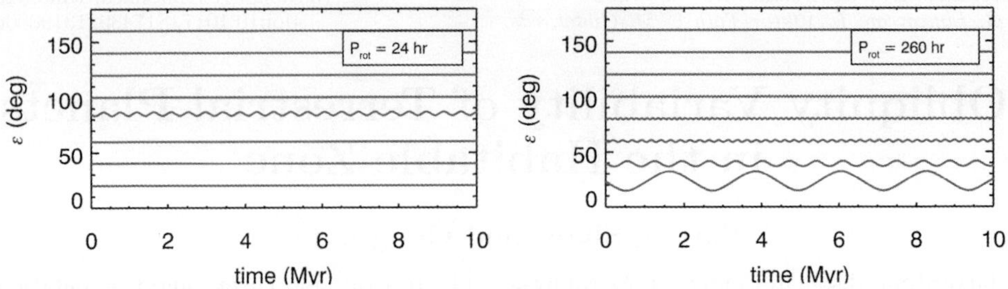

Figure 1. Evolution of obliquity over 10 Myr for range of initial obliquity values for Kepler-186f with spin period 24h (left) and 260h (right). Reproduced from Shan & Li (2018).

we run N-body simulations of each system to derive orbital elements over 10 Myr (Shan & Li 2018). These are in turn fed into the secular Hamiltonian of obliquity variation (e.g., Laskar *et al.* 1993). To determine the precession coefficient for the HZ planets of interest, we assume similar internal structure as the Earth such that we can scale the Earth's dynamical ellipticity. We examine the obliquity time series for a few representative spin periods, as shown in the Figure 1 for Kepler-186f. Retrograde obliquities ($\epsilon > 90°$) are always stable, whereas prograde ($\epsilon \leqslant 90°$) obliquities can undergo variations. Spin period affects the number and locations of the unstable zones. The baseline case – an Earthlike 24h-rotator – remains stable for all obliquities $\lesssim 80°$. The results for Kepler-62f are qualitatively similar.

The fundamental reason for obliquity variations is spin-orbit resonance (Laskar *et al.* 1993; Touma & Wisdom 1993; Lissauer *et al.* 2012). We use an analytic framework to compute the resonance locations as a function of rotation periods, moon properties and system architecture. We find that only one (two) dominant oscillation mode(s) exists for Kepler-186f (62f), making obliquity variations for these planets unlikely except under fine-tuned circumstances (i.e., at very specific rotation periods or within very limited range of obliquity values). The analytic frequencies allows us to characterize obliquity variability in a large parameter space. Using Monte Carlo trials, we also verify that our results are robust to measurement errors in stellar and planet masses. Additional undetected planets and moons in the system could modify the resonance structures, which could have either a stabilizing or destabilizing effect to existing obliquity.

References

Barclay, T., Burke, C. J., Howell, S. B., *et al.* 2013, *ApJ*, 768, 101
Borucki, W. J., Agol, E., Fressin, F., *et al.* 2013, *Science*, 340, 587
Chen, J., & Kipping, D. 2017, *ApJ*, 834, 17
Laskar, J., Joutel, F., & Robutel, P. 1993, *Nature*, 361, 615
Li, G., & Batygin, K. 2014, *ApJ*, 790, 69
Lissauer, J. J., Barnes, J. W., & Chambers, J. E. 2012, *Icarus*, 217, 77
Quintana, E. V., Barclay, T., Raymond, S. N., *et al.* 2014, *Science*, 344, 277
Shan, Y. & Li, G. 2018, *AJ*, 155, 237
Torres, G., Kipping, D. M., Fressin, F., *et al.* 2015, *ApJ*, 800, 99
Touma, J., & Wisdom, J. 1993, *Science*, 259, 1294
Ward, W. R. 1974, *J. Geophys. Res.*, 79, 3375
Williams, D. M., & Kasting, J. F. 1997, *Icarus*, 129, 254

Origins: From the Protosun to the First Steps of Life
Proceedings IAU Symposium No. 345, 2018
B. G. Elmegreen, L. Viktor Tóth & M. Güdel, eds.

doi:10.1017/S174392131900142X

Planetary Systems in Star Clusters: the dynamical evolution and survival

F. Flammini Dotti[1,2], Maxwell Xu Cai[3], Rainer Spurzem[4,5,6] and M.B.N. Kouwenhoven[1,2]

[1]Department of Mathematical Sciences, Xi'an Jiaotong-Liverpool University, 111 Ren'ai Rd., Suzhou Dushu Lake Science and Education Innovation District, Suzhou Industrial Park, Suzhou 215123, P.R. China
email: flammini.francesco@xjtlu.edu.cn

[2]Department of Mathematical Sciences, University of Liverpool, Liverpool L69 3BX, UK

[3]Leiden Observatory, Leiden University, PO Box 9513, 2300 RA, Leiden, Netherlands

[4]National Astronomical Observatories and Key Laboratory of Computational Astrophysics, Chinese Academy of Sciences, 20A Datun Rd., Chaoyang District, 100012, Beijing, China

[5]Kavli Institute for Astronomy and Astrophysics at Peking University, 5 Yiheyuan Rd., Haidian District, 100871, Beijing, China

[6]Zentrum für Astronomie der Universität Heidelberg, Astronomisches Rechen-Institut, Mönchhofstr. 12-14, 69120 Heidelberg, Germany

Abstract. Most stars form in crowded stellar environments. Such star forming regions typically dissolve within ten million years, while others remain bound as stellar groupings for hundreds of millions to billions of years, and then become the open clusters or globular clusters that are present in our Milky Way galaxy today. A large fraction of stars in the Galaxy hosts planetary companions. To understand the origin and dynamical evolution of such exoplanet systems, it is necessary to carefully study the effect of their environments. Here, we combine theoretical estimates with state-of-the-art numerical simulations of evolving planetary systems similar to our own solar system in different star cluster environments. We combine the REBOUND planetary system evolution code, and the NBODY6++GPU star cluster evolution code, integrated in the AMUSE multi-physics environment. With our study we can constrain the effect of external perturbations of different environments on the planets and debris structures of a wide variety of planetary systems, which may play a key role for the habitability of exoplanets in the Universe.

A large fraction of stars in the Galaxy hosts planetary companions (e.g., Mayo *et al.* 2018). To understand the origin and evolution of planetary systems, it is necessary to study the influence of their surroundings. Most field stars form in dense gaseous environments, which are similar to, but slightly less massive and concentrated than the progenitor of the longer-lived open clusters. In such environments, planetary systems may be perturbed or disrupted following close encounters with neighbouring stars (e.g., Spurzem *et al.* 2009; Hao *et al.* 2013; Zheng *et al.* 2015; Cai *et al.* 2017). Most of these young groups of stars disperse within $10 - 50$ Myr (e.g., Lada & Lada 2003), after which their member stars become part of the field star population, while others remain bound for hundreds of millions to billions of years. Here, we present the results of our numerical study of the evolution of complex planetary systems (Solar-like Systems) in star clusters, in which we model the effects of nearby cluster member stars on existing planetary systems Fig 1. We aim to study a different variety of surrounding environments, to obtain a comprehensive picture of the conditions under which external perturbations are important or negligible. The evolution of planetary systems is determined by (i) secular evolution due to planet-planet interactions, (ii) gravitational encounters with nearby stars, and (iii) the cumulative effect of distant encounters. The external contributions are

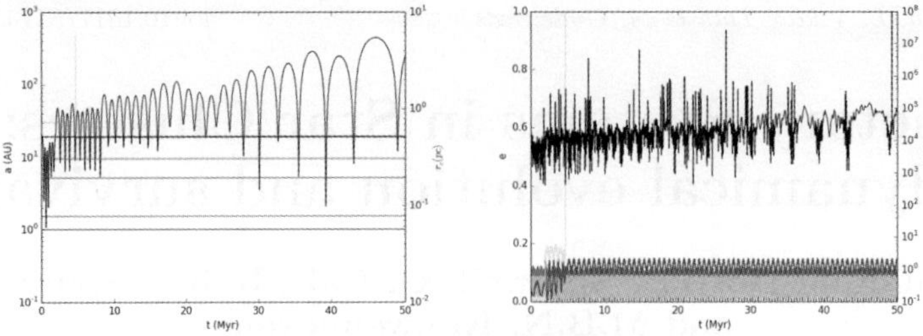

Figure 1. A typical example of a planetary system evolving under the influence of neighbouring stars (Flammini Dotti *et al.*, submitted). *Left:* semi-major axis, a, and distance from the cluster centre, r_c. *Right:* eccentricity, e, and distance to the nearest neighbour star, d_n. The planetary systems were evolved using the `LonelyPlanets` code (Cai *et al.* 2015, 2017; Flammini Dotti *et al.*, submitted), which combines the stellar dynamics (`NBODY6++GPU`; Wang *et al.* 2016) with that of the planetary systems (`REBOUND`; Rein and Liu. 2012) in the `AMUSE` framework (Pelupessy *et al.* 2013). We model star clusters with $N = 500 - 10^4$ stars and radii $r_{\rm vir} = 1$ pc in a Solar-orbit tidal field. Stars are drawn from the Kroupa (2001) IMF, with masses of $0.1 - 25\,M_\odot$. Stars in the mass range $0.9 - 1.1\,M_\odot$ are assigned a planetary system similar to our own Solar system (Earth, Mars, Jupiter, Saturn, Uranus and Neptune). All models are integrated for 50 Myr.

primarily determined by the local stellar density: planetary merger rates and escape rates increase strongly with density. Most systems remain stable for at least 50 Myr, although a substantial fraction experiences orbital reconfigurations, planet-planet mergers, and escape events. As a results of its high mass, Jupiter often serves as a barrier that separates dynamics in the inner and outer planetary system, and hence to some degree protects the terrestrial planets from external perturbations. In systems where Jupiter itself is strongly perturbed, however, the terrestrial planets are unlikely to remain in the habitable zone. Amongst the planetary systems that remain intact, we do not find differences between those that remain in the cluster, and those that have escaped into the Galactic field. On the other hand, we find a notable correlation between the evolution of planetary systems and the velocity with which a star escapes from the cluster: planetary systems amongst ejected stars often experienced highly destructive events, while evaporated stars remain mostly intact. The fraction of ejected planets is comparable for all models, while the fraction of planet-planet mergers tends to increase with the local stellar density in the star cluster. For further details we refer to Flammini Dotti *et al.*, submitted.

Acknowledgements

MBNK acknowledges the National Natural Science Foundation of China (grant 11573004) and the XJTLU Research Development Fund (RDF-16-01-16). RS and MXC acknowledge the support of the DFG priority program SPP 1992 "Exploring the Diversity of Extrasolar Planets" (SP 345/20-1), and of Silk Road Project at NAOC Beijing.

References

Cai, M. X. Meiron, Y., Kouwenhoven, M. B. N., Assmann, P., & Spurzem, R., 2015, *ApJS*, 219, 12
Cai, M. X., Kouwenhoven, M. B. N., Portegies Zwart, S. F., & Spurzem, R., 2017, *MNRAS*, 470, 4337
Hao, W., Kouwenhoven, M. B. N., & Spurzem, R., 2013, *MNRAS*, 433, 867
Lada C. J., & Lada E. A., 2003, *ARA&A*, 41, 57
Mayo A. W., Vanderburg A., *et al.* 2018, *AJ*, 155, 25
Pelupessy, F. I., van Elteren, A., de Vries, N., *et al.* 2013, *A&A*, 557, A84
Rein, H., & Liu, S.-F. 2012, *A&A*, 537, A128
Spurzem, R., Giersz, M., Heggie, D. C., & Lin, D. N. C., 2009, *ApJ*, 697, 458
Wang, L., Spurzem, R., Aarseth, S., *et al.* 2016, *MNRAS*, 458, 1450
Zheng, X. C., Kouwenhoven, M. B. N., & Wang, L., 2015, *MNRAS*, 453, 2759

Origins: From the Protosun to the First Steps of Life
Proceedings IAU Symposium No. 345, 2018
B. G. Elmegreen, L. Viktor Tóth & M. Güdel, eds.

doi:10.1017/S1743921319001431

Rising magnetic flux tubes as a source of IR-variability of the accretion disks of young stars

Sergey A. Khaibrakhmanov[1,2], Alexander E. Dudorov[2] and Andrey M. Sobolev[1]

[1]Kourovka Astronomical Observatory, Ural Federal University,
620000, 51 Lenina str., Ekaterinburg, Russia
email: khaibrakhmanov@csu.ru

[2]Theoretical Physics Dept., Chelyabinsk State University,
454001, 129 Bratiev Kashirinykh str., Chelyabinsk, Russia

Abstract. We investigate dynamics of slender magnetic flux tubes (MFT) in the accretion disks of young stars. Simulations show that MFT rise from the disk and can accelerate to 20-30 km/s causing periodic outflows. Magnetic field of the disk counteracts the buoyancy, and the MFT oscillate near the disk's surface with periods of 10-100 days. We demonstrate that rising and oscillating MFT can cause the IR-variability of the accretion disks of young stars.

Keywords. Accretion, accretion disks, instabilities, MHD, ISM: jets and outflows, ISM: magnetic fields, infrared: ISM

1. Introduction

Young stars with accretion disks born with large-scale magnetic field, according to theory of fossil magnetic field (see Dudorov & Khaibrakhmanov 2015). We developed an MHD model of the accretion disks with magnetic field (Dudorov & Khaibrakhmanov 2014, Khaibrakhmanov *et al.* 2017). Simulations have shown that Ohmic dissipation and ambipolar diffusion limit generation of the magnetic field in regions of low ionization fraction. In the regions where ionization fraction is high, magnetic field is frozen in gas, and generation of strong toroidal magnetic field is possible. Runaway growth of the magnetic field can be limited by magnetic buoyancy instability. This instability leads to formation of slender magnetic flux tubes (MFT, see Parker 1979), that rise from the disk and carry away the excess of magnetic flux (see Khaibrakhmanov & Dudorov 2017). In this work, we present our model of MFT dynamics in the accretion disks of young stars (see details in Khaibrakhmanov *et al.* 2018; Dudorov *et al.* 2019).

2. Model description

We consider geometrically thin, optically thick accretion disk in hydrostatic equilibrium. Vertically stratified polytropic disk is considered. We investigate the dynamics of the unit length cylindrical MFT as a part of magnetic flux ring. Dynamics of slender MFT in the z-direction under the action of buoyant and drag forces is investigated. Turbulent drag is considered inside the disk, and aerodynamic drag above the disk. Heat exchange between the MFT and external gas is calculated in frame of radiative diffusion approximation.

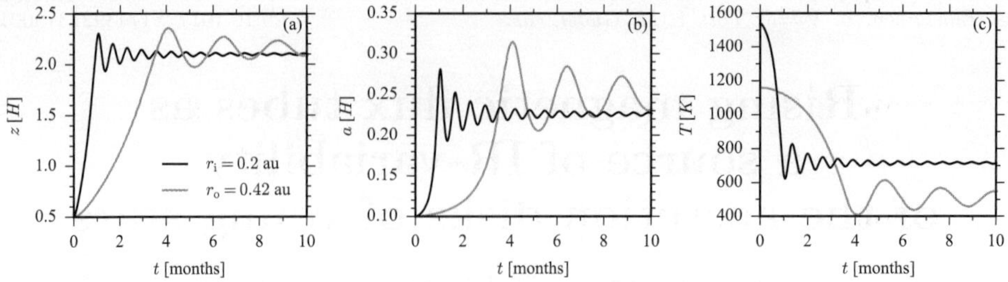

Figure 1. Dependences of z-coordinate (panel (a)), radius (panel (b)) and temperature (panel (c)) of the MFT in the disk of star J11092266-7634320 from Chameleon I cluster. Two representative distances are considered (0.2 au and 0.42 au). H is the scale height of the disk.

3. Results and discussion

In absence of exernal magnetic field, the MFT rapidly accelerates, then rises with slowly increasing velocity, decelerates a little and acquires nearly steady velocity above the surface of the disk. The MFT with stronger magnetic field move faster. The MFT with plasma beta of 0.01 can reach velocity up to $20 - 30$ km s^{-1}. The MFT expands during rise. We conclude that rising MFT form outflowing magnetized corona of the disk.

External magnetic field changes dynamics of the MFT. The MFT rise from the disc and then oscillate near its surface, where intensities of external and internal magnetic fields are equal, and the buoyant force vanishes. The oscillations periods are of 10-100 days depending on the magnetic field strength. Ultimately the oscillations decay. The MFT pulsates during the oscillations. It also exhibits temperature oscillations.

We apply our numerical results for interpretation of IR-variability of T Tauri stars from Chameleon I cluster found by Flaherty, *et al.* (2016). As an example, in Fig. 1 we plot dependences of z-coordinate, radius and temperature on time for the MFT in the disk of star J11092266-7634320. Fig. 1 shows that radius and temperature of MFT experience oscillations at height $z \sim 2.1\,H$ decaying in time, where H is the scale height. Maximal temperature variations are of 200 K. Corresponding periods are 1 and 4 months at $r = 0.2$ au and 0.42 au, respectively. The MFT has large optical depth in IR range, therefore their temperature variations should appear as variability of the IR-brightness of the disk. Observed IR-variability period for star J11092266-7634320 (32 days) can be explained by oscillating MFT with initial plasma beta 1 and radius $0.1\,H$.

Acknowledgements

This investigation is supported by Russian Foundation for Basic Research (project 18-02-01067) and by the Ministry of Science and High Education (the basic part of the State assignment, RK no. AAAA-A17-117030310283-7).

References

Dudorov, A. E., & Khaibrakhmanov, S. A. 2014, *Ap&SS*, 352, 103

Dudorov, A. E., & Khaibrakhmanov, S. A. 2015, *Adv. Sp. Res.*, 55, 843

Dudorov, A. E., Khaibrakhmanov, S. A., & Sobolev, A. M. 2019, *MNRAS*, 487, 5388

Flaherty, K. M., DeMarchi, L., Muzerolle, J., Balog, Z., Herbst, W., Megeath, S. T., Furlan, E., & Gutermuth, R. 2016, *ApJ*, 833, 104

Khaibrakhmanov, S. A., Dudorov, A. E., Parfenov, S. Yu., & Sobolev, A. M. 2017, *MNRAS*, 464, 586

Khaibrakhmanov, S. A., & Dudorov, A. E. 2017, *Phys. of Particles and Nuclei Lett.*, 14, 882

Khaibrakhmanov, S. A., Dudorov, A. E., & Sobolev, A. M. 2018, *Research in Astronomy and Astrophysics*, 18, 90

Parker, E. 1979, *Cosmical magnetic fields: Their origin and their activity* (Oxford: Clarendon Press), p. 858

Origins: From the Protosun to the First Steps of Life
Proceedings IAU Symposium No. 345, 2018
B. G. Elmegreen, L. Viktor Tóth & M. Güdel, eds.

doi:10.1017/S1743921319001443

Large-scale magnetic field of the accretion disks of T Tauri stars

Alexander E. Dudorov[1], Sergey A. Khaibrakhmanov[1,2], Sergey Yu. Parfenov[2] and Andrey M. Sobolev[2]

[1]Theoretical Physics Dept., Chelyabinsk State University, 454001, 129 Bratiev Kashirinykh str., Chelyabinsk, Russia
email: dudorov@csu.ru

[2]Kourovka Astronomical Observatory, Ural Federal University, 620000, 51 Lenina str., Ekaterinburg, Russia

Abstract. The large-scale magnetic field in the accretion disks of young stars is investigated. Main features of our magnetohydrodynamical (MHD) model of the accretion disks and typical simulation results are presented. We discuss the role of MHD effects, ionization structure, magnetic field geometry and strength of the accretion disks.

Keywords. Accretion, accretion disks, instabilities, MHD, ISM: magnetic fields

1. Introduction

Stars with accretion disks born with large-scale magnetic field, according to theory of fossil magnetic field (see Dudorov & Khaibrakhmanov 2015). Efficiency of angular momentum transport mechanisms in the accretion disks depends on the strength and geometry of the magnetic field (see Papaloizou & Lin 1995), which are still poorly known.

In this paper, we present an MHD model of the accretion disks of young stars developed by Dudorov & Khaibrakhmanov (2014), Khaibrakhmanov *et al.* (2017). We calculate the strength and geometry of the magnetic field in the accretion disks of T Tauri stars and make synthetic polarization maps of the dust continuum emission to predict the appearance of the disks with magnetic field.

2. Model description

We consider a stationary geometrically thin and optically thick disk of a T Tauri star. The disk's inner boundary $r_{\rm in}$ is the stellar magnetosphere, the outer boundary is the contact boundary with the ISM. Disk vertical structure is calculated assuming hydrostatic equilibrium. The model includes equations of Shakura & Sunyaev (1973), induction equation taking into account Ohmic diffusion (OD) and ambipolar diffusion (AD), magnetic buoyancy and the Hall effect, equations of thermal and collisional ionization taking into account cosmic rays, X-rays, radionuclides, radiative recombinations and recombinations on dust grains. In the present work, we additionally include dissociative recombinations, charge transfer reactions and grain charges following Tassis & Mouschovias (2005). Evaporation of dust grains is included in the model.

Below we discuss the results of simulation for stellar mass $M = 1\,M_{\odot}$, accretion rate $\dot{M} = 3\times10^{-8}\,M_{\odot}\,{\rm yr}^{-1}$, turbulence parameter $\alpha = 0.01$, dust grain radius $0.5\,\mu$m. Other parameters are adopted according to Khaibrakhmanov *et al.* (2017).

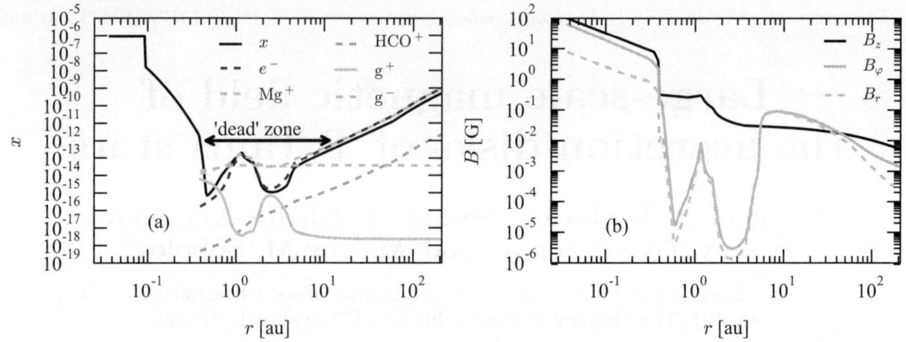

Figure 1. Panel (a): radial profiles of fractional abundances of the charged species (x is the ionization fraction). Panel (b): radial profiels of magnetic field components.

3. Results and discussion

In Fig. 1(a) we plot radial profiels of the fractional abundances of charged species and of the total ionization fraction x. The profile of the ionization fraction is non-monotonic. Thermal ionization operates, and magnetic field is frozen into gas at $r < 0.5$ au. The region of low ionization fraction, $x < 10^{-12}$ ('dead' zone) is located at $0.5 < r < 30$ au. The small peak $x \sim 10^{-12}$ near $r \approx 1$ au is due to evaporation of dust grain mantles. The ionization fraction is high due to cosmic rays ionization at $r > 30$ au. Electrons, metal ions and negatively charged grains are the main charge carriers inside the 'dead' zone.

In Fig. 1(b) we plot the radial profiels of the magnetic field components. The magnetic field geometry is quasi-azimuthal at $r < 0.5$ au. Magnetic buoyancy hinders growth of B_φ in this region. The magnetic field is quasi-uniform due to OD inside the 'dead' zone. It is quasi-radial near the borders of the 'dead' zone due to the Hall effect. AD hinders growth of B_r outside the 'dead' zone, and the magnetic field geometry is quasi-azimuthal. The upper boundary of the 'dead' zone is determined by AD and lies below the disk's surface. Surface density of the active layer above the 'dead' zone is of $5 - 10\,\mathrm{g\,cm^{-2}}$.

Magnetic field strength is characterized by plasma $\beta \sim 10 - 100$ near the inner boundary of the disk, $\beta \sim 10^{-5}$ at $r = 3$ au, and $\beta \sim 1 - 10$ near the outer boundary (240 au).

We made the synthetic maps of polarized continuum dust emission using the accretion disk structure. It is considered that the polarization comes from dust grains alignment with the magnetic field direction in the disk (see details in Khaibrakhmanov *et al.* 2017). The polarization maps shows that the 'dead' zones can be observed as regions with lowest polarization degree in the polarization maps.

Acknowledgements

This investigation is supported by Russian Foundation for Basic Research (project 18-02-01067) and by the Ministry of Science and High Education (the basic part of the State assignment, RK no. AAAA-A17-117030310283-7).

References

Dudorov, A. E., & Khaibrakhmanov, S. A. 2014, *Ap&SS*, 352, 103
Dudorov, A. E., & Khaibrakhmanov, S. A. 2015, *Adv. Sp. Res.*, 55, 843
Khaibrakhmanov, S. A., Dudorov, A. E., Parfenov, S. Yu., & Sobolev, A. M. 2017, *MNRAS*, 464, 586
Papaloizou, J. C. B, & Lin, D. N. C 1995, *ARAA*, 24, 337
Shakura, N. I., & Sunyaev, R. A. 1973, *A&A*, 24, 337
Tassis, K., & Mouschovias, T. Ch. 2005, *ApJ*, 618, 769

Origins: From the Protosun to the First Steps of Life
Proceedings IAU Symposium No. 345, 2018
B. G. Elmegreen, L. Viktor Tóth & M. Güdel, eds.

doi:10.1017/S1743921319001467

Near-Earth object population and formation of lunar craters during the last billion of years

Sergei I. Ipatov[1], **Ekaterina A. Feoktistova**[2] **and Vladimir V. Svetsov**[3]

[1]Vernadsky Institute of Geochemistry and Analytical Chemistry of RAS, 119991, 19 Kosygin st., Moscow, Russia
email: siipatov@hotmail.com

[2]P.K. Sternberg Astronomical Institute, M.V. Lomonosov Moscow State University, 119899, 13 Universitetsky Pr., Moscow, Russia

[3]Institute of Dynamics of Geospheres of RAS, 119334, 38 Leninskii pr., Moscow, Russia

Abstract. We compared the number of lunar craters with diameters greater than 15 km with age less than 1.1 Gyr in the region of the Oceanus Procellarum with the estimates of the number of craters made based on the number of near-Earth objects and on the characteristic times elapsed before collisions of near-Earth objects with the Moon. Our estimates allow the increase of the number of near-Earth objects after a recent catastrophic disruption of a large main-belt asteroid. However, destruction of some old craters and variations in orbital distribution of near-Earth objects with time could allow that the mean number of near-Earth objects during the last billion years could be close to the present value.

Keywords. methods: n-body simulations, Moon, solar system: formation

---

We analyzed diameters of lunar craters with age less than T_{OP}=1.1 Gyr in the region of the Oceanus Procellarum. Data on the craters were taken from the database of the lunar craters of the P.K. Sternberg Astronomical Institute. The ratio r_{OP} of the area of the considered region to the full surface of the Moon is 0.176. In the above region, the number N_{obs} of the craters with a diameter $D > 15$ km (the same number is with $D \geqslant 16$ km) is 52. Results of our analysis of the number of craters of different sizes in the above region are in accordance with the earlier conclusion (Hazards 2010) that the number of impactors with a diameter $d > d_p$ is proportional to $d_p^{-2/3}$. The mean lunar impact velocity of asteroids was calculated to be 18.3 km/s by Minton, Richardson & Fassett (2015) and 19.7 km/s by Stuart & Binzel (2004). Werner & Ivanov (2015) concluded that the final rim crater diameter, D_v (in km), for a vertical impact is

$$D_v = 4(d_p \cdot U^{0.58})^{0.91}, \tag{1}$$

where d_p is the diameter of the impactor, and U is the impact velocity in km/s. The formula (1) was obtained for complex craters with a diameter $D > 15$ km. The size of a crater depends also on the impact angle θ approximately as $(\sin\theta)^{1/3}$ (Melosh H. J. 1989). Taking into account that the impact probability density is equal to $\sin 2\theta$, for the mean (over angles) value of the diameter of the crater, the coefficient 4 in (1) should be changed for 3.3. Based on the formula (1) with the coefficient 3.3, we obtain that the average diameter D_v of the crater that is produced by impactors with a diameter $d_p = 1$ km is 15.3 km and 15.7 km, for U equal to 18.3 km/s and 19.7 km/s, respectively.

The total number N_{1km} of NEOs (near-Earth objects) with diameter $d > 1$ km is estimated to be about 920. The ratio k_{ECO} of the number of NEOs to the number of

ECOs (Earth-crossing objects) was estimated by Ipatov & Mather (2004) to be about 1300/756≈1.72. The similar ratio (6718/3906≈1.72) was considered for the objects discovered before June 1, 2010 (Hazards 2010). Characteristic times T_E elapsed before collisions of an Apollo object and of an Aten object with the Earth were estimated in Ipatov & Mather (2004) to be about 164 and 15 Myr, respectively. T_{Eecos}, which is the value of T_E for all ECOs, equaled to 67 Myr. Before 2004 the obtained values of T_{Eecos} were greater. For example, T_{Eecos}=100 Myr in Ipatov (2001). The smaller than in previous papers values of T_{Eecos} in Ipatov & Mather (2004) were due to several Atens with small inclinations discovered at the beginning of 2000th. For the increase of the inclination of the Aten object 2000 SG344 from its present value equaled 0.1° to 1°, the values of T_{Eecos} were obtained to be equal to 97 Myr for all ECOs. Such difference in the values of T_{Eecos} illustrates the contribution to the mean value of T_{Eecos} of rare objects with high probabilities of collisions with the Earth. Studies of dynamics of initially Jupiter-crossing objects showed (Ipatov & Mather 2004) that the probability of a collision of one object with the Earth can be greater than the sum for thousands of other objects in similar initial orbits. The ratio p_{EM} of probabilities of collisions of NEOs with the Earth to that with the Moon is considered to be about 22 (Leuvre & Wieczorek 2011).

The number of impacts of NEOs with a diameter $d > 1$ km onto the considered lunar region during time T_{OP} can be estimated to be

$$N_{est} = N_{1km} \cdot r_{OP} \cdot T_{OP}/(T_{Eecos} \cdot k_{ECO} \cdot p_{EM}). \tag{2}$$

At N_{1km}=920, $k_{ECO} = 1.72$, $r_{OP} = 0.176$, $T_{OP} = 1100$ Myr, $T_{Eecos} = 100$ Myr, $p_{EM} = 22$, we have $N_{est} \approx 47$. For N_{obs}=52 the ratio N_{obs}/N_{est} is close to 1.1. Bottke, Vokrouhlicky & Nesvorny (2007) suggested that a recent catastrophic disruption of a large main-belt asteroid 160 Myr ago could increase the present value of N_{1km} compared to its mean value for a whole 1 Gyr interval. For such disruption, the mean value of N_{1km} should be smaller than 920, and the number N_{est} of craters with $D > 15$ km should be smaller than 52. As we noted above, Ipatov & Mather (2004) obtained $T_{Eecos} = 67$ Myr. Some ECOs earlier also could have small inclinations. In Ipatov & Mather (2004) 110 Atens were considered. If inclinations are randomly distributed in the range from 0 to 22°, one of them would have inclination between 0 and 0.2°, i.e. 0.1° on average. If we consider $T_{Eecos} = 67$ Myr for $T_{OP} = 1.1$ Gyr and $N_{1km} = 920$, then $N_{est} \approx 70$ and in order to have $N_{est} = N_{obs} = 52$ in the formula (2), we need to consider the mean value of N_{1km} during T_{OP} to be smaller than 920 by a factor of $70/52 \approx 1.35$, that is in accordance with the asteroid breakup considered in Bottke, Vokrouhlicky & Nesvorny (2007). Not all craters with a large age could survive. So the actual value of N_{obs} could be greater than its present value. For greater values of $N_{est} = N_{obs}$, in the formula (2) we need to consider greater values of N_{1km} or smaller values of $T_{Eecos} \cdot k_{ECO}$. In principle, our estimates can also allow that the mean number of NEOs during the last billion years was about the same as that in the present time.

References

Bottke, W., Vokrouhlicky, D., & Nesvorny, D. 2007, *Nature*, 449, 48

Hazards 2010, B. M. Shustov, L. V. Rykhlova (eds.), *Hazards due to asteroids and comets: Yesterday, today, tomorrow* (Moscow: Fizmatlit), 384 p., in Russian

Ipatov, S. I. 2001, *Adv. Space Res.*, 28, 1107

Ipatov, S. I. & Mather, J. C. 2004, *Annals of the New York Academy of Sciences*, 1017, 46

Melosh, H. J. 1989, *Impact cratering: A geologic process* (New York: Oxford Univ. Press), p. 245

Minton, D. A., Richardson, J. E., & Fassett, C. I. 2015, *Icarus*, 2015, 247, 172

Leuvre, M. L. & Wieczorek, M. A. 2011, *Icarus*, 2011, 214, 1

Stuart, J. S., Binzel, R. P. 2004, *Icarus*, 170, 295

Werner, S. C., Ivanov, B. A. 2015, in: G. Schubert (ed.), *Treatise on Geophysics* (Second Edition), 10, 327

Origins: From the Protosun to the First Steps of Life
Proceedings IAU Symposium No. 345, 2018
B. G. Elmegreen, L. Viktor Tóth & M. Güdel, eds.

doi:10.1017/S1743921319001480

A Hydrodynamic Modelling of Atmospheric Escape and Absorption Line of WASP-12b

N. K. Dwivedi[1], M. L. Khodachenko[1,2], I. F. Shaikhislamov[3], A. G. Berezutsky[3], I. B. Miroshnichenko[3], L. Fossati[1], H. Lammer[1], Y. Sasunov[1], K. G. Kislyakova[4], C. P. Johnstone[4] and M. Güdel[4]

[1]Space Research Institute, Austrian Academy of Sciences, Graz, Austria
email: `Navin.Dwivedi@oeaw.ac.at`

[2]Skobeltsyn Institute of Nuclear Physics, Moscow State University, Moscow, Russia

[3]Institute of Laser Physics, SB RAS, Novosibirsk, Russia

[4]Department of Astronomy, University of Vienna, Vienna, Austria

Abstract. Self-Consistent 2D modelling of stellar wind interaction with the upper atmosphere of WASP-12b has been performed. The two case-scenarios of the planetary material escape and interaction with the stellar wind, namely the 'blown by the wind' (without the inclusion of tidal force) and 'captured by the star' (with the tidal force) have been modelled under different stellar XUV radiations and stellar wind parameters. In the first scenario, a shock is formed around the planet, and the planetary mass loss is controlled completely by the stellar radiation energy input. In the second scenario, the mass loss is mainly due to the gravitational interaction effects. The dynamics of Mg_{II} and related absorption were modelled with three sets of different stellar wind parameters and XUV flux values.

Keywords. Dynamical evolution and stability, gaseous planets, hot Jupiters, etc.

---

1. Introduction

WASP-12b is a close-orbit, short period hot Jupiter for which Fossati *et al.* (2010) have detected an extra absorption in the Mg_{II} resonance line cores. Because of the close orbital distance of WASP-12b, XUV radiation and stellar wind (SW) result in the escape (Khodachenko *et al.* 2007) of the ionized planetary atmosphere and associated mass loss. The main objective of the present work is to analyse the two scenarios namely the formation of bow-shock (scenario 1) and diffused cloud or torus (scenario 2) due to the planetary material escape to simulate the extra absorption in Mg_{II} resonance line (Fossati *et al.* 2010) and mass loss.

2. Multi-fluid Simulation results

A 2D model developed in (Shaikhislamov *et al.* 2014, Shaikhislamov *et al.* 2018) is adopted for the specific case of WASP-12b. The summary of the modelling parameters used for the simulations of the scenarios 1 and 2 as well as the calculated upper atmosphere mass loss rate are given in Table 1. Figure 1 illustrates the density distribution of the main interacting constituents and figure 2 shows simulated Mg_{II} absorption line, respectively for both scenarios.

3. Summary

The simulation results demonstrate that in the 'blown by the wind regime' when a slow SW condition and no tidal force are considered in the model, we observe a very strong

Table 1. Adopted parameters and mass loss rate.

Scenario	F_{XUV} $[ergcm^{-2}s^{-1}]$	n_{sw} $[cm^{-3}]$	V_{sw} $[kms^{-1}]$	T_{sw} $[MK]$	**Mass loss rate** $[10^{10}gs^{-1}]$
Blown by the wind	5	$1 \cdot 10^6$	226	1.4	9.91
Blown by the wind	10	$1 \cdot 10^6$	226	1.4	11.91
Blown by the wind	20	$1 \cdot 10^6$	226	1.4	20.49
Captured by the star regime	5	$1 \cdot 10^4$, $3 \cdot 10^4$	417	3.17	145.38, 222.98
Captured by the star regime	10	$1.5 \cdot 10^4$, $3 \cdot 10^5$	417	3.17	169.95, 197.64
Captured by the star regime	20	$1.5 \cdot 10^4$, $3 \cdot 10^5$	417	3.17	186.53, 225.29

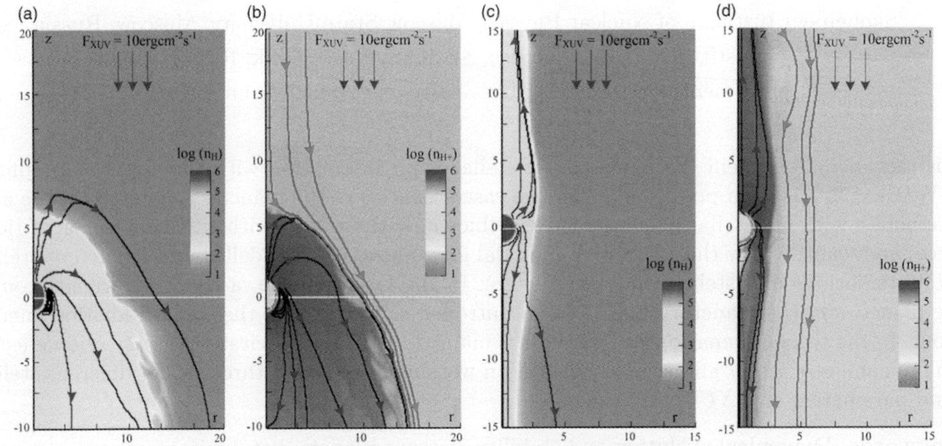

Figure 1. The density distribution of planetary atoms (a, c), ions and stellar protons (b, d) respectively in 'blown by the wind (a, b)' and 'captured by the star (c, d)' regimes.

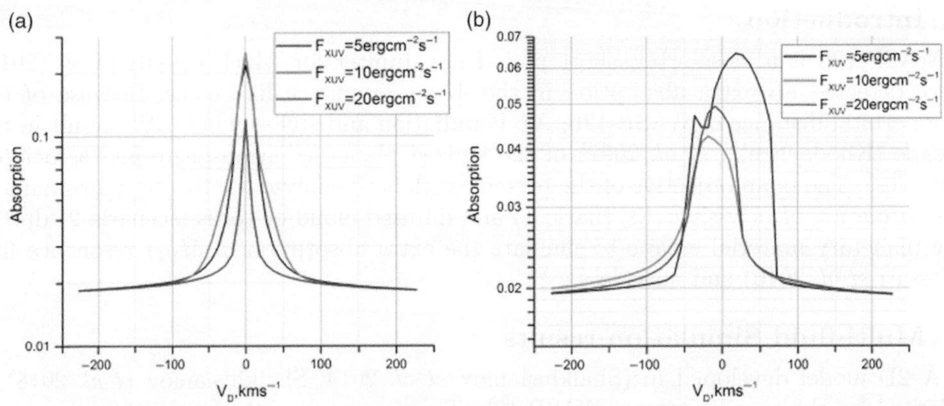

Figure 2. Absorption line profile of the Mg_{II} h line, respectively in 'blown by the wind (a)' and 'captured by the star (b)' regimes.

absorption of Mg_{II}. However, in the 'captured by the star' regime, we found that the change in the XUV radiation fluxes does not make a significant change in the absorption profile of Mg_{II} line. In the case of 'captured by the star' regime, the calculated absorption shows a good agreement with the observed value of absorption in Mg_{II} line which is 3–4 percent.

References

Fossati, L., Haswell, C. A, Froning, C. S., Hebb, L., Holmes, S., Kolb, U., Helling, Ch., Carter, A., Wheatley, P., Cameron, A. C., Loeillet, B., Pollacco, D., Street, R., Stempels, H. C., Simpson, E., Udry, S., Joshi, Y. C., West, R. G., Skillen, I., & Wilson, D. 2010, *APJ Letts.*, 714, L222–L227

Khodachenko, M. L., Ribas, I., Lammer, H., Griemeier, J M., Leitne, M., Selsis, F., Eiroa, C., Hanslmeier, A., Biernat, H. K., Farrugia, C. J., & Rucker, H. O. 2007, *AsBio*, 7, 167

Shaikhislamov, I. F., Khodachenko, M. L., Sasunov, Yu L., Lammer, H., Kislyakova, K. G., & Erkaev, N. V. 2014, *ApJ*, 795, 132

Shaikhislamov, I. F., Khodachenko, M. L., Lammer, H., Fossati, L., Dwivedi, N., Güdel, M., Kislyakova, K. G., Johnstone, C. P., Berezutsky, A. G., Miroshnichenko, I. B., Posukh, V. G., Erkaev, N. V., & Ivanov V. A. 2018, *APJ*, 866, 47

Origins: From the Protosun to the First Steps of Life
Proceedings IAU Symposium No. 345, 2018
B. G. Elmegreen, L. Viktor Tóth & M. Güdel, eds.

doi:10.1017/S1743921319001509

Formation, Initial Evolution and Galactic Dynamic Evolution of Young Star Clusters

Annibal Hetem

Center for Engineering, Modelling and Applied Social Sciences - UFABC,
Rua Santa Adélia, 11111, Santo André, SP, Brazil
email: annibal.hetem@ufabc.edu.br

Abstract. We present an analysis of a sample of clusters of young stars in order to investigate the inherent properties of clustering and dynamic evolution of stellar components, based on fractal statistics. In addition, we present the application of new mathematical and numerical techniques with potentiality for use in models of filamentary structures.

Keywords. stars: pre–main-sequence, ISM: clouds, open clusters and associations: general.

1. Introduction

In previous works, we measured the parameter $\mathcal{Q}$, proposed by Cartwright & Whitworth (2004) for a set of clusters and the results were correlated with other properties of the spatial distributions determined through the King profile and show that almost half of these groups have a relation with the fractal dimension of their parental cloud.

The statistical parameter $\mathcal{Q}$ was measured for the clusters and its correlations with the estimated fractal dimension for the projected near clouds were presented. There are also indications of the presence of substructures similar to those observed in the surrounding clouds. However, other clusters have a radial distribution that does not coincide with the structure of the clouds. These properties may lead to conclusions about the initial conditions of clusters formation (cold collapse or hot collapse), initial evolution (bound or not) and its expected galactic dynamic evolution (crossing time). Figure 1(left) shows the minimal spanning tree, smallest circle and convex hull for NGC 2244, as a visual example of $\mathcal{Q}$ estimation. These studies may give us information about the history of the influence of the Galaxy on clusters and how they were affected by their passage through their structures, as can be seen in Figure 1(right).

2. Genetic Algorithm used to select members of the clusters

In our previous work the dynamical evolution of the clusters was evaluated by Gregorio-Hetem *et al.* (2015) on the basis of the variation of core radius as a function of age for the sample from Santos-Silva & Gregorio-Hetem (2012) compared with σ Ori, Associations and Young Massive Clusters of Milky Way (MW), and clusters of other galaxies of the Local Group (LG) and outside the LG (Portegies Zwart *et al.* 2010). In this poster we presented another method to estimate the membership of the stars in a cluster, based on Bayesian modelling and genetic algorithm by using proper motion provided by the GAIA Data Release 2.

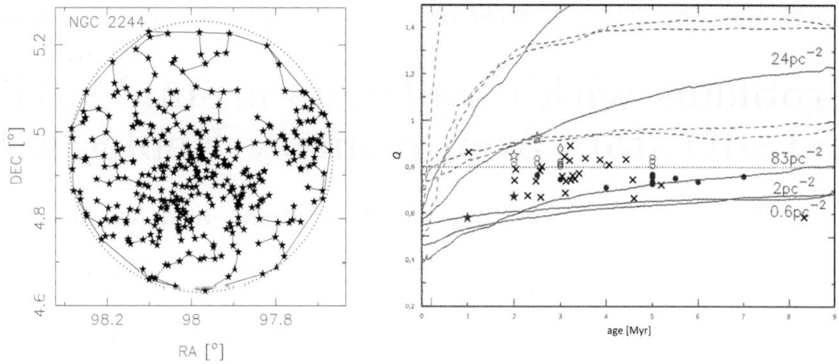

Figure 1. *Left:* Minimal spanning tree, smallest circle and convex hull for NGC 2244. *Right:* Distribution of $\mathcal{Q}$ as a function of age for our sample and other clusters. The results (crosses) are compared to the simulations with feedback (solid lines) and without feedback (dashed lines), indicating the initial densities (number of stars per pc^2) for some of the models adapted from Parker & Dale (2013) and Gregorio-Hetem *et al.* (2015).

3. Estimates of the deviations and error bars

We adopted two methods to estimate the errors associated with the parameter $\mathcal{Q}$. First, the standard deviation was estimated with bootstrapping (Efron 1979), which is independent of model or calculations, is not based on asymptotic results and is simple enough to be automatized. The second method was based on traditional error propagation techniques. The error in a derived measure is obtained straight forward from errors in measurements. It is based on real measurements (observations) and can be implemented in an algorithm if formalism is known.

4. Conclusions

Traditional error propagation techniques can achieve correct estimates of deviations but is strongly dependent on the number of stars. Standard deviation estimated with bootstrapping is independent of number of stars ($\Delta\mathcal{Q} \sim 0.7$), but overestimates deviations for high number of stars (N>200), underestimates deviations for small number of stars (N<100), and gives wrong estimates of deviations for few stars (N<50). Fractal statistics is not indicated to study clusters with 50 stars or less.

Acknowledgements

This work was supported by FAPESP Proc. No. 2017/19458-8 (AH).

References

Cartwright, A., & Whitworth, A. P. 2004, *MNRAS*, 348, 589

Efron, B. 1979, *Bootstrap Methods: Another Look At the Jackknife, The Annals of Statistics*, 7, 126.

Gregorio-Hetem, J., Hetem, A., Santos-Silva, T., & Fernandes, B. 2015, *MNRAS*, 448, 2504

Gower J. C., & Ross G. J. S. 1969, *Appl. Stat.*, 18, 54

Hetem, A. & Lépine, J. R. D. 1993, *A&A*, 270, 451

Megiddo, N. 1983, *SIAM J. Comput.*, 12, 759

Parker, R. J. & Dale, J. E. 2013, *MNRAS*, 432, 986

Portegies Zwart, S. F., McMillan, S. L. W., & Gieles, M., 2010, *ARA&A*, 48, 431

Santos-Silva, T. & Gregorio-Hetem, J. 2012, *A&A* 547, A107

Origins: From the Protosun to the First Steps of Life
Proceedings IAU Symposium No. 345, 2018
B. G. Elmegreen, L. Viktor Tóth & M. Güdel, eds.

doi:10.1017/S1743921319001522

Problems with exoplanets around sdBV and sdO stars from primary Kepler field

Jerzy Krzesinski[1,2], A. Blokesz[1], W. Ogłoza[1] and M. Dróżdż[1]

[1]Mt. Suhora Observatory, Pedagogical University of Cracow,
ul. Podchorążych 2, 30-084 Cracow, Poland
email: jk@oa.uj.edu.pl

[2]Astronomical Observatory, Jagiellonian University, ul. Orla 171, PL-30-244 Cracow, Poland

Abstract. The existence of exoplanets around evolved objects is one of the most interesting subjects from the viewpoint of planetary system evolution and its fate. What happens to the exoplanets engulfed in the host star envelope during red giant branch (RGB) phase? Can planets survive this evolutionary stage of the host star? Here, we are showing that at least some of the exoplanetary candidates recently found around a couple of sdBV stars, KIC 5807616 and KIC 10001893, might not be exoplanets after all. One "exoplanetary signal" visible in the light curve FT of KIC 10001893 can be just a frequency combination of stellar pulsation modes, while others are likely artifacts. Similarly, low frequency signals found in KIC 5807616 light curve FT, are beating frequencies of stellar oscillations, rather than resulting from the exoplanetary radiation. We also analyzed frequency and amplitude changes of the signal around 0.256 c/d (~3.9 day) visible in the light curve FT of the KIC 10449976 sdO star. Our simulations show that it is difficult to reproduce the observed signal frequency variations by the weather changes in the exoplanet atmosphere.

Keywords. exoplanets, sdB, sdO

---

1. Introduction

Two pulsating blue subdwarfs (sdBVs) and a helium one (sdO) from the extreme horizontal branch (EHB) of the HR diagram were recently reported to harbor candidates for exoplanetary systems. The exoplanetary signals were found in the low frequency regions of sdBVs and sdO light curve Fourier Transforms (FTs). Some of the detections were done below or just above the standard 4σ detection threshold of their FT amplitude spectra, but supported by episodes of stronger signal occurrences during shorter sets of Kepler time series data. The signals were discovered below the cut-off frequency for gravity modes (Hansen *et al.* 1985) of the KIC 5807616 (Charpinet *et al.* 2011) and KIC 10001893 (Silvotti *et al.* 2014) stars. Based on data available at that time, in both stars the stellar nature of the signals was excluded and exoplanetary reflection/radiation effects were inferred to explain observed low FT frequency signals. A signal found in KIC 10449976 sdO star (Jeffery *et al.* 2013, Bear & Soker 2014) light curve was not firmly established as of the exoplanetary or stellar origin, but the pulsation nature or exoplaneatry reflection/radiation effects were suggested. Then, the weather in the exoplanet atmosphere was proposed as a cause of the observed signal amplitude and frequency variations (Bear & Soker 2014).

2. Data Analysis

Based on analysis of Q5-Q8 quarters of Kepler data, two planets in an extreme planetary system were suggested to orbit KIC 5807616 sdBV star (Charpinet *et al.* 2011).

The star itself is a typical sdB, but the exoplanets around it would be very hot Earth-sized remnants of former Jupiter mass planets. The "exoplanetary signals" were found in the star light curve FT at $F_2 = 33.755$ and $F_1 = 48.204\,\mu$Hz (Charpinet *et al.* 2011). Latter, using full Q5-Q17 data, Krzesinski (2015) determined slightly different values for $F_2 = 33.839$ and $F_1 = 48.182\,\mu$Hz signals in the low FT frequency region of KIC 5807616 light curve and prewhitened the full set of pulsation frequencies for this star. He also suggested that signals visible at the low FT frequency region might be actually of stellar, not the exoplanetary origin. Our analysis shows that F_2 signal fits the combination frequency of two 201.667 and 167.848 μHz gravity modes of moderate (836 and 125 ppm) amplitudes, while F_1 can be a beating frequency of 167.693 and 119.526 μHz pulsation frequencies (239 and 49 ppm respectively) of the star, with the accuracy better than 0.021 μHz.

Similar analysis was performed for the second candidate for extreme exoplanetary system around KIC 10001893 where three signals were found in the low FT frequency region at $k_1 = 52.683$, $k_2 = 35.578$ and $k_3 = 14.261\,\mu$Hz (Silvotti *et al.* 2014). Using the newest list of pulsation frequencies identified by Uzundag *et al.* (2017), as well as, frequencies prewhitened by us from pure SC (Q3.2, Q6-Q7) and LC (Q1-Q17) data, we performed analysis of pulsation frequency combinations. It appears, that allowing for n and m natural numbers > 1 in $n \cdot f_i - m \cdot f_k$ frequency combinations we can match all three low FT frequency signals with one or more combination frequencies. However, a simple beating frequency ($n = m = 1$) can be fit only to k_3 signal with an accuracy better than 0.0028 μHz (two gravity modes 238.983 and 224.719 μHz, 91 ppm and 54 ppm respectively). k_1 and k_2 seem to be just artifacts depending on data reduction methods.

There are no distinct features visible in the KIC 10449976 light curve FT, but some weak signal between 2.9–3.2 μHz. Jeffery *et al.* (2013) were not able to identify strictly the source of variations. Bear & Soker (2014) suggested, the weather changes in the exoplanet atmosphere could shift maximum amplitude of the flux within 1–3 days. Our simulations could not confirm that variation of the signal are caused by the weather changes in the atmosphere of an exoplanet orbiting the star. We found however, that observed FT pattern can be closely simulated by 435 day period phase modulation of the signal frequency. The source of the 435 day phase modulation is unknown. We also have shown that single night ground observations of the star did not reveal any short period (minutes) pulsations.

Summary: Low FT frequency signals observed in two sdBV stars are not necessarily signatures of exoplanetary systems. We suggest that the FT detection threshold should be increased up to 5σ level to avoid spurious or artifact signals to be taken for analysis.

Acknowledgements

This work was supported by the National Science Centre, Poland.
The project registration number: 2017/25/B/ST9/00879.

References

Bear, E., & Soker, N. 2014, *MNRAS*, 437, 1400
Charpinet, S., Fontaine, G., Brassard, P., *et al.* 2011, *Nature*, 480, 496
Hansen, C. J., Winget, D. E., & Kawaler, S. D. 1985, *ApJ*, 297, 544
Jeffery, C. S., Ramsay, G., Naslim, N., *et al.* 2013, *MNRAS*, 429, 3207
Krzesinski, J. 2015, *A&A*, 581, A7
Silvotti, R., Charpinet, S., Green, E., *et al.* 2014, *A&A*, 570, A130
Uzundag, M., Baran, A. S., Østensen, R. H., *et al.* 2017, *MNRAS*, 472, 700

Origins: From the Protosun to the First Steps of Life
Proceedings IAU Symposium No. 345, 2018
B. G. Elmegreen, L. Viktor Tóth & M. Güdel, eds.

doi:10.1017/S1743921319001546

RCW98: a dust enshrouded HII region

Soňa Ehlerová[1], Lenka Zychová[2], Jan Palouš[1] and Richard Wünsch[1]

[1]Astronomical Institute, Academy of Sciences of the Czech Republic,
Boční II 1401, 141 00 Praha 4, Czech Republic
email: ehlerova@asu.cas.cz

[2]Institute of theoretical physics and astrophysics, Masaryk University,
Kotlářská 267/2, 611 37 Brno, Czech Republic

Abstract. RCW 98 is an HII region with an active star formation in its immediate neighbourhood. At least three early-type stars ionize it and it contains the bright rimmed cloud. Our observations of this region in the optical regime shows that RCW 98 is a complex and dynamical region. We have created a simple model of the dust distribution in this object, but it is not fully consistent with the optical observations. We also study the effects of ionizing stars on the dust.

Keywords. ISM: bubbles, ISM: individual (RCW 98)

1. Introduction

RCW 98 is an HII region inside the infrared bubble S73 (Churchwell *et al.* 2006) with another smaller infrared bubble S74 inside. There are three known early type stars in the region (Pinheiro, *et al.* 2010): LSS 3423 (O9.5V star, considered to be the main ionizing star), ALS19480 (O9.5V star, with a high extinction) and 2MASS J15554264-5439018 (B2.5V star). RCW 98 contains a bright rimmed cloud SFO75: Sugitani & Ogura (1994), Urquhart *et al.* (2009). It is a place of the active star formation; see e.g. Sharma *et al.* (2016) and Urquhart *et al.* (2009); with many potential young stellar objects. Inside the molecular cloud an ATLASGAL clump AGAL327.568-00.852, Csengeri *et al.* (2014), is located.

All three main stars of RCW 98 are listed in the Gaia DR2 catalogue (Gaia Collaboration (2018): LSS 3423 = DR2 5884682031137168768, ALS19480 = DR2 5884681069064466176, 2MASS J15554264-5439018 = DR2 5884681275222918272. Their parallaxes are 0.165 *mas*, 0.355 *mas* and 0.388 *mas*, respective distances (Bailer-Jones *et al.* 2018) are 4.2 *kpc* (lower limit: 3.0 *kpc*, upper limit: 6.4 *kpc*), 2.6 *kpc* (lower limit: 2.3 *kpc*, upper limit: 3.2 *kpc*) and 2.4 *kpc* (lower limit: 2.2 *kpc*, upper limit: 2.7 *kpc*). Distances around 2.5 *kpc* are in agreement with previous distance estimates, the Gaia distance of the brightest star LSS 3423 is very different. Since LSS 3423 is a binary, its Gaia DR2 distance is probably incorrect.

2. HII region and the dust

We observed several selected points inside RCW 98 by the spectrograph FEROS at the 2.2m MPG/ESO telescope (ESO programs 097.A-9039(C) and 099.A-9039(C)). Radial velocities of the HII gas (the $H\alpha$ line) agree with measured radial velocities of the molecular cloud, they lie in the interval $v_{lsr} \in (-44, -35)$ kms^{-1}. $H\alpha$ and $[NII6584]$ line profiles are shown in Fig. 1 (right) for two of the measured points. The location $P1$ lies further away from the star forming activities and is connected to the smoother S73 bubble. The location $P2$ lies inside the small infrared bubble S74, which is directly connected to the cloud. Line profiles are different from each other, $P1$ is simple while $P2$

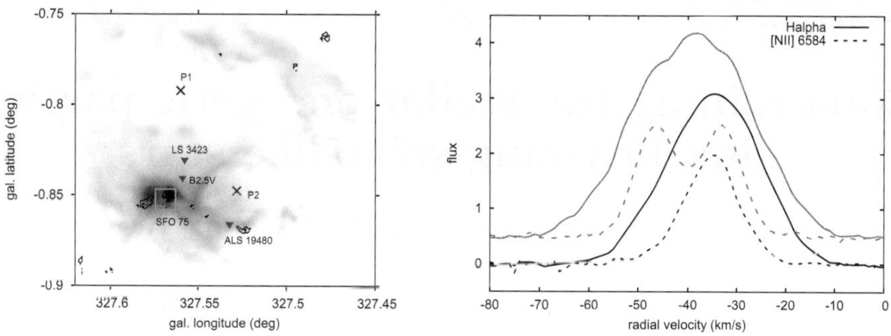

Figure 1. Left: the far-infrared image of RCW98 (Herschel, PACS blue) with superimposed contours of the calculated optical thickness. Triangles are early-type stars in the region, square is the position of the bright rimmed cloud SFO75, crosses denote positions of measurements shown in the right panel. Right: $H\alpha$ and $[NII6584]$ emission from two positions inside the HII region: $P1$ (darker lines) and $P2$ (lighter lines). $P2$ values are artificially shifted upwards by 0.5.

is more complex with a shape, which can be explained by the presence of two denser ionised walls surrounding the more rarefied (ionised) gas, with the velocity separation of these walls 14 kms^{-1}. Using this velocity separation as twice the expansion velocity we can calculate the kinematic age of the structure (Weaver *et al.* 1977; Ehlerová & Palouš 2018): $t_{kin} \simeq 0.7\ Myr$.

Using HERSCHEL observations of this region (all 5 filters) we have calculated the temperature and optical thickness of the emitting dust. The average temperature of the dust in the studied region is 28 K with the minimum value of 17 K and the maximum value 42 K. However, the model of one-temperature component is probably too simplistic for this complex region. Only the star ALS19480 was found to lie in the dust filament, surrounded by a local increase in the temperature. Other two stars, including the brightest LSS 3423, have no visible effect on their surrounding in our calculated maps of temperature and optical thickness. Moreover, we were not able to construct a reasonable picture, which would explain both the dust and ionized gas properties.

Acknowledgements

This work is supported by the grant RVO:67985815 and by the GAČR grant 19-15008S.

References

Bailer-Jones, C. A., Rybizki, J., Fouesneau, M., *et al.* 2018, *AJ*, 156, 58
Churchwell, E., Povich, M. S., Allen, D., *et al.* 2006, *ApJ*, 649, 759
Csengeri, T., Urquhart, J. S., Schuller, F., *et al.* 2014, *A&A*, 565, 75
Ehlerová, S., & Palouš, J. 2018, *A&A*, 619, 101
Gaia Collaboration 2018, *A&A*, 616, 1
Pinheiro, M. C., Copetti, M. V., & Oliviera, V. A. 2010, *A&A*, 521, 26
Sharma, S., Pandey, A. K., Borissova, J., *et al.* 2016, *AJ*, 151, 126
Sugitani, K., & Ogura, K. 1994, *ApJS*, 92, 163
Urquhart, J. S., Thompson, M. A., Morgani, L. K., *et al.* 2009, *A&A*, 450, 625
Weaver, R., McCray, R., Castor, J., *et al.* 1997, *ApJ*, 218, 931

Origins: From the Protosun to the First Steps of Life
Proceedings IAU Symposium No. 345, 2018
B. G. Elmegreen, L. Viktor Tóth & M. Güdel, eds.

doi:10.1017/S174392131900156X

Constraining the stellar energetic particle flux in young solar-like stars

Ch. Rab[1], M. Padovani[2], M. Güdel[3], I. Kamp[1], W-F. Thi[4] and P. Woitke[5]

[1]Kapteyn Astronomical Institute, University of Groningen, P.O. Box 800, 9700 AV Groningen, The Netherlands
email: rab@astro.rug.nl

[2]University of Vienna, Dept. of Astrophysics, Türkenschanzstr. 17, 1180 Wien, Austria

[3]INAF-Osservatorio Astrofisico di Arcetri, Largo E. Fermi, 5 - 50125 Firenze, Italy

[4]MPE, Giessenbachstrasse 1, 85748 Garching, Germany

SUPA, School of Physics & Astronomy, University of St. Andrews, St. Andrews KY16 9SS, UK

Abstract. Anomalies in the abundance measurements of short lived radionuclides in meteorites indicate that the protosolar nebulae was irradiated by a large number of energetic particles ($E \gtrsim 10$ MeV), often called solar cosmic rays. The particle flux of the contemporary Sun cannot explain these anomalies, but, similar to T Tauri stars, the young Sun was more active and probably produced enough high energy particles. However, the stellar particle (SP) flux of young stars is essentially unknown. We model the impact of high-energy ionization sources on the chemistry of the circumstellar environment (disks and envelopes). The model includes X-ray radiative transfer and makes use of particle transport models to calculate the individual molecular hydrogen ionization rates. We study the impact on the chemistry via the ionization tracers HCO^+ and N_2H^+. We argue that spatially resolved observations of those molecules combined with detailed models allow for disentangling the contribution of the individual high-energy ionization sources and to put constraints on the SP flux in young stars.

Keywords. stars: pre–main-sequence, (stars:) circumstellar matter, stars: activity, astrochemistry, radiative transfer, methods: numerical

1. Introduction & Methods

Our Sun can accelerate particles to energies similar to galactic cosmic rays ($E \gtrsim 10$ MeV), hence this stellar energetic particles (SP) are also called solar cosmic rays. Abundance measurements of short-lived radionuclides in meteorites indicate that the protosolar nebulae experienced an about 10^5 times higher SP flux than the Sun produces nowadays (see e.g. Feigelson *et al.* 2002, Gounelle *et al.* 2006). Similar to T Tauri stars (young solar analogues) the young Sun was likely more active and was probably able to produce the SP flux required to explain the abundance anomalies measured in meteorites. This scenario is also supported by the observed high X-ray luminosities of T Tauri stars (Güdel *et al.* 2007). However, contrary to X-rays, the SP flux in T Tauri stars cannot be directly measured, and is not well constrained.

We present 2D radiation-thermo chemical models to study the impact of SPs on the ionization structure of the circumstellar environment. We use ProDiMo (PROtoplanetary DIsk MOdel, Woitke *et al.* 2009, Kamp *et al.* 2017) to calculate the temperature structure, chemical abundances and to produce synthetic observables. We include the main H_2 ionization sources: X-rays, galactic cosmic-rays (CR) and SPs. To calculate the H_2 ionization rate we use X-ray radiative transfer (Rab *et al.* 2018) and make use of particle

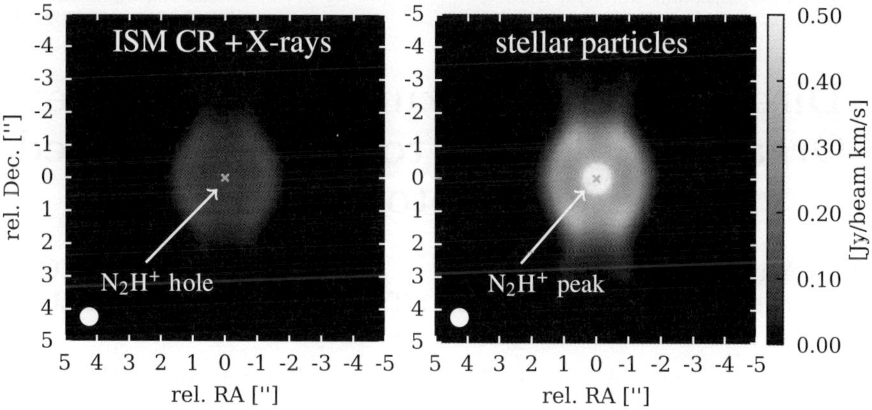

Figure 1. Synthetic $N_2H^+ J = 3-2$ integrated intensity maps for an envelope model (Class 0/I) without SP irradiation (left panel) and the same model with SPs (right panel).

transport models (Padovani *et al.* 2009). This allows us to study the individual contributions of the ionization agents to the total ionization rate and on the chemistry of observable molecules such as HCO^+ and N_2H^+.

2. Results & Conclusions

In Rab *et al.* (2017), we show that SPs can be the dominant H_2 ionization source in the upper layers of a protoplanetary disk. Differently to X-rays and CRs, SPs cannot penetrate down to the midplane as they originate close to the star and are not scattered towards the midplane. With observations of two molecules, HCO^+, which traces the upper disk layers and N_2H^+, which traces deeper layers of the disk, it is therefore possible to measure the impact of SPs and to constrain the SP flux in young stars.

Herschel observations of Ceccarelli *et al.* (2014) for the deeply embedded Class 0 protostar OMC-2 FIR 4 show very low HCO^+/N_2H^+ line ratios. They explained their observations with a strongly elevated H_2 ionization rate of $10^{-14} - 10^{-12}\,s^{-1}$, produced by a high flux of SPs. We used the same approach as for our disk models but now for an envelope structure (see Rab *et al.* 2017a for details) to simulate the conditions for embedded sources. We find that indeed a high SP flux is required to reproduce the low HCO^+/N_2H^+ ratios observed by Ceccarelli *et al.* (2014). In Fig. 1, we present ALMA simulations for this model showing that spatially resolved N_2H^+ observations should clearly show the signatures of SP ionization in embedded sources. We plan to also include different particle transport methods in our model (e.g. Rodgers-Lee *et al.* 2017) to test if they produce yet other observable signatures.

References

Ceccarelli, C., Dominik, C., López-Sepulcre, A., Kama, M. *et al.* 2014, *ApJ* (Letters), 790, L1
Feigelson, E. D., Garmire, G. P., & Pravdo, S. H. 2002, *ApJ*, 572, 335–349
Gounelle, M., Shu, F. H., Shang, H., Glassgold, A. E. *et al.* 2006, *ApJ*, 640, 1163–1170
Güdel, M., Briggs, K. R., Arzner, K., Audard, M., Bouvier, J. *et al.* 2007, *A&A*, 468, 353–377
Kamp, I., Thi, W.-F., Woitke, P., Rab, C., Bouma, S., & Ménard, F. 2017, *A&A*, 607, A41
Padovani, M., Galli, D. & Glassgold, A. E. 2009, *A&A*, 501, 619–631
Rab, Ch., Elbakyan, V., Vorobyov, E., Güdel, M., Dionatos, O. *et al.* 2017a, *A&A*, 604, A15
Rab, Ch., Güdel, M., Padovani, M., Kamp, I., Thi, W.-F. *et al.* 2017, *A&A*, 603, A96
Rab, Ch., Güdel, M., Woitke, P., Kamp, I., Thi, W.-F., Min, M. *et al.* 2018, *A&A*, 609, A91
Rodgers-Lee, D., Taylor, A. M., Ray, T. P., & Downes, T. P. 2017, *MNRAS*, 472, 26–38
Woitke, P., Kamp, I., & Thi, W.-F. 2009, *A&A*, 501, 383

Origins: From the Protosun to the First Steps of Life
Proceedings IAU Symposium No. 345, 2018
B. G. Elmegreen, L. Viktor Tóth & M. Güdel, eds.

doi:10.1017/S1743921319001571

Direct Infall Signatures and Complex Organic Molecules toward an Isolated Embedded Protostar BHR 71

Yao-Lun Yang[1], Neal J. Evans II[1,2,3], Aaron Smith[4], Joel D. Green[5,1], Jes K. Jørgensen[6], Jeong-Eun Lee[7], Tyler L. Bourke[8], John J. Tobin[9] and Susan Terebey[10]

[1]Department of Astronomy, The University of Texas at Austin, USA
email: yaolunyang.astro@gmail.com

[2]Korea Astronomy and Space Science Institute, Korea

[3]Humanitas College, Kyung Hee University, Korea

[4]Department of Physics, Massachusetts Institute of Technology, USA

[5]Space Telescope Science Institute, USA

[6]Centre for Star and Planet Formation, Niels Bohr Institute and Natural History Museum of Denmark, University of Copenhagen, Denmark

[7]Department of Astronomy & Space Science, Kyung Hee University, Korea

[8]SKA Organization, Jodrell Bank Observatory, UK

[9]National Radio Astronomy Observatory, USA

[10]Department of Physics and Astronomy, California State University at Los Angeles, USA

Abstract. Our ALMA observations of HCO^+ and HCN show such redshifted absorption toward an isolated core, BHR 71. Both lines show a similar redshifted absorption profile. We also found emissions of complex organic molecules (COMs) around 345 GHz from a compact region centered on the continuum source, which is barely resolved with a beam of 0.″27, corresponding to ~50 AU.

Keywords. ISM: kinematics and dynamics, astrochemistry, and stars: formation

1. Introduction

Gas infall transforms dense cores into protostars. While interferometric observations of molecular spectra best probe the gas kinematics at the densest part of the prorotstellar envelope, only a few direct measurements of infall exist to constrain the model of infall. ALMA observations of optically-thick molecular transitions exclusively probe the line-of-sight velocity structure within a pencil beam toward protostars, resulting in a red-shifted absorption against the continuum, which is an unambiguous signature of infall (Pineda *et al.* 2012; Evans *et al.* 2015).

Several protostars harbor complex organic molecules (COMs), named as "hot corinos", suggested by Ceccarelli (2004). With the high resolution and sensitivity of ALMA, the emission of COMs not only represents the chemical complexity of the protostars but also probes the dynamics at the disk-forming regions (e.g. Oya *et al.* 2017).

2. Results

In our ALMA Cycle 4 observation of BHR 71 (PI: Y.-L. Yang, 2016.0.00391S), we detect strong redshifted absorption against the continuum from the HCO^+ $J = 4 \rightarrow 3$ and HCN $J = 4 \rightarrow 3$ lines (Figure 1) with a synthetic beam size of 0.″39×0.″27.

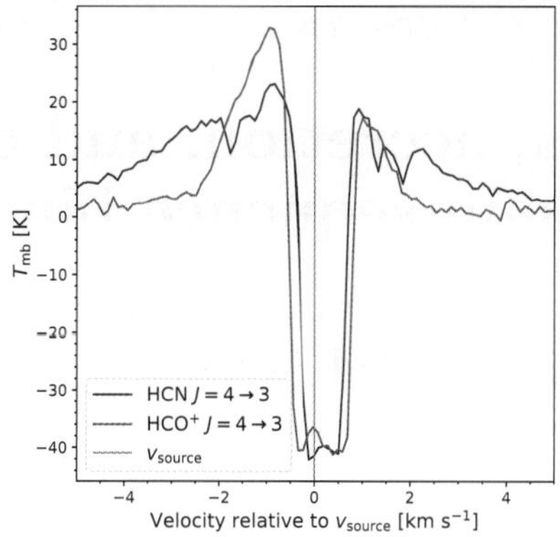

Figure 1. The infall signature detected from the HCO^+ $J=4 \rightarrow 3$ (blue) and HCN $J=4 \rightarrow 3$ (red) lines extracted from a $0.''52 \times 0.''39$ ellipse with PA=125°.

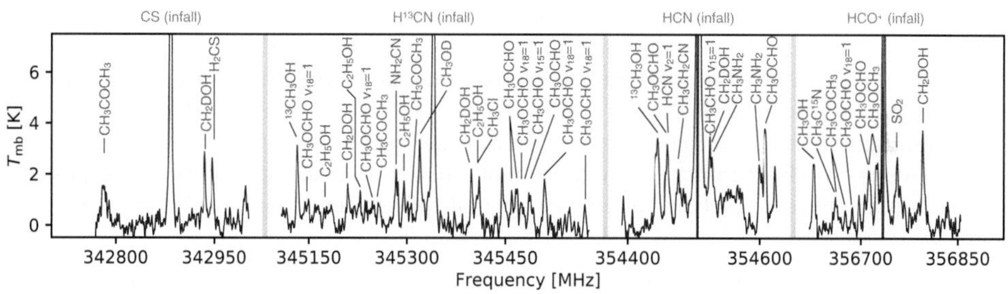

Figure 2. The full ALMA spectra at Band 7, showing numerous weak and narrow emission features identified as the emission of complex organic molecules. The annotated texts show the identified species.

In the same observation, we detect more than 40 weak narrow emission features over the entire ALMA spectra (Figure 2). We identified 14 species of COMs toward the compact center of BHR 71, discovering the "hot corino" nature of BHR 71. The COMs emit from an averaged 90 AU×70 AU region at the center, similar to the size of continuum emission, 100 AU×80 AU, suggesting that the COMs locate at the high temperature (>100 K) region around the central protostar (Yang *et al.* 2017).

References

Ceccarelli, C. 2004, in: Johnstone, D., Adams, F. C., Lin, D. N. C., Neufeld, D. A., & Ostriker, E. C., *Star Formation in the Interstellar Medium: In Honor of David Hollenbach*, ASP Conference Proceedings, Vol. 323. (San Francisco: ASP) p. 195

Evans, N. J., Di Francesco, J., Lee, J.-E., Jørgensen, J. K., Choi, M., Myer, P. C., & Mardones, D. 2015, *ApJ*, 814, 22

Oya, Y., Sakai, N., Watanabe, Y., Higuchi, A. E., Hirota, T., López-Sepulcre, A., Sakai, T., Aikawa, Y., Ceccarelli, C., Lefloch, B., Caux, E., & Vastel, C. 2017, *ApJ*, 837, 174

Pineda, J. E., Maury, A. J., Fuller, G. A., Testi, L., García-Appadoo, D., Peck, A. B., Villard, E., Corder, S. A., van Kempen, T. A., Turner, J. L., Tachihara, K., & Dent, W. 2012, *A&A*, 544, L7

Yang, Y.-L., Evans, N. J., Green, J. D., Jørgensen, J. K., & Dunham, M. M., 2017, *ApJ*, 835, 259

Origins: From the Protosun to the First Steps of Life
Proceedings IAU Symposium No. 345, 2018
B. G. Elmegreen, L. Viktor Tóth & M. Güdel, eds.

doi:10.1017/S1743921319001583

Spots, flares, accretion, and obscuration in the pre-main sequence binary DQ Tau

Á. Kóspál[1,2], P. Ábrahám[1], G. Zsidi[1], K. Vida[1], R. Szabó[1], A. Moór[1] and A. Pál[1]

[1]Konkoly Observatory, Research Centre for Astronomy and Earth Sciences, Hungarian Academy of Sciences, Konkoly-Thege Miklós út 15-17, H-1121, Budapest, Hungary
`kospal@konkoly.hu`

[2]Max Planck Institute for Astronomy, Königstuhl 17, D-69117 Heidelberg, Germany

Abstract. DQ Tau is a young low-mass spectroscopic binary, consisting of two almost equal-mass stars on a 15.8 day period surrounded by a circumbinary disk. We analyzed DQ Tau's light curves obtained by Kepler K2, the Spitzer Space Telescope, and ground-based facilities. We observed variability phenomena, including rotational modulation by stellar spots, energetic stellar flares, brightening events around periastron due to increased accretion, and short dips due to temporary circumstellar obscuration. The study on DQ Tau will help in discovering and understanding the formation and evolution of other real-world examples of "Tatooine-like" systems. This is especially important because more and more evidence points to the possibility that all Sun-like stars were born in binary or multiple systems that broke up later due to dynamical interactions.

Keywords. stars: pre–main-sequence, accretion disks, circumstellar matter

---

Pre-main sequence stars are strongly linked with their circumstellar environment, causing variability at a wide range of wavelengths and timescales. The four main origins of photometric variability are variable accretion, rotational modulation due to hot or cold stellar spots, variable line-of-sight extinction, and stellar flares. The second mission of the Kepler spacecraft (K2) provided high precision photometry for many young stars in the Taurus star forming region in 2017 (Campaign 13), enabling us to search for young stellar variability, disentangle the different effects, and determine their relative contributions.

DQ Tau consists of two almost identical $0.6\,M_{\odot}$ stars that orbit each other in an eccentric orbit with a period of 15.8 days. At periastron they approach each other within $12.5\,R_{\odot}$. The binary is surrounded by a circumbinary protoplanetary disk, from which gas and dust are accreting onto the stars. The system is seen close to pole-on (Czekala *et al.* 2016), it displays quasi-periodic optical variability due to pulsed accretion (Tofflemire *et al.* 2017), and millimeter flares and elevated X-ray activity near periastron due to a combination of magnetic and dynamic effects (Getman *et al.* 2011; Kóspál *et al.* 2011).

Kepler K2 observed DQ Tau in 2017, providing an 80-day-long uninterrupted monitoring of the white-light brightness with a cadence of 1 min (Fig. 1). For a few weeks at the beginning of the K2 monitoring, we could observe DQ Tau from the ground using the 90 cm Schmidt telescope of Konkoly Observatory (Hungary), with $BV(RI)_C$ filters. We complemented these with mid-infrared photometry at 3.6 and 4.5 μm using the Spitzer Space Telescope during the final 11 days of the K2 campaign with a cadence of 20 hours.

The K2 data displays strong periodicity with $P = 3.017 + 0.004$ days, consistently with the stellar rotational period. The phase-folded multi-filter optical data could be fitted with a model containing three spots that are 400 K cooler than the stellar photosphere and together cover 50% of the stellar surface (assuming that only one star is spotted).

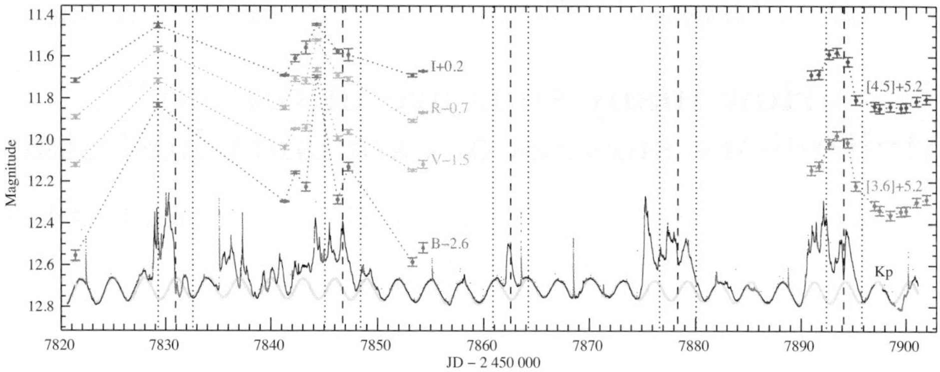

Figure 1. Light curves of DQ Tau at different wavelengths (Kóspál *et al.* 2018).

After subtracting the rotational modulation (gray curve in Fig. 1), we searched for short, flare-like brightening events (fast rise and exponential decay). We identified 40 such events (highlighted with red in Fig. 1), which last for 100–200 minutes. The energy released in the flares is between 4.4×10^{32} – 1.2×10^{35} erg, more powerful than usual for main sequence late-type stars, but typical for young low mass stars. The energy distribution of the flares suggests non-thermal emission mechanism. The flares occur randomly; we saw no correlation either with the rotational period (probably a consequence of very extended active regions that are always visible), or with the binary's orbital period (suggesting that the flares are single-star flares happening just above the stellar surface rather than between the two stars where their magnetospheres reconnect and separate)

After subtracting the rotational modulation and the flares, the remaining variability mostly consists of complex brightening events clustered around the periastrons, probably caused by increased accretion rate. This can be explained by pulsed accretion: the stars gravitationally perturb the inner edge of the circumbinary disk during each apoapsis passage and pull some material from the disk that eventually lands on the binary components. The strongest peaks reach $5\times10^{-8}\,M_{\odot}$/yr, when the accretion luminosity goes up to $0.8\,L_{\odot}$. The total mass accreted in each orbital cycle is typically about $1.2\times10^{-10}\,M_{\odot}$. Our Spitzer monitoring revealed that the increased accretion luminosity during the periastron events warmed up the inner disk temporarily by about 100 K.

DQ Tau shows short dips of <0.1 mag in its light curve (highlighted with blue in Fig. 1), reminiscent of the well-known "dipper phenomenon" observed in many low-mass young stars, probably caused by dusty material lifted up from the inner edge of the disk. A rather small amount of mass ($7\times10^{-15}\,M_{\odot}$) is enough to explain the dips in DQ Tau.

This project has received funding from the European Research Council (ERC) under the European Unions Horizon 2020 research and innovation programme under grant agreement No 716155 (SACCRED, PI: Á. Kóspál). The authors acknowledge the Hungarian National Research, Development and Innovation Office grants OTKA K-109276, OTKA K-113117, NKFIH K-115709, as well as Lendület LP2018-7/2018. KV is supported by the Bolyai János Research Scholarship of the Hungarian Academy of Sciences.

References

Czekala, I., Andrews, S. M., Torres, G., *et al.* 2016, *ApJ*, 818, 156

Getman, K. V., Broos, P. S., Salter, *et al.* 2011, *ApJ*, 730, 6

Kóspál, Á., Salter, D. M., Hogerheijde, *et al.* 2011, *A&A*, 527, A96

Kóspál, Á., Ábrahám, P., Zsidi, G., *et al.* 2018, *ApJ*, 862, 44

Tofflemire, B. M., Mathieu, R. D., Ardila, D. R., *et al.* 2017, *ApJ*, 835, 8

Origins: From the Protosun to the First Steps of Life
Proceedings IAU Symposium No. 345, 2018
B. G. Elmegreen, L. Viktor Tóth & M. Güdel, eds.

doi:10.1017/S1743921319001595

How many suns are in the sky? Multiplicity surveys of exoplanet host stars

M. Mugrauer[1], C. Ginski[2], N. Vogt[3], R. Neuhäuser[1] and C. Adam[4]

[1]Astrophysical Institute and University Observatory Jena,
Schillergäßchen 2, D-07745 Jena, Germany
email: markus@astro.uni-jena.de

[2]Leiden Observatory,
Leiden University, PO Box 9513, 2300, RA Leiden, The Netherlands

[3]Instituto de Física y Astronomía,
Facultad de Ciencias, Universidad de Valparaíso, Av. Gran Bretaña, 1111, Valparaíso, Chile

[4]Instituto de Astronomía,
Universidad Católica del Norte, Av. Angamos 0610, Antofagasta, Chile

Abstract. In order to determine the true impact of stellar multiplicity on the formation and evolution of planets, we initiated direct imaging surveys to search for (sub)stellar companions of exoplanet host stars on close orbits, as their gravitational impact on the planet bearing disk at first and on formed planets afterwards is expected to be maximal. According to theory these are the most challenging environments for planet formation and evolution but might occur quite frequently in the milky way, due to the large number of multiple stars within our galaxy. On this poster we showed results, obtained so far in the course of our AO and Lucky-imaging campaigns of exoplanet host stars, conducted with NACO/ESO-VLT for southern and with AstraLux/CAHA2.2m for northern targets, respectively. In addition, we introduced our new high contrast imaging survey with SPHERE/ESO-VLT to search for close companions of southern exoplanet host stars, and presented some first results.

Keywords. binaries (including multiple): close, visual, stars: imaging, planetary systems, techniques: high angular resolution

---

Theoretical studies predict that a stellar companion of a star alters the formation process of its planets and the evolution of their orbits. As observations can test these theoretical predictions, we have already searched for wide stellar companions of exoplanet host stars using Seeing-limited IR imagers with large fields of view e.g. MAGIC/CAHA 2.2 m, SofI/ESO-NTT, and UFTI/UKIRT. In the course of these surveys, beside several stellar companions, HD 3651 B the first directly imaged brown dwarf companion of an exoplanet host star was detect (Mugrauer *et al.* 2006). Furthermore, these studies revealed several differences between the properties of exoplanets in star systems and of those orbiting single stars (Mugrauer *et al.* 2007a). With AO and speckle imaging we could also detect close companions of our targets, e.g. γ Cep B (about $0.4\,M_{\odot}$, a$\sim$20 au), which was directly imaged by us for the first time using the 8.2 m Subaru, and the CAHA 3.5 m telescope (Neuhäuser *et al.* 2007), or Gl 86 B, a white dwarf companion (about $0.6\,M_{\odot}$, a$\sim$20 au), whose true nature was revealed by us with NACO/ESO-VLT imaging and spectroscopy (Mugrauer & Neuhäuser 2005), proving that planets can survive the post main sequence evolution of a nearby star, as predicted by theory. Beside close binaries also several triple stars with exoplanets were detected (see e.g. Mugrauer *et al.* 2007b). Today more than 100 multiple stars with exoplanets are known and 35 (27 doubles, 8 triples) of them were

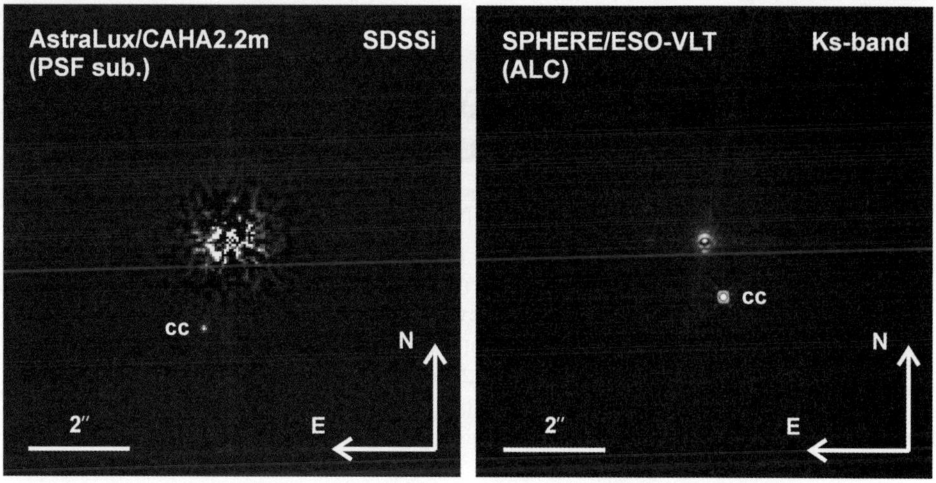

Figure 1. Two examples of close companion-candidates (cc) of exoplanet host stars, recently detected by us in our imaging surveys, carried out with AstraLux/CAHA 2.2 m, and SPHERE/ESO-VLT. The PSF of the bright exoplanet host star (centered in each image) is subtracted in the AstraLux image, while the SPHERE data is taken in coronagraphic mode to suppress the flux of the bright primary star. The companionship of the detected candidates is checked with follow-up imaging observations, which are currently ongoing.

detected in our surveys. This demonstrates the efficiency of these projects and shows that the multiplicity of the exoplanet host stars is significantly underestimated (see also Mugrauer & Dinçel 2016, and Mugrauer *et al.* 2014).

Regarding the impact of stellar multiplicity on the planet formation and evolution, the closest star systems with exoplanets are most intriguing, as the gravitational perturbation of a close stellar companion on the planet bearing disk at first and on formed planets later on is expected to be maximal. However, surveys for such close stellar companions of exoplanet host stars were conducted so far only with small samples sizes. In order to perform a homogeneous multiplicity study of exoplanet host stars with a large enough sample size to draw statistically significant conclusions for the whole exoplanet population, we use AstraLux/CAHA 2.2 m for northern, and NACO as well as SPHERE at the ESO-VLT for southern targets, to discover and characterize close stellar companions (proj. separation > 13 au) of exoplanet host stars and to clarify their multiplicity status (see Fig. 1, Mugrauer & Ginski 2015, Ginski *et al.* 2016, and Ginski *et al.* 2019).

References

Ginski, C., *et al.* 2019, in preparation
Ginski, C., Mugrauer, M., Seeliger, M., *et al.* 2016, *MNRAS*, 457, 2173
Mugrauer, M., & Dinçel, B. 2016, *AN*, 337, 627
Mugrauer, M., & Ginski, C. 2015, *MNRAS*, 450, 3127
Mugrauer, M., Ginski, C., & Seeliger, M. 2014, *MNRAS*, 439, 1063
Mugrauer, M., Neuhäuser, R., & Mazeh, T. 2007, *A&A*, 469, 755
Mugrauer, M., Seifahrt, A., & Neuhäuser, R. 2007, *MNRAS*, 378, 1328
Mugrauer, M., Seifahrt, A., Neuhäuser, R., & Mazeh, T. 2006, *MNRAS*, 373, L31
Mugrauer, M., & Neuhäuser, R. 2005, *MNRAS*, 361, L15
Neuhäuser, R., Mugrauer, M., Fukagawa, M., Torres, G., & Schmidt, T. 2007, *A&A*, 462, 777

Origins: From the Protosun to the First Steps of Life
Proceedings IAU Symposium No. 345, 2018
B. G. Elmegreen, L. Viktor Tóth & M. Güdel, eds.

doi:10.1017/S1743921319001601

Multiplicity study of T Tauri stars in the Lupus star forming region

M. Mugrauer[1], C. Ginski[2], N. Vogt[3] and R. Neuhäuser[1]

[1]Astrophysical Institute and University Observatory Jena,
Schillergäßchen 2, D-07745 Jena, Germany
email: markus@astro.uni-jena.de

[2]Leiden Observatory,
Leiden University, PO Box 9513, 2300, RA Leiden, The Netherlands

[3]Instituto de Física y Astronomía,
Facultad de Ciencias, Universidad de Valparaíso, Av. Gran Bretaña, 1111, Valparaíso, Chile

Abstract. We carried out a high contrast imaging search for (sub)stellar companions of young pre-main sequence stars in the Lupus star forming region. For this project we utilized NACO/ESO-VLT, operated at the Paranal observatory. On this poster, we presented the results of this survey. In several observing campaigns we could obtain diffraction limited deep IR imaging data and detected faint co-moving companions around our targets, whose astro- and photometry was determined in all observing epochs. The co-moving companions found in our survey exhibit angular separations in the range between about 0.1 and a few arcsecs, i.e. projected separations between about 10 and a few hundreds of au, at the average distance of our targets of about 140 pc. Beside several new binary and triple star systems, whose multiplicity was revealed in this survey, also faint co-moving companions in the substellar mass regime could be identified close to some of our targets.

Keywords. binaries (including multiple): close, visual, stars: pre–main-sequence, low-mass, brown dwarfs, techniques: high angular resolution

---

As T Tauri stars are young only few Myr old pre main-sequence stars, they are the most favorable targets for a direct imaging search for (sub)stellar companions, because their companions are also young, and therefore self luminous, due to still ongoing contraction.

Among all known T Tauri stars in the Lupus star forming region, we have selected those as targets, which were not observed with NACO/ESO-VLT before, exhibit well known and sufficiently high proper motions, and are bright enough to be imaged at a high Strehl-ratio with NACO/ESO-VLT in the Ks-band but do not saturate the used infrared detector, most important to obtain accurate astro- and photometric measurements of the detected (sub)stellar companions.

In total, more than 60 mainly weak-lined T Tauri stars are selected as targets for our multiplicity study and deep high contrast AO imaging data could be taken in multiple observing epochs (Mugrauer *et al.* 2019). Examples for co-moving (sub)stellar companions, detected in the course of our multiplicity study around the observed targets, are shown in Figs. 1 and 2.

Reference

Mugrauer, M., *et al.* 2019, in preparation

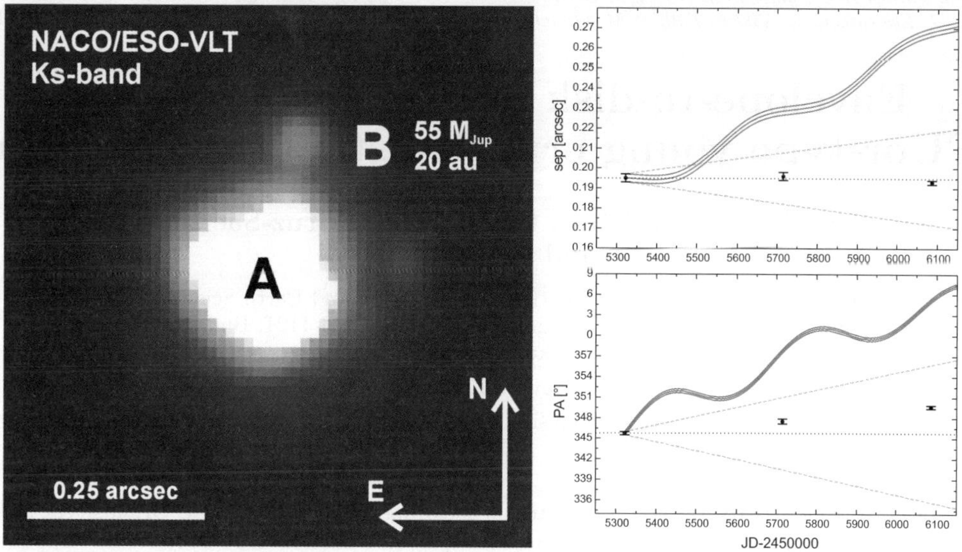

Figure 1. As example for the companions detected in our project, we show here a new brown dwarf companion imaged in three observing epochs close (only 20 au of projected separation) to one of our targets in the Lupus star forming region. As it is shown in the two plots on the right the companion clearly shares a common proper motion with its primary star and even orbital motion is significantly detected in position angle. The expected maximal change of separation and position angle due to orbital motion is shown as straight blue dashed lines. The expected change of both quantities for a non-moving background source is illustrated as red wobbled lines. Based on its infrared photometry the newly found co-moving companion exhibits a mass of about 55 M_{Jup}.

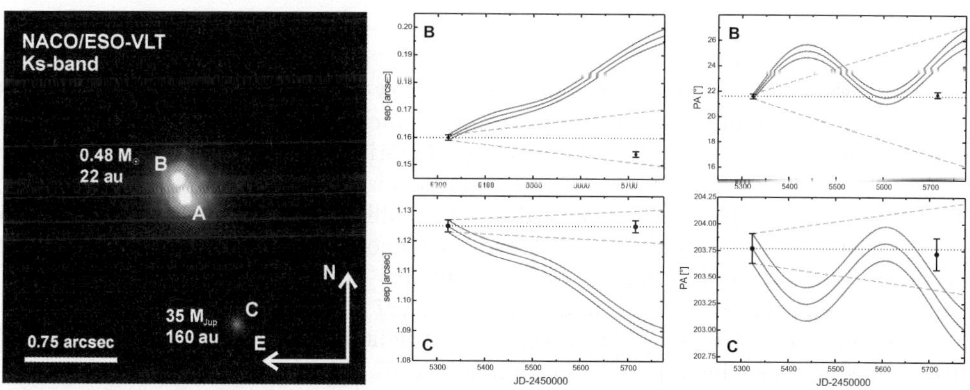

Figure 2. Here we show another example for our targets with detected co-moving companions, all identified in our high contrast imaging survey of T Tauri stars in the Lupus star forming region. This target exhibits a close bright companion (B) at only 22 au of projected separation, which is clearly co-moving to its primary star. The infrared photometry of this companion is consistent with a low-mass star of about 0.48 $M_{\odot}$, whose orbital motion could also be detected in separation (see the relative astrometry of the companion, illustrated in the plots on the top). In addition to this close stellar companion we identified a further but much fainter companion (C) of the star at a projected separation of about 160 au. The detected companion is clearly co-moving with our target (as shown in the plots at the bottom) and its infrared photometry is consistent with a brown dwarf with a mass of about 35 M_{Jup}. Hence, this T Tauri star, previously only known as a single star in the Lupus star forming region, is actually a hierarchical triple system with a wide substellar companion.

Origins: From the Protosun to the First Steps of Life
Proceedings IAU Symposium No. 345, 2018
B. G. Elmegreen, L. Viktor Tóth & M. Güdel, eds.

doi:10.1017/S1743921319001613

Envelope-to-disk mass transport in the FUor-type young eruptive star V346 Normae

Á. Kóspál[1,2], P. Ábrahám[1], O. Fehér[1], F. Cruz-Sáenz de Miera[1] and M. Takami[3]

[1]Konkoly Observatory, Research Centre for Astronomy and Earth Sciences, Hungarian Academy of Sciences, Konkoly-Thege Miklós út 15-17, H-1121, Budapest, Hungary
email: kospal@konkoly.hu

[2]Max Planck Institute for Astronomy, Königstuhl 17, D-69117 Heidelberg, Germany

[3]Institute of Astronomy and Astrophysics, Academia Sinica, PO Box 23-141, 106, Taipei, Taiwan

Abstract. Having disk-to-star accretion rates on the order of 10^{-4} $M_{\odot}$/yr, FU Orionis-type stars (FUors) are thought to be the visible examples for episodic accretion. FUors are often surrounded by massive envelopes, which replenish the disk material and enable the disk to produce accretion outbursts. We observed the FUor-type star V346 Nor with ALMA at 1.3 mm continuum and in different CO rotational lines. We mapped the density and velocity structure of its envelope and analyzed the results using channel maps, position-velocity diagrams, and spectro-astrometric methods. We discovered a pseudo-disk and a Keplerian disk around a 0.1 $M_{\odot}$ central star. We determined an infall rate from the envelope onto the disk of 6×10^{-6} $M_{\odot}$/yr, a factor of few higher than the quiescent accretion rate from the disk onto the star. This hints for a mismatch between the infall and accretion rates as the cause of the eruption.

Keywords. stars: pre–main-sequence, accretion disks, circumstellar matter

---

A long-standing open issue of the paradigm of low-mass star formation is that most protostars are less luminous than theoretically predicted. One possible solution is that the accretion process is episodic. The young outbursting FU Ori-type stars (FUors) are thought to be the visible examples for objects in the high accretion state. Mass infall from FUor envelopes may play a fundamental role in refilling the disk, and triggering instabilities causing accretion outbursts. However, our knowledge on the envelope dynamics and on the envelope-to-disk mass transfer is still very limited.

V346 Nor is an embedded FUor in the Sandqvist 187 dark cloud, at a distance of 700 pc. It erupted between 1976 and 1980, stayed in the bright state until 2008, rapidly faded in 2010, and is now showing a partial re-brightening. We modeled the multi-epoch optical-infrared spectral energy distributions and found a peak accretion rate of 10^{-4} $M_{\odot}$/yr in 1992, and a drop of accretion by at least a factor of 100 in 2010, suggesting that the quiescent accretion rate is less than about $10^{-7} - 10^{-6}$ $M_{\odot}$/yr (Kóspál *et al.* 2017a).

We observed V346 Nor with ALMA in the J = 2–1 line of ^{12}CO, ^{13}CO, C^{18}O, and 1.32 mm continuum in one single setting in Band 6. We used a combination of 12 m array configurations, as well as data from the 7 m array and Total Power antennas. We had two runs: in 2014–15 the beam size was $1''.0$ with a continuum sensitivity of 70 μJy/beam, while in 2017–18 the beam size was $0''.1$ with a continuum sensitivity of 35 μJy/beam.

We detected a fairly compact continuum source coinciding with the near-infrared stellar location (Fig. 1). The source is marginally resolved, with a deconvolved FWHM of $0''.21\times0''.15$. We interpret this as an inclined disk-like object, with a radius of 50 au. The position angle is NW-SE, same as the CO rotating disk. The central source is surrounded

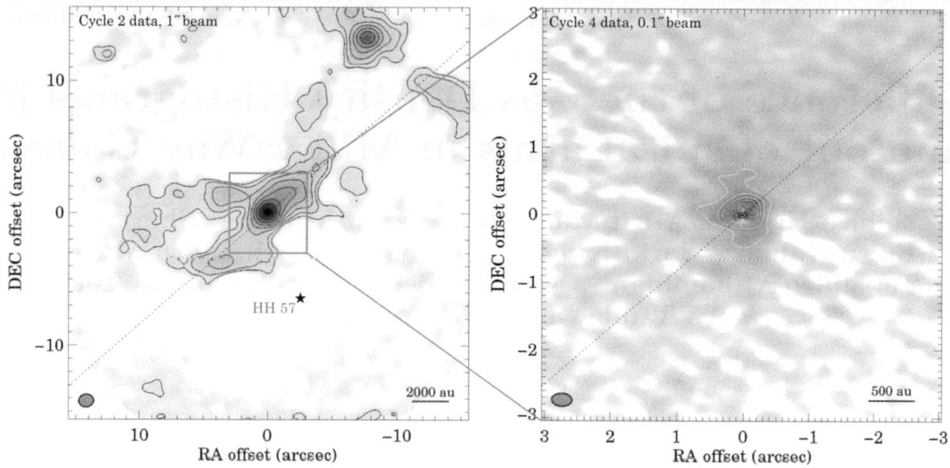

Figure 1. ALMA 1.32 mm continuum map of V346 Nor (left panel from Kóspál *et al.* 2017b).

by more extended, fainter emission out to 6300 au. The dust mass in the compact source within 50 au is $3{\times}10^{-4}\,M_{\odot}$.

Results from our spectro-astrometric analysis of the $C^{18}O$ data cube indicates that from 350 to 700 au, the radial velocity profile is consistent with a pseudo-disk (infalling-rotating motion with angular momentum conservation). The inner 350 au resembles a Keplerian disk around a $0.1\,M_{\odot}$ central star. The total gas mass within 350/700 au is $0.01/0.03\,M_{\odot}$.

Both the ^{12}CO and ^{13}CO profiles exhibit high-velocity line wings. The redshifted emission shows a parabola opening towards the northeast with a relatively wide opening angle of about 80°. The blueshifted emission forms a narrower ellipse extending towards the southwest with an opening angle of about 40°. The geometry is reminiscent of an outflow cavity, where emission is coming from the swept-up material in the cavity walls. The Herbig-Haro object HH 57 is situated along the axis of the southwestern CO-emitting ellipse, close to its farther edge.

We calculated the infall rate using an analytic formula from Momose *et al.* (1998). The resulting infall rate from the envelope onto the disk is $6{\times}10^{-6}\,M_{\odot}/\mathrm{yr}$, which is a factor of few higher than the quiescent accretion rate from the disk onto the star. It is a hint for a mismatch between the infall and accretion rates. This is the first observational support for such mismatch in a FUor, previously invoked to explain FUor outbursts.

Acknowledgements

This project has received funding from the European Research Council (ERC) under the European Union's Horizon 2020 research and innovation programme under grant agreement No 716155 (SACCRED, PI: Á. Kóspál). This paper makes use of the following ALMA data: ADS/JAO.ALMA#2013.1.00870.S, ADS/JAO.ALMA#2016.1.00209.S. ALMA is a partnership of ESO (representing its member states), NSF (USA) and NINS (Japan), together with NRC (Canada) and NSC and ASIAA (Taiwan) and KASI (Republic of Korea), in cooperation with the Republic of Chile. The Joint ALMA Observatory is operated by ESO, AUI/NRAO and NAOJ.

References

Momose, M., Nagayoshi, O., Ryohei, K., *et al.* 1998, *ApJ*, 504, 314
Kóspál, Á., Ábrahám, P., Westhues, Ch. & Haas, M. 2017a, *A&A*, 597, L10
Kóspál, Á., Ábrahám, P., Csengeri, T. *et al.* 2017b, *ApJ*, 843, 45

Origins: From the Protosun to the First Steps of Life
Proceedings IAU Symposium No. 345, 2018
B. G. Elmegreen, L. Viktor Tóth & M. Güdel, eds.

doi:10.1017/S1743921319001649

Difference of the Gas Density Histograms in and out of spiral arms in Milky Way Galaxy

Toshihiro Handa[1], Miru Maebata[1], Takeru Murase[1], Shinji Fujita[2], Mikito Kohno[2], Nario Kuno[3], Mika Kuriki[3], Mitsuhiro Matsuo[4], Tetsuhiro Minamidani[4], Atsushi Nishimura[2], Kazufumi Torii[4], Yuya Tsuda[5] and Tomofumi Umemoto[4]

[1]School of Science & Engineering, Kagoshima university,
Korimoto 1-21-35, Kagoshima, Kagoshima 890-0065, Japan,
email: handa@sci.kagoshima-u.ac.jp

[2]Department of Astrophysics, Nagoya University,
Furo-cho, Chikusa-ku, Nagoya, Aichi 464-8602, Japan.

[3]Graduate School of Pure and Applied Sciences, University of Tsukuba,
Tennodai 1-1-1, Tsukuba, Ibaraki 305-8571, Japan.

[4]Nobeyama Radio Observatory, National Astronomical Observatory of Japan,
Minamimaki 462-2, Minamisaku, Nagano 384-1305, Japan.

[5]Meisei University,
Hodokubo, Hino, Tokyo 191-0042, Japan.

Abstract. "Gas Density Histogram (GDH)" is an observational counterpart of the probability density function (PDF) of the gas density of interstellar medium (ISM). We used ^{12}CO data in $(l, b) = (29°, 0)$ region from "FOREST unbiased galactic imaging survey with Nobeyama 45-m telescope (FUGIN)", which is a large coverage survey in three CO (1-0) lines. Using the kinetic distance, we estimated the volume density of the voxel from the observed column density. The resultant GDHs of the inter-arm regions show lognormal or lognormal-like, but those in the spiral arm regions show flat-top shape.

Keywords. ISM: evolution, radio lines: ISM

1. Introduction

To address the density structure of ISM we investigate the GDH in the Milky Way. It is cumulative volumes as a function of gas density in an assigned region and an observational counterpart of PDF, if the density structure is steady. Although observational PDF is estimated for some galactic objects, they are 2 dimensional, *i.e.* sky coverage vs column density. Here, we show a 3D study in the galactic scale using the kinetic distance.

2. Data & Method

We investigate a $2° \times 2°$ region around $(l, b) = (29°, 0°)$ from the ^{12}CO (1-0) data of FUGIN survey (Umemoto *et al.* 2017). The spatial and velocity resolutions are 20″ and 1.3 km s^{-1}, respectively. The grid spacing is 8.5″. The typical noise level is 0.44 K in $T_{\rm mb}$. To reduce the noise level we integrated the data over 8 velocity channels before making GDHs; the velocity range we used is $26.75 \leqslant v_{\rm LSR} \leqslant 99.55$ km s^{-1} by every 5.2 km s^{-1}. The number of voxels we used is, therefore, $481 \times 481 \times 15$.

The column density of ISM is derived from ^{12}CO (1-0) intensity using the conversion factor $X_{\rm CO}$. We used $X_{\rm CO} = 1.8 \times 10^{20}$ cm^{-2} (K km s^{-1})$^{-1}$ (Dame *et al.* 2001). The

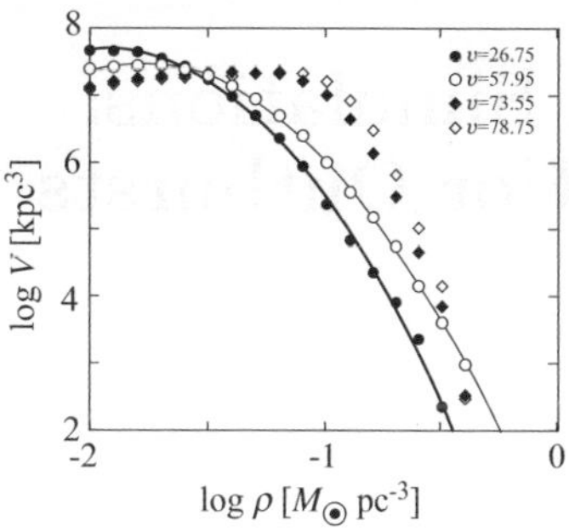

Figure 1. Gas Density Histograms from ^{12}CO (1-0) data around $(l, b) = (29°, 0°)$. Four different symbols show the GDHs at a different LSR velocity *i.e.* distance from the Sun. GDHs at $v_{\rm LSR} = 26.75$ and 57.95 km s^{-1} show lognormal (parabolic in this space), and two solid lines are the best fit curves. The others are flat-top and rapidly decrease in the denser end.

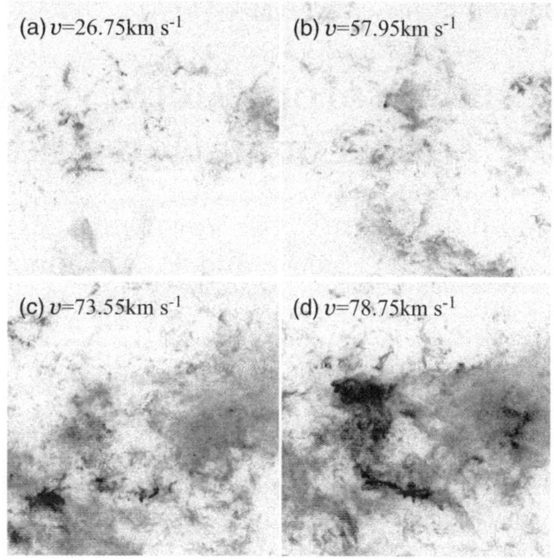

Figure 2. Four typical chanel maps in ^{12}CO (1-0). The area is 2° square centered at $(l, b) = (29°, 0°)$. The LSR velocity is shown in each panel. The GDHs at velocity shown in panels (a) and (b) are type A, and those in panels (c) and (d) are type B.

line-of-sight (LoS) depth of a voxel is derived using kinetic distance based on the flat rotation model with $R_0 = 8.5$ kpc and $V_0 = 220$ km s^{-1} (Kerr & Lynden-Bell 1986). The volume density of each voxel is the column density derived by this LoS depth, and the volume of each voxel is calculated using this LoS depth and the observed solid angle. For each velocity channel map we accumulate the volume for each volume density range, which gives the GDH at the corresponding distance, although it is an average of near and far locations. The low density limit is given by the noise level and LoS depth correspondent velocity range. It is about $10^{-2} M_\odot$ pc^{-3} for this region.

3. Results & Discussion

Out of 15 channels 3 GDHs are lognormal-like (type A), but the others are flat top or with a bump beyond the peak (type B). The typical GDHs at 4 velocities are shown in Fig. 1. Type A is expected for turbulence dominated region where gas density is randomly fluctuated (Vazquez-Semadeni 1994). Type A is found at the inter-arm regions in the $l - v$ diagram, where only few small molecular clouds are found (Fig. 2). Type B is found at the spiral-arm regions, where many dense cores and extended gas with filamentary structure are found. This suggests that molecular gas density is controlled by different mechanism in and out of the spiral arm. In the arm cloud-cloud collision and/or gravitational collapse may work to make dense clouds. We will apply this method for other regions.

References

Dame T. M. *et al.* 2001, *ApJ*, 547, 792
Kerr F. J. & Lynden-Bell D. 1986, *MNRAS*, 221, 1023
Vazquez-Semadeni E. 1994, *ApJ*, 423, 681
Umemoto T. *et al.* 2017, *PASJ*, 69, 78

Origins: From the Protosun to the First Steps of Life
Proceedings IAU Symposium No. 345, 2018
B. G. Elmegreen, L. Viktor Tóth & M. Güdel, eds.

doi:10.1017/S1743921319001650

Global Protoplanetary Disk Simulations: Dead Zone Formation and FUor Outbursts

Kundan Kadam[1,*], E. Vorobyov[2], Zs. Regály[1], Á. Kóspál[1,2] and P. Ábrahám[1]

[1]Konkoly Observatory, MTA CSFK, Budapest, Hungary
[*]email: kundan.kadam@csfk.mta.hu

[2]Max Planck Institute for Astronomy, Heidelberg, Germany

Abstract. We conducted global hydrodynamic simulations of protoplanetary disk evolution with an adaptive Shakura-Sunyaev α prescription to represent the layered disk structure, and starting with the collapse phase of the molecular cloud. With the canonical values of model parameters, self-consistent dead zones formed at the scale of a few au. The instabilities associated with the dead zone and corresponding outbursts, similar to FUor eruptions, were also observed in the simulations.

Keywords. planetary systems: protoplanetary disks, hydrodynamics, turbulence

1. Introduction

T Tauri-type stars in young stellar clusters show an apparent "luminosity problem", where their luminosity is observed to be about an order of magnitude lower as compared to that predicted by theoretical models (Kenyon *et al.* 1990). Variable young stars called FUors (prototype- FU Orionis) indeed show brightening by a factor of ~ 100, which can last for several decades (Hartmann & Kenyon 1996). Magneto-rotational instability (MRI) and gravitational instability (GI) are two of the prominent mechanisms for angular momentum transport in a protoplanetary disk. Gammie (1996) suggested that the midplane temperature of a protoplanetary disk at ~ 1 au is low enough to suppress the MRI, and the accretion occurs only through a thin surface layer due to cosmic ray ionization. Armitage *et al.* (2001) demonstrated that the transport of gas due to GI in the outer regions may lead to pile up of material at the magnetically inactive dead zone. The increased viscous heating in the this zone can eventually trigger MRI if the midplane temperature reaches a critical value ($T_{\rm crit}$) when the alkali metals are ionized, giving rise to the gravo-magneto (GM) instability in the disk.

2. Results

We performed global hydrodynamic simulations of protoplanetary disks in the thin disk limit, including the gravitational collapse phase of the cloud core (for details about the hydrodynamic code, see Vorobyov & Basu 2015). The kinematic viscosity was parametrized using a variable Shakura, & Sunyaev (1973) α parameter- $\alpha_{\rm eff} = (\Sigma_a\alpha_a + \Sigma_d\alpha_d)/(\Sigma_a + \Sigma_d)$, where the subscript 'a' denotes the gas surface density of the MRI active layer, while 'd' denotes that of the dead zone. This prescription of the angular momentum transport is consistent with the expectations from a layered disk model.

With the canonical values of layered disk parameters ($T_{\rm crit} = 1300$K, $\Sigma_{\rm a} = 100$ g/cm^2), self-consistent dead zones formed with the variable α prescription; while such zones did

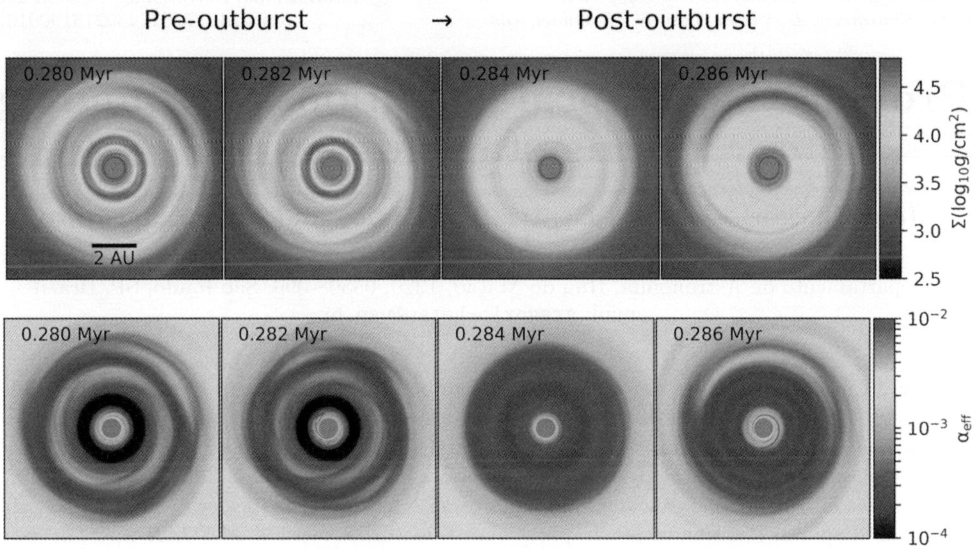

Figure 1. Evolution of the surface density and effective α-parameter in the inner 10 au box, showing the progression of the dead zone across an outburst.

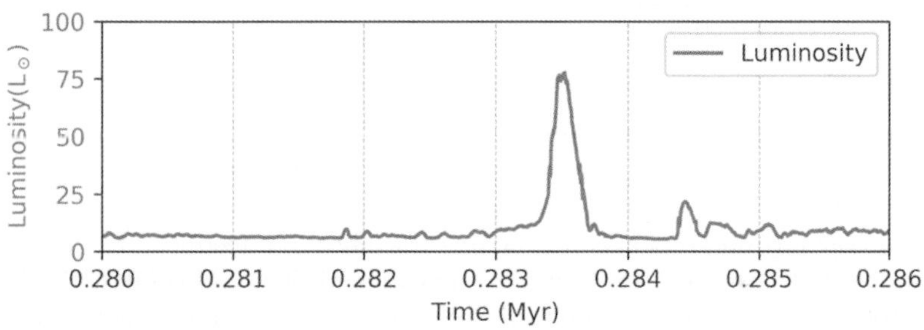

Figure 2. Total luminosity of the outbursting system, depicting an eruption similar to FUors.

not form in (an otherwise identical) fully MRI active disk. The dead zone formed in the region between $1 - 5$ au has high surface density rings and low effective viscosity (Figure 1, first two columns).

Since the dead zone can not effectively transport angular momentum, the gas continued to accumulate in its vicinity from the outer regions. The central disk temperature eventually reached T_{crit} through viscous heating triggering MRI, and thus the GM instability. The dead zone disappeared (see Figure 1, last two columns) and the accumulated material was accerted onto the star. An associated eruption was observed, which can be considered as FUor-like outburst. The eruption typically lasted few hundred years, while the total luminosity of the system increased to $\sim 100 L_{\odot}$ (Figure 2).

References

Armitage, P. J., Livio, M., & Pringle, J. E. 2001, *MNRAS*, 324, 705
Gammie, C. F. 1996, *ApJ*, 457, 355
Hartmann, L., & Kenyon, S. J. 1996, *ARA&A*, 34, 207
Kenyon, S. J., Hartmann, L. W., Strom, K. M., & Strom, S. E. 1990, *AJ*, 99, 869
Shakura, N. I., & Sunyaev, R. A. 1973, *A&A*, 24, 337
Vorobyov, E. I., & Basu, S. 2015, *ApJ*, 805, 115

Origins: From the Protosun to the First Steps of Life
Proceedings IAU Symposium No. 345, 2018
B. G. Elmegreen, L. Viktor Tóth & M. Güdel, eds.

doi:10.1017/S1743921319001674

Probing the physical conditions surrounding young star clusters

Jane Gregorio-Hetem, Beatriz Fernandes and Felipe Navarete

Universidade de São Paulo IAG/USP,
Departamento de Astronomia, Rua do Matão, 1226, 05508-090, São Paulo, SP, Brazil
email: gregorio-hetem@usp.br

Abstract. The scenario of star cluster formation can be better understood based on the detailed study of the dynamical conditions of the associated gas, clustering properties and effects of ionizing sources, among others. Some of these characteristics are explored in our ongoing work based on observations with the *SOAR* and *T80-S* telescopes. With these data we have obtained a complete multi-band photometric catalogue of selected clusters that we characterize using color-color diagrams and flux ratios. In particular, this work is focused on *SOAR/Spartan* (near-infrared) observations of the Canis Major star-forming region.

Keywords. stars: pre–main-sequence, open clusters and associations: general.

1. Introduction

The early evolution of star clusters can be deciphered by comparing the cinematic and structural parameters of stars distributions with dynamical conditions of their local interstellar medium. This study can bring us valuable information about the clustering properties of star-forming regions and the possible effects of feedback from massive stars.

One example of this scenario is found in the Canis Major (CMa) OB1/R1 Association, a wide ($\sim 10^o \times 10^o$) complex of molecular clouds, emission and reflection nebulae that show evidence of star formation induced by supernova events (Fernandes *et al.* 2019). About two hundred emission-line stars, most having spectral type B, were reported by Shevchenko *et al.* (1999) in the area of about ($\sim 2^o \times 2^o$) covering several reflection nebulae, called CMa R1 region, which is located at d $\sim$ 1kpc (for a review, see Gregorio-Hetem 2008).

The most prominent nebula belonging to CMa R1 is Sh2-296, which shows an arc-shaped structure associated with more than 300 low-mass young stars (Fernandes *et al.* 2015). However, less than 10% of them shows evidence of circumstellar disks, which is quite rare when compared with most of the star-forming regions younger than 3 Myr (Haisch *et al.* 2001, Hernández *et al.* 2008, Fedele *et al.* 2010) suggesting that some external factor accelerated the disk dissipation.

2. Characterization of pre-main sequence stars

Our long-term goal is to study the young stellar population in different Galactic regions, and investigate the influence of the environment and feedback from massive stars (ionization, winds) in the formation and evolution of star clusters and circumstellar disks.

The present work is focused on the census of the CMa young stellar population, improved by means of the infrared (IR) characterization.

Following Navarete *et al.* (2015), we performed three observing runs with the *Spartan* camera at the 4m *SOAR* telescope (Cerro Pachón, Chile) to acquire near-IR images (H_2,

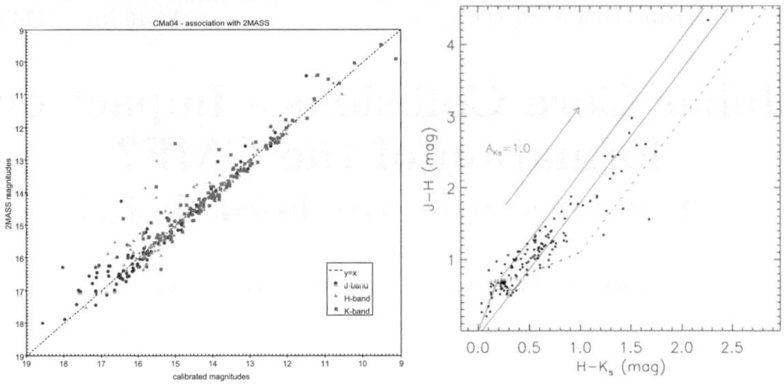

Figure 1. *SOAR/Spartan* data for one of the CMa R1 fields. *Left:* Calibrations of JHK magnitudes. *Right:* Color-color diagram.

Brγ) that are often used as probes of extended and filamentary structures and could bring us some light on the conditions of the molecular and atomic gas associated with our targets.

Figure 1 shows the calibration (based on 2MASS data) of the *Spartan* JHK magnitudes, and the color-color diagram of the sources detected in one of the CMa fields. Our near-IR observations greatly increased the number of detected sources (when compared with the previously known objects having good 2MASS photometry, i.e., AAA flags).

3. Partial results and conclusion

In the field around Z CMa we detected 219 near-IR sources, 99 of them have 2MASS counterpart with good photometry. This means that 120 are new sources, not previously identified or that had an improvement on their photometric measurements. In two other fields, we detected a total of 281+144 sources, revealing 178+97 new objects.

These near-IR results have significantly improved the census of low-mass stars in CMa R1. However, H_2 and Brγ images do not reveal filamentary or extended structures in the fields observed with *Spartan*. Next steps of this ongoing project is to study the ionized gas based on Hα, SII and OIII observations obtained with *SAMI* (*SOAR Adaptative Module Image*) for CMa R1 and other young clusters.

Acknowledgements

This work was supported by FAPESP Procs. 2018/05866-0 (JGH), 2015/24946-6 (JGH, BF), and 2017/18191-8 (FN). Based on observations obtained at the Southern Astrophysical Research (SOAR) telescope, which is a joint project of the Ministério da Ciência, Tecnologia, Inovações e Comunicações (MCTIC) do Brasil, the U.S. National Optical Astronomy Observatory (NOAO), the University of North Carolina at Chapel Hill (UNC), and Michigan State University (MSU).

References

Fedele, D., van den Ancker, M. E., Henning, Th., *et al.* 2010, *A&A* 510, 72
Fernandes, B., Gregorio-Hetem, J., Montmerle, T, & Rojas, G. 2015, *MNRAS* 448, 119
Fernandes, B., Montmerle, T., Santos-Silva, T., & Gregorio-Hetem, J. 2019, *(accepted)*
Gregorio-Hetem, J. 2008, in: B. Reipurth (ed.), *Handbook of Star Forming Regions*, Vol. 2, pg. 1
Haisch, K. E., Lada, E. A., & Lada, C. J. 2001, *ApJ* 553, L153
Hernández, J., Hartmann, L., Calvet, N., *et al.* 2008, *ApJ* 686, 1195
Navarete, F., Damineli, A., Barbosa, C., & Blum, R. 2015, *MNRAS* 450, 4364
Shevchenko, V. S., Ezhkova, O. V., Ibrahimov, M. A. *et al.* 1999, *MNRAS* 310, 210

Origins: From the Protosun to the First Steps of Life
Proceedings IAU Symposium No. 345, 2018
B. G. Elmegreen, L. Viktor Tóth & M. Güdel, eds.

doi:10.1017/S1743921319001698

Prestellar Core Collisions - Impact on the formation of the CMF? A case study on FeSt 1-457

Gabor I. Herbst-Kiss[1] and Joao Alves[2]

[1]Department of Astrophysics, University of Vienna, Türkenschanzstrasse 17, 1180 Vienna, Austria

[2]Radcliffe Institute for Advanced Study, Harvard University, 10 Garden Street, Cambridge, MA 02138, USA
email: gabor.imre.kiss@univie.ac.at

Abstract. The initial mass function (IMF) is a profoundly studied subject, however its origin is still unclear and heavily disputed. The Core Mass Function (CMF) has a remarkable resemblance to a shifted IMF along the mass axis of a factor of 3. This CMF has been observed amongst others in the Pipe Nebula, a calm molecular cloud at approximately 130 pc. We study the origin of the CMF under the assumption that collisions and merging of prestellar cores shape the CMF. We present our preliminary results of core collisions for the well known FeSt 1-457.

Keywords. ISM: clouds, kinematics and dynamics

---

1. Introduction

In the timeline of star formation, a molecular cloud has to lose 6 – 7 orders of magnitude of its initial specific angular momentum. This loss happens in distinct stages due to various processes (e.g. magnetic braking, fragmentation and turbulence). The first entities formed out of fragmenting filaments are structures of several Jeans masses. Often these objects have small separations and could collide in regions of high core density. Indeed colliding clouds, filaments and prestellar cores have been already identified (Frau *et al.* 2015; Alves & Burkert 2009). To study the effects of these collisions on the loss of angular momentum, we use in this specific example truncated Plummer spheres in a variety of initial conditions. The resulting v_{lsr} gradient of the 2D projected core, the oscillation patterns and dissipated material after the merge will help in the identification and interpretation of real observed cores.

2. Motivation

Several prestellar Cores in the Pipe Nebula show a quasi stable state. The observed cloud cores from Rathborne *et al.* (2009) and Roman-Zuniga *et al.* (2010) are distributed like a shifted IMF. Further studies have shown that the v_{lsr} of the cores can be split into two groups with 2-4 km/s and 6-7 km/s indicating filamentary structures oriented E-W and N-S (Frau *et al.* 2015). The authors propose a filamentary collision and therefore objects especially in the so called bowl of the Pipe Nebula are of great interest for possible colliding candidates.

3. Methods and preliminary results

We use the Herschel column density maps and data from the literature for core selection. The criteria for the selected cores are an elongated shape or a close core within the distance of a Jeans length. First we analyze each single core and subsequently test collision scenarios with the smoothed particle hydrodynamics code called GANDALF (Hubber *et al.* 2018). The simulations were run with and without moderate turbulence and a Boss Bodenheimer test was performed. The cores 94 and 96 fall into the higher velocity filament with 6.3 and 5.87 km/s respectively and are located at the rim of the bowl. By extrapolating the velocity gradient of FeSt 1-457 along the northeast-southwest direction of 1.8 km/s (Aguti *et al.* 2007) to the local surrounding structure we find that the cores collide within 10^5 yrs starting with their current separation of 0.1 pc. This collision produces a clear shock wave between the two cores, which was not observed by any author.

The small velocity dispersion of the line broadening and the small difference in v_{lsr} would speak for a fragmentary origin. A Boss-Bodenheimer test with the current rotation of 10^{-14} radians implemented from the observed small velocity gradient of core 96 could not fit the current observations. The literature is divided whether the core is stable or unstable, but Aguti *et al.* (2007) and Juarez *et al.* (2016) agree that it is at the rim of stability. However, we calculate that FeSt 1-457 exceeds the critical Bonnor Ebert mass and should collapse and therefore star formation will set in before the two cores (94 and 96) can merge by pure gravitational interaction. A simulation of two equally sized prestellar cores colliding with velocity of 1.8 km/s did not instantly collapse after the collision but a slower motion of drifting apart with slight preference to dispersal was observed. We notice here that no background gas is implemented as external pressure in the simulations, although it could have a considerable effect on the pulsation of the cores (Aguti *et al.* 2007). The initial setup for this simulation were two 2 $M_\odot$ cores with plummer density profile at a separation of 0.1 pc. Our study has shown that two prestellar cores with plummer density profile could produce a 4 $M_\odot$ core with an elongated bullet-shaped companion. This structure is however not stable and collapses without any further support.

4. Conclusion and future implications

We use the two-fluid version of GANDALF adding a dust component to study the observable dynamics. Especially in the case of the bullet companion the shown diverging behaviour of larger dust grains from gas dynamics is of great interest.

References

Aguti, E. D., Lada, C. J., Bergin, E. A., Alves, J. F., & Birkinshaw, M. 2007, *ApJ*, 665, 457

Alves, J., & Burkert, A. 2009, *ApJ*, 695, 1308

Frau, P., Girart, J. M., Alves, F. O., Franco G. A. P., Onishi, T. & Roman-Zuniga, C. G.2015, *AA*, 574, 6

Hubber, D. A., Rosotti, G. P., & Booth, R. A. 2018, *MNRAS*, 473, 1603

Juarez, C., Girart, J. M., Frau, P., Palau, A., Estalella, R., Morata, O., Alves, F. O., Beltran, M. T., & Padovani, M. 2017, *AA*, 597, 74

Rathborne, J. M., Lada, C. J., Muench, A. A., Alves, J. F., Kainulainen, J., & Lombardi, M. 2009, *ApJ*, 699, 742

Roman-Zuniga, C. G., Alves, J. F., Lada, C. J., & Lombardi, M. 2010, *ApJ*, 725, 2232

Origins: From the Protosun to the First Steps of Life
Proceedings IAU Symposium No. 345, 2018
B. G. Elmegreen, L. Viktor Tóth & M. Güdel, eds.

doi:10.1017/S1743921319001728

Observations of H_2O and OH masers in star-forming regions

Georgij Rudnitskij[1], Nuriya Ashimbaeva[1], Olga Bayandina[2], Pierre Colom[3], Evgeny Lekht[1], Mikhail Pashchenko[1], Nadezhda Shakhvorostova[2], Alexander Tolmachev[4] and Irina Valtts[2]

[1]Lomonosov Moscow State University, Sternberg Astronomical Institute, Moscow, 119234 Russia
email: gmr@sai.msu.ru
[2]Astrospace Center, Lebedev Institute of Physics, Russian Academy of Sciences, Moscow, 117997 Russia
[3]LESIA, Observatoire de Paris-Meudon, Meudon, 92195 France
[4]Pushchino Radio Astronomy Observatory, Astrospace Center, Lebedev Institute of Physics, Russian Academy of Sciences, Pushchino, 142290 Russia

Abstract. We report long-term observations of H_2O and OH maser emission sources at wavelengths of 1.35 and 18 cm associated with star-forming regions. Strong quasi-periodic flares of maser emission have been observed. Several sources (in particular, G25.65+1.05, IRAS 16293−2422, Cep A) have displayed strong flares in the H_2O line, when their peak flux density raised by a few orders of magnitude above the quiet state. Possible causes of this are discussed.

Keywords. stars: formation, stars: planetary systems: protoplanetary disks, radio lines: stars.

Cosmic maser sources associated with protostellar and/or young stellar objects in star-forming regions often demonstrate strong variability. In particular, this is typical of masers emitting in the 6_{16}–5_{23} rotational line of the water-vapor molecule at a wavelength of 1.35 cm. Since 1980, we have been monitoring a sample of 60 H_2O masers in star-forming regions at the 22-meter radio telescope of the Pushchino Radio Astronomy Observatory, Russia. The same sources are also monitored in the OH lines at $\lambda = 18$ cm on the Nançay radio telescope, France. Some of them are well-known masers, such as Ori A, W43M3, W51, W75N. Strong quasi-periodic flares of H_2O maser emission have been observed by Colom *et al.* (2018); Kurtz *et al.* (2018); Shakhvorostova *et al.* (2018); Volvach *et al.* (2018) in the source G25.65+1.05 (IRAS 18316−0602), maximum peak flux density reaching scores of kilojanskys, whereas the source in its quiescent state had hardly 300 Jy (Fig. 1). This source was observed on 29.09.2017 with a record angular resolution on a baseline of 9 Earth diameters by the space-ground RadioAstron interferometer just during its flare, when its peak flux density reached 13 kJy. It displayed a single unresolved feature smaller than 0.1 mas, see Kurtz *et al.* (2018).

Another example is IRAS 16293−2422, a young binary system, the nearest one to the Earth with two collimated molecular outflows, Girart *et al.* (2018). In 2018 the source also had a strong flare in the H_2O line (Fig. 2).

Possible causes of this are discussed, among them consecutive excitation of the maser by a propagating shock wave and processes related to turbulence and gravitational instability in the circumstellar protoplanetary disks.

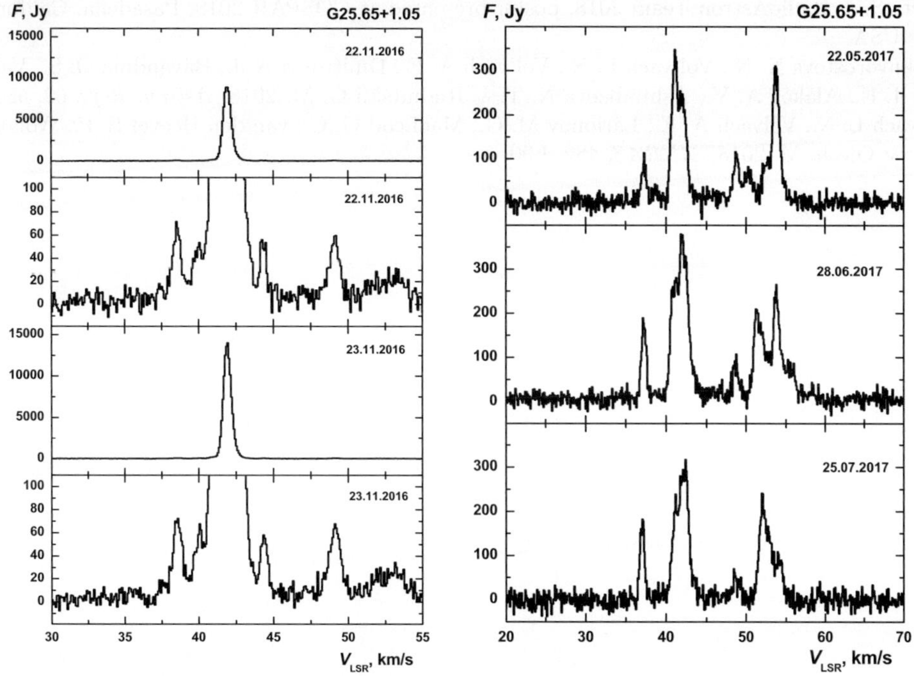

Figure 1. The flare of the source G25.65+1.05 (IRAS 18316−0602) in 2016 (left) and the source in its quiescent state in 2017 (right).

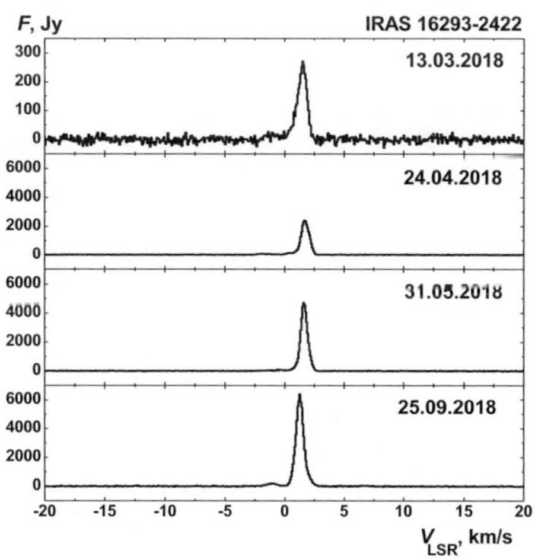

Figure 2. The flare of the source IRAS 16293−2422 in 2018.

The complete version of this poster in the electronic form (together with some OH data) can be found at:
`http://comet.sai.msu.ru/~gmr/IAUS345_24122.pdf`

References

Colom P., Lekht E. E., Pashchenko M. I., Rudnitskii G. M., & Tolmachev A. M. 2018, *Astron. Rep.*, 62, 440

Girart J. M., Estalella R., Palau A., Torrelles J. M. & Rao R. 2014, *ApJ*, 780, L11

Kurtz S. & RadioAstron Team 2018, poster presented at COSPAR 2018, Pasadena, California, USA

Shakhvorostova N. N., Vol'vach L. N., Vol'vach A. E., Dmitrotsa A. I., Bayandina O. S., Val'tts I. E., Alakoz A. V., Ashimbaeva N. T. & Rudnitskii G. M. 2018, *Astron. Rep.*, 62, 584

Volvach L. N., Volvach A. E., Larionov M. G., MacLeod G. C., van den Heever S. P., Wolak P., & Olech, M. 2018, *MNRAS*, 482, L90

Origins: From the Protosun to the First Steps of Life
Proceedings IAU Symposium No. 345, 2018
B. G. Elmegreen, L. Viktor Tóth & M. Güdel, eds.

doi:10.1017/S1743921319001741

HCL1 and HCL2 - low mass star formation in violent and quiet environments

L. Viktor Tóth[1,2], Orsolya Fehér[2,3], Mika Juvela[4], Julien Montillaud[5] and Sándor Pintér[6]

[1]Department of Astronomy, Eötvös Loránd University, Pázmány Péter sétány 1/A, H-1117 Budapest, Hungary
email: l.v.toth@astro.elte.hu

[2]Konkoly Observatory of the HAS, 15-17 Konkoly Thege Miklós út, H-1121 Budapest, Hungary

[3]IRAM, 300 Rue de la Piscine, 38406, Saint Martin d'Hères, France

[4]Department of Physics, PO Box 64, 00014 University of Helsinki, Finland

[5]Institut UTINAM - UMR 6213 - CNRS - Univ Bourgogne Franche Comte, France, OSU THETA, 41bis avenue de l'Observatoire, 25000 Besançon, France

[6]National University of Public Service, Budapest, Hungary

Abstract. Recently a large number of Galactic cold clumps were located with the Planck all-sky survey. Our radio line observations have revealed the distribution and physical properties of the interstellar medium in dozens of PGCC sources. Clumps can be affected by many external effects. HCL1 (a.k.a. L1251) and HCL2 (which contains also TMC-1) are examples of low mass star forming clouds in violent and quiet environments.

Keywords. stars: formation, radio lines: ISM

1. Introduction

The Solar System may have formed in a violent environment but solar mass stars can also form without external triggering. Heiles (1968) traced cold interstellar gas with OH detections at positions of high visual extinction. Since then the name "Heiles Cloud 2" (HCL2) identifies the most massive molecular cloud in Taurus as was shown e.g. by a large scale $C^{18}O$ survey by Onishi *et al.* (1996). It is however not recognized these days, that Heiles' Cloud 1, where he also detected normal (i.e. not non-thermal) OH lines, is L1251, one of the prominent molecular clouds of the Cepheus region. Here we compare HCL1 and HCL2 as examples of low mass star formation in different environments.

2. Results and discussion

L1251 is at the boundary of the Cepheus Flare GMC (Lebrun 1986, Kiss *et al.* 2006). It faces an old SNR bubble as the overlaid soft X-ray contours show in the CO map of Grenier *et al.* (1989). The Cepheus Planck FIR, sub-mm polarization pattern (Planck collaboration *et al.* 2015) shows a large scale structure, and the elongated L1251 cloud (see recently Levshakov *et al.* 2016) looks unrelated.

HCL1 has two Planck Galactic Cold Clumps (PGCC, Planck collaboration 2011, 2015, Montillaud *et al.* 2015) and a number of low and intermediate mass YSOs mostly around it (Tóth & Walmsley 2011). HCL2 is a "bush" type cluster of PGCCs, as revealed by the minimum spanning tree (MST) method, with a group of low mass YSOs (Tóth *et al.* 2004, 2017). HCL2 is embedded in a complex of clouds, with a B field pattern that may be the result of colliding HI flows (Ballesteros-Paredes *et al.* 1999). The Taurus filaments

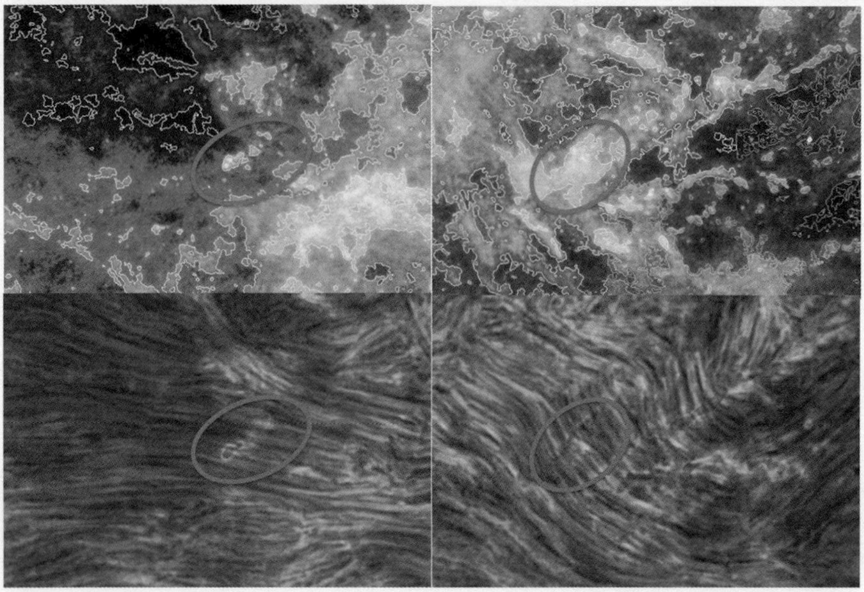

Figure 1. FIR continuum and polarization images of the environment of L1251 (left) and HCL2 (right). The positions of the clouds are marked with red ellipses. Top row: Composite of Red: Planck 857GHz; Green: IRIS 100μm; Blue: IRIS 12μm continuum images. Bottom row: B field pattern based on Planck polarization data (LIC convolved to 10 arcmin).

are perpendicular to the B field while striations (fibers outside filaments) are parallel to B field, based on CO observations and optical polarization (Goldshmith *et al.* 2008) and Planck FIR and sub-mm polarization data.

Acknowledgements

Based on observations with the 100-m telescope of the MPIfR (Max-Planck-Institut für Radioastronomie) at Effelsberg. Herschel is an ESA space observatory with science instruments provided by European-led Principal Investigator consortia and with important participation from NASA. This research was supported by the Hungarian OTKA grant NN111016, the Balaton program 40470VL and TÉT 2017-2.2.5-TÉT-FR-2017-00027.

References

Ballesteros-Paredes, J. *et al.* 1999, *ApJ*, 527, 285
Fehér, O. *et al.* 2016, *A&A*, 590, A75
Goldshmith, P. *et al.* 2008, *ApJ*, 680, 428
Grenier, I. *et al.* 1989, *ApJ*, 347, 231
Heiles, C. *et al.* 1968, *ApJ*, 151, 919
Kiss, Z. T. *et al.* 2006, *A&A*, 453, 923
Lebrun, F. *et al.* 1986, *ApJ*, 306, 16
Levshakov, S. A. *et al.* 2016, *A&A*, 586, 126
Montillaud, J. *et al.* 2015, *A&A*, 584, 92
Onishi, T. *et al.* 1996, *ApJ*, 465, 815
Planck collaboration *et al.* 2011, *A&A*, 536, A23
Planck collaboration *et al.* 2015, *A&A*, 576, A104
Tóth, L. V. *et al.* 1995, *Ap&SS*, 233, 175
Tóth, L. V. & Walmsley, C. M 1996, *A&A*, 311, 981
Tóth, L. V. *et al.* 2004, *A&A*, 420, 533
Tóth, L. V. *et al.* 2017, *IAUS*, 316, 133

Origins: From the Protosun to the First Steps of Life
Proceedings IAU Symposium No. 345, 2018
B. G. Elmegreen, L. Viktor Tóth & M. Güdel, eds.

doi:10.1017/S1743921319001753

A double core in the Auriga-California Molecular Cloud

Sarolta Zahorecz[1,2,3], Daniel Molnar[4], Alex Kraus[5] and Toshikazu Onishi[1]

[1]Department of Physical Science, Graduate School of Science, Osaka Prefecture University, 1-1 Gakuen-cho, Naka-ku, Sakai, Osaka 599-8531, Japan
email: s.zahorecz@p.s.osakafu-u.ac.jp

[2]Chile Observatory, National Astronomical Observatory of Japan, National Institutes of Natural Science, 2-21-1 Osawa, Mitaka, Tokyo 181-8588, Japan

[3]Eötvös Loránd University, Department of Astronomy, Pázmány Péter sétány 1/A, 1117, Budapest, Hungary

[4]Osservatorio Astronomico di Cagliari, Via della Scienza, 5, 09047 Cuccuru Angius, Selargius CA, Italy

[5]Max Planck Institut fúr Radioastronomie, Auf dem Hügel 69, 53121, Bonn, Germany

Abstract. Planck cold clump G163.82-8.44 is part of the Auriga-California Molecular Cloud. It was observed with Herschel PACS and SPIRE instruments as part of the Herschel open time key programme Galactic Cold Cores. Follow-up ground-based molecular line observation of NH_3 was performed to the densest part of the filament with the Effelsberg-100m telescope. We detected two different velocity components with a separation of 0.5 km/s. We performed radiative transfer modeling with two 3-dimensional spheres to characterise the temperature and density of the dense cores. We have found that the temperatures of the two cores are almost the same, 10.8 K and 11.1 K and their mass and size ratios are 1:10 and 1:5, respectively.

Keywords. radiative transfer, molecular data, ISM: structure, stars: formation

1. Introduction

Far-infrared colours help us exploring the structures of the nearby cold interstellar matter: large shells seen as far-infrared loops (Könyves *et al.* 2007 using *IRAS* maps), molecular cloud complexes such as HCL2 and isolated clouds (see Tóth *et al.* 2004 and Tóth *et al.* 2002 respectively using ISOPHOT data). Large-scale surveys are needed to study cold, dense interstellar matter in the whole Galaxy. The *Planck* (Tauber *et al.* 2010) all-sky survey has provided us a perfect database to identify the coldest structures of the interstellar matter in the Galaxy. The follow-up observations of these Planck objects with the European Infrared Space Observatory *Herschel* gave a unique opportunity to study in great detail the density and temperature structure of these objects at high angular resolution. In the Herschel open time key programme Galactic cold cores project (Juvela *et al.* 2010, 2012) we mapped ~390 cold Planck objects on ~120 fields with the *Herschel* PACS and SPIRE instruments (100-500 μm). Detailed study of these Planck objects were performed based on the Herschel data to study the physical properties, column density structures, filaments of these regions (see e.g. Montillaud *et al.* 2015, Rivera Ingraham *et al.* 2016, 2017, Juvela *et al.* 2018).

One of our fields is G163.82-8.44 in the Auriga-California Molecular Cloud, it contains more than 30 Planck cold clumps (Planck Collaboration *et al.* 2011) found in large groups similarly to other fields in our sample, see e.g. Tóth *et al.* (2017). G163.82-8.44 is one of the longest and most striking filaments in the sample. Very clear striations are apparent in the Herschel images. It is located at a distance of 450 pc (Montillaud *et al.* 2015). We used the column density maps based on *Herschel* SPIRE data calculated by Juvela *et al.* (2012) to identify dense cores on the field and select the densest ones for molecular line follow-up observations.

2. Ammonia observation and radiative transfer modeling

We observed the NH_3 (1,1) and (2,2) inversion transition lines in the densest region with $N(H_2)>2\times10^{22}$ cm^{-2} in G163.82-8.44 with the Effelsberg 100-m telescope. The total on-source integration time was ~4 hours. We used the 13 mm primary receiver of the telescope with frequency switch mode. The half-power beamwidth at this frequency is 40", 0.085 pc at the distance of 450 pc. Data were reduced with the GILDAS package. See more details of the observations and data reduction in the description of the NH_3 survey by Tóth *et al.* (in prep). The observations revealed two velocity components with a separation of 0.5 km/s. Based on the ammonia observations, the two dense cores have kinetic temperatures, effective radii and volume densities of 8.7 K, 10.1 K; 0.071 pc, 0.123 pc; 7.18×10^3 cm^{-3}, 9.59×10^3 cm^{-3}, respectively. The estimated masses give a mass ratio of 7:1.

We used a non-LTE Monte-Carlo radiative transfer simulation (CPPSIMU, Juvela 1997) for the radiative transfer modeling of NH_3 line emission to determine the physical properties of the dense cores based on the best signal-to-noise ratio spectra in the centre part of the cores. The system is described with two spherically symmetric density distributions with Plummer-like profile. We successfully reproduced the spectra in the centre part of the cores with the model. We have found that the temperatures of the two cores are almost the same, 10.8 K and 11.1 K and their mass and size ratios are 1:10 and 1:5, respectively. A detailed description of the model with using all observations will be presented in Zahorecz *et al.* (in prep).

Acknowledgements

Based on observations with the 100-m telescope of the MPIfR (Max-Planck-Institut fr Radioastronomie) at Effelsberg, the authors thank their high level of technical support and assistance. This research was supported by the NAOJ ALMA Scientific Research Grant Number 2016-03B.

References

Juvela, M. 1997, *A&A*, 322, 943
Juvela, M., *et al.* 2012, *A&A*, 518, L93
Juvela, M., *et al.* 2012, *A&A*, 541, A12
Juvela, M., *et al.* 2018, *A&A*, 614, A83
Könyves, V., *et al.* 2007, *A&A*, 463, 1227
Montillaud, J., *et al.* 2015, *A&A*, 584, A92
Planck Collaboration, *et al.* 2011, *A&A*, 536, A23
Rivera-Ingraham, A., *et al.* 2016, *A&A*, 591, A90
Rivera-Ingraham, A., *et al.* 2017, *A&A*, 601, A94
Tauber, J. A., *et al.* 2010, *A&A*, 520, A1
Tóth, L. V., *et al.* 2002, *A&A*, 395, 663
Tóth, L. V., *et al.* 2004, *A&A*, 420, 533
Tóth, L. V., *et al.* 2017, *IAUS*, 316, 133

Origins: From the Protosun to the First Steps of Life
Proceedings IAU Symposium No. 345, 2018
B. G. Elmegreen, L. Viktor Tóth & M. Güdel, eds.

doi:10.1017/S1743921319001765

Deuteration of formaldehyde - an important precursor of hydrogenated complex organic molecules - during star formation in our Galaxy

Sarolta Zahorecz[1,2], Izaskun Jimenez-Serra[3], Leonardo Testi[4], Katharina Immer[5], Francesco Fontani[6], Paola Caselli[7], L. Viktor Toth[8,9], Ke Wang[4] and Toshikazu Onishi[1]

[1]Department of Physical Science, Graduate School of Science, Osaka Prefecture University, 1-1 Gakuen-cho, Naka-ku, Sakai, Osaka 599-8531, Japan
email: s.zahorecz@p.s.osakafu-u.ac.jp

[2]Chile Observatory, National Astronomical Observatory of Japan, National Institutes of Natural Science, 2-21-1 Osawa, Mitaka, Tokyo 181-8588, Japan

[3]School of Physics and Astronomy, Queen Mary University of London, Mile End Road, E1 4NS London, United Kingdom

[4]European Southern Observatory, Karl-Schwarzschild-Str. 2, 85748, Garching, Germany

[5]Joint Institute for VLBI ERIC, Postbus 2, 7990 AA Dwingeloo, The Netherlands

[6]INAF-Osservatorio Astrofisico di Arcetri, L.go E. Fermi 5, 50125 Firenze, Italy

[7]Max-Planck-Institut für Extraterrestrische Physik, Giessenbachstrasse 1, 85748 Garching, Germany

[8]Eötvös Loránd University, Dep., Pázmány P. sétány 1/A, 1117, Budapest, Hungary

[9]Konkoly Observatory, Research Centre for Astronomy and Earth Sciences, Hungarian Academy of Sciences, H-1121 Budapest, Konkoly Thege Miklós út 15-17, Hungary

Abstract. Formaldehyde (H_2CO) and its deuterated forms can be produced both in the gas phase and on grain surfaces. However, the relative importance of these two chemical pathways is unclear. Our recent single dish observation of formaldehyde and its deuterated species suggests that they form mostly on grain surfaces although some gas-phase contribution is expected at the warm HMPO stage. Since the single dish beam is larger, and since these high-mass star-forming regions are clustered and complex, it is however unclear whether the emission arises from the protostellar sources or from starless/pre-stellar cores associated with them. Therefore, interferometric observations are needed to separate the emission originating from the small and dense cores, to disentangle their formation routes and then being able to use them as powerful diagnostic tools of the physical and chemical properties of high-mass star forming regions.

Keywords. astrochemistry, molecular data, ISM: molecules, stars: formation

1. Introduction

H_2CO is crucial in any chemical networks because it is very abundant and it is an important precursor of hydrogenated complex organic molecules. Our single dish observation of high-mass star-forming cores in the Galaxy in different evolutionary stages suggests that formaldehyde and its deuterated species form mostly on grain surfaces although some gas-phase contribution is expected at the warm HMPO stage (Zahorecz *et al.* 2017). Resolving the emitting region of a molecule is fundamental to

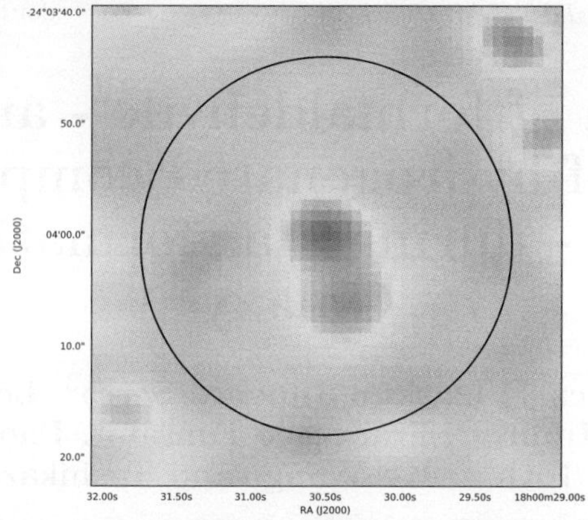

Figure 1. Integrated intensity map of $H_2^{13}CO$ $3_{1,2}$-$2_{1,1}$ emission based on ALMA ACA observations toward G5.89+0.39. Two cores are revealed according to the two velocity components found in our previous APEX observations. Black circle indicates the single dish beam size of our previous observations.

understand its chemistry. Only higher angular resolution observations allow us to shed light on the physical conditions of the region traced by the selected molecule, and based on them the molecules can finally be grouped in families depending on their emission morphology and chemical paths can be found by simply correlating their distribution.

2. ALMA observations of formaldehyde in G5.89+0.39

G5.89-0.39 is a shell-like ultra-compact H II region (diameter of ∼4") found at 3 kpc (Sato *et al.* 2014). Sub-arcsecond observations revealed at least five dust condensations (Su *et al.* 2009). Our single dish APEX (Atacama Pathfinder EXperiment) SEPIA receiver (Swedish-ESO PI receiver for APEX; Billade *et al.* 2012) observations were centred on the SMA-N dust condensation. Two velocity components were detected within the APEX beam of 34", see details in (Zahorecz *et al.* 2017).

We performed ALMA (Atacama Large Millimeter/submillimeter Array) ACA (Atacama Compact Array) observations as part of the 2017.1.01157.S. project. Two cores corresponding to the two velocity components are resolved, see on Figure 1. Detailed analysis will help us to better understand the deuteration process of formaldehyde and the chemical properties of these high-mass star-forming cores (Zahorecz *et al.* in prep).

Acknowledgements

This paper makes use of the following ALMA data: ADS/JAO.ALMA#2017.1.01157.S. ALMA is a partnership of ESO (representing its member states), NSF (USA) and NINS (Japan), together with NRC (Canada), MOST and ASIAA (Taiwan), and KASI (Republic of Korea), in cooperation with the Republic of Chile. The Joint ALMA Observatory is operated by ESO, AUI/NRAO and NAOJ. This research was supported by the NAOJ ALMA Scientific Research Grant Number 2016-03B.

References

Billade, B. *et al.* 2012, *IEEE Trans. Terahertz Sci. Tech.*, 2, 208
Sato, M. *et al.* 2014, *ApJ*, 793, 72
Su, Y.-N., *et al.* 2009, *ApJ*, 704, L5
Zahorecz, S., *et al.* 2017, *A&A*, 602, L3

Origins: From the Protosun to the First Steps of Life
Proceedings IAU Symposium No. 345, 2018
B. G. Elmegreen, L. Viktor Tóth & M. Güdel, eds.

doi:10.1017/S1743921319001790

Impact of the interstellar medium on processes on the Earth

Nikolai G. Bochkarev

Sternberg Astronomical Institute, Moscow State University,
13 Universitetskij prosp. Moscow 119991, Russia
email: boch@sai.msu.ru

Abstract. The paper discusses possible impacts of the interstellar medium (ISM) on processes on Earth, first of all those, which may affect the Earth biosphere.

Keywords. astrobiology, molecular processes, ISM: bubbles ISM: clouds (ISM:) cosmic rays (ISM:) dust, ISM: evolution, ISM: magnetic fields

ISM is a multicomponent dynamic matter. Cosmic rays (CR) is what most strongly affects processes on Earth and its biosphere. Brief reviews of physical processes due to the CR impact on life on Earth is given by Medvedev & Melott (2007) and Atri & Melott (2014). Here we consider mainly the astronomical aspects and influence of long time variations of ISM parameters near and inside Solar system (SS).

SS moves through the interstellar (IS) matter (ISM) around the center of the Galaxy along a quasi-circular orbit distorted by the spiral arms gravitational fields (Gies & Helsel 2005), and rises a little (70 pc) above or sinks under the galactic plane. Presently the Sun is located in the ISM region known as Local ISM (LISM), not far from the edge of the superbubble (Bochkarev, 1990) formed by the Sco-Cen stellar association, from which stellar winds and supernova explosions, in a cumulative effort, have blown away most of ISM to a distance of about 170 pc. Inside the blown away gas remain fragments of structures (small clouds, filaments, etc).

ISM in the Sun close vicinity, but outside heliosphere has a temperature 8000 K, a number density of a few tenths particle per cubic centimeter, ionization degree 50%. That is, the SS is located in the warm HI region but very close to the hot gas. Once inside heliosphere, ISM neutral atoms gradually get ionized by the Sun UV radiation and charge exchange reactions with solar wind ions take place. Once electrically charged, the particles are carried back by the solar wind forming the "pickup ion" flow. Approaching the two-layer shock wave inner edge, those ions accelerate at the wave front forming the anomalous CR component. Unlike galactic CR the anomalous CR are sub-relativistic. They interact more intensively with atoms and molecules, in particular with those in planetary atmosphere outer layers (if manage to reach them).

Statistics of ISM cloud size and density distribution show that the Sun in its 4.5 billion year long travel along the Galaxy have about 130 times (7-8 time per galactic year) crossed ISM clouds with number density $n > 100\ cm^{-3}$ and about 16 times those with $n > 1000\ cm^{-3}$ (Talbot & Newman 1977; Pavlov *et al.* 2005a).

CR effect on the ozone amount in the Earth atmosphere. The present-day heliosphere radius is ~ 100 a.u. When the SS enters an ISM cloud, heliosphere contracts down to $1-2$ a.u., which practically does not change the parameters of the CR penetrating it. But the energy of the anomalous CR accelerating at the heliosphere outer border increases by hundreds of times. As a result, the Earth upper atmosphere (at least at the polar regions) is irradiated by the anomalous CR hundreds of times more intensively, leading, in the now-a-day nitrogen-oxygen Earth atmosphere, to formation of ozone disrupting nitrogen oxide molecules and subsequent significant ozone layer thinning. The characteristic time for ozone concentration to achieve a new stationary state is about 5-10 years (Pavlov *et al.* 2005a). Results of computation of ozone model distribution over the Earth globe surface for today heliosphere size, and for the case of heliosphere being strongly compressed by the surrounding IS gas of a concentration of 100 cm^{-3} are given in Pavlov *et al.* (2005a). The ozone mass is shown to diminish by 2-4 times in sub-polar regions but only by 40% in the equatorial ones. Reversals of the Earth magnetic field polarity and magnetic excursions, when the magnetic field dipole component weakens drastically for 1-10 kyr, approaching zero, are well documented. They occur, in the average, once in 200 kyr, though tens of millions of years long periods of stable polarity are known to have taken place. The typical time of the SS remaining inside a giant molecular cloud is of the order of 1 Myr. Over this time-interval the Earth magnetic field polarity may suffer several alternations, the ozone protection of the biosphere weakening for several kyrs all over the Earth surface, including the equatorial regions, which, naturally, undermines the biosphere and may lead to a massive extinction of species.

Galactic CR flux variations. Low energetic galactic CR (GCR) are distributed unevenly in the Galaxy. This is seen even from the fact that a significant part of gamma radiation comes from molecular clouds. CR being very dense there produce gamma rays that interact with ISM. Molecular clouds are located mainly in the Galaxy spiral arms. That is why the content of low energy CR is higher there.

The SS in its orbital motion, passes through areas with different CR density. Comparison of such phenomena on Earth as successive cooling and warming epochs, great glaciations, mass extinction of species with the periods when the Sun crosses the spiral arms and with galactic CR flow variations, suggests our planet climate dependence on these phenomena (Shaviv 2003). It should be reminded that as the Sun evolves, solar wind and solar CR parameters change, which affects, in particular, the size of the heliosphere and, consequently, the penetration of GCR into the SS.

The role of the Galaxy evolution. The SS age makes a significant part of our Galaxy age (~13 Gyr). During that time, the physical conditions in the ISM that surrounds the SS and through which it moves could have experienced considerable changes.

IS dust influence. IS dust can also influence the biosphere through various processes. First, if dense enough it would on the way down to the planet surface absorb sunlight. This mechanism was first proposed By F.Hoyle in his science-fiction novel "The Black Cloud" and described quantitatively in Pavlov *et al.* (2005b). Besides, if a planet upper atmosphere is dusty enough, electromagnetic radiation absorption by the dust leads to this layer heating, also affecting the planet's climate. For the biosphere of planets with a scarce stock of nutrients, organic molecules frozen in the dust particles and carried by them down to the planet surface could make an important additional source of nutrients.

References

Atri D., Melott A. 2014, *Astroparticle Phys.*, 53, 186
Bochkarev, N. G. 1990, *The local interstellar medium* (Moscow: Nauka)
Gies D. R. & Helsel J. W. 2005, *ApJ*, 626, 844
Medvedev M. V. & Melott A. L. 2007, *ApJ*, 664, 879
Pavlov, A. A., Pavlov, Anatoli, K., Mills, Michael J. *et al.* (2005a) *GeoRL* 32, L01815
Pavlov A. A., Toon O. B., Pavlov A. K., Bally J. & Pollard D. (2005b), *GeoRL*, 32, L03705
Shaviv N. J. (2003) *New Astronomy*, 8, 39
Talbot R. J., Newman M. J. 1977, *ApJS*, 34, 295

Origins: From the Protosun to the First Steps of Life
Proceedings IAU Symposium No. 345, 2018
B. G. Elmegreen, L. Viktor Tóth & M. Güdel, eds.

doi:10.1017/S1743921319001935

Magnetic geometry and activity of cool stars

Sudeshna Boro Saikia, Theresa Lüftinger and Manuel Guedel

University of Vienna, Department of Astrophysics, Tuerkenschanzstrasse 17, 1180, Vienna, Austria
email: sudeshna.boro.saikia@univie.ac.at

Abstract. Stellar magnetic field manifestations such as stellar winds and EUV radiation are the key drivers of planetary atmospheric loss and escape. To understand how the central star influences habitability, it is very important to perform detailed investigation of the star's magnetic field. We investigate the surface magnetic field geometry and chromospheric activity of 51 sun-like stars. The magnetic geometry is reconstructed using Zeeman Doppler imaging. Chromospheric activity is measured using the Ca II H& K lines. We confirm that the Sun's large-scale geometry is dominantly poloidal, which is also true for slowly rotating stars. Contrary to the Sun, rapidly rotating stars can have a strong toroidal field and a weak poloidal field. This separation in field geometry appears at Ro=1. Our results show that detailed investigation of stellar magnetic field is important to understand its influence on planetary habitability.

Keywords. star: activity, star: chromospheres, star: rotation, star: magnetic fields

1. Introduction

Stellar activity and magnetic field are known to play an important role in the evolution of a habitable planetary system. Our central star, the Sun, has a magnetic field of moderate strength that interacts with the orbiting planets, initiating different reactions in their atmospheres. We investigate the vector magnetic field and associated chromospheric activity of 51 sun-like stars to provide observational constraints for different space weather models.

2. Data collection

Majority of the stars were observed as part of the BCool project (Marsden *et al.* 2014, Petit *et al.* in prep). Few data were also taken from separate multi-epoch studies (Hackman *et al.* 2016; Rosén *et al.* 2016; Alvarado-Gómez *et al.* 2015). The rotation periods for these stars were taken from Wright *et al.* (2011) when not available in the original literature, to determine the rossby number (Ro). Ro is defined as the rotation period divided by the convective overturn time, τ_c, which is determined from the empirical relationship of Noyes *et al.* (1984).

3. Results

Figure 1 (Left) shows the chromospheric activity of the stars in our sample as a function of $B-V$. The large-scale geometry in terms of the poloidal field, axisymmetry, and the mean magnetic field B is shown. The grey lines separate the active and inactive stars, while the intermediately active stars lie in the middle. Majority out of the 51 stars lie in the active side of the figure, which is not surprising as stronger the activity (or field strength), easier is the detection. Active stars do not have any strong inclination towards purely poloidal (red) or toroidal (blue) large-scale field. Strong poloidal and toroidal field are both detected in these stars. The axisymmetry and B also show a wide range. Where

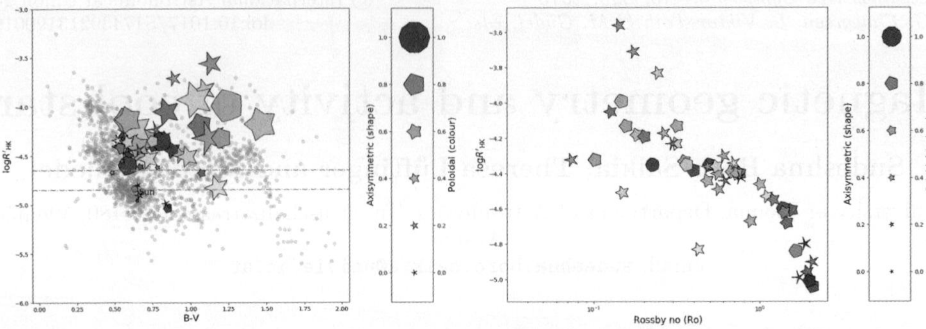

Figure 1. *Left*:Chromospheric activity vs $B-V$. The large-scale poloidal field, axisymmetry and mean field strength (B) of the total field is shown for 51 sun-like stars (the coloured symbols). Colour represents the magnetic energy of the poloidal field (strength increases from blue to dark red), shape represents the axisymmetry (increases from star to hexagonal shape) and symbol size increases with increasing B. Gray circles in the background are the chromospheric activity of ~4000 stars with no large-scale field information (Boro Saikia *et al.* 2018). The grey lines show the intermediate activity range. *Right*: Chromospheric activity of the 51 stars vs Rossby number Ro. The symbols are the same as on the Right, except the symbol sizes are not scaled to B.

as the inactive stars, including the Sun, show a predominantly poloidal field structure but the axisymmetry can vary from one star to the other.

As magnetic field generation and amplification is influenced by the rotation of the star, Fig. 1 (Left) was plotted as a function of Ro (Fig. 1, Right). Chromospheric activity shows a linear trend with Ro for slowly rotating stars. For very fast rotators the trend is more scattered. At Ro$\geqslant$1, slowly rotating stars like our Sun have a dominant poloidal field.

4. Conclusion

Our results show that the large-scale magnetic field geometry of a sun-like star could be very different from the Sun. Additionally the large-scale geometry and Ro has a tight relationship, especially for older slowly rotating stars. The strong poloidal geometry at Ro$\geqslant$1 suggests that, as stars get older and rotate slower their dynamos operate in a similar manner. This might not be true for younger stars as suggested by the wide variety in field geometry. Hence it is very important to investigate the magnetic nature of central star to understand the associated space weather in different exoplanetary systems.

References

Alvarado-Gómez, J. D., Hussain, G. A. J., Grunhut, J., *et al.* 2015, *A&A*, 582, A38

Boro Saikia, S., Marvin, C. J., Jeffers, S. V., *et al.* 2018, *A&A*, 616, A108

Hackman, T., Lehtinen, J., Rosén, L., Kochukhov, O., & Käpylä, M. J. 2016, *A&A*, 587, A28

Marsden, S. C., Petit, P., Jeffers, S. V., *et al.* 2014, *MNRAS*, 444, 3517

Noyes, R. W., Hartmann, L. W., Baliunas, S. L., Duncan, D. K., & Vaughan, A. H. 1984, *ApJ*, 279, 763

Rosén, L., Kochukhov, O., Hackman, T., & Lehtinen, J. 2016, *A&A*, 593, A35

Wright, N. J., Drake, J. J., Mamajek, E. E., & Henry, G. W. 2011, *ApJ*, 743, 48

Origins: From the Protosun to the First Steps of Life
Proceedings IAU Symposium No. 345, 2018
B. G. Elmegreen, L. Viktor Tóth & M. Güdel, eds.

doi:10.1017/S1743921319001947

Nitrogen-included Carbonaceous Compounds (NCC): Laboratory-synthesized organics as the probable candidate for the carrier of the UIR bands observed in dusty classical novae

Izumi Endo[1], Itsuki Sakon[1], Takashi Onaka[1], Andrew L. Helton[2], Ryan M. Lau[3], Seiji Kimura[4], Setsuko Wada[4], Nanako Ogawa[5], Naohiko Ohkouchi[5], and Yoko Kebukawa[6]

[1]University of Tokyo 7-3-1 Hongo, Bunkyo-ku Tokyo 113-0033, Japan

[2]SOFIA Science Center, USRA, NASA Ames Research Center MS N232-12, Moffett Field, CA 94035, USA

[3]Jet Propulsion Laboratory, California Institute of Technology, 4800 Oak Grove Drive, Pasadena, CA 91109, USA

[4]The University of Electro-Communications 1-5-1 Chofugaoka, Chofu Tokyo 182-8585, Japan

[5]Japan Agency for Marine-Earth Science and Technology 2-15, Natsushima-cho, Yokosuka-city Kanagawa 237-0061 Japan

[6]Yokohama National University 795 Tokiwadai, Hodogaya-ku, Yokohama, Kanagawa 2408501, Japan

Abstract. The unidentified infrared (UIR) bands have been ubiquitously observed in various astrophysical environments and consist of a series of emission features arising from aromatic and/or aliphatic C-C and C-H bonds (Allamandola *et al.* 1989). Therefore, their carriers are thought to be related to interstellar organics. However, our knowledge on the true carriers of the UIR bands is still limited. Recently Kwok & Zhang (2011) has proposed Mixed Aromatic Aliphatic Organic Nanoparticles, which contains hetero atoms in addition to conventional hydrocarbon models, as a more realistic interpretation of the band carriers. The challenges toward identifying the carriers of the UIR bands are still ongoing. Past studies have shown that the UIR bands observed around classical novae, which characterized by the presence of broad feature around $8\mu m$ (Helton *et al.* 2011), are somewhat different from those observed in other astrophysical environment. Here we report the success of experimentally synthesizing the organics called Nitrogen-included Carbonaceous Compounds (NCC; Sakon *et al.* 2015) whose infrared properties can reproduce the UIR bands observed in classical novae.

Keywords. (ISM:) dust, extinction, ISM: lines and bands, infrared: ISM

---

1. Experiment and Result

Nitrogen-included Carbonaceous Compounds (NCC) is synthesized by rapidly cooling the plasma gas produced from nitrogen gas and hydrocarbon dust (e.g. QCC; Sakata *et al.* 1983, PAHs; Allamandola *et al.* 1989) via 2.45GHz microwave discharge. Figure. 1 shows the comparison of the infrared absorption spectra of NCC and QCC with the UIR bands observed in dusty classical nova V2361 Cyg. The band profiles and the peak positions of features in the mid-infrared wavelength range are remarkably similar between the infrared absorption spectrum of NCC and the UIR bands in V2361 Cyg. The earlier

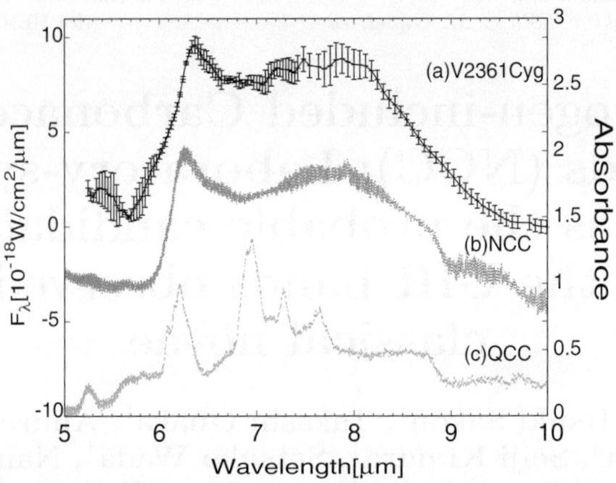

Figure 1. Comparison of (a)the UIR band observed in classical nova V2361Cyg at 116 days after outburst (Helton *et al.* 2011) with the infrared absorption spectra of (b)NCC and (c)QCC

works suggest that nitrogen inclusion into carbonaceous dust can naturally occur in the process of dust condensation in circumstellar environment (Hudgins *et al.* 2005) and that Nitrogen abundance in nova ejecta is higher than other circumstellar environment (e.g. Munari *et al.* 2008). Further more, the synthesis method of NCC simulates the circumstellar conditions of novae, where nitrogen-rich nova wind react with the existing circumstellar hydrocarbon dust. We, therefore, suggest that NCC be a powerful candidate of the carriers of the UIR bands in V2361 Cyg.

2. Properties of NCC

To investigate the chemical properties of NCC, we have carried out X-ray Absorption Near Edge Structure (XANES) measurements in the energy range of carbon (280-300eV) and nitrogen (395-415eV). From the results of the Nitrogen XANES (N-XANES) analysis, we have found that nitrogen atoms in NCC are contained in the form of imine (C=N) and amine (C-N) structures. In the infrared spectrum, the features arising from imine structures should appear around 6-7μm and are not clearly identified in the absorption spectrum of NCC mainly because of the blending with the 6.25μm feature arising from aromatic C-C. On the other hands, the feature arising from aromatic and aliphatic amine structure apear in 7.4-8.2μm and 8.2-9.8μm, respectively. We, therefore, concluded that the broad 8μm feature in the absorption spectrum of NCC is contributed by the amine structures. The present result suggest that nitrogen-inclution into conventional hydrocarbon model is significant for the better understanding of the carriers of the UIR bands.

References

Allamandola, L. J., Tielens, A. G. G. M., & Barler, J.R. 1989, *ApJS*, 71, 733

Helton, L. A., Woodward, C. E., & Gehrz R. D. 2011, EAS Pub. Ser., 46, 407

Hudgins, D. M., Bauschlicher, Jr., C. W., & Allamandola, L. J. 2005, *ApJ*, 632, 316

Kwok, S., & Zhang, Y. 2011, *Nature*, 479, 80

Munari, U., Siviero, A., Henden, A., Cardarelli, G., Cherini, G., Dallaporta, S., Dalla V. G., Frigo, A., Jurdana-Sepič, R., Moretti, S., Ochner, P., Tomaselli, S., Tomasoni, S., Valisa, P., Navasardyan, H., & Valentini, M. 2008, *A&A*, 492, 145

Sakata, A., Wada, S., Okutsu, Y., Shintani, H., & Nakada, Y. 1983, *Nature*, 301, 493

Sakon, I., Sato, K., Onaka, T., Kimura, S., Wada, S., Ogawa, N., Ohkouchi, N., & Pathak, A. 2015, Asian Journal of Physics, 24, 8, 1143

Origins: From the Protosun to the First Steps of Life
Proceedings IAU Symposium No. 345, 2018
B. G. Elmegreen, L. Viktor Tóth & M. Güdel, eds.

doi:10.1017/S1743921319001959

Evidences of asymmetry in properties of L4 & L5 Jupiter Trojans

Ivan Slyusarev[1,2], Daniella Glezina[1] and Irina Belskaya[1,2]

[1]Dept. of Astronomy & Space Informatics, V.N. Karazin Kharkiv National University, 4 Svobody Sq., 61022, Kharkiv, Ukraine
email: i.slyusarev@karazin.ua

[2]Institute of Astronomy, V.N. Karazin Kharkiv National University, 35 Sumskaya St., Kharkiv, Ukraine

Abstract. We search for possible differences in rotational frequencies, diameters, albedos, and orbital parameters between Trojans belonging to the L4 and L5 swarms using our own observations and literature data. With increasing number of observational data it becomes evident that the L4 and L5 populations have very similar distributions of most parameters with an exception of orbital inclination distribution.

Keywords. Trojans, asteroid families, orbital inclinations, asymmetry, proper elements

1. Introduction

In the second half of 2000's, after the emergence of new class of models of evolution of the early Solar system in the second half of 2000s (Nesvorny *et al.* 2013 and ref. therein) has increased interest to Jupiter Trojans. Mechanism of Trojans capture and the place of their origin (the population is either primordial or was captured from the outer part of Solar system during gas giants migration) were the significant component of these models. Therefore, investigation of this population is necessary to test these models. Below we investigate one of the most intriguing features of the Jupiter Trojans population related to the observed asymmetry between the L4 and L5 groups.

2. Search for asymmetries in dynamical and physical properties

From the beginning of the study of Jupiter Trojans, there is a well-known difference in number of objects between L4 and L5 groups. At present, according to MPC, there are more than 7000 Trojans in both swarms, moreover L4 population outnumbers the L5 in about 1.9 times. Except this well-known asymmetry between L4 and L5 swarm, in our previous work (Slyusarev & Belskaya 2014) we found a significant difference in orbit inclinations distribution. Here we search for possible differences in physical and dynamical properties using the increasing set of observational data. We examined such properties as albedo and diameters (Grav *et al.* 2011); inclinations, eccentricities, and absolute magnitudes (from MPCORB database); lightcurve amplitudes and periods (MPC Asteroid Lightcurve Database). We did not find statistically significant differences in the distributions of these parameters, except diameters (within fraction of asteroids smaller than 20 km) and inclinations. Distributions of inclinations show drastic differences in L4 and L5 (Fig. 1a). After peak at 8 degrees of inclination, the distribution of L4 Trojans decreases exponentially, while there is a wide plateau in L5 group. As it was firstly noted in (Di Sisto *et al.* 2014), the main contribution to this asymmetry is made by bodies smaller than 30 km in diameters. We validated this assumption for smaller diameters

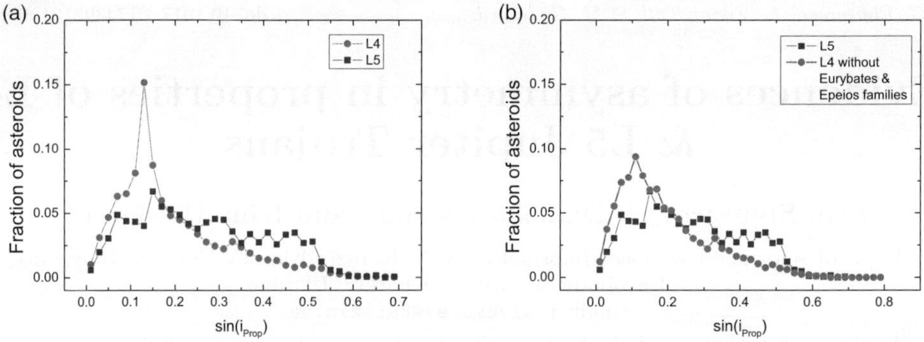

Figure 1. Distribution of sine of proper inclinations in L4 and L5 swarms; a) all asteroids, b) L4 after elimination Eurybates and Epeios families.

and found the same shape of asymmetry for asteroids smaller than 20 km. Since such small asteroids are a product of collisional history, we assume that families in Trojan swarms could be the key to understanding the reason of asymmetries because they are also formed due to collisions. That is why we also built distributions of sine of the proper inclinations and saw similar kind of asymmetry (Fig. 1). We have checked if the elimination of family members influences the asymmetry. Firstly, we extracted asteroids of Eurybates family, the only robust family among Trojans (Broz & Rozehnal 2011), and found that a strong peak in L4 significantly decreased (Fig. 1b). We also checked the contribution of Ennomos family in L5 and Epeios family in L4 (Broz & Rozehnal 2011 and Vinogradova 2015). Elimination of Ennomos family members didn't affect the distribution, while those of Epeios family did. We see that families cannot wholly explain the asymmetries between the swarms: differences still present in the sin(i) intervals of 0.07-0.16 and 0.25-0.55. (Fig. 1b).

3. Conclusions

We have compared the distributions of albedos, diameters, lightcurve amplitudes, rotational frequencies, eccentricity, and inclinations of Trojan asteroids of L4 and L5 swarms. All these parameters, except inclinations, showed no significant differences. We found that asteroids smaller than 20 km in diameter mainly contribute to asymmetry inclinations distribution. Taking into account that small asteroids are the product of collisions, we checked the contribution of the most reliable Jupiter Trojans families to this asymmetry and found significant influence of Eurybates and Epeios families. However, families solely cannot explain all the differing features of the distributions. This fact gives us the reason to assume that some asymmetries of L4 and L5 swarms could be primordial.

References

Nesvorny, D., Vokrouhlicky, D., Morbidelli, A. 2013, *Ap.J.*, 768, 8
Slyusarev I. G. & Belskaya I. N. 2014, *Solar System Res.*, 48, 139
Vinogradova T. A. 2015, *MNRAS*, 454, 2436
Grav T. *et al.* 2011, *Ap.J.*, 742, 49
Di Sisto R., Ramos X. S., & Beauge C. 2014, *Icarus*, 243, 287
Broz M. & Rozehnal J. 2011, *MNRAS*, 414, 565

Origins: From the Protosun to the First Steps of Life
Proceedings IAU Symposium No. 345, 2018
B. G. Elmegreen, L. Viktor Tóth & M. Güdel, eds.

doi:10.1017/S1743921319001960

Physical conditions in the early Solar system and life origin: compatible models

Mariya Ragulskaya[1], **Elizaveta Khramova**[1] **and Vladimir Obridko**[1]

[1]IZMIRAN
108840, Moscow, Russia
email: ra_mary@mail.ru

Abstract. The article discusses the physical conditions in the early Solar system and on Earth, determining the origin, selection and development of the first living systems. The role of the young Sun dynamics, cosmic rays, magnetic fields and other protective shells of the Earth in the formation of the biosphere is emphasized. The selection of a single genetic code, ancient methods of long-term storage of energy and adaptive technologies of the first living systems occurred under the influence of cosmological and geophysical factors. A hypothesis was suggested that the accumulation of energy in polyphosphates without the participation of solar radiation could have ensured the survival of the primary biosphere in the conditions of the low luminosity of the young Sun.

Keywords. origin of life, physical conditions on early Earth, cosmic rays, biosphere, etc.

1. Overview

The form and structure of the heliosphere, the intensity of galactic and solar cosmic rays near the Earth and other processes in the solar system (SS) are determined by the magnetic field of the Sun and the position of the SS in the Galaxy. The magnetic field simulation of the early Sun (O.Cohen *et al.* 2012, Pipin 2017) revealed that the intensity of SCR was more, but the intensity of GCR was less than the current values. The intensity of the GCR could change by an order during the passage of the SS through the arms of the galaxy.

Young Sun (YS) had a circulation period of ~5–8 days, a mass of ~103% of the modern. There was no ordered periodicity. The first solar cycles (~5 years) appeared ~2.5 billion years ago (Pipin 2017). The intensity of sporadic radiation exceeded the current one by several orders. The intense UV radiation of YS exerted a significant influence on the biosphere formation. The luminosity of YS was less by 30%. Then the surface of the Earth must be frozen. However, according to geological data, the temperature of the Earth's surface exceeded the modern. The paradoxs solution can lie in other reflection and absorption by the early Earth. The existence of biological films on the oceans surface is reported. They significantly altered the albedo of the planet and the composition of the atmosphere. Also, the early Earth could be closer to the Sun due to the migration of Jupiter and Saturn. The geomagnetic field (GF) protects the biosphere from intense cosmic rays. The formation of the GF occurred 2 billion years ago according to dynamo theory. According to paleomagnetic data there already was a magnetic field of the Earth 4.2 billion years ago. Its values were comparable with modern. The reason for the emergence of such a field could be the process of formation of the Earth-Moon system as well as tidal interaction of proto-Earth with migrating giant planets.

The influence of space weather factors on the biosphere are observed to the present day. In the modern biosphere there are objects similar to the most ancient living structures. Inorganic polyphosphates (the so-called "fossil molecules") accompanying biological objects at all stages of evolution. For example, they are present in human hemoglobin or yeast cells. Polyphosphates are responsible for one of the most ancient ways to store energy in cells. *Saccharomyces cerevisiae* strain Y-517 was studied by daily measurements in 2000-2012 (Gromozova *et al.* 2014, Ragulskaya 2017). The statistical distributions for space factors in random days and in the days with the relevant type of metachromasia was considered.

The main result of the analysis is that the distribution of MTC=3 considerably depends of galaxy cosmic rays intensity variations during 2006–2010; it means that galaxy cosmic rays intensity variations are the main biotropical agent of the space weather for sell structures during solar activity minimum.

2. Implications

Thus during a minimum of solar activity, a significant contribution to the dynamics of cellular structures is made by GCR. At the maximum, the influence of solar activity and variations of the geomagnetic field are expressed. Apparently the effect of GCR is not directly manifested. It can occur because changes in the oxygen, ozone and nitrogen content in the Earth's atmosphere during the solar activity cycle. Or can occur because a change in the amount of oxygen, phosphorus and iron in the blood of mammals and humans.

Probably the impact of cosmic rays on cellular structures has been evident since the origin of life and the evolutionary adaptation of the first ecosystems. The impact of galactic cosmic rays and magnetic fields on organic molecules begins else in galactic molecular clouds, during the pre-biological evolution of organic matter in the Universe. A key role in interstellar molecular chemistry is played by ion-neutral (or ion-molecular) reactions. These are the reactions between the ionized and neutral reagents, with the primary ionization provided by the particles of cosmic rays. A hypothesis was proposed that the accumulation of energy in polyphosphates without the participation of solar radiation could have ensured the survival of the primary biosphere in the conditions of the low luminosity of the young Sun in this report.

3. Acknowledgment

The research was supported by a Russian-Azerbaijani grant from the Russian Fund for Basic Research 18-52-06002 Az_a and Presidium of the Russian Academy of Sciences Program №17.

References

Cohen, O., Drake, J. J., & Kota J. 2012, *The Astophysical Journal*, 760, 85

Pipin, V. V. 2017, *MNRAS*, 466, 3007

Gromozova, E., Voychuk, S., Grigoriev, P., Vishnevsky, V., & Ragulskaya, M. 2012, *Sun and Geosphere*, 7, 117

Ragulskaya, M., & Obridko, V. 2017, *Radiophysics and radioastronomy*, 22/4, 276

Origins: From the Protosun to the First Steps of Life
Proceedings IAU Symposium No. 345, 2018
B. G. Elmegreen, L. Viktor Tóth & M. Güdel, eds.

doi:10.1017/S1743921319001972

Exploration of the molecular gas content of young debris disks

A. Moór[1], Á. Kóspál[1,2], P. Ábrahám[1], and N. Pawellek[1,2]

[1]Konkoly Observatory, Research Centre for Astronomy and Earth Sciences, H-1121
Konkoly-Thege Miklós út 15-17, Budapest, Hungary
email: moor.attila@csfk.mta.hu

[2]Max Planck Institute for Astronomy, Königstuhl 17, D-69117 Heidelberg, Germany

Abstract. Thanks especially to the ALMA interferometer, there are several new detections of CO gas in debris disks. Based on our own and archival ALMA observations, we found that the presence of CO gas in dust-rich debris disks around young (10–50 Myr) A-type stars is common. Interestingly, similarly dust-rich debris disks around young F-K type stars exhibit significantly lower gas incidence. The main difference between the two subsamples is related to a special population of gaseous debris disks whose total CO gas quantity is comparable to that of the less massive Herbig Ae disks. The origin of gas in these CO-rich debris systems is not fully clarified yet.

Keywords. (stars:) circumstellar matter, radio lines: stars.

While the evolution of dust between the protoplanetary and debris stages is rather well studied, the parallel evolution of gas remained poorly explored (Wyatt *et al.* 2015). In order to better understand this aspect of the transition and the early phase of debris disks' evolution, using ALMA data we explored the CO content of a sample of young (10–50 Myr), cold (<140 K), dust-rich ($L_{\rm disk}/L_* > 5 \times 10^{-4}$) debris systems within 150 pc. By supplementing our own ^{12}CO measurements (for 20 targets) with literature (Lieman-Sifry *et al.* 2016, Marino *et al.* 2016, Matrà *et al.* 2017) and archival ALMA line observations, we gathered data for 44 objects, thereby compiling a nearly complete volume-limited sample of such disks. Their analysis have brought the following results.

Detection rate of CO gas. 11 out of the 17 disks that surround A-type stars harbor CO gas. The high detection rate (~65%) implies that the presence of CO gas in dust-rich debris disks around young A-type stars is common. Out of the 27 disks that surround F, G, or K-type stars, only two (7.4%) show detectable CO emission. By applying Fisher's exact test we found that the observed difference in the proportions of gas-bearing systems in the two subsamples is statistically significant (see also in Moór *et al.* 2017).

Evolution of gas and dust from the protoplanetary to debris phase. By carrying out ^{13}CO and C^{18}O line observations for seven gas-bearing debris disks – all surround A-type stars – with ALMA, we demonstrated that their ^{12}CO emission is optically thick and their CO mass is a factor of 30–250× higher than that calculated from ^{12}CO measurements alone using an optically thin assumption. In Figure 1 we compared the dust and CO content of the eleven gas-bearing debris disks around A-type stars from our sample with those of known Herbig Ae systems. Consistently with previous findings (e.g. Wyatt *et al.* 2015) we found a pronounced drop in dust mass from primordial to debris disks. On the contrary, the CO gas mass of debris disks show a significantly larger spread and

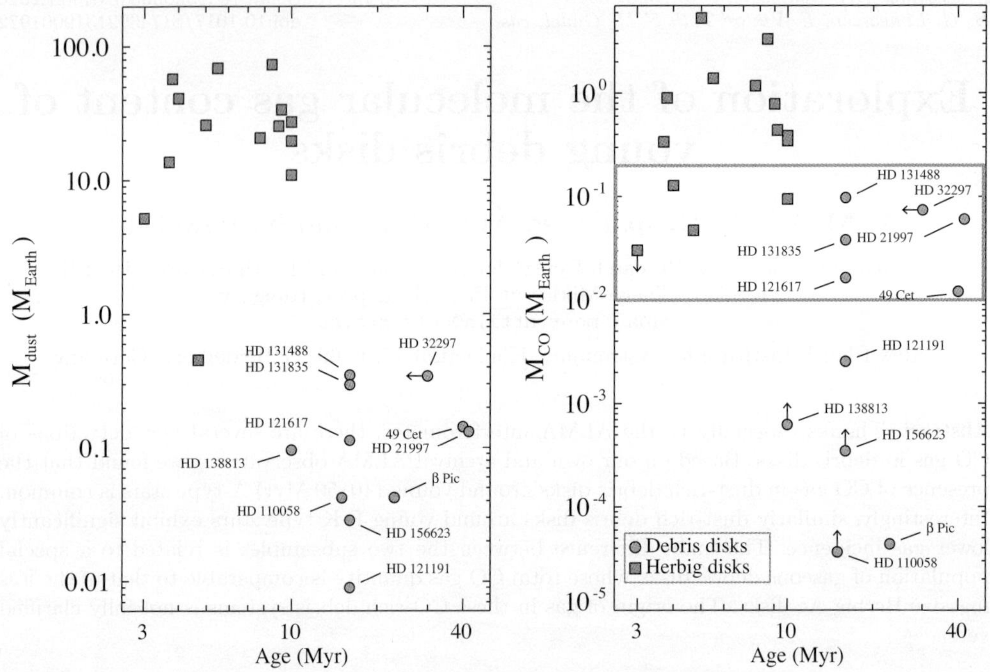

Figure 1. Dust and CO masses of the eleven gas-bearing debris disks around young A-type stars from our volume-limited sample and of some nearby Herbig Ae stars as a function of age. In the right panel the red rectangle highlights those six debris disks whose CO mass is comparable to that of Herbig Ae systems.

there are at least six systems whose CO mass is comparable to that of Herbig Ae disks (see the region highlighted by red in Figure 1).

Origin of gas in debris disks around young A-type stars. Currently there are two viable scenarios to explain the existence of CO gas at protoplanetary level in the debris disks around young A-type stars. One possibility is that they exhibit a hybrid nature: while their dust is secondary and derived from the erosion of planetesimals, the gas component is predominantly leftover from the primordial phase and is dominated by H_2 molecules that provide strong shielding to CO gas (Kóspál *et al.* 2013). Alternatively, Kral *et al.* (2018) recently demonstrated that in debris disks with sufficiently high gas production from icy planetesimals via collisions, shielding by neutral carbon atoms – photodissociation products of the released CO and CO_2 gas – can explain the long lifetime of CO molecules. The currently available observational results are not conclusive enough on the gas origin in these systems. To clarify this question, further observations and application of new approaches are needed (e.g., Kral *et al.* 2018).

References

Kóspál, Á., Moór, A., Juhász, A., *et al.* 2013, *ApJ*, 776, 77
Kral, Q., Marino, S., Wyatt, M. C., Kama, M., & Matrà, L. 2018, arXiv:1811.08439
Lieman-Sifry, J., Hughes, A. M., Carpenter, J. M., *et al.* 2016, *ApJ*, 828, 25
Marino, S., Matrà, L., Stark, C., *et al.* 2016, *MNRAS*, 460, 2933
Matrà, L., Dent, W. R. F., Wyatt, M. C., *et al.* 2017, *MNRAS*, 464, 1415
Moór, A., Curé, M., Kóspál, Á., *et al.* 2017, *ApJ*, 849, 123
Wyatt, M. C., Panić, O., Kennedy, G. M., & Matrà, L. 2015, *Ap&SS*, 357, 103

Origins: From the Protosun to the First Steps of Life
Proceedings IAU Symposium No. 345, 2018
B. G. Elmegreen, L. Viktor Tóth & M. Güdel, eds.

doi:10.1017/S1743921319001996

Interaction of solid bodies with atmospheres of protoplanets

Ernst A. Dorfi[1] **and Florian Ragossnig**[2]

[1]Institut für Astrophysik, Universität of Wien,
Türkenschanzstr. 17, A-1180 Wien, Austria
email: ernst.dorfi@univie.ac.at

[2]email: florian.ragossnig@univie.ac.at

Abstract. During the early stages of planet formation accretion of small bodies add mass to the planet and deposit their energy kinetic energy. Caused by frictional heating and/or large stagnation pressures within the dense and extended atmospheres most of the in-falling bodies get destroyed by melting or break-up before they impact on the planet's surface. The energy is added to the atmospheric layers rather than heating the planet directly. These processes can significantly alter the physical properties of protoplanets before they are exposed with their primordial atmospheres to the early stellar source when the protoplanetary disk becomes evaporated.

Keywords. solar system: formation, celestial mechanics, hydrodynamics

---

1. Introduction

Planets are embedded in a protostellar disk when they form. The outermost layers of their atmospheres are in direct contact with variable boundary conditions during the early phases of disk evolution. IR-observations indicate that the typical disk life times are less than 10^7 years before the gas is evaporated. To study the time-dependent evolution of atmospheres of earth-like planets (e.g. Stökl *et al.* 2015) we perform computations in spherical symmetry, based on solutions of the equations of radiation hydrodynamics (RHD) including convective energy transport, a realistic equation of state and recent dust opacities. Such protoatmospheres are influenced by several time-dependent processes like the evaporation of the protoplanetary disk and changes of the stellar irradiation. Typical atmospheric masses are around $10^{-2}\,M_\mathrm{p}$ and therefore not comparable with current atmospheres of earth-like planets. Here we concentrate on the bombardment by a population of solid bodies ranging over a wide scale in mass and diameter. During the passage through the atmosphere they are heated by friction, can melt and are partially or fully eroded. Hence, depending on the initial conditions the larger bodies will reach the planetary surface and deposit their remaining kinetic energy. Smaller bodies are eroded and provide a height-dependent source of energy as well as an additional source of dust particles modifying the atmospheric properties by changing the opacity. Since the thick envelopes can insulate the hot planetary surface the cooling rate and the further evolution of the young planets are modified.

2. Results

Depending on the mass of the planet the atmospheres react on the accretion rate and the size distribution of the planetesimals. In Fig. 1 the velocities are plotted for bodies (radii between $a = 1\,\mathrm{cm}$ and $a = 100\,\mathrm{km}$) falling on a protoplanet with $M_\mathrm{p} = 1\,M_\oplus$. Only

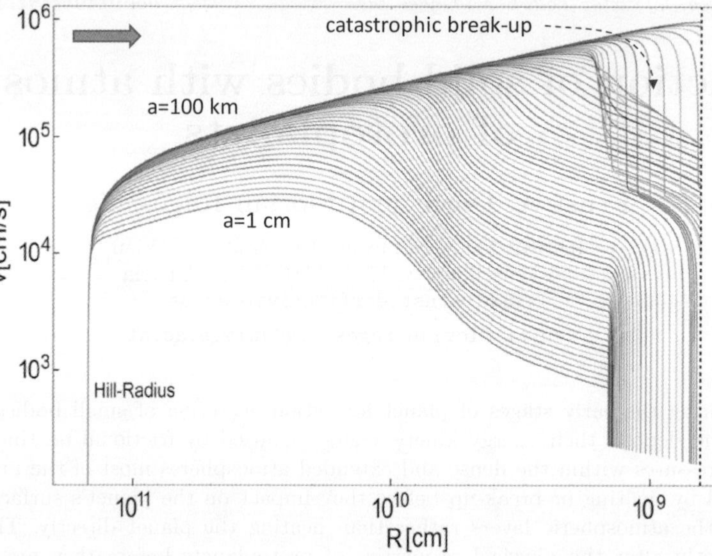

Figure 1. The inward velocity of planetesimals for initial radii between $a = 1$ cm and $a = 100$ km as a function distance R. In the case of an earth-sized planet only the most massive bodies can reach the surface (vertical dashed line) because of melting and/or break-up.

km-sized bodies are able to impact on the surface of the planet (plotted as vertical dashed line) thereby heating the surface by deposition of their remaining kinetic energy. Since all accreted bodies are orbiting the protosun, we can use the escape velocity as initial inward velocity at the Hill-radius. The motion is controlled by friction, heating and ongoing evaporation. If the stagnation pressure becomes comparable to the material strength, a simple break-up model is included which cuts in half the mass of a planetesimal (as seen in upper right corner of Fig. 1). At these stages only rather solid planetesimals above 1 km size can reach the surface, all other will deposit their kinetic energy well above the planet. Solid bodies of sizes up 10 km are stopped or melt by frictional dissipation and explode into smaller fragments during their flight. The larger the mass of the planetary body the better the accreted bodies get stopped in the dense and extended atmospheres of these newly born protoplanets.

3. Conclusions

As shown by the computations a significant amount of gravitational energy is indeed dissipated into the atmosphere (especially for larger planetary cores) which consequently can not contribute to the atmospheric planetary luminosity. Furthermore, we examine that planetesimal infall-events for cores with $M_{\rm p} \geqslant 2\, M_{\oplus}$ actually result in a local cooling of the atmosphere which is totally in contradiction with the classical model. More details (and references therein) of these computations are published in Ragossnig *et al.* (2018).

References

Ragossnig, F., Stökl, A., Dorfi, E. A., Johnstone, C. P., Steiner, D., Güdel, M., 2018, *A&A*, 618, A19

Stökl, A., Dorfi, E. A., & Lammer, H. 2015, *A&A*, 576, A87

Origins: From the Protosun to the First Steps of Life
Proceedings IAU Symposium No. 345, 2018
B. G. Elmegreen, L. Viktor Tóth & M. Güdel, eds.

doi:10.1017/S174392131900200X

"KAGONMA" NH_3 mapping observations of molecular clouds with Nobeyama 45m telescope.

T. Murase[1], T. Handa[1], M. Maebata[1], Z. Yang[1], Y. Hirata[1], K. Ishizaki[1], K. Sunada[2], M. Nakano[3], T. Omodaka[1], S. Shihara[1], K. Wada[1], T. Umemoto[4], M. Matsuo[4], T. Minamidani[4], K. Torii[4], N. Kuno[5,6], S. Fujita[4,5], M. Kohno[7], M. Kuriki[5], A. Nishimura[5] and Y. Tsuda[8]

[1]Faculty of Science, Kagoshima University, 1-21-35 Korimoto, Kagoshima 890-0065, Japan.
email: takerun.charvel@gmail.com

[2]Mizusawa VERA Observatory, National Astronomical Observatory of Japan, 2-21-1 Osawa, Mitaka, Tokyo 181-8588, Japan.

[3]Faculty of Education, Oita University, 700 Dannoharu, Oita 870-1192, Japan.

[4]Nobeyama Radio Observatory, National Astronomical Observatory of Japan, Minamimaki 462-2, Minamisaku, Nagano 384-1305, Japan.

[5]Graduate School of Pure and Applied Sciences, University of Tsukuba, Tennodai 1-1-1, Tsukuba 305-8577, Japan.

[6]Tomonaga Center for the History of the Universe, University of Tsukuba, 1-1-1 Tennodai, Tsukuba, Ibaraki 305-8571, Japan.

[7]Department of Astrophysics, Nagoya University, Furo-cho, Chikusa-ku, Nagoya 464-8602, Japan.

[8]Meisei University, Hodokubo, Hino, Tokyo 191-0042, Japan.

Abstract. In the collapsing phase of a molecular cloud, the molecular gas temperature is a key to understand the evolutionary process from a dense molecular cloud to stars. In order to know this, mapping observations in NH_3 lines are required. Therefore, we made them based on the FUGIN (FOREST Unbiased Galactic plane Imaging survey with Nobeyama 45m telescope). The 6 maps were observed in NH_3 (J,K) – (1,1), (2,2), (3,3) and H_2O maser lines and obtained temperature maps; some show temperature gradient in a cloud. Additionally 72 cores were observed. These candidates were called as KAGONMA or KAG objects as abbreviation of KAgoshima Galactic Object survey with Nobeyama 45-M telescope in Ammonia lines. We show the results of two regions in W33 and discuss their astrophysical properties.

Keywords. stars: formation, ISM: clouds, radio lines: ISM

1. Observations

We observed W33 star formation region (KAG64) and CMa OB1 east filament (KAG71) with the Nobeyama 45-m radio telescope since 2016 December. The NH_3 (J,K) = (1,1), (2,2), (3,3) and H_2O maser lines were observed simulaneously. We observed 703 positions with a 37.5" grid along the equatorial coordinates using the position switch method. The pointing accuracy was checked using known H_2O maser sources and was better than 5". We integrated at least 20 min at each position. The typically system noise temperature was between 100 and 300 K. The rms noise level was typically 0.035 K in $T_{\rm a}^*$ at each observed position.

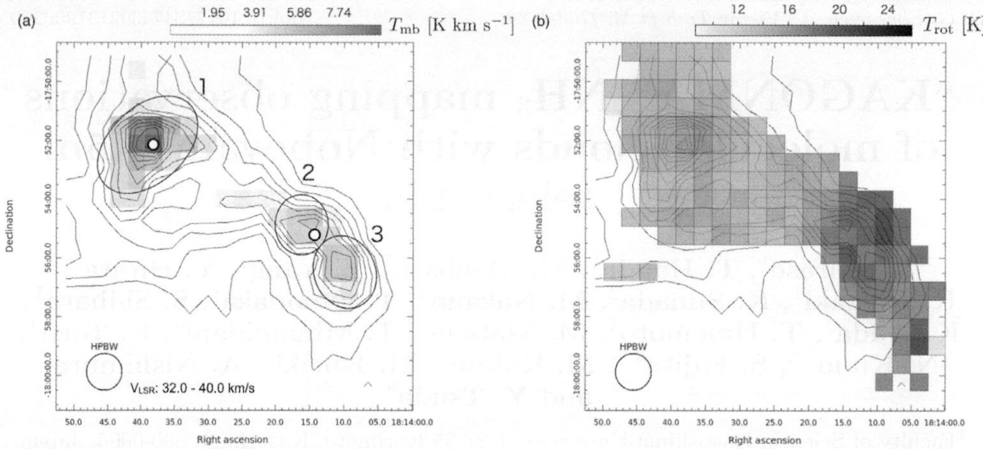

Figure 1. (panel a) The integrated intensity map in $NH_3(J, K) = (1, 1)$ line of KAG 64. Contour interval is 0.675 K in T_{mb}. Three clumps we defined are enclosed with black ellipses. Two circles show the positions of H_2O masers found in our observations. (panel b) The rotation temperature distribution of KAG 64. Contours are the same as in panel a.

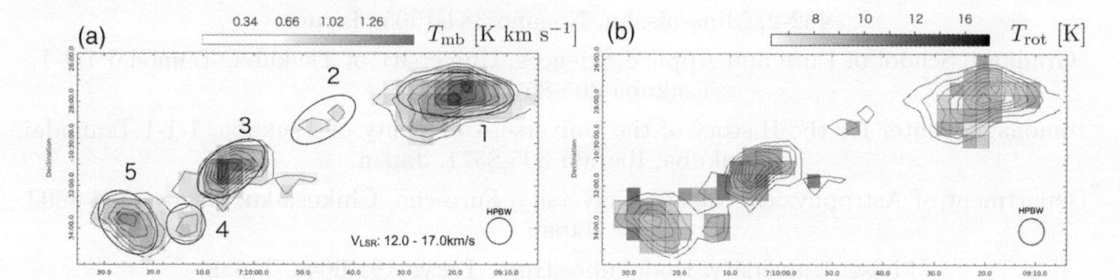

Figure 2. (panel a) The integrated intensity map in $NH_3(J, K) = (1, 1)$ line of KAG 71. Contour interval is 0.175 K in T_{mb}. Five clumps we defined are enclosed with black ellipses. (panel b) The rotation temperature distribution of KAG 71.

2. Results

We show the preliminary results of two sources; KAG64 (W33 star formation region) and KAG71 (CMa OB1 east filament). The optical depth, the rotational temperature and the column density were estimated using KAGONMA data. These physical parameters were estimated for positions with over-3σ detection in the NH_3 (J,K) = (1,1) line.

KAG64, the 3 clumps were detected in NH_3 (J,K)=(1,1) emission. Based on our observations the western and eastern parts of this region show a clear rotational temperature difference. Furthermore, H_2O maser lines were observed in each part. Radio recombination lines (RRLs) were detected between clump 2 and clump 3, where also the ammonia rotational temperature peaks.

In KAG71, we found 5 clumps in NH_3 (J,K)=(1,1) emission. The rotational temperature values show that these clumps are almost the same between 11 and 15 K. KAG71 is well traced also in 250 μm continuum emission and dust temperature is consistent to our results (Elia *et al.* 2013).

References

Elia, D. *et al.* 2013, *ApJ*, 772, 45.

Origins: From the Protosun to the First Steps of Life
Proceedings IAU Symposium No. 345, 2018
B. G. Elmegreen, L. Viktor Tóth & M. Güdel, eds.

doi:10.1017/S1743921319002011

Dust Trapping and Coagulation in Protoplanetary Disks

Ya-Ping Li

Shanghai Astronomical Observatory, Chinese Academy of Sciences, 80 Nandan Road, Shanghai 200030, China
email: leeyp2009@gmail.com

Abstract. In this work, we carry out two-fluid (gas+dust) hydrodynamical simulations on a large family of models in order to study the dust coagulation and the dust-gas dynamical processes in protoplanetary disks. Our theoretical effort is guided by the observational results of disks in nearby star forming regions at sub-millimeter and millimeter (mm) wavelengths. By a systematic comparison with the continuum emission at several mm bands from ALMA observations, we find that ringed structures are predicated in the unresolved faint disks for those with mm spectral indexes as low as about 2.0. Our parameter exploration can also be used to constrain the fragmentation velocity, one key parameter of the dust coagulation model, and some other disk parameters.

Keywords. planetary systems: protoplanetary disks — planets: rings — submillimeter

1. Introduction

The discovery of thousands of exoplanets over the last couple of decades has clearly shown that the birth of planets is a very efficient process in nature (e.g., Burke *et al.* 2015). The commonly accepted theory is that planets form in young disks orbiting Pre-Main Sequence stars through the agglomeration of small dust particles into km-sized "planetesimals", which are massive enough to gravitationally attract other solids in the disk (see review by Chiang & Youdin 2010). Grain growth from sub micrometer sizes, which are the typical sizes of dust in the interstellar medium (ISM), to millimeter and centimeter sizes is thus the first step toward the formation of planetesimals inside young circumstellar disks. However, models of the evolution of disks solids with density and temperature both decreasing with distance from the central star have predicted radial drift timescales which are too short (e.g., only $100 \sim 1000$ orbits) to form planetesimals (see review by Johansen *et al.* 2014). It is generally believed that some disk substructures have to be invoked to slow down the dust radial drift because of the positive pressure gradient in the inner edge of the bump.

Motivated by the low mm spectral indexes $\alpha_{\rm mm}$ observed for a large sample of protoplanetary disks (PPDs) in several nearby star forming regions (Ricci *et al.* 2012; Ansdell *et al.* 2018; Andrews *et al.* 2018), we take a local maximum substructure for the gas surface density profile, which is assumed to be generated by a gap in the disk viscosity. By performing detailed simulations for two fluids (gas+dust) hydrodynamics, we study the dust trapping and coagulation processes in the disks and then calculate the dust continuum emission at several mm bands with detailed radiative transfer (Li *et al.* 2019, in preparation). Our main goal is to interpret the low spectral indexes observed for many young disks, and understand the link between the spectral slope and the dust dynamics in the disks. Based on this diagnostics, we can constrain some fundamental parameters for

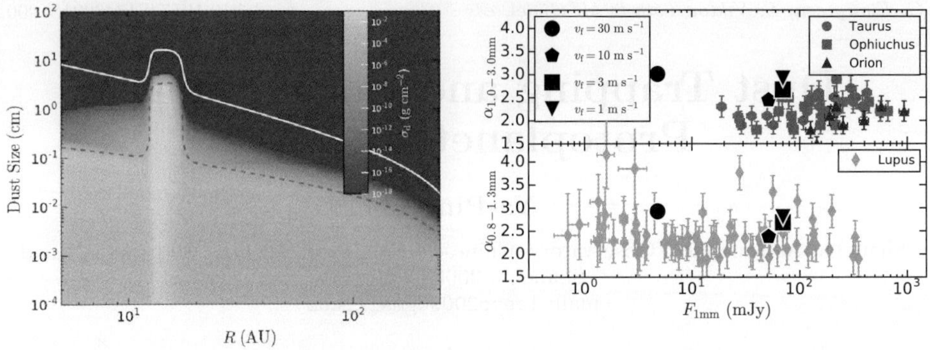

Figure 1. Left: dust surface density distribution for different dust species at the time 2.1 Myr for a typical model. The red dashed line indicates the fragmentation barrier, which represents the maximum size of the particles before they reach fragmentation velocities. The white solid line denotes the grain size corresponding to a Stokes number of unity. Right: 1 mm flux $F_{1\mathrm{mm}} - \alpha_{\mathrm{mm}}$ plot for models with different fragmentation velocities. The observed fluxes are all normalized with a disk distance of 140 pc and extrapolated to 1 mm based on the observed spectral indexes (Ricci *et al.* 2012; Ansdell *et al.* 2018). The spectral indexes are calculated between 1.0 mm and 3.0 mm (upper panel), and between 0.89 mm and 1.33 mm (lower panel).

the dust coagulation model and shed new insight into the mechanism of dust trapping and growth.

2. Results

We present detailed hydrodynamical simulations for PPDs to study the two-fluid co-evolution with self-consistently considering dust coagulation/fragmentation processes (for previous version of the code, see Li *et al.* 2005; Birnstiel *et al.* 2010a). We then compare the dust continuum emissions produced from *RADMC-3D* (Dullemond *et al.* 2012) with multi-wavelength data. By assuming a viscosity gap in the inner region of the disk, a gas bump can be generated, which results in an effectively dust trapping in the same region. With a systematically parameter study for different model parameters (e.g., the fragmentation velocity v_{f}, gas surface density, disk global viscosity α_0, viscosity gap parameters, disk scale height, and disk gas profile index), we can quantify the role of the viscosity gap in the mm dust continuum spectral index.

Our main conclusions are summarized as follows (Li et al. 2019, in preparation). 1) Gas bump is necessary to produce a lower spectral index. Such a gas bump could facilitate the trapping of the dust particle and also make efficient dust size growth (left panel of Figure 1). 2) Our model with a reasonable v_{f} ($3 \sim 10$ m s^{-1}) can reproduce a low α_{mm} close to 2.0 (right panel of Figure 1). The tentatively negative relation between α_{mm} and $F_{1\mathrm{mm}}$ obtained from Ansdell *et al.* (2018) for the faint PPD sources can be explained by a relatively low v_{f} ($\lesssim 10$ m s^{-1}) with a variation of dust mass surface density. This is simply because the optical depth becomes smaller as the dust mass (mm flux) decreases. 3) α_{mm} is not very sensitive to the gas bump width, gap depth, disk scale height, and disk profile index. Even though one ringed structure in the outer region of the disk can produce a slightly brighter disk, our one ringed-structure model can hardly reproduce the bright end disks with the low spectral index from previous surveys. We expect that multiple rings can be responsible for such very bright disks.

References

Andrews, S. M., Huang, J., Pérez, L. M., *et al.* 2018, arXiv e-prints, arXiv:1812.04040.
Ansdell, M., Williams, J. P., Trapman, L., *et al.* 2018, *ApJ*, 859, 21

Birnstiel, T., Dullemond, C. P., & Brauer, F. 2010a, *A&A*, 513, A79
Burke, C. J., Christiansen, J. L., Mullally, F., Seader, S., Huber, D. *et al.* 2015, *ApJ*, 809, 8
Chiang, E., & Youdin, A. N. 2010, *Annual Review of Earth and Planetary Sciences*, 38, 493
Dullemond, C. P., Juhasz, A., Pohl, A., *et al.* 2012, *ASCL*, ascl:1202.015
Johansen, A., Blum, J., Tanaka, H., *et al.* 2014, *Protostars and Planets VI*, 547-570
Li, H., Li, S., Koller, J., *et al.* 2005, *ApJ*, 624, 1003
Li, Y.-P., Li, H., Ricci, L., *et al.* 2019, *ApJ*, 878, 39
Ricci, L., Trotta, F., Testi, L., *et al.* 2012, *A&A*, 540, A6

Origins: From the Protosun to the First Steps of Life
Proceedings IAU Symposium No. 345, 2018
B. G. Elmegreen, L. Viktor Tóth & M. Güdel, eds.

doi:10.1017/S1743921319002023

Constraints of habitability for the young Earth in a highly eccentric orbit

Elke Pilat-Lohinger[1], Kristina G. Kislyakova[1,2], Helmut Lammer[2], Colin P. Johnstone[1], David Bancelin[1] and Ákos Bazsó[1]

[1]Dept. of Astrophysics, University of Vienna
Türkenschanzstrasse 17, A-1180 Wien, Austria
emails: elke.pilat-lohinger@univie.ac.at, kristina.kislyakova@univie.ac.at, colin.johnstone@univie.ac.at, david.bancelin@univie.ac.at, akos.bazso@univie.ac.at

[2]Space Research Institute, Austrian Academy of Science
Schmiedlstrasse 6, 8042 Graz, Austria
email: helmut.lammer@oeaw.ac.at

Abstract. Thousands of planets outside the Solar system have been discovered, with exoplanets in different environments. Since we cannot expect to find an exoplanetary system fully resembling our Solar System, we consider a Solar System type configuration where the Earth moves in an eccentric orbit. We focus on young Earth 1 billion years ago, when the Sun's extreme UV (EUV) flux was about 5 times higher than the current radiation. In case of eccentric motion of Earth, strong variations of the EUV flux would influence the evolution of the planet's atmosphere (EUV radiation of 50 times the current EUV flux would be possible). Taking into account a certain amount of Nitrogen in the atmosphere of such a young Earth, we study the non-thermal loss of N2 over a long time interval. We therefore investigate to what extent eccentric motion will influence the conditions of habitability of a terrestrial planet.

Keywords. Young Earth, eccentric motion, EUV flux, Nitrogen loss

---

1. Introduction

Pilat-Lohinger *et al.* (2008a) & (2008b) investigated the dynamical evolution of various solar system like configurations. These studies showed that the Sun's habitable zone (HZ) is mainly perturbed by the gravitational interaction of Jupiter and Saturn. Uranus and Neptune do not have a significant influence due to their lower masses. Consequently, it is possible to reduce the planetary system of Jupiter-Saturn like configurations. In Pilat-Lohinger *et al.* (2008a) it can be seen that not all Jupiter-Saturn pairs would lead to a nearly circular orbit for an Earth at 1 au. Contrariwise, it was found that strong variations in the orbital eccentricity could arise with a maximum eccentricity up to nearly 0.7 (see Pilat-Lohinger (2015)) which might change the conditions for habitability significantly. Especially, the evolution of the young Earth could be endangered when the young Sun is still more active, causing a higher extreme UV (EUV) flux which might spoil the planet's atmosphere.

2. EUV flux and Nitrogen loss for an Earth in eccentric motion

Tu *et al.* (2015) showed that the early Sun can follow different activity tracks which might influence the evolution of a planet since the planets form during this phase. If we consider an early time of the solar system (e.g. at 3.5 Gyrs) where the activity tracks of the Sun converged already and the atmosphere of Earth contained Nitrogen. Considering the case that strong gravitational perturbations cause eccentricity motion of the Earth

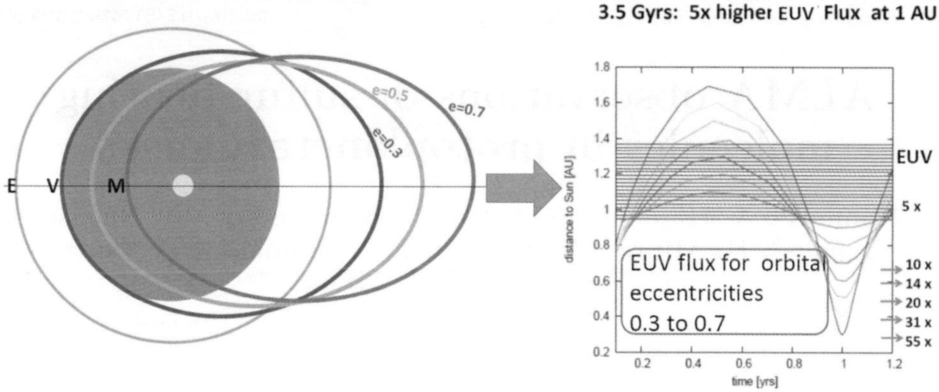

Figure 1. Eccentricity motion of an Earth-like planet at 1 au: left panel shows that the planet enters the "red" region with higher EUV flux when $e > 0.3$. Right panel shows the distance of the planet to the Sun over one orbital period for different eccentricities represented by different colours: 0.1 (red), 0.2 (green), 0.3 (blue), 0.4 (magenta), 0.5 (light blue), 0.6 (yellow) and 0.7 (black). The blue area lables the HZ (as defined by Kasting *et al.* (1993)). The EUV flux inside the red area (left panel) is indicated at the right side of this panel.

we studied the Nitrogen-loss due to the high EUV flux when the planet approaches the Sun. Simulations of exopheric clouds assuming a Nitrogen dominated atmosphere (Kislyakova *et al.* (2018)) based on the study by Tian *et al.* (2008) indicated the following features: (i) A higher EUV flux heats up the atmosphere efficiently, (ii) the lower atmosphere expands to high altitudes then consequently (iii) the exobase is located at very high altitudes. Consequently, high exobase altitudes lead to intense interactions between the stellar wind and the atmosphere so that severe consequences for the evolution of this planet can be expected. A detailed study thereto is in progress.

Acknowledgements

The authors acknowledge the support of the Austrian Science Fund (FWF) NFN project S11601-N16 "Pathways to Habitability: From Disks to Active Stars, Planets to Life" and related sub-projects S11608-N16 (EP-L, DB, ÁB), S11607-N16 (KGK, HL) and S11604-N16 (CPJ).

References

Kasting, J. F., Whitmire, D. P., & Reynolds, R. T. 1993 *Icarus*, 101, 108

Kislyakova, K. G., Holmström M., Odert, P., Lammer, H., Erkaev, N. V., Dorfi, E., & Güdel, M. 2018 *A&A*, 2018 in prep.

Pilat-Lohinger, E., Süli, A., Robutel, P., & Freistetter, F. 2008, *ApJ*, 681, 1639

Pilat-Lohinger, E., Robutel, P., Süli, A., & Freistetter, F. 2008, *CeMDA*, 102, 83

Pilat-Lohinger 2015, *IJAsB*, 14, 145

Tian, F., Kasting, J. F., Liu, H.-L., & Roble, R. G. 2008, *Journal of Geophys. Research (Planets)*, 113, E05008

Tu, L., Johnstone, C. P., Güdel, M., & Lammer, H. 2015 *A&A*, 577, L3

Origins: From the Protosun to the First Steps of Life
Proceedings IAU Symposium No. 345, 2018
B. G. Elmegreen, L. Viktor Tóth & M. Güdel, eds.

doi:10.1017/S1743921319002072

ALMA observations of sulfur-bearing molecules in protoplanetary disks

H. Nomura[1], A. Higuchi[2], N. Sakai[2], S. Yamamoto[3], M. Nagasawa[4], K.K. Tanaka[5], H. Miura[6], T. Nakamoto[1], H. Tanaka[5], T. Yamamoto[7], C. Walsh[8] and T.J. Millar[9,10]

[1]Department of Earth and Planetary Sciences, Tokyo Institute of Technology, Japan
[2]RIKEN Cluster for Pioneering Research, Japan
[3]Department of Physics, The University of Tokyo, Japan
[4]School of Medicine, Kurume University, Japan
[5]Department of Astronomy, Tohoku University, Japan
[6]Graduate School of Natural Sciences, Nagoya City University, Japan
[7]Institute of Low Temperature Science, Hokkaido University, Japan
[8]School of Physics and Astronomy, University of Leeds, UK
[9]ARC, School of Mathematics and Physics, Queen's University Belfast, UK
[10]Leiden Observatory, Leiden University, The Netherlands

Abstract. It is thought that protoplanets formed in protoplanetary disks excite the orbital motion of the surrounding planetesimals, and the bow shocks caused by the highly excited planetesimals heat their icy component evaporating into gas. We have performed model calculations to study the evolution of molecular abundances of the evaporated icy component, which suggests sulfur-bearing molecules can be good tracers of icy planetesimal evaporation. Here we report the result of our ALMA observations of sulfur-bearing molecules towards protoplanetary disks. The lines were undetected but the obtained upper limits of the line fluxes and our model calculations give upper limits of the fractional abundances of $x(H_2S) < 10^{-11}$ and $x(SO) < 10^{-10}$ in the outer disk. These results are consistent with the molecular abundances in comets in our Solar system.

Keywords. astrochemistry, planetary systems: protoplanetary disks, ISM: molecules

---

Protoplanetary disks are the natal birth place of planets and ALMA observations are now revealing the physical and chemical structure of planet forming regions in disks. ALMA long baseline observations have detected gaps and rings which suggest possible formation of giant planets in disks. Meanwhile, molecular lines are useful tracers of physical and chemical properties of the disks.

Theoretically, it is thought that once (proto)planets are formed in disks, eccentricities of planetesimals are increased due to gravitational interactions. As a result, bow shocks are formed around the icy planetesimals that evaporate via shock heating. The evaporation rates and orbital evolution of such icy planetesimals under various conditions of disks and (proto)planets have been investigated (Tanaka *et al.* 2013; Nagasawa *et al.* 2014, 2018). For example, if a Jovian planet exists at an orbital radius of around 20 au, planetesimals can be excited to a velocity of $3 - 6$ km s^{-1} inside a disk radius of 20 au. The excited velocity is high enough for bow shocks to evaporate icy planetesimals and low enough that the evaporated molecules are not dissociated in the hot gas caused by the bow shock.

We have performed model calculations of time evolution of the abundances of evaporated icy molecules, using a gas-phase chemical reaction network with desorption and

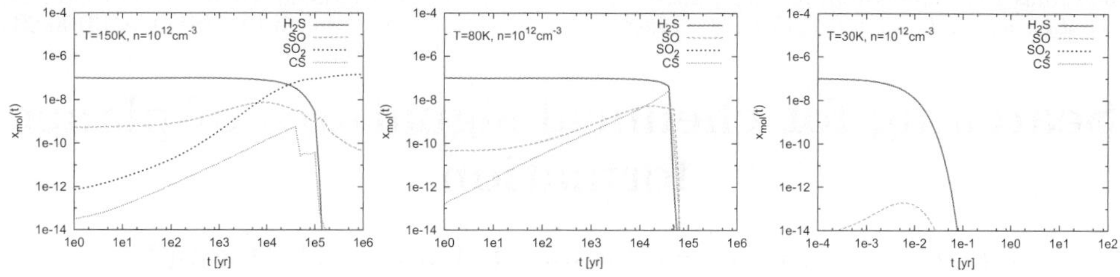

Figure 1. Time evolution of the fractional abundances of molecules evaporated from icy planetesimals. The behaviors are different depending on the dust temperature in the disk.

adsorption of molecules from/onto grains. Figure 1 shows the resulting time evolution of the abundances of sulfur(S)-bearing species at different temperatures. S-bearing molecules are known as shock tracers and can be abundant in the gas-phase only after they are evaporated from the ice. For example, emission from SO molecule has recently been observed around protostellar disks surrounded by an infalling envelope, suggesting the existence of a shock at the centrifugal radius (e.g., Sakai *et al.* 2017, Miura *et al.* 2017). In our model, the S-bearing parent molecules in ice are assumed to be H_2S and OCS. The figure shows that after the evaporation from ice, H_2S is destroyed by gas-phase reactions to form SO and then SO_2 with a timescale of 10^{4-5} yrs. Since the freeze-out temperature of SO_2 ($T_d \sim 130$ K) is higher than those of H_2S and SO ($T_d \sim 60$ K), SO_2 formed via the gas-phase reactions is quickly frozen out on grains except for in the very inner disk where the dust temperature is high ($T_d > 130$ K). In the cold outer disk ($T_d <$ 60 K) even H_2S and SO are frozen out on grains quickly. Thus, we expect H_2S and SO to become abundant in the region with 60 K $< T_d <$ 130 K, which corresponds to the giant planet forming region in disks, after the evaporation of icy planetesimals.

Based on our model calculations, we have executed ALMA observations of the 217 GHz H_2S $2_{2,0} - 2_{1,1}$ and 220 GHz SO $6_5 - 5_4$ lines in Band 6 towards 10 T Tauri disks in the Taurus molecular clouds on 1, 17 and 27 of August 2016 in Cycle 3 (2015.1.01207.S), aiming to observe molecules evaporated from icy planetesimals as a tracer of (proto)planet formation in these disks. The lines were undetected but we obtained the upper limits of the line flux densities of 12 – 19 mJy and 18 – 25 mJy for the H_2S and SO lines, respectively.

We have performed radiative transfer calculations of the molecular lines by using the density and temperature profiles of a disk model and assuming the gas-phase abundances of H_2S and SO in the outer disk, considering non-thermal desorption of molecules from icy grains. This is an analogue of the recent ALMA observations of methanol towards a protoplanetary disk, which suggests non-thermal desorption of methanol from ice in the outer disk (Walsh *et al.* 2016). By comparing the model calculations and the upper limits of the line flux densities in our ALMA observations, we have derived the upper limits of the fractional abundances with respect to hydrogen nuclei of $x(H_2S) < 10^{-11}$ and $x(SO) < 10^{-10}$. The obtained upper limits of the fractional abundances are consistent with those observed towards comets in our Solar system.

References

Miura, H., Yamamoto, T., Nomura, H. *et al.* 2017, *ApJ*, 839, 47
Nagasawa, M., Tanaka, K. K., Tanaka, H., *et al.* 2014, *ApJL*, 794, L7
Nagasawa, M., Tanaka, K. K., Tanaka, H., *et al.* 2018, *ApJ*, in press (arXiv:1812.05250)
Sakai, N., Oya, Y., Higuchi, A. *et al.* 2017, *MNRAS*, 467, L76
Tanaka, K. K., Yamamoto, T., Tanaka, H. *et al.* 2013, *ApJ*, 764, 120
Walsh, C., Loomis, R., Oberg, K.I. *et al.* 2016, *ApJL*, 823, L10

Origins: From the Protosun to the First Steps of Life
Proceedings IAU Symposium No. 345, 2018
B. G. Elmegreen, L. Viktor Tóth & M. Güdel, eds.

doi:10.1017/S1743921319002114

Searching for chemical signatures of planet formation

Ch. Rab[1], G.A. Muro-Arena[2], I. Kamp[1], C. Dominik[2], L.B.F.M. Waters[2,3], W-F. Thi[4] and P. Woitke[5]

[1]Kapteyn Astronomical Institute, University of Groningen, P.O. Box 800, 9700 AV Groningen, The Netherlands
email: rab@astro.rug.nl

[2]Anton Pannekoek Institute for Astronomy, University of Amsterdam, Science Park 904, 1098 XH Amsterdam, The Netherlands

[3]SRON Netherlands, Sorbonnelaan 2, 3584 CA Utrecht, The Netherlands

[4]MPE, Giessenbachstrasse 1, 85748 Garching, Germany

[5]SUPA, School of Physics & Astronomy, University of St. Andrews, North Haugh, St. Andrews KY16 9SS, UK

Abstract. High spatial resolution observations with ALMA and VLT/SPHERE show gaps and rings in continuum emission of protoplanetary disks, possibly indicating ongoing planet formation. However, it is still unclear if the gas follows the dust distribution. We present radiation thermo-chemical models for the disk of HD 163296 to study the impact of dust and gas gaps on the chemistry and molecular line emission. We compare a model with only dust gaps to a model that also has gas gaps. In both models, rings and gaps are visible in (sub)mm molecular line emission. Due to chemistry, certain molecules are sensitive to dust gaps where others are more sensitive to gas depletion. Observations of multiple molecules might allow to accurately determine the degree of gas depletion within the dust gaps, information crucial to discriminate between gap formation theories (e.g. planets, ice lines).

Keywords. stars: pre–main-sequence, (stars:) planetary systems: protoplanetary disks, astrochemistry, radiative transfer, methods: numerical

---

1. Introduction & Methods

High spatial resolution observations with ALMA and VLT/SPHERE reveal gaps and rings in dust continuum images of protoplanetary disks (e.g. Andrews *et al.* 2016, Avenhaus *et al.* 2018). Although those gaps and rings are very prominent in dust observations, they are not yet clearly detected in the gas. An accurate measurement of the gas column densities and the possible gas depletion within the dust gaps is crucial to discriminate between different gap formation scenarios. If planets produced the gaps, one would expect strong gas depletion and gas gaps, other formation scenarios such as dust evolution, molecular ice lines or deadzones predict no or only shallow gas gaps (e.g. Isella *et al.* 2016, Birnstiel *et al.* 2015, Zhang *et al.* 2015, Pinilla *et al.* 2016).

Here we present a model for the HD 163296 protoplanetary disk to study in detail the impact of the observed dust gaps and possible gas gaps on the chemistry and on molecular line emission. We used the radiation thermo-chemical disk model ProDiMo (PROtoplanetary DIsk MOdel, e.g. Woitke *et al.* 2009, Kamp *et al.* 2017) to self-consistently calculate dust and gas temperatures, molecular abundances and synthetic observables. The models are based on the HD 163296 ProDiMo model presented in Woitke *et al.* (2019),

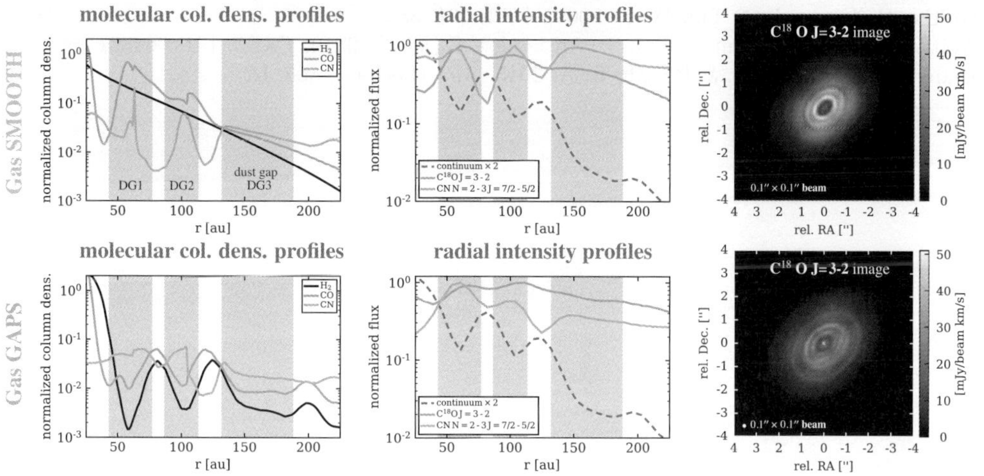

Figure 1. Two models for the HD 163296 disk, one without gas gaps (SMOOTH, *top row*) and one with gas gaps (GAPS, *bottom row*). The gray boxes mark the dust gaps present in both models. *First column:* normalized radial molecular column density profiles (H_2 traces the total gas column); *second column:* the corresponding radial intensity profiles (normalized to the peak intensity); *third column:* integrated intensity map for $C^{18}O$ $J = 3 - 2$.

but was refined by Muro-Arena *et al.* (2018) to fit new VLT/SPHERE and ALMA continuum observations that show prominent dust gaps and rings.

2. Results & Conclusions

In Fig. 1, we compare two models, one with a smooth gas surface density profile (SMOOTH) and one with gas gaps (GAPS). As an example, we show the column densities and observables for the molecules CO and CN. The molecular column density profiles show gaps and rings in both models. In the SMOOTH model this is solely due to the impact of dust gaps on the chemistry (e.g. temperature, radiation field). In the models CO quite nicely follows the total gas surface density (see also Facchini *et al.* 2018), but also in the SMOOTH model gaps are seen in the $C^{18}O$ line image. The $C^{18}O$ line is optically thick and therefore sensitive to temperature changes caused by the dust gaps (see also van der Marel *et al.* 2018). In the SMOOTH model, CN gaps and rings appear due to dust depletion and the column density actually peaks within the dust gaps, but in the gas GAPS model these peaks are washed out, making CN a good tracer of dust gaps without gas depletion. This example shows that self-consistent dust and gas modeling including chemistry is required to accurately infer gas column densities (gas gaps) from molecular line observations and that observations of multiple molecules will allow for a more accurate measurement of gas depletion within dust gaps.

References

Andrews, S. M.,Wilner, D. J., Zhu, Z., Birnstiel, T. *et al.* 2016, *ApJ* (Letters), 820, L40
Avenhaus, H., Quanz, S. P.,Garufi, A., Perez, S., Casassus, S. *et al.* 2018, *ApJ*, 863, 44
Birnstiel, T., Andrews, Sean M., Pinilla, P., & Kama, M. 2015, *ApJ*, 813
Facchini, S., Pinilla, P., van Dishoeck, E. F. & de Juan Ovelar, M. 2018, *A&A*, 612
Isella, A., Guidi, G., Testi, L., Liu, S., Li, H., Li, S. *et al.* 2016, *Phys. Rev. Lett.*, 117, 251101
Kamp, I., Thi, W.-F., Woitke, P., Rab, C., Bouma, S., & Ménard, F. 2017, *A&A*, 607, A41
Muro-Arena, G. A., Dominik, C., Waters, L. B. F. M., Min, M. *et al.* 2018, *A&A*, 614, A24
Pinilla, P., Flock, M., Ovelar, M. de Juan & Birnstiel, T. 2016, *A&A*, 596
van der Marel, N., Williams, J. P., & Bruderer, S. 2018, *ApJ* (Letters), 867, L14

Woitke, P., Kamp, I., & Thi, W.-F. 2009, *A&A*, 501, 383
Woitke, P., Kamp, I., Antonellini, S., Anthonioz, F., *et al.*, 2019, *PASP*, 131, 064301
Zhang, K., Blake, G. A., & Bergin, E. A. 2015, *ApJ* (Letters), 806, L7

Origins: From the Protosun to the First Steps of Life
Proceedings IAU Symposium No. 345, 2018
B. G. Elmegreen, L. Viktor Tóth & M. Güdel, eds.

doi:10.1017/S174392131900214X

Infrared Spectroscopic Studies of Gases in the Circumstellar Environments of Young Stellar Objects

Benjamin A. Sargent[1], William J. Forrest[2], Joel D. Green[1], Edward J. Montiel[3] Curtis DeWitt[3] and Matthew J. Richter[3]

[1]Space Telescope Science Institute, 3700 San Martin Drive, Baltimore, MD 21218, USA
email: sargent@stsci.edu

[2]Department of Physics and Astronomy, University of Rochester, Rochester, NY 14627, USA

[3]USRA/SOFIA, NASA Ames Research Center, MS 232-12, Bldg. 232, Rm. 130-31, Moffett Field, CA 94035-000, USA

Abstract. The building blocks of planets in planet-forming ("protoplanetary") disks are assembled early in the lifetime of a young star. The gas disks are relatively short-lived, with a half-life of about 3 million years, as chemical reactions modify the reservoir of material from the natal molecular cloud. *Spitzer Space Telescope* Infrared Spectrograph (IRS) spectra of protoplanetary disks around T Tauri stars show emission from H_2O and absorption from other gases, sometimes consistent with formaldehyde, H_2CO, and other times consistent with formic acid, HCOOH, in the 5-7.5 μm region. SOFIA-EXES spectra of YSOs that follow up on these *Spitzer*-IRS studies are presented. How the gaseous features observed between 5-7.5 μm relate to those at other wavelengths is discussed. This work suggests that water and organic molecules, which are crucial for life as we know it, are present in the habitable zones of stars at a very early age [of 1-3 million years].

Keywords. infrared: stars

Though many previous studies found evidence for water from *Spitzer*-IRS high resolution (R $\sim$ 600) spectra (e.g., Carr & Najita 2008), we find that *Spitzer*-IRS low resolution (R $\sim$ 90) spectra can identify water emission as well (note Pascucci *et al.* 2009 and Teske *et al.* 2011 identified HCN and C_2H_2 in *Spitzer*-IRS low resolution spectra). Models of low resolution *Spitzer*-IRS spectra of protoplanetary disks around young stars suggest the presence of H_2O, H_2CO, and HCOOH in these protoplanetary disks (Sargent *et al.* 2014). We are seeking high resolution infrared spectroscopic confirmation of the presence of these molecules (e.g., Roueff *et al.* 2006 found absorption at 3.6 μm in the spectrum of protostar W33A from H_2CO).

For the SOFIA-EXES medium resolution (Richter *et al.* 2018) spectrum of the young stellar object (YSO) IRAS 04278+2253, the spectral model shown in Figure 1 (at *left*) assumes a cool (300K) cloud of H_2O in front of a warm blackbody, where the H_2O has sufficient column density so that the absorption features at 5.782, 5.789, 5.802, 5.809, 5.818, 5.823, and 5.826 μm are clearly visible in the plot. The SOFIA-EXES spectra of IRAS 04278+2253 (not telluric-corrected) can largely be explained by telluric absorption and the H_2O model. There are features not explained by telluric absorption or the H_2O model, but it is not clear what their carrier is. The SOFIA-EXES low-resolution spectrum of DL Tau (Figure 1, *right*) is plagued by fringing. We have attempted to remove the fringes with a simple sine wave, following Cami *et al.* (2000), but this was not successful. A more sophisticated means of removing fringes is needed.

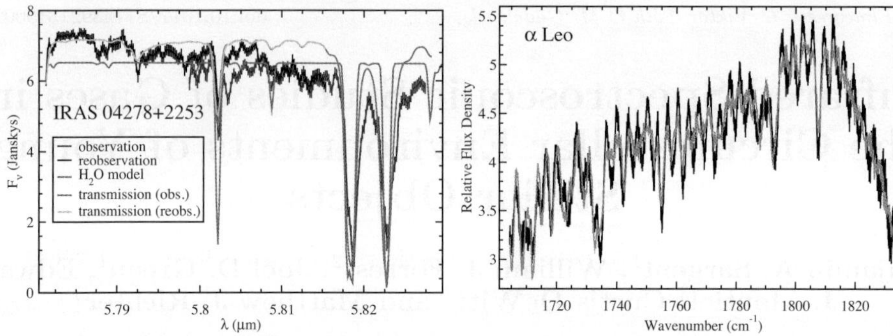

Figure 1. Left, SOFIA-EXES medium-resolution spectra of IRAS 04278+2253 from program 04_0180, with the original (March 2016) spectrum in blue; the re-observation (January 2017) in black; 300K HITEMP (Rothman *et al.* 2010) H_2O model in red; and the transmission curves for Earth's atmosphere in green and orange. **Right**, SOFIA-EXES low-resolution spectrum of α Leo, the telluric calibrator for the DL Tau observation (program 04_0188). The signal from the observation is in black, and an attempt at removing the fringes with a simple sine wave (following Cami *et al.* 2000) is in magenta.

Initial reductions of archival Keck-NIRSPEC 2-4 μm spectra of a few of our stars do not reveal any H_2CO lines at 3.6 μm; however, the lines may be variable, or the H_2CO may be located in the disks such that absorption is seen at 5.7 μm but not at 3.6 μm. An interesting future project would be to measure and model the emission and absorption from gases in a larger sample of *Spitzer*-IRS spectra of T Tauri stars.

This work is based in part on observations made with the *Spitzer Space Telescope*, which is operated by the Jet Propulsion Laboratory, California Institute of Technology under NASA contract 1407. This work is also based [in part] on observations made with the NASA/DLR Stratospheric Observatory for Infrared Astronomy (SOFIA). SOFIA is jointly operated by the Universities Space Research Association, Inc. (USRA), under NASA contract NAS2-97001, and the Deutsches SOFIA Institut (DSI) under DLR contract 50 OK 0901 to the University of Stuttgart. B.A.S. acknowledges SOFIA funding from subcontract SOF 04-0180 Sargent and subcontract SOF 04-0188 Sargent of prime contract NAS2-97001.

References

Cami, J., *et al.* 2000, *A&A*, 360, 562
Carr, J. S., & Najita, J. R. 2008, *Science*, 319, 1504
Pascucci, I., *et al.* 2009, *ApJ*, 696, 143
Richter, M. J., *et al.* 2018, *Journal of Astronomical Instrumentation*, 7, 1840013
Rothman, L., *et al.* 2010, *JQSRT*, 111, 2139
Roueff, E., *et al.* 2006, *A&A*, 447, 963
Sargent, B. A., *et al.* 2014, *ApJ*, 792, 83
Teske, J. K., *et al.* 2011, *ApJ*, 734, 27

Origins: From the Protosun to the First Steps of Life
Proceedings IAU Symposium No. 345, 2018
B. G. Elmegreen, L. Viktor Tóth & M. Güdel eds.

doi:10.1017/S1743921319002230

Chemical modeling of FU Ori protoplanetary disks

Tamara Molyarova[1], Vitaly Akimkin[1], Dmitry Semenov[2], Péter Ábrahám[3], Thomas Henning[2], Ágnes Kóspál[2,3], Eduard Vorobyov[4] and Dmitri Wiebe[1]

[1]Institute of Astronomy, Russian Academy of Sciences, 119017, 48 Pyatnitskaya st., Moscow, Russia
email: molyarova@inasan.ru

[2]Max Planck Institute for Astronomy, Königstuhl 17, 69117 Heidelberg, Germany

[2]Konkoly Observatory, Research Centre for Astronomy and Earth Sciences, Hungarian Academy of Sciences, Konkoly-Thege Miklós út 15-17, 1121 Budapest, Hungary

[4]Research Institute of Physics, Southern Federal University, Stachki 194, Rostov-on-Don, 344090, Russia

Abstract. Luminosity outbursts of the FU Ori type stars, which have a magnitude of $\sim 100\,L_\odot$ and last for decades, may affect chemical composition of the surrounding protoplanetary disk. Using astrochemical modelling we analyse the changes induced by the outburst and search for species sensitive to the luminosity rise. Some changes in the disk molecular composition appear not only during the outburst itself but can also retain for decades after the end of the outburst. We analyse main chemical processes responsible for these effects and assess timescales at which chemically inert species return to the pre-outburst abundances.

Keywords. astrochemistry, planetary systems: protoplanetary disks, stars: pre–main-sequence

---

Young stars of the FU Ori type are objects experiencing sudden luminosity outburst, associated with increased accretion rate (Dunham & Vorobyov 2012). These outbursts are thought to be episodic and could happen to many young protostars (Audard *et al.* 2014). Anomalous abundances of specific molecules in disks around quiescent stars may indicate past luminosity outbursts (Rab *et al.* 2017). Our goal is to point out the chemical species useful for this purpose.

We use ANDES astrochemical model by Akimkin *et al.* (2013), adjusted to the case of variable accretion rate. We consider disk with the mass $M_{\rm disk} = 0.01\,M_\odot$ and characteristic radius $R_{\rm c} = 100\,{\rm au}$ around a young sun-like star $(1\,M_\odot,\ 0.9\,L_\odot)$. Dust size distribution is described by a power-law with the index of -3.5, $a_{\rm min} = 5\cdot10^{-7}\,{\rm cm}$ and $a_{\rm max} = 2.5\cdot10^{-3}\,{\rm cm}$. The outburst is simulated as an increased accretion luminosity lasting for $\approx 50\,{\rm yr}$, with the profile shown in Fig. 1. The detailed description of this model as well as chemistry of different disk models can be found in Molyarova *et al.* (2018).

The patterns of chemical behavior during the outburst are shown in Figure 1. Most of gas-phase species grow in abundance during the outburst due to the thermal evaporation of their ices (the first three panels). For CO the increase is not very significant due to the chemical depletion of its ice (Molyarova *et al.* 2017, Bosman *et al.* 2018). Abundant volatiles, such as CH_3OH and C_3H_4, follow closely the luminosity profile and immediately return to their quiescent abundances after the end of the outburst. The species shown in the last three panels of Fig. 1 are found to stay overabundant after the outburst for the longest timescales, from decades to thousand years.

Different mechanisms are responsible for time-dependent chemistry of the selected species. While H_2CO, NH_2OH, CH_3CHO, $HCOOCH_3$, and CH_3OCH_3 are evaporated

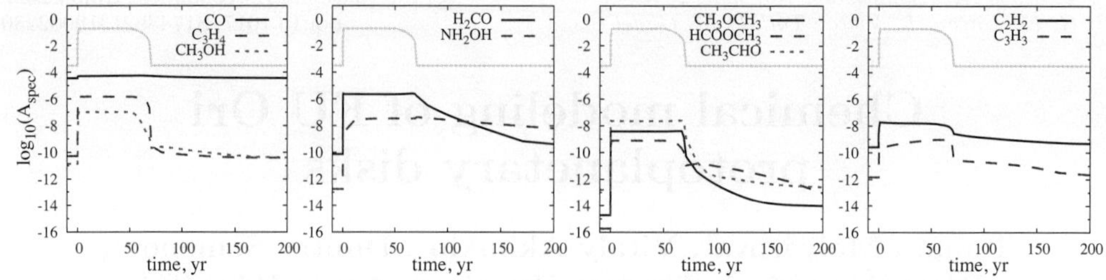

Figure 1. Disk total abundances (with respect to H_2) of selected species changing with time. The solid gray line denotes the luminosity profile (in $1000\,L_\odot$).

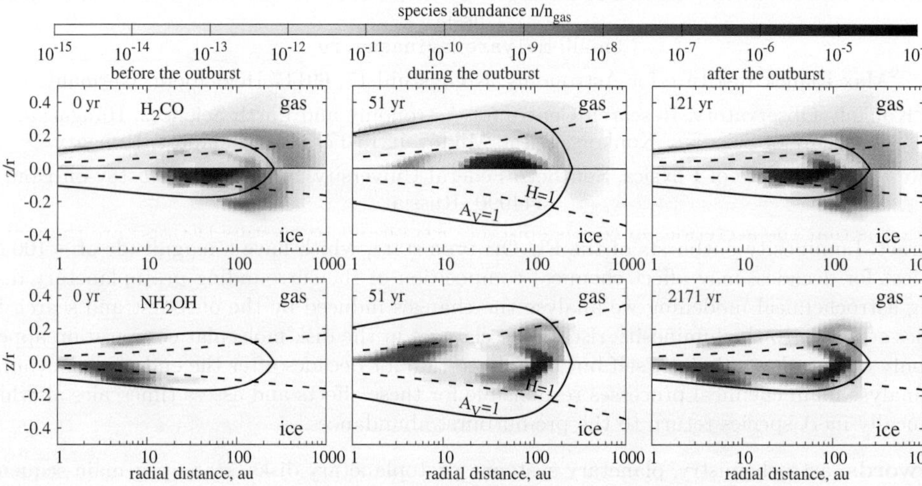

Figure 2. 2D distribution of H_2CO and NH_2OH in gas (upper half-plots) and in ice (lower half-plots) at specified time moments after the beginning of the outburst. Solid and dash black lines show the surface of $A_v = 1$ and scale height equal to unity, correspondingly.

from ice, C_3H_3 and C_2H_2 are formed directly in the gas. Fig. 2 shows spatial distribution of H_2CO and NH_2OH, which are the most promising post-outburst tracers. Their abundances are determined mostly by the evaporation, although NH_2OH is also actively formed on ice during the outburst. The times of returning to the quiescent values are defined by freeze-out timescales, which are long in the outer disk regions.

The abundances of H_2CO and NH_2OH jump for several orders of magnitude during the outburst and stay elevated for $10^2 - 10^3$ yr after it. This makes them the most suitable candidates for recent outburst tracers.

Acknowledgment

The study is supported by RFBR grant 17-02-00644.

References

Akimkin, V., Zhukovska, S., Wiebe, D., *et al.* 2013, *ApJ*, 766, 8
Audard, M., Ábrahám, P., Dunham, M. M., *et al.* 2014, *Protostars and Planets VI*, 387
Bosman, A. D., Walsh, C., and van Dishoeck, E. F. 2018, *A&A*, 618, A182
Dunham, M. M., & Vorobyov, E. I. 2012, *ApJ*, 747, 52
Molyarova, T., Akimkin, V., Semenov, D., *et al.* 2017, *ApJ*, 849, 130
Molyarova, T., Akimkin, V., Semenov, D., *et al.* 2018, *ApJ*, 866, 46
Rab, C., Elbakyan, V., Vorobyov, E., *et al.* 2017, *A&A*, 604, A15

Origins: From the Protosun to the First Steps of Life
Proceedings IAU Symposium No. 345, 2018
B. G. Elmegreen, L. Viktor Tóth & M. Güdel, eds.

doi:10.1017/S1743921319002242

Disks around FUor-type young eruptive stars with ALMA

Fernando Cruz-Sáenz de Miera[1], Ágnes Kóspál[1,2], Péter Ábraham[1], Hauyu Baobab Liu[3] and Michihiro Takami[4]

[1]Konkoly Observatory, Budapest, Hungary
email: cruzsaenz.fernando@csfk.mta.hu
[2]MPIA, Heidelberg, Germany
[3]ESO, Garching, Germany
[4]ASIAA, Taipei, Taiwan

Abstract. A long-standing problem of the general paradigm of low-mass star formation is the "luminosity problem": protostars are less luminous than theoretically predicted. One possible solution is that the accretion process is episodic. FU Orionis-type stars (FUors) are thought to be the visible examples for objects in the high accretion state and it is still debated what physical mechanism triggers the phenomenon. For many of these objects their disk properties are still largely unknown so we conducted a deep, high spatial resolution (down to 20 au) ALMA Band 6 (1.3 mm) dust continuum survey of a sub-sample of known FUors. Here we present preliminary results of our survey, including the mass, size and spectral slope of each disk.

Keywords. stars: pre–main-sequence, stars: circumstellar matter, techniques: interferometric

1. Introduction

Young eruptive stars are low-mass pre-main sequence stars showing large outbursts in visible light (ΔV $\sim$ 5). See Audard *et al.* (2014) for a review of these objects. The cause for these events is thought to be a temporary increase of accretion rate going from 10^{-8} to 10^{-4} $M_{\odot}$ year^{-1}. The duration of these outbursts is ~100s years; therefore, they contribute significantly to the final stellar mass. The outbursts clear the material surrounding the star, helping the transition from the embedded Class I to disk-only Class II. Recent studies have shown that FUor disks are smaller and more massive than previously thought (Cieza *et al.* 2018, and Liu *et al.* 2018).

2. Observations & Analysis

ALMA Band 6 with two continuum windows plus ^{12}CO, ^{13}CO and C^{18}O using the 7m and 12m arrays, with baselines from ~9m to ~3600m. Here we present only the continuum results. Self-calibration of phases and amplitudes for targets with high enough SNR. Measured radii, inclination and position angles by fitting a 2D Gaussian model convolved with the synthesized beam. Estimated disk masses assuming 100:1 gas-to-dust ratio, $\kappa = 2.2$ cm^2 g^{-1} and $T_{dust} = 50$ K. The α spectral index was calculated using the two continuum windows.

3. Results

The L1551 IRS 5 image shows the two circumstellar disks if the system, N and S, separated by 51 au, and a resolved circumbinary ring (CBR), extending from ~100 to ~200 au. The circumstellar disks have radii of 9.4 au and 6.9 au for N and S, respectively.

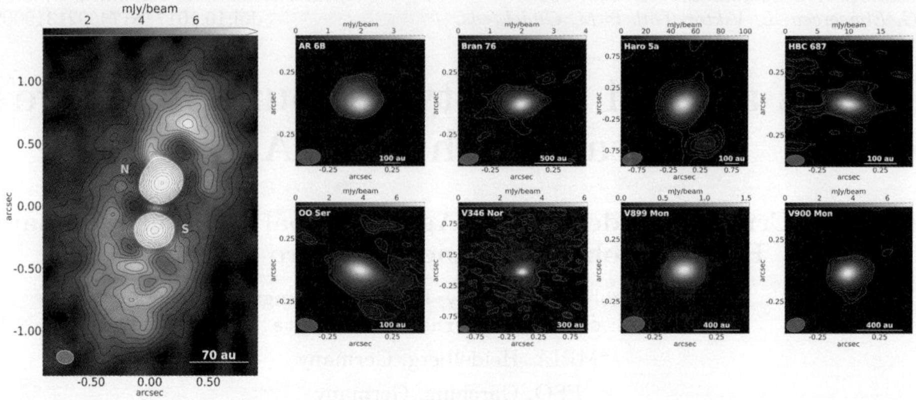

Figure 1. Continuum maps for L1551 IRS 5, AR6B, Bran 76, Haro 5a, HBC 687, OO Ser, V346 Nor, V899 Mon and V900 Mon.

N, S and the CBR have inclinations of 47, 39 and 45°, and PAs of 148, 150 and 150°, respectively, i.e. they are aligned, as previously determined by Lim *et al.* (2016). N shows extension towards the north which appears to connect it to the CBR. The CBR has a spectral index $\alpha = 3.3$; N and S have $\alpha \approx 2$; the rest of the sample has typical α value between 2.2 and 3.8, with the exception of AR 6B with $\alpha \approx 1.4$. The other FUors have disks ranging from 20 to 50 au with the exception of AR 6B with a radius of 70 au. The masses of all the circumstellar disks in our sample range from 10^{-4} to 10^{-3} $M_\odot$.

4. Conclusions

AR 6B, Bran 76, OO Ser and V899 Mon show dust extensions at a $\sim$10σ level. Haro 5a shows a faint companion 0.7" to the southwest. V346 Nor shows a much larger and fainter envelope. The values we report are 2–3 orders of magnitude below the values reported by Cieza *et al.* (2018); however, because of high optical depths and the uncertainty on the dust parameters, our mass estimates should be taken as lower limits. Difference of spectral index between circumbinary ring and circumstellar disks in L1551 IRS 5 can be an indication of grain growth or of high optical depth in the circumstellar disks. The other FUor disks show a wide range of values for the spectral index without a clear trend. A more detailed analysis of L1551 IRS 5 will be presented in Cruz-Sáenz de Miera *et al.* (in preparation) and for the full sample in Kóspál *et al.* (in preparation). This project has received funding from the European Research Council (ERC) under the European Union's Horizon 2020 research and innovation programme under grant agreement No 716155 (**SACCRED**, PI: Á. Kóspál).

References

Audard, M., Ábrahám, P., Dunham, M. M., Green, J. D., Grosso, N., Hamaguchi, K., Kastner, J. H., Kóspál, Á., Lodato, G., Romanova, M. M., Skinner, S. L., Vorobyov, E. I. & Zhu, Z. 2005, *Protostars and Planets VI*, pp. 387–410

Cieza, L. A., Ruíz-Rodríguez, D., Perez, S., Casassus, S., Williams, J. P., Zurlo, A., Principe, D. A., Hales, A., Prieto, J. L., Tobin, J. J., Zhu, Z. & Marino, S. 2018, *MNRAS*, 474, 4347

Lim, J., Yeung, P. K. H., Hanawa, T., Takakuwa, S., Matsumoto, T. & Saigo, K. 2016, *ApJ*, 826, 153

Liu, H. B., Dunham, M. M., Pascucci, I., Bourke, T. L., Hirano, N., Longmore, S., Andrews, S., Carrasco-González, C., Forbrich, J., Galván-Madrid, R., Girart, J. M., Green, J. D., Juárez, C., Kóspál, Á., Manara, C. F., Palau, A., Takami, M., Testi, L. & Vorobyov, E. I. 2018, *A&A*, 612, A54

Origins: From the Protosun to the First Steps of Life
Proceedings IAU Symposium No. 345, 2018
B. G. Elmegreen, L. Viktor Tóth & M. Güdel, eds.

doi:10.1017/S1743921319002254

Pre- and protostellar cores in the Rosette Nebula

R. Bőgner[1], T. Csengeri[2], M. Wienen[3], N. Schneider[4], J. Montillaud[5] and L. V. Tóth[1,6]

[1]Department of Astronomy, Eötvös Loránd University, 1/A Pázmány Péter sétány, 1117 Budapest, Hungary
email: r.bogner@astro.elte.hu

[2]Max-Planck-Institut für Radioastronomie, 69 Auf dem Hügel, 53121 Bonn, Germany

[3]College of Engineering, Mathematics and Physical Sciences, University of Exeter, North Park Road, EX4 4QL Exeter, UK

[4]I. Physikalische Institut, University of Cologne, 77 Zuelpicher Str., 50937 Cologne, Germany

[5]Institut Utinam, CNRS UMR 6213, OSU THETA, Université de Franche-Comté, 41 bis avenue de l'Observatoire, 25000 Besançon, France

[6]Konkoly Observatory, Research Centre for Astronomy and Earth Sciences, Hungarian Academy of Sciences, 15-17 Konkoly-Thege Mikló út, 1121, Budapest, Hungary

Abstract. Recent theories on the formation of the Solar System turned the attention to the study of low mass cloud cores in massive star forming regions. The Rosette Molecular Cloud is a well-known star forming area having highly filamentary structure with dense cores covering a wide range of masses. These pre- and protostellar cores were observed by Herschel and key core properties were derived from its data. With the Effelsberg 100m telescope a sample of these cores with masses ranging between 3-40 $M_\odot$ were observed in ammonia inversion lines. In this work we are examining the correlations between these two datasets with the aim of gaining insight of the processes behind the star formation of the region.

Keywords. stars: formation, radio lines: ISM

1. Introduction

The earliest stages of high-mass star formation is an actively researched area. High-mass stars are rarer according to the initial mass function, they are found at larger distances than low-mass star forming regions and evolve on a much shorter timescale in complex and dynamic environments. They originate in clusters and since a large fraction of all stars form in such environments, the very first stages of high mass star formation can represent the initial condisitons for star formation in general (Sridharan *et al.* 2005). The density of high-mass star forming regions is around the critical NH_3 volume density ($\sim 10^4\,\mathrm{cm}^{-3}$), therefore NH_3 is suitable for investigating the parameters of high-mass star forming cores without the surrounding lower density structures.

The Rosette Nebula at a distance of 1.6 kpc is an emission nebula where the central OB cluster NGC 2244 is blowing a cavity into the surrounding molecular cloud (Schneider *et al.* 2010). The role of triggered star formation in this area is debated: Cambrésy *et al.* (2013) claimed that the evolution of the Rosette complex is not governed by the influence of its OB star population, and that simply the morphological appearance of an active region is not sufficient to conclude much about the triggering role.

With its wavelength coverage the Herschel satellite could detect massive young stellar objects at all evolutionary stages (Montillaud *et al.* 2015) and was able to detect the

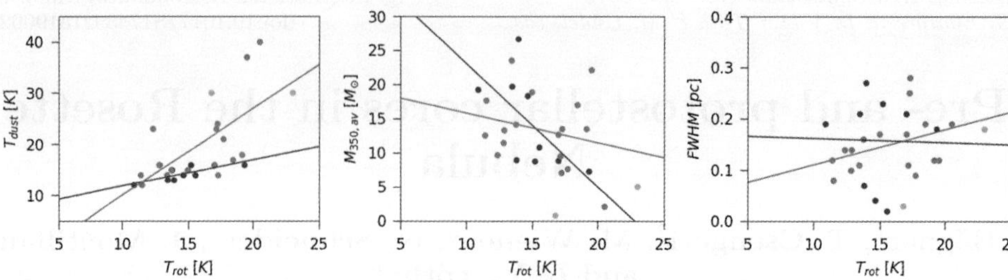

Figure 1. Correlation plots of rotational and dust temperature, mass and physical size of the cores, respectively. Green: warm-starless, blue: prestellar, red: protostellar cores. Weak correlation can be seen between the rotational and the dust temperatures by types.

precursors of O and B stars. Cores in early evolutionary phases of the Rosette Molecular Cloud were observed by Herschel and discussed by Motte *et al.* (2010) and Hennemann *et al.* (2010). The Effelsberg NH_3 observations make it possible to calculate properties of the molecular gas associated with the cores, complementing Herschel data.

2. Results and discussion

Single-point NH_3 observations were carried out in 2011 with the Effelsberg 100m telescope. 28 of the observed cores had detectable NH_3 (1,1) and (2,2) lines with an average rms level of 0.01 K. Data reduction was performed with the methods of Wienen *et al.* (2012). Rotational and kinetic temperatures, thermal linewidths and NH_3 column densities were calculated.

From Herschel data Motte *et al.* (2010) derived physical properties of the cores (dust temperature, mass, FWHM size) and classified them using Spitzer 24 m data into warm-starless, pre- and protostellar sources, our sample having 2, 9 and 17 sources, respectively.

Acknowledgements

This work is based on observations with the 100-m telescope of the MPIfR (Max-Planck-Institut für Radioastronomie) at Effelsberg. Herschel is an ESA space observatory with science instruments provided by European-led Principal Investigator consortia and with important participation from NASA. We thank the French ministry of foreign affairs (French embassy in Budapest) and the Hungarian national office for research and innovation (NKFIH) for financial support (Balaton program 40470VL/2017-2.2.5-TÉT-FR-2017-00027). This research is supported by the Hungarian OTKA grant NN111016. R.B. is grateful to the Physics Doctoral School of Eötvös Loránd University, Erasmus+, the French government, CampusFrance and the IAU for supporting her work. T.Cs. acknowledges support from the *Deutsche Forschungsgemeinschaft, DFG* via the SPP (priority programme) 1573 'Physics of the ISM'.

References

Sridharan, T. K., Beuther, H., Saito, M., Wyrowsky, F., & Schilke, P. 2005, *ApJ*, 634L, 57S
Motte, F., Zavagno, A., Bontemps, S., *et al.* 2010, *A&A*, 518, L77
Hennemann, F., Motte, F., Bontemps, S. *et al.* 2010, *A&A*, 518, L84
Schneider, N., Motte, F., Bontemps, S. *et al.* 2010, *A&A*, 518, L83
Wienen, M., Wyrowski, F., Schuller *et al.* 2012, *A&A*, 544, A146
Cambrésy, L., Marton, G., Fehér, O., Tóth, L. V., Schneider, N. 2013, *A&A*, 557, A29
Montillaud, J., Juvela, M., Rivera-Ingraham, A. *et al.* 2015, *A&A*, 584, A92

Origins: From the Protosun to the First Steps of Life
Proceedings IAU Symposium No. 345, 2018
B. G. Elmegreen, L. Viktor Tóth & M. Güdel, eds.

doi:10.1017/S1743921319002266

The TOP-SCOPE survey of Planck Galactic Cold Clumps: The 200 brightest compact sources of Planck

R. Bőgner[1], T. Liu[2], D. J. Eden[3] and L. V. Tóth[1,4]

[1]Department of Astronomy, Eötvös Loránd University,
1/A Pázmány Péter sétány, 1117 Budapest, Hungary
email: r.bogner@astro.elte.hu

[2]Korea Astronomy and Space Science Institute,
776 Daedeokdaero, Yuseong-gu, Daejeon 34055, Republic of Korea

[3]Astrophysics Research Institute, Liverpool John Moores University,
IC2, Liverpool Science Park, 146 Brownlow Hill, Liverpool L3 5RF, UK

[4]Konkoly Observatory of the HAS, 15-17 Konkoly Thege Miklós út, 1121 Budapest, Hungary

Abstract. The Planck all-sky submillimetre observations have made it possible to study Galactic cold clumps in diverse environments, to probe dust properties and to examine the earliest stages of star formation. The TOP-SCOPE joint survey program aims to statistically study the evolution of molecular clouds and the initial conditions of star formation in a wide variety of environments. In this work we carry out an investigation of the 200 brightest compact sources detected by Planck.

Keywords. surveys, ISM: clouds, ISM: abundances, stars: formation

1. Introduction

Regarding star formation both the analysis of individual star-forming regions and survey-driven research have become of huge importance. From the results of Planck an all-sky catalogue of Galactic cold clump candidates (PGCC) have been made (Planck Collaboration 2016). TOP-SCOPE is a joint follow-up survey program targeting around 2000 PGCCs in CO isotopologues and around 1000 PGCCs in 850 micrometer continuum emission using ground-based radio telescopes.

The TRAO Observations of Planck cold clumps (TOP - PI: Liu, T.) is a Key Science Program of the Taeduk Radio Astronomy Observatory surveying the J=1-0 transitions of 12CO and 13CO toward 2000 PGCCs. The goals are to find dense CO clumps, study the universality of filaments in the cold interstellar medium and cloud evolution, investigate the role of magnetic fields, turbulence and gravity and to examine the dynamical effects of stellar feedback and cloud collisions on star formation (Liu *et al.* 2018).

The Scuba-2 Continuum Observations of Pre-protostellar Evolution (SCOPE - PI: Liu, T.) is one of the large programs at the JCMT of the East Asia Observatory, a survey of 1000 PGCCs at 850 μm. The main goal of the survey is to investigate dense clumps in different environments, study the roles of filaments in dense core formation, examine dust properties and detect rare populations of dense clumps (Liu *et al.* 2018).

2. Data

The TOP200 sample consists of the 200 brightest sources of the Planck Compact Source Catalog that can be observed by telescopes from the northern hemisphere. These are not

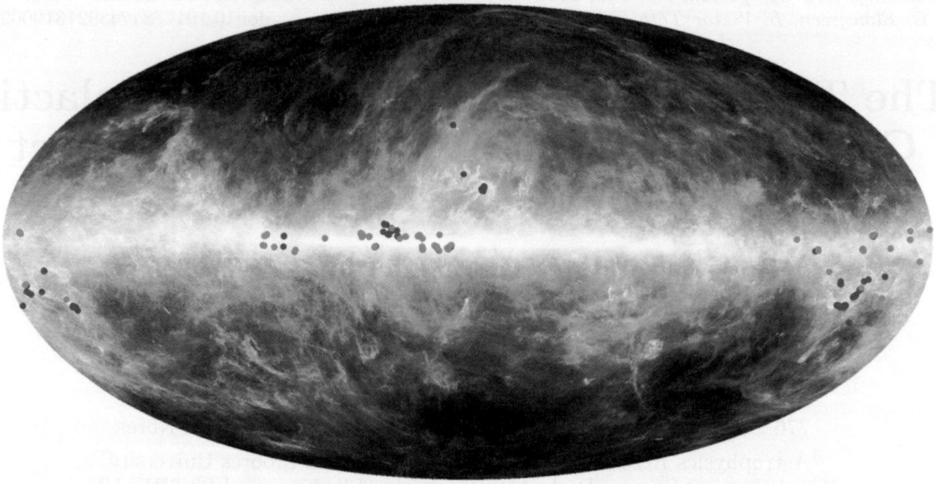

Figure 1. Distribution of the TOP200 sources on the Planck 857 GHz all-sky map in histogram-equalized colour mapping. Blue dots: TOP sources, red dots: sources in the SCOPE fields.

necessarily PGCCs, most of them are in star forming regions: the sample contains more evolved sources, starless core candidates and protostellar cores as well as UCHII regions. 80 of the sources have been observed by the SCOPE survey.

The primary aim of the project is to make a catalog and a statistical study of the sources as in the work of Eden *et al.* (2017). Using Herschel and JCMT observations we are investigating the dust properties of the cores, making column density and dust temperature maps. We are also investigating the young stellar objects in the vicinity of the sources to examine their influence on such star-forming sites. We aim to compare the properties of this sample to PGCCs to see how star formation efficiency, clustering behavior and core masses evolve with time.

Acknowledgements

The James Clerk Maxwell Telescope is operated by the East Asian Observatory on behalf of The National Astronomical Observatory of Japan; Academia Sinica Institute of Astronomy and Astrophysics; the Korea Astronomy and Space Science Institute; the Operation, Maintenance and Upgrading Fund for Astronomical Telescopes and Facility Instruments, budgeted from the Ministry of Finance (MOF) of China and administrated by the Chinese Academy of Sciences (CAS), as well as the National Key R&D Program of China (No. 2017YFA0402700). Additional funding support is provided by the Science and Technology Facilities Council of the United Kingdom and participating universities in the United Kingdom and Canada. Herschel is an ESA space observatory with science instruments provided by European-led Principal Investigator consortia and with important participation from NASA. This research was supported by the Hungarian OTKA grant NN111016. R.B. is grateful to the Physics Doctoral School of Eötvös Loránd University, Erasmus+, the French government, CampusFrance and the IAU for the support.

References

Planck Collaboration 2016, *A&A*, 594, A28
Liu, T. *et al.* 2018, *ApJS*, 234, 28L
Eden, D. J. *et al.* 2017, *MNRAS*, 469, 2163E

Origins: From the Protosun to the First Steps of Life
Proceedings IAU Symposium No. 345, 2018
B. G. Elmegreen, L. Viktor Tóth & M. Güdel, eds.

doi:10.1017/S1743921319002588

Time-variability and disk geometry in Herbig Ae/Be disks

R. Szakáts[1], **P. Ábrahám**[1] **and Á. Kóspál**[1,2]

[1]Konkoly Observatory, Research Centre for Astronomy and Earth Sciences, Hungarian Academy of Sciences, Konkoly-Thege Miklós út 15-17, H-1121, Budapest, Hungary
email: szakats.robert@csfk.mta.hu

[2]Max Planck Institute for Astronomy, Königstuhl 17, D-69117 Heidelberg, Germany

Abstract. The origin of optical-infrared variability in young, intermediate mass Herbig Ae/Be stars is linked to their circumstellar disk. Therefore, variability could serve as a diagnostic tool to constrain the structure and dynamics of the (inner) disk. Here we discuss this diagnostic potential, and report some preliminary results from our coordinated $BVRIJHK_S$ and Spitzer monitoring observations of nine Herbig Ae stars. We aim to understand the response of the inner disks thermal emission on the changing stellar irradiation, and to separate it from UX Orionis-type fading events, which also provide information on the disk. This project is a pilot study for the era of time domain astronomy of young stars, opened by Kepler K2, Gaia, ASAS-SN, TESS, Spitzer, WISE, and JWST.

Keywords. stars: pre–main-sequence, circumstellar matter

1. Motivation: linking variability with disk structure

The three main physical processes causing variability are all connected to the disk: (a) fluctuating mass accretion from the disk changes the luminosity of the central source (optical light curve), and leads to variable irradiation of the disk (infrared light curve); b) variable emitting surface of the disk, e.g. by temporary dust clouds above the disk surface, increase the thermal infrared emission; and c) UX Ori (UXor)-type eclipses, when dust clumps in the inner disk pass through the line-of-sight. Radiative transfer models quantitatively predict the wavelength dependent light curves (Kesseli *et al.* 2016). Observationally, we looked for a correlation between the variability amplitude at $6 - 8\ \mu$m (Kóspál *et al.* 2012) and the disk flaring (Meeus *et al.* 2001). We found higher amplitudes in flatter (Group IIa) disks. Plotting the same variability amplitudes wrt. the radius of the interferometric emitting area (Menu *et al.* 2015) implies that variability in disks with more centrally concentrated brightness distribution is stronger. These results imply a physical link between variability and disk structure, and suggest that variability may provide information about the disk structure and dynamics on the smallest spatial scales, a few au from the star.

2. The Konkolyvar project

In order to decide which physical processes are responsible for the observed variability in a particular system, and to deduce disk properties, simultaneous optical-infrared observations are needed. We carried out a 2-week-long quasi-simultaneous optical–infrared photometric monitoring program with daily cadence, in 2009 Oct–Nov. We targeted nine

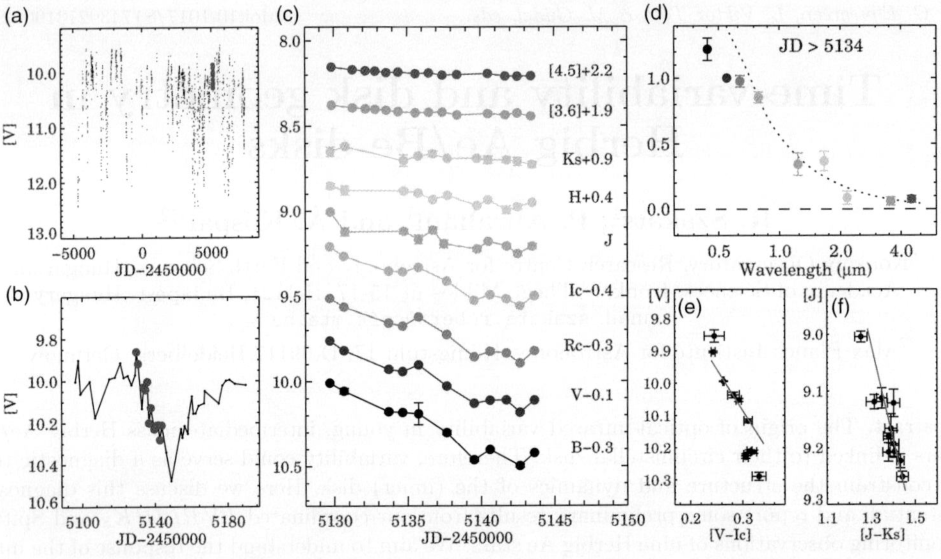

Figure 1. Multiwavelength light curves for BF Ori from the Konkolyvar project.

Herbig Ae stars and collected observations from Piszkéstető Observatory (Hungary), Teide Observatory (Spain), and at 3.6 and 4.5 μm from the Spitzer Space Telescope.

3. The physical origin of the variability, and the response of the disk

As an example, Fig. 1 presents our observations for BF Ori, where panel *c* shows the multiwavelength light curves. Comparison with earlier/later V-band observations (panel *b*) outlines that our first two epochs coincided with a small brightness peak, while later the source entered a deep minimum. Panels *d*, *e*, and *f* imply that the fading occurred along the extinction path, and was probably a typical UXor event. The faintest data points exhibit optical colors deviating from the extinction curve: here we see the well-known blueing of UXors, caused by scatter light.

We also compared the Spitzer 3.6/4.5 μm data with the I-band light curve. If the whole inner disk was obscured in an UXor event, shapes of the I-band and Spitzer light curves would be identical, apart from a lower amplitude in the mid-infrared according to the extinction law (e.g., Cardelli *et al.* 1989). This is the case for BF Ori at JD > 2455135, but the first few points exhibit the response of the disk with unexpectedly high amplitude for the small V-band brightness peak. Analysing the other stars from the Konkolyvar sample we found large variety: in RR Tau most variability is related to extinction changes, while in UX Ori the disk emission is virtually unrelated to the optical flux changes, because in the second part of our campaign the optical flux increases while the mid-infrared flux drops. The detailed analysis of these results in terms of disk geometry and dynamics is ongoing.

Acknowledgements

This project has received funding from the European Research Council (ERC) under the European Union's Horizon 2020 research and innovation programme under grant agreement No 716155 (SACCRED). This work is based in part on observations made with the Spitzer Space Telescope, which is operated by the Jet Propulsion Laboratory, California Institute of Technology under a contract with NASA.

References

Cardelli, J. A., Clayton, G. C., & Mathis, J. S. 1989, *ApJ*, 345, 245
Kesseli, A. Y., Petkova, M. A., Wood, Kenneth *et al.* 2016, *ApJ*, 828, 42
Kóspál, Á., Ábrahám, P., Acosta-Pulido, J. A. *et al.* 2012, *ApJ Suppl. Ser.*, 201, 11
Meeus, G., Waters, L. B. F. M., Bouwman, J. *et al.* 2001, *A&A*, 365, 476
Menu, J., van Boekel, R., Henning, Th. *et al.* 2015, *A&A*, 581, A107

Origins: From the Protosun to the First Steps of Life
Proceedings IAU Symposium No. 345, 2018
B. G. Elmegreen, L. Viktor Tóth & M. Güdel, eds.
doi:10.1017/S1743921319002825

Konkoly optical catalog of young stars for the Gaia Photometric Science Alerts

E. Varga-Verebélyi, M. Kun, E. Szegedi-Elek, P. Ábrahám, J. Varga, Cs. Kiss, Á. Kóspál, G. Marton and L. Szabados

Konkoly Observatory, Konkoly Thege Miklós út 15-17, H-1121 Budapest, Hungary
email: verebelyi.erika@csfk.mta.hu

Abstract. Our aim is to present a new and so far most complete catalog of optically selected young stars. The basis of this work is an extensive literature search for young stars in all the known nearby (< 2 kpc) star forming regions, included in the Handbook of Star Forming Regions Reipurth (2008a,b), and in 67 additional catalogs. We collected data on known young, pre-main-sequence stars detected in optical bands. The catalog contains the celestial coordinates, object names, names of the enclosing star forming region, identification methods, distances, and other information (e.g., references, binarity) for 15208 young stellar objects. It is already in use by the Gaia Photometric Science Alerts Team to identify variable young stars in the Gaia data. Our catalog was cross-correlated with the Gaia DR2 and we obtained flux and distance estimations for 86% of the stars.

Keywords. catalogs, stars: pre–main-sequence, techniques: miscellaneous

---

The Gaia Photometric Science Alerts† is an all-sky photometric transient survey, based on the measurements of the Gaia space observatory (Gaia Collaboration *et al.* 2016). The team, responsible for the alerts, runs a dedicated data processing pipeline to look for transient events in the Gaia data and alert immediately if there are noticeable changes in brightness of detected stars.

During the alert identification the objects are classified into basic transient types (e.g., supernova, cataclysmic variable, young stellar object), and the alerts are published online to promote follow-up observations. This identification process is not very effective when it comes to young stars, as it is optimized for transient detection (e.g., supernovae). In contrast, the variability patterns of young stars are quite diverse, and more complex algorithms are needed for an efficient event detection. In order to support the more efficient detection and classification of variable young stars among the alerts, we compiled a catalog of optically confirmed young stellar objects (YSOs), which is so far the most complete catalog of this kind. There are some other YSO catalogs existing like Marton *et al.* (2016) or Tóth *et al.* (2014), but most of them on the one hand, call their sources only YSO candidates since there are serious amount of contamination and uncertainty in the classifications on the other hand, these based on infrared detections and many of the sources do not have optical counterparts. In contrast, our catalog includes known young stars, primarily identified at optical wavelengths, from numerous literature sources: 67 individual catalogs and the two volumes of the Handbook of Star Forming Regions Reipurth (2008a, 2008b).

We collected the YSO lists from the original catalogs, and loaded them into a database using a common data format. Then we removed the duplicates using a 1.5 arcsec searching

† http://gsaweb.ast.cam.ac.uk/alerts

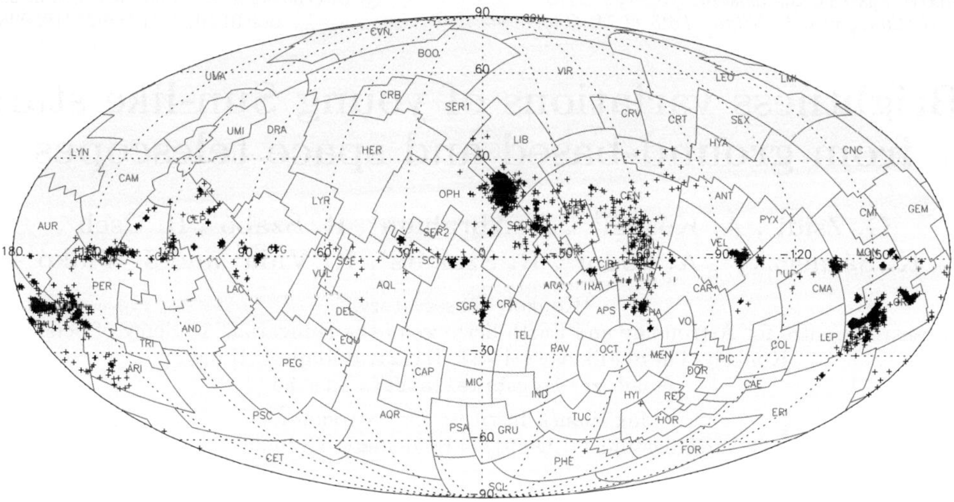

Figure 1. Full sky map of the objects from the Konkoly optical catalog of young stars. The background map shows the boundaries of the constellations.

radius, and flagged the known multiple systems. The catalog contains the object names, celestial coordinates, names of the embedding star forming region, identification methods, references, and binarity flags for 15208 YSOs, which are all confirmed young stars. Fig 1 shows the positions of the sources on the galactic sky.

We cross-correlated our catalog with the Gaia Data Release 2 (DR2), Gaia Collaboration *et al.* (2018). As a result, our catalog also contains optical (mean G) magnitudes for 87% and distance estimates for 79% of the objects. The catalog is already in use in the form presented here by the Gaia Photometric Science Alerts Team to help identify YSOs among the many variable sources picked up by the alert algorithm. Nevertheless, we constantly keep the catalog up to date until we publish it soon.

Acknowledgement

This project has been supported by ESA Contract No. 4000121377/17/NL/CBi. We acknowledge ESA Gaia, DPAC and the Photometric Science Alerts Team (http://gsaweb.ast.cam.ac.uk/alerts). This research made use of the SIMBAD database, the VizieR catalogue access tool, and the cross-match service provided by CDS, Strasbourg, France.

References

Gaia Collaboration, Prusti, T., de Bruijne, J. H. J., *et al.* 2016, *A&A*, 595, A1

Gaia Collaboration, Brown, A. G. A., Vallenari, A., *et al.* 2018, *A&A*, 616, A1

Marton, G., Tóth, L. V., Paladini, R., *et al.* 2016, *MNRAS*, 458, 3479

Reipurth, B. 2008a, Handbook of Star Forming Regions, Volume I., ASP Monograph Publications, Vol. 4. Ed. Bo Reipurth

Reipurth, B. 2008b, Handbook of Star Forming Regions, Volume II., ASP Monograph Publications, Vol. 5. Ed. Bo Reipurth

Tóth, L. V., Marton, G., Zahorecz, S., *et al.* 2014, *PASJ*, 66, 17

Origins: From the Protosun to the First Steps of Life
Proceedings IAU Symposium No. 345, 2018
B. G. Elmegreen, L. Viktor Tóth & M. Güdel, eds.

doi:10.1017/S1743921319002837

Brightness variations of young Sun-like stars from ground-based and space telescopes

G. Zsidi[1], Á. Kóspál[1,2], P. Ábrahám[1], R. Szabó[1], B. Cseh[1], K. Sárneczky[1], Á. Sódor[1], R. Szakáts[1], K. Vida[1] and J. Vinkó[1]

[1]Konkoly Observatory,
Research Centre for Astronomy and Earth Sciences of the Hungarian Academy of Sciences,
Konkoly-Thege Miklós út 15-17, H1121, Budapest, Hungary
email: zsidi.gabriella@csfk.mta.hu

[2]Max Planck Institute for Astronomy,
Königstuhl 17, D-69117, Heidelberg, Germany

Abstract. Young stellar objects often show photometric variability, which is well examined at optical wavelengths, but more and more infrared data are also available. The wavelength dependence of the variability carries information on the physical cause of the changing brightness. Here, we examine seven T Tauri-type stars known for their large amplitude variability selected from the Campaign 13 field of the Kepler K2 mission. We complemented the K2 light curves by multifilter optical monitoring observations made with the 90 cm Schmidt telescope of Konkoly Observatory, and by 3.6 and 4.5 μm infrared photometry with a 20 hours cadence using the Spitzer Space Telescope. We found that the wavelength dependence of the observed variability is not consistent with changing interstellar extinction. We suggest that the brightness changes are due to variable accretion, causing a variable illumination of the inner disk.

Keywords. stars: pre–main-sequence, circumstellar matter

T Tauri stars are young Sun-like stars, which are still being built up by accreting mass from the surrounding circumstellar disk made of gas and dust. They often show photometric variability (e.g. Herbst *et al.* 1994, Morales-Calderón *et al.* 2011), which may be due to non-steady accretion from the circumstellar disk onto the star, but other processes, such as rotating cold or hot spots, or eclipses caused by dust clouds moving in and out of the line-of-sight can also change the observed brightness. The observations at different wavelengths trace different regions and processes in the system, and they are all connected through the disk (Sicilia-Aguilar *et al.* 2016).

We selected seven T Tauri-type stars (Haro 6-10, HBC 393, HL Tau, XZ Tau, UZ Tau E, DR Tau and DQ Tau) from the Campaign 13 field of the Kepler K2 mission, which covered part of the Taurus star forming region. We obtained high precision optical photometry with 1 minute cadence between 2017 March 8 and May 27. In order to extend the wavelength coverage, we complemented the K2 data with 3.6 and 4.5 μm infrared observations at a 20 hours cadence using the Spitzer Space Telescope between 2017 May 17 and May 28. We also carried out monitoring observation in B, V, R_C and I_C bands with the 90 cm Schmidt telescope of Konkoly Observatory (Hungary) between 2017 March 8 and April 10. Here, we present results on two of our targets: DR Tau and HL Tau.

DR Tau is a highly accreting classical T Tauri star. The K2 light curve (Kp magnitudes are plotted in Fig. 1) shows variability on hourly timescale. Stochastic variations with peak-to-peak amplitude of 1.4 mag appear on this light curve. In contrast, HL Tau is a

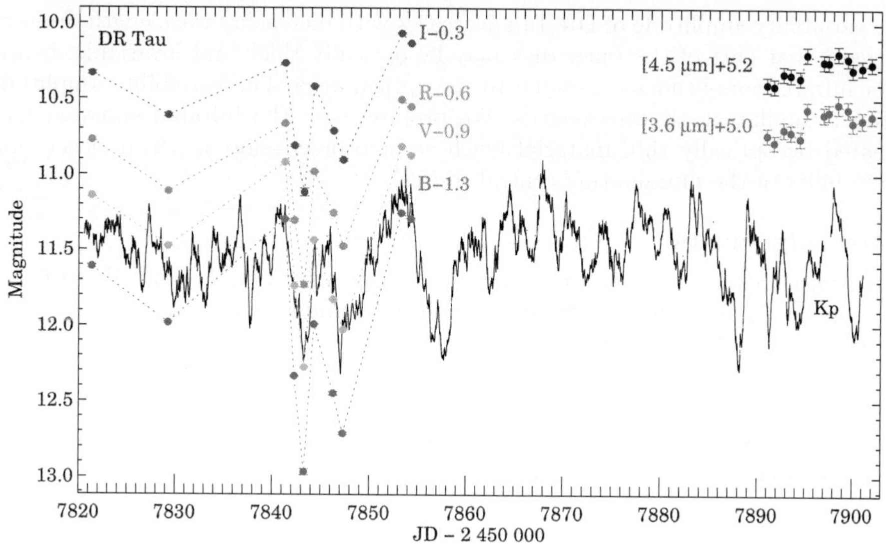

Figure 1. Multiwavelength light curves of DR Tau.

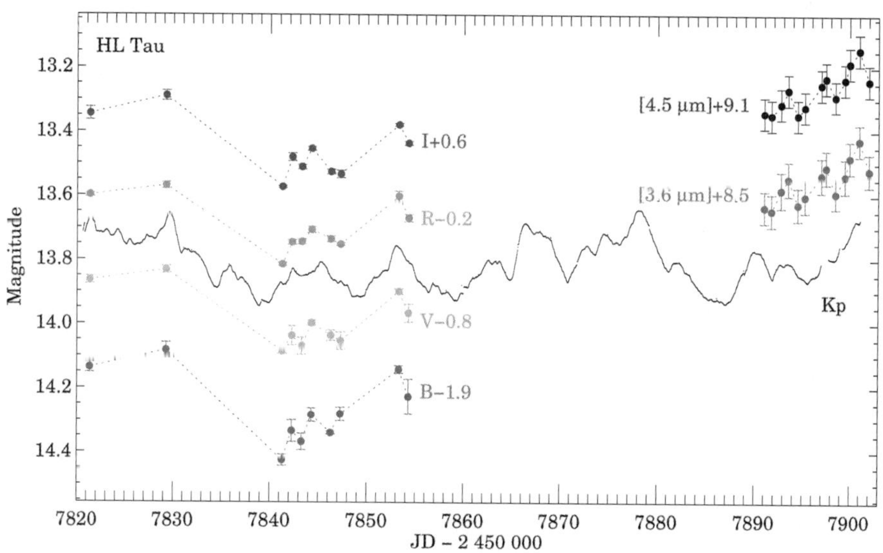

Figure 2. Multiwavelength light curves of HL Tau.

deeply embedded classical T Tauri star. Its K2 light curve (Fig. 2) shows small amplitude variations on a timescale of a few days with peak-to-peak amplitude of 0.3 mag.

Our ground-based multifilter photometry revealed that the shape of the light curves are similar at all wavelengths, although the amplitude of the variability may differ. By plotting the data points as a function of the Kp magnitude, the plots could be well fitted with a line whose slope gives the variability amplitude relative to the Kp band. By comparing the measured amplitudes to the standard interstellar extinction law (Cardelli *et al.* 1989), we can exclude variable extinction as the cause of the variability. A possible explanation is variable accretion. The accretion hot spot on the star causes the optical variability, while the changing illumination causes variability of the inner disk emission observable at the Spitzer wavelengths.

The variability amplitude of DR Tau decreases with increasing wavelength, from which we suggest that part of the inner disk may be optically thick and invariable, hence the smaller infrared amplitudes compared to the optical ones. The variability amplitudes of HL Tau is similar at all wavelengths. We propose that the infrared emission may be dominated by optically thin material (such as halo or envelope), which closely follows the variability of the illuminating central source.

Acknowledgements

This project has received funding from the European Research Council (ERC) under the European Union's Horizon 2020 research and innovation programme under grant agreement No 716155 (SACCRED).

References

Herbst, W., Herbst, D. K., Grossman, E. J., & Weinstein, D. 1994, *AJ*, 108, 5
Morales-Calderón, M, Stauffer, J. R., Hillenbrand, L. A. *et al.* 2011, *ApJ*, 733, 50
Sicilia-Aguilar, A., Banzatti, A., Carmona, A. *et al.* 2016, *PASA*, 33, 59
Cardelli, J. A., Clayton, G. C. & Mathis, J. S. 1989, *ApJ*, 345, 245

Origins: From the Protosun to the First Steps of Life
Proceedings IAU Symposium No. 345, 2018
B. G. Elmegreen, L. Viktor Tóth & M. Güdel, eds.

doi:10.1017/S1743921319002953

Modeling the Transmission Spectra of WASP-31b

J. Chouqar[1], M. L. Morales[2], A. Daassou[1], A. Jabiri[1] and Z. Benkhaldoun[1]

[1]Oukaïmeden Observatory LPHEA Cadi Ayyad University BP 2390 Marrakech Morocco
email: jamila.chouqar@ced.uca.ma

[2]Department of Astronomy, Harvard-Smithsonian Center for Astrophysics, Cambridge, MA 02138, USA

Abstract. Wasp-31b is a planet of 0.48 Jupiter masses and 1.55 Jupiter radii, with orbital period of 3.4-days around a metal-poor, late-F-type, V = 11.7 dwarf star. The planet has a large atmospheric scale height that makes it a good target for transmission spectroscopy. Sing *et al* (2014) presented an optical and near-IR transmission spectrum of the atmosphere of WASP-31b obtained with the HST and show the presence of a strong potassium line. In contrast, Gibson *et al.* (2017) reports a spectrum of the atmosphere of WASP-31b, obtained with the FORS2 instrument on the VLT and find that there is no strong potassium line. Here, we take those two datasets and, using models, we try to find a case where both solutions are correct by considering different cloud scenarios.

Keywords. planets and satellites: atmospheres - planets and satellites: composition - planets and satellites: individual (WASP-31b)

1. Introduction

Many forefront facilities such as the Hubble Space Telescope and ESOs Very Large Telescope are being pointed at exoplanets to try to find out what their atmospheres are made of. WASP-31b is one of the exoplanets for which its atmosphere has been probed with the HST and VLT. The observed transmission spectrum of WASP-31b is mostly flat, implying that the planet has a cloudy atmosphere, but while one point from previous HST data (the green-circle data point) indicates the presence of a strong Potassium line, the new data from the VLT (small yellow circle) is incompatible with the HST result, showing no strong Potassium line. So how this discrepancy can be explained?

2. ATMO Model Grid

Using a 1D radiative-convective-chemical equilibrium model termed ATMO, Goyal, *et al.* (2017) generated a thousands of models for WASP-31b with varying C-to-O ratio, Temperature, Metallicity and various levels of cloud and haze coverage. We fitted all those models to the two sets of data, the Sing *et al.* (2015) data & the Gibson, *et al.* (2016) one, to find wich ones fit the data well. The two best fit models found have hazes and clouds, both have high metallicities, one has a 2xC/O ratio then the other and a temperature of about 1425k hotter then the other but none of those models can reproduce the strong potassium absorption in the transmission spectrum of WASP-31b observd with HST. See Fig. 2.

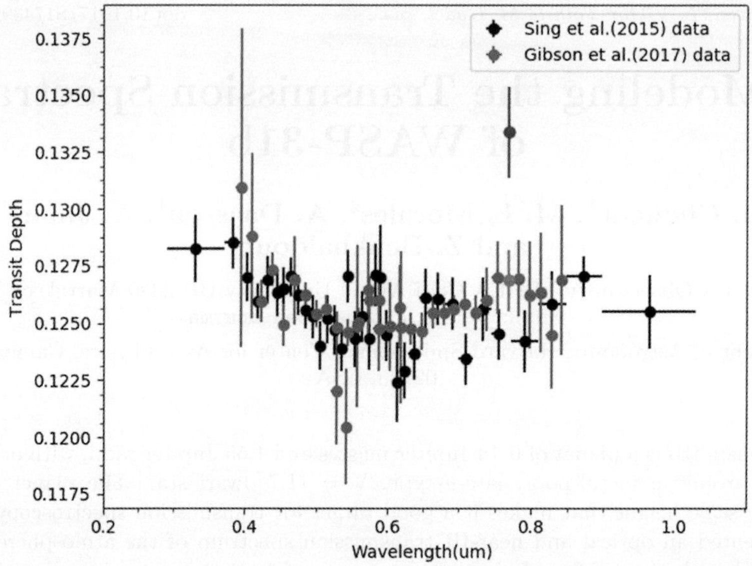

Figure 1. Transmission spectrum of WASP-31b. The black points are the STIS results reported from Sing *et al.* (2015) while The Brouwn one are the FORS2 results reported from Gibson *et al.* (2017).

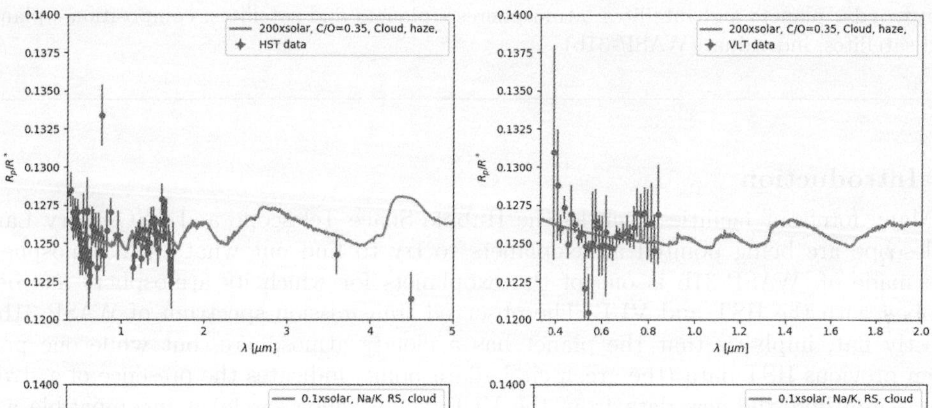

Figure 2. In the upper panel, the best fit models found for the atmosphere of WASP-31b using Atmo code. The best fit models found with Exo_Transmit for WASP-31b are shown in the bottom panel.

Table 1. Table showing the entire parameter space of the Exo_Transmit grid. The temperature is with respect to the planetary equilibrium temperature (T_{eq}).

	Temperature (K)	Metallicity (x solar)	N_a/K	Haze RS augmentation factor	Cloud cloud-top pressure in (Pa)
Sing *et al.* (2015)	1300	0.1	0.1	1	5000
Gibson et al. (2016)	1300	0.1	0.1	1	4800

3. Exo_Transmit Model Grid/Results

Using Exo_Transmit code (Kempton *et al.* (2017)), we generated more then 2000 models for the atmosphere of WASP-31b with varying the C-to-O ratio, the metallicity and Na/K ratio. We simulated those spectra with clouds by varying the cloud-top pressure. With this code, we fitted the two sets of data, the Sing *et al.* (2015) data & the Gibson, *et al.* (2016) one with the same parameters of the planet, see Table.1, exept for where the cloud deck is. As we see in Fig. 2, the two sets of data are consistent with low Sodium abundance $\frac{N_a}{K}$= 0.1 and with an atmospheric model of 0.1xsolar metallicity. The two observations (Sing *et al.* (2015) data & the Gibson, *et al.* (2016)) can be reproduced with the same atmospheric parameters for the planet, just changing the height of the cloud deck, with cloud-top pressure of 5000 pa and 4800 pa respectivelly. Because we are looking at the terminator of the planet, it could be possible that the cloud structure in the terminator was different in each observation epoch. So that means that the presence of a thik cloud in the epoch of Gibson observations obscured the absorption line of Potassium could explain the lack of Potassium feature in the transmission spectra of VLT data.

4. Conclusion

Briefly, the absence of the strong Potassium line in the epoch of Gibson observations could be explained by the presence of a tick cloud in that epoch.

References

E. M. Kempton *et al.* 2017, *PASP*, 129, 044402
J. M. Goyal, *et al.* 2018, *MNRAS*, 474, 5158
N. P. Gibson, *et al.* 2017, *MNRAS*, 467, 4591
Sing D. K., *et al.* 2015, *MNRAS*, 446, 2428

Origins: From the Protosun to the First Steps of Life
Proceedings IAU Symposium No. 345, 2018
B. G. Elmegreen, L. Viktor Tóth & M. Güdel, eds.

doi:10.1017/S1743921319002965

Searching for methylamine in Orion-KL using ALMA archival data

Harumi Minamoto[1], Yoko Oya[2], Hirota Tomoya[3] and Hideko Nomura[1]

[1]Department of Earth and Planetary Sciences, Tokyo Institute of Technology,
2-12-1 Ookayama, Meguro-ku, Tokyo 152-8551, Japan,
email: `minamoto.h@eps.sci.titech.ac.jp`

[2]Department of Physics, The University of Tokyo,
7-3-1 Hongo, Bunkyo-ku, Tokyo 113-0033, Japan, email: `oya@taurus.phys.s.u-tokyo.ac.jp`

[3]National Astronomical Observatory of Japan,
2-21-1 Osawa, Mitaka, Tokyo 181-8588, Japan, email: `tomoya.hirota@nao.ac.jp`

Abstract. Methylamine (CH_3NH_2) is the simplest amine and thought to be a potential interstellar precursor to the amino acid glycine (NH_2CH_2COOH). It is confirmed by the experimental work and in terms of exploration in the Solar system, CH_3NH_2 has been detected in two comets. However, in molecular clouds, a robust detection of CH_3NH_2 has been reported only for Sgr B2(N) so far, while a variety of complex organic molecules have been detected by radio observations in many star-forming regions. To search for CH_3NH_2, we used the ALMA Cycle 2 archival data toward Orion Kleinmann-Low nebula (Orion-KL) at Band 6 and found 5 candidate emission at the hot core region. By using the rotation diagram method, we evaluated its tentative column density and rotational temperature to be 4.9×10^{14} cm^{-2} and 102 K, respectively.

Keywords. astrobiology, astrochemistry, ISM: individual (Orion-KL)

1. Introduction

The simplest amine methylamine (CH_3NH_2) has been suggested as a precursor to the amino acid glycine (NH_2CH_2COOH). Theoretical and laboratory studies have demonstrated that glycine is formed on icy grain surfaces from CH_3NH_2 and CO_2 under UV irradiation (Holtom *et al.* 2005). However, in molecular clouds, a robust detection of CH_3NH_2 has been reported only for Sgr B2(N) (Halfen *et al.* 2013) so far, while a variety of complex organic molecules have been detected by radio observations.

Orion Kleinmann-Low nebula (Orion-KL) is known as the nearest high-mass star-forming region (at $d=$ 388 $\pm$ 5 pc, Kounkel *et al.* 2017) and one of the most prolific sources of line emission of a variety of complex organic molecules. Because of its richness in chemical composition, it has also been regarded as a suitable target for molecular line survey in the interstellar medium. Although there is a report of tentative detection of CH_3NH_2 (Pagani *et al.* 2017), detailed analysis has not been carried out, and so we conducted further verification of this species.

2. Results

We used the ALMA Cycle 2 archival data toward Orion-KL (#2013.1.00533.S) in Band 6. In Orion-KL, we found 5 candidate features (Table 1) of CH_3NH_2 lines in Hot core. Figure 1(a) shows one example of the intensity distribution maps. We extracted the spectrum from Hot core and show it in Figure 1(b).

Table 1. Observed rotational transitions of CH_3NH_2 in Orion-KL.

Fequency [GHz]	Sμ^2 [D^2]	E$_u$ [K]	Transition (J, K_a, Γ)	Comments
215.670	53.92	111.48	9, 2, E_{1-1} $\rightarrow$ 9, 1, E_{1+1}	Partially blended
217.758	129.88	182.05	12, 2, B_2 $\rightarrow$ 12, 1, B_1	
229.908	27.37	92.71	8, 2, A_2 $\rightarrow$ 8, 1, A_1	
235.735	82.06	92.76	8, 2, B_2 $\rightarrow$ 8, 1, B_1	
244.887	49.54	48.09	5, 2, B_1 $\rightarrow$ 5, 1, B_2	

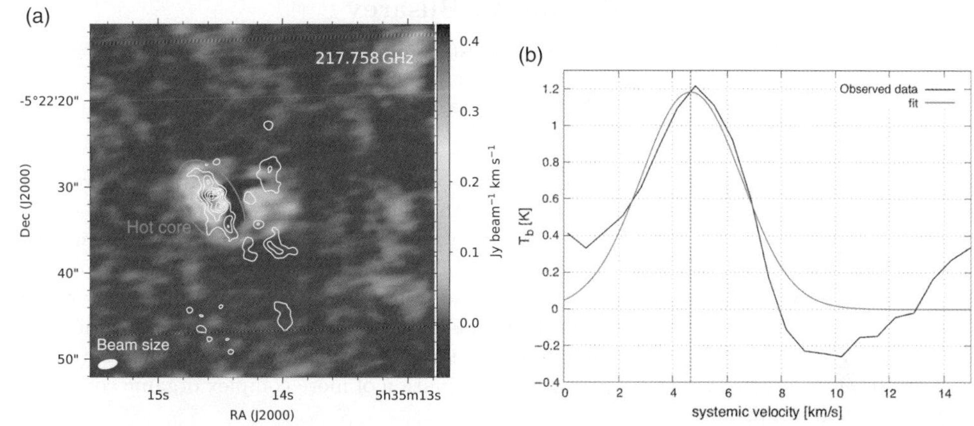

Figure 1. (a) Intergrated intensity map of the line at 217.758 GHz. The white contour represents the 1.3 mm continuum map (Hitota *et al.* 2015) and the black cross marks Hot core center where the spectrum analyzed in this work are extracted. (b) Spectrum of the CH_3NH_2 line at 218.758 GHz observed in Hot core center (black) and the result of the Gaussian fitting (red).

The systemic velocity (V_{LSR}) and the width (FWHM, ΔV) of the line in Figure 1(b) are derived by Gaussian fitting as $V_{LSR} = 4.7 \pm 0.1$ km s^{-1} and $\Delta V = 4.2 \pm 0.2$ km s^{-1}. Since the typical values of Hot core are V_{LSR} of 4$\sim$ 6 km s^{-1} and ΔV of 5$\sim$15 km s^{-1} (Favre *et al.* 2011a), our results are consistent within the range of errors.

Assuming the local thermodynamic equilibrium (LTE) condition and optically thin emission, we plotted the rotation diagram using the observed lines in Table 1. The 215.670 GHz transition seems to be contaminated by other molecular line emission, and so we used the other 4 lines for the rotation diagram analysis.

We derived a rotational temperature of $T_{rot} = 102^{+26}_{-17}$ K with a column density of $N_{tot} = 4.9^{+3.7}_{-19} \times 10^{14}$ cm^{-2} tentatively. For more robust detection, it is necessary to identify more emission lines by the observation with high sensitivity.

References

Favre, C., Despois, D., Brouillet, N., *et al.* 2011a, *A&A*, 532, A32
Halfen, D. T., Ilyushin, V. V., & Ziurys, L. M. 2013, *ApJ*, 767, 66
Hirota, T., Kim, M. K., Kurono, Y., & Honma, M. 2015, *ApJ*, 801, 82
Holtom, P. D., Bennett, C. J., Osamura, Y., *et al.* 2005, *ApJ*, 626, 940
Kounkel, M., Hartmann, L., Loinard, L., *et al.* 2017, *ApJ*, 834, 142
Pagani, L., Favre, C., Goldsmith, P. F., *et al.* 2017, *A&A*, 604, A32

Origins: From the Protosun to the First Steps of Life
Proceedings IAU Symposium No. 345, 2018
B. G. Elmegreen, L. Viktor Tóth & M. Güdel, eds.

doi:10.1017/S174392131900303X

Discovering periodic sublimation on main-belt primitive asteroids near perihelion and its possible astrobiological significance

Vladimir V. Busarev

Moscow Lomonosov State University, Sternberg Astronomical Institute,
13 Universitetskij prospect, Moscow 119992, Russian Federation
email: busarev@sai.msu.ru

Abstract. Discovered periodic sublimation activity on four main-belt primitive asteroids led us to conclusions about possible origin of those and similar bodies (or their parent bodies) near or beyond the snow line in the early Solar System making incorporated in them a considerable water ice stock. Water differentiation of the bodies owing to ^{26}Al decay and their internal thermal evolution might have created conditions for water soluble organics and prebiotic compounds formation. Subsequent longtime periodic changing temperature and other physico-chemical parameters (due to spinning and moving around the Sun) in the near-surface layers of primitive asteroids have led probably to formation of more complex organic compounds of astrobiological significance.

Keywords. asteroids, spectroscopic methods, sublimation of ices, activity of the Sun, solar wind, astrobiology

---

Based on the data of reflectance spectroscopy, we have discovered simultaneous sublimation activity on four main-belt primitive asteroids 145 Adeone, 704 Interamnia, 779 Nina, and 1474 Beira being simultaneously near their perihelion distances in September 2012 (Busarev *et al.* 2015; Busarev *et al.* 2018a). As follows from Mie theory and numeric modeling (Hansen & Travis 1974), observed maxima in reflectance spectra of the asteroids in the near-UV or visible spectral ranges point to dispersion of light in a subtle regular coma of frozen H_2O micron-sized particles around each of the asteroids.

We have confirmed sublimation activity three of the asteroids (Nina, Interamnia and Adeona) at their next passages of perihelion in 2016-2018 using data obtained at four observatories and, correspondingly, established periodicity of this process for asteroids with considerable ellipticity of the orbit (Busarev *et al.* 2018b). Additionally, we noticed that last sublimation activity of the asteroids was less intense then previous one in 2012, which we connect with different solar activity (more strong in 2012) and its direct influence to extent and density of the sublimation coma (Busarev *et al.* 2018b). All considerable factors of solar wind (fluxes of protons, electrons and many other charged particles, as well as strong gradients of electric and magnetic fields) (e.g., McComas *et al.* 2003; Owens & Forsyth 2013) should influence the surface and sublimation coma of primitive asteroids.

The results led us to assumptions about origin of those and similar primitive bodies (or their parent bodies) near or beyond the snow line in the early Solar System making incorporated in them a considerable water ice stock. Water differentiation of such stone-ice bodies owing to ^{26}Al (and similar short-lived isotopes) decay and their internal thermal evolution within the first about 5 millions years might have created conditions for formation of hydrated silicates, water soluble and other organic compounds (e.g., Busarev *et al.* 2003; Busarev 2012). Subsequent longtime periodic changing temperature and other physico-chemical parameters in the near-surface layers of primitive asteroids (due to their spinning and moving around the Sun) have led probably to growing more complex organic compounds. As concluded from the study of kinetic principles of genesis and evolution of protobiopolymeric molecules (Varfolomeev *et al.* 2018), the thermocycles could provide the thermodynamic feasibility of formation of monomer-monomer bond and kinetic selection of more regular (developing) complexes. Presently, it is considered commonly accepted (Varfolomeev *et al.* 2018 and references herein) that action of radiation, temperature, electric and magnetic fields and so on to simple molecules (CO_2,hydrocarbons, HCN, etc.) under prebiological conditions can synthesize low-molecular-weight compounds (sugars, amino acids, nitrogen bases). As known, such compounds are found in carbonaceous chondrites (e. g., Dodd 1981) which are possible fragments of primitive asteroids. Thus, organic compounds formed with origin and evolution of primitive asteroids and delivered to Earth at the time of its heavy bombardment could form the basis for subsequent appearance of macromolecular systems (peptides, proteins, nucleic acids, etc.) and terrestrial life in the end.

References

Busarev, V. V. 2012, *Asteroids, Comets, Meteors (ACM) 2012*, Abstract #6017

Busarev, V. V., Dorofeeva, V. A., & Makalkin, A. B. 2003, *Earth, Moon and Planets*, 92, 345–357

Busarev ,V. V., Barabanov, S. I., Rusakov, V. S., Puzin, V. B., & Kravtsov, V. V. 2015, *Icarus*, 262, 44–57

Busarev ,V. V., Makalkin, A. B., Vilas, F., Barabanov, S. I., & Scherbina, M. P. 2018a, *Icarus*, 304, 83–94

Busarev ,V. V., Scherbina, M. P., Barabanov, S. I., Irsmambetova, T. R., Kokhirova, G. I., Khamroev, U. Kh., Khamitov, I. M., Bikmaev, I. F., Gumerov, R. I., Irtuganov, E. N., & Melnikov, S. S. 2018b, *Ninth Moscow Solar System Symposium (9MS3), Space Research Inst., Moscow*, Abstract 9MS3-SB-08

Dodd, R. T. 1981, *Meteorites A petrologic-chemical synthesis*, Cambridge: Cambridge Univ. Press, 368 p.

Hansen, J. E., & Travis, L. D. 1974, *Space Sci. Rev.*, 16, 527–610.

McComas, D. J., Elliot, H. A., & Schwadron, N. A. 2003, *Geoph. Res. Let.*, 30, No.10, 1517

Owens, M. J., & Forsyth, R. J. 2013, *Living Rev. Solar Phys.*, 10, 5

Varfolomeev, S. D., Tsybenova, S. B., Bykov, V. I., & Goldberg, V. M. 2018, *Doklady Physical Chemistry*, 481, Part 2, 117–120

Origins: From the Protosun to the First Steps of Life
Proceedings IAU Symposium No. 345, 2018
B. G. Elmegreen, L. Viktor Tóth & M. Güdel, eds.

doi:10.1017/S1743921319003041

An UXor among FUors: extinction-related brightness variations of the young eruptive star V582 Aur

P. Ábrahám[1], Á. Kóspál[1,2], M. Kun[1], O. Fehér[1], G. Zsidi[1,3] and J. A. Acosta-Pulido[4,5]

[1]Konkoly Observatory, Research Centre for Astronomy and Earth Sciences, Hungarian Academy of Sciences, Konkoly-Thege Miklós út 15-17, H-1121, Budapest, Hungary
[2]Max Planck Institute for Astronomy, Königstuhl 17, D-69117 Heidelberg, Germany
[3]Eötvös Loránd University, Dept. Astronomy, Pázmány Péter sétány 1/A, H-1117 Budapest
[4]Instituto de Astrofísica de Canarias, Avenida Vía Láctea, E-38205 La Laguna, Tenerife, Spain
[5]Departamento de Astrofísica, Universidad de La Laguna, E-38205 La Laguna, Tenerife, Spain
email: abraham@konkoly.hu

Abstract. The early evolution of Sun-like stars may be interspersed by energetic FU Orionis (FUor) type accretion outbursts. We analysed eight years of photometric and spectroscopic variability of V582 Aur, a bona fide FUor, in outburst. While the accretion rate derived from near-infrared measurements was constant, radical brightness changes occurred due to dust clumps crossing the line of sight. The brightness minima resemble the variability patterns of the UXor phenomenon. Orbiting density enhancements or short-lived clumps moving in and out of the line-of-sight may explain these observations. Our message is that during FUor outbursts the inner disk is a dynamically active place, affecting the initial conditions for planet formation.

Keywords. accretion disks, stars: pre–main-sequence, circumstellar matter

1. Our target

V582 Aur is a pre–main-sequence star that brightened by 4-5 magnitudes in 1986, and is still in its high state. The brightening turned out to be an FU Orionis-type outburst, when mass accretion from the inner disk onto the protostar increases by 100–1000× for decades, due to a still debated physical instability (Hartmann & Kenyon 1996). According to Gaia DR2, the distance of V582 Aur is $2.4^{+0.7}_{-0.4}$ kpc (Bailer-Jones *et al.* 2018), which implies that the outburst luminosity is 500–1050 $L_\odot$, powerful enough to modify the density structure, mineralogy, and chemistry of the inner disk. Changes in both the accretion rate and the disk structure are expected to lead to optical-infrared variations, that has indeed been observed in V582 Aur. Here we analyze its multiwavelength light curves to learn about the physical processes in the inner disk during a FUor eruption.

2. Time evolution of the outburst

Multiepoch optical and infrared observations of V582 Aur between 2010 and 2018 are plotted in Fig. 1. The figure outlines a relatively well defined constant high flux level of the star, that is modulated by two deep minima, in 2012 and in 2016–18. In the first event the system recovered within a year, but the second dip is still ongoing. The source brightened until 2018 February, then entered a new fading period, but new data from 2018 autumn show a higher flux again. The mid-infrared flux evolution is not correlated with the optical-near-IR light curves.

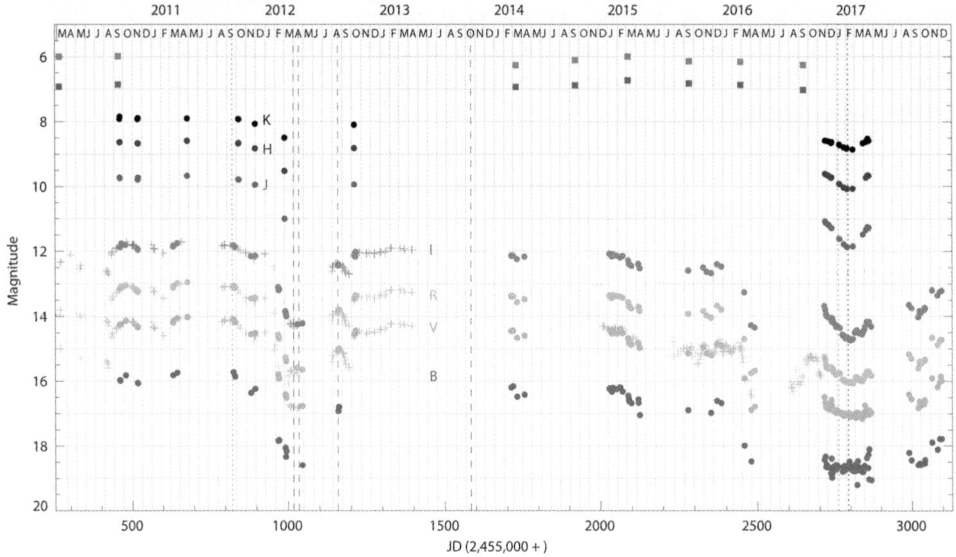

Figure 1. Light curves of V582 Aur. Filled dots are from this work, obtained with the following telescopes: Schmidt & RCC at Konkoly Observatory, IAC-80 and TCS at Teide Observatory, and the WHT on La Palma. Plus signs are from Semkov *et al.* (2013), and from ASAS-SN (Shappee *et al.* 2014, Kochanek *et al.* 2017). Mid-IR data points are from WISE (filled squares).

3. Accretion or extinction?

We modeled the near-IR spectral energy distribution at several epochs by assuming that the emission of the FUor outburst is dominated by a steady, optically thick, and geometrically thin viscous accretion disk, and fitted the circumstellar reddening and accretion rate. We found that while the accretion was relatively constant over the whole period, the extinction increased during the brightness minima. These results imply that the observed variability was caused by extinction changes (Ábrahám *et al.* 2018).

4. Nature of the obscuring material

The two deep minima in 2012 and 2016–18 may suggest an orbiting dust clump periodically occulting the inner part of the system. The different lengths and profiles of the two minima might indicate a viscous spreading of the dust particles along the orbit. With an orbital radius of 2.8 au, extinction of $\Delta A_V = 7.3$ mag, and dust mass of $\sim 0.004\ M_\oplus$, it would be a minor density fluctuation within the disk. Alternatively, dust could be freshly created by colliding planetesimals, or be part of a giant ($M \sim 2 \times 10^6$ Hale-Bopp) exocomet visiting the inner system on an excentric orbit. The mid-infrared flux peak in 2015 (Fig. 1) may mark the periastron of the exocomet's orbit (Ábrahám *et al.* 2018).

Acknowledgements

This project has received funding from the European Research Council (ERC) under the European Union's Horizon 2020 research and innovation programme under grant agreement No 716155 (SACCRED).

References

Ábrahám, P., Kóspál, Á., Kun, M. *et al.* 2018, *ApJ*, 853, 28

Bailer-Jones, C. A. L., Rybizki, J., Fouesneau, M., Mantelet, G., & Andrae, R. 2018, *AJ*, 156, 58

Hartmann, L., & Kenyon, S. J. 1996, *ARAA*, 34, 207
Kochanek, C. S., Shappee, B. J., Stanek, K. Z. *et al.* 2017, *PASP*, 129, 104502
Semkov, E. H., Peneva, S. P., Munari, U. *et al.* 2013, *A&A*, 556, A60
Shappee, B. J., Prieto, J. L., Grupe, D. *et al.* 2014, *ApJ*, 788, 48

Origins: From the Protosun to the First Steps of Life
Proceedings IAU Symposium No. 345, 2018
B. G. Elmegreen, L. Viktor Tóth & M. Güdel, eds.

doi:10.1017/S1743921319003053

Possibility to locate the position of the H_2O snowline in protoplanetary disks through spectroscopic observations

Shota Notsu[1], Hideko Nomura[2], Catherine Walsh[3], Mitsuhiko Honda[4], Tomoya Hirota[2], Eiji Akiyama[5], Takashi Tsukagoshi[2], Alice S. Booth[3] and T. J. Millar[6]

[1]Leiden Observatory, Leiden University, P.O. Box 9513, 2300 RA Leiden, The Netherlands JSPS (Japan Society for the Promotion of Science) Overseas Research Fellow
email: notsust@strw.leidenuniv.nl

[2]National Astronomical Observatory of Japan, 2-21-1 Osawa, Mitaka, Tokyo 181-8588, Japan

[3]School of Physics and Astronomy, University of Leeds, Leeds, LS2 9JT, UK

[4]Faculty of Biosphere-Geosphere Science, Okayama University of Science, 1-1 Ridai-chou, Kita-ku, Okayama 700-0005, Japan

[5]Institute for the Advancement of Higher Education, Hokkaido University, Kita 17, Nishi 8, Kita-ku, Sapporo, Hokkaido 060-0817, Japan

[6]Astrophysics Research Centre, School of Mathematics and Physics, Queen's University Belfast, University Road, Belfast, BT7 1NN, UK

Abstract. Observationally locating the position of the H_2O snowline in protoplanetary disks is crucial for understanding planetesimal and planet formation processes, and the origin of water on the Earth. In our studies, we conducted calculations of chemical reactions and water line profiles in protoplanetary disks, and identified that ortho/para-$H_2{}^{16}O$, $H_2{}^{18}O$ lines with small Einstein A coefficients and relatively high upper state energies are dominated by emission from the hot midplane region inside the H_2O snowline. Therefore, through analyzing their line profiles the position of the H_2O snowline can be located. Moreover, because the number density of the $H_2{}^{18}O$ is much smaller than that of $H_2{}^{16}O$, the $H_2{}^{18}O$ lines can trace deeper into the disk and thus they are potentially better probes of the exact position of the H_2O snowline in disk midplane.

Keywords. astrochemistry— protoplanetary disks— ISM: molecules— sub-millimeter: planetary systems— stars: formation—

Observationally locating the position of the H_2O snowline in protoplanetary disks is crucial for understanding the planetesimal and planet formation processes, and the origin of water on the Earth. The velocity profiles of emission lines from disks are usually affected by doppler shift due to keplerian rotation. Therefore, the line profiles are sensitive to the radial distribution of the line emitting regions. However, water lines which have been observed by previous infrared spectroscopic observations (*Spitzer* and *Herschel*, see, e.g., van Dishoeck *et al.* 2014) mainly traced the disk surface and the cold water resevoir outside the H_2O snowline. Thus, they are not good direct tracer of the H_2O snowline.

In our studies (Notsu *et al.* 2016, Notsu *et al.* 2017, Notsu *et al.* 2018), on the basis of our calculations of disk chemical structures and water line profiles, we proposed how to identify the position of the H_2O snowline directly by analyzing the Keplerian profiles of water lines which can be obtained by high dispersion spectroscopic observations across

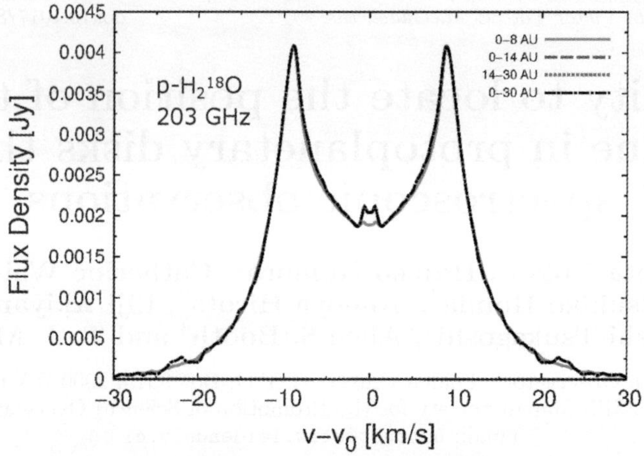

Figure 1. The profile of para-$H_2{}^{18}O$ 203 GHz line for the Herbig Ae disk. In this line profile, we ignore dust emission and adopt a disk inclination, $i = 30$ deg and the distance to the object, $d = 140$ pc. The line profile from inside 8 au (=the inner high temperature region) is displayed with *red solid line*, that from within 14 au (~within the H_2O snowline) is *blue dashed line*, that from 14-30 au (~outside the H_2O snowline) is *green dotted line*, and that from the total area inside 30 au is *black dashed dotted line*.

a wide range of wavelengths (from mid-infrared to sub-millimeter, e.g., ALMA, SPICA). We selected candidate water lines to locate the H_2O snowline based on specific criteria. We concluded that lines which have small Einstein *A* coefficients ($A_{ul}=10^{-6} \sim 10^{-3}$ s^{-1}) and relatively high upper state energies ($E_{up} \sim$ 1000K) trace the hot water reservoir within the H_2O snowline, and can locate the position of the H_2O snowline (see Figure 1). In these candidate water lines, the contribution of the optically thick hot midplane inside the H_2O snowline is large compared with that of the outer optically thin surface layer. This is because the intensities of lines from the optically thin region are proportional to the Einstein *A* coefficient. Moreover, the contribution of the cold water reservoir outside the H_2O snowline is also small, because lines with high excitation energies are not emitted from the regions at a low temperature.

The position of the H_2O snowline of a Herbig Ae disk exists at a larger radius compared with that around less massive and cooler T Tauri stars. Therefore, it is expected to be easier to observe the candidate water lines, and thus identify the location of the H_2O snowline, in Herbig Ae disks. In addition, since the number densities of the ortho- and para-$H_2{}^{18}O$ molecules are about 1/560 times smaller than their ^{16}O analogues, they trace deeper into the disk than the ortho-$H_2{}^{16}O$ lines, and lines with relatively smaller upper state energies (~ a few 100K) can also locate the position of the H_2O snowline. Thus these $H_2{}^{18}O$ lines are potentially better probes of the position of the H_2O snowline at the disk midplane, depending on the dust optical depth (Notsu *et al.* 2018).

There are several candidate water lines that trace the position of the H_2O snowline in ALMA Bands 5 – 10. Finally, we have proposed the water line observations for a Herbig Ae disk HD163296 in ALMA Cycle 3, and partial data were delivered. We constrain the line emitting region (the location of the H_2O snowline) and the mm dust opacity from the observations (Notsu *et al.* 2019).

References

Notsu, S., Akiyama, E., Booth, A., *et al.* 2019, *ApJ*, 875, 96 (paper IV)
Notsu, S., Nomura, H., Walsh, C., *et al.* 2018, *ApJ*, 855, 62 (paper III)

Notsu, S., Nomura, H., Ishimoto, D., Walsh, C., Honda, M., Hirota, T., & Millar, T. J. 2017, *ApJ*, 836, 118 (paper II)

Notsu, S., Nomura, H., Ishimoto, D., Walsh, C., Honda, M., Hirota, T., & Millar, T. J. 2016, *ApJ*, 827, 113 (paper I)

van Dishoeck, E. F., Bergin, E. A., Lis, D. C., & Lunine, J. I. 2014, *in Protostars and Planets VI, ed. H. Beuther et al. (Tucson, AZ: Univ. Arizona Press)*, 835

Author index